ELECTROPLATING

FREDERICK A. LOWENHEIM

ELECTROPLATING

Sponsored by the American Electroplaters' Society

McGRAW-HILL BOOK COMPANY

New York St. Louis San Francisco Auckland Bogotá Düsseldorf
Johannesburg London Madrid Mexico Montreal New Delhi Panama
Paris São Paulo Singapore Sydney Tokyo Toronto

Library of Congress Cataloging in Publication Data

Lowenheim, Frederick Adolph, date.
Electroplating.

Bibliography: p.
Includes index.
1. Electroplating. I. American Electroplaters' Society. II. Title.
TS670.L67 671.7'32 77-11993
ISBN 0-07-038836-9

34567890 ***FGRFGR*** 87654321

The editors for this book were Jeremy Robinson and Ruth Weine, the designer was Elliot Epstein, and the production supervisor was Frank P. Bellantoni. It was set in Times Roman VIP by University Graphics, Inc.

Printed and bound by Fairfield Graphics, Inc.

Contents

Foreword

While the protective value of a metallurgically applied coating of tin on iron was recognized as early as the fifth century B.C., it was not until the advent of the voltaic cell that the electrodeposition of metals became practical. In the formative years, beginning in the early 1800s, the process of electrodeposition was an art practiced by only a few individuals, including such early scientists as Sir Humphry Davy and Michael Faraday, whose well-known laws of electrolysis are the foundation by which electrodeposition processes are governed.

The first practical application of electrodeposited metal occurred about 1840. Because of the investigations of countless individuals over the ensuing years, the art of electrodeposition slowly evolved into a more scientifically based technology; unlike its modest beginnings, it now requires practitioners to have a solid foundation in the fundamentals governing the electrodeposition process.

The primary goals of the American Electroplaters' Society (founded in 1909) deal with the educational needs of its membership. Our objectives have been achieved through the sponsorship of technical meetings, symposia, and courses, as well as through dissemination of information in the AES monthly journal, *Plating and Surface Finishing*. In the Society's continuing efforts to advance the science and practice of electrodeposition and surface finishing, Dr. Frederick A. Lowenheim, one of the world's authorities in this area, was commissioned to write a book intended for use as a classroom text as well as a research source. It was felt that such a book would provide a much-needed reference in the field of electroplating and metal finishing.

Therefore this book is offered as a mark of the concern of the American Electroplaters' Society for the educational advancement of its members and is dedicated to providing expanded knowledge in this field. It is to this end that this comprehensive text is tendered.

RICHARD G. BAKER, CHAIRMAN
AES Technical Education Board

HARRY J. LITSCH, CHAIRMAN
AES Education Committee

Preface

This book had its origins in the decision by the American Electroplaters' Society (AES) to sponsor courses at the Junior College or College level in Metal Finishing throughout the country, with the purpose of granting certification to those successfully completing such a course. Although some of the local Branches of the Society have sponsored such courses, they were independent of each other and there could be no assurance that they were equivalent in content, nor that those taking the courses were receiving equivalent instruction. The AES wished to expand the program and to ensure that they would be at least approximately equivalent in content and in difficulty, so that all "graduates," no matter where they received their instruction, could truly claim to be familiar with the fundamentals of the technology. A further aim was to so structure the courses that they would be deserving of due credit toward a degree from the institution where they were taken.

It was soon realized that no satisfactory textbook in English for such a course was available, and the author was honored with the request to prepare such a text; this book is the result.

Electroplating and metal finishing is both an art and a science. Although based on several technologies and sciences, including chemistry, physics, chemical and electrical engineering, metallurgy, and perhaps others, it retains in some ways the aspects of an art, in which experience is the only teacher. In fact, of course, all the sciences have elements of art which can be learned only by experience; all the reading of textbooks on chemistry will not produce a chemist. *A fortiori,* no text on electroplating will produce an expert electroplater; there is no substitute for experience and what is somewhat inelegantly termed know-how. Nevertheless, the sponsors of this book, as well as the author, believe that experience alone is not enough, especially in our modern technological society; that a background in fundamentals—know-why in addition to know-how—is a basic requirement. Accordingly, this book is offered as an introduction to electroplating, with the full realization that without a follow-up in an operating finishing shop or research laboratory one's knowledge will necessarily be incomplete.

It is hoped that the book will be useful to at least two classes of readers: those with no background in the subject, who want an introduction which will be followed by practice; and those with some practical experience who may want to improve their skills by a deeper understanding of the science and technology behind their day-to-day practice.

No one book, of course, can claim to offer complete instruction in all aspects of any subject as wide-ranging and as encompassing as metal finishing. Readers must recognize the necessity to supplement their use of this book with reading in allied fields. It is hoped that the bibliography which constitutes the final chapter will aid in directing the reader to such supplemental reading, but even that cannot claim to be complete. It goes without saying that awareness of current literature as presented in journals, and which no hardcover book can possibly cover, is an absolute requirement.

In general, this book follows more or less conventional outlines for covering the topics that must be considered in presenting the groundwork of the electroplating art. In one respect, however, the author flatters himself that he has departed from convention: In the chapters on the individual plating metals, an attempt is made to provide some background in the provenance, economic importance, and chemistry of each metal concerned. This is done with the thought that the finisher will more intelligently approach his task if he or she knows not only how to electroplate a given metal, but also something about the production, uses, and chemistry of the metal being processed. I know of no other text in the field that attempts this approach, and I hope it proves useful.

This has not been a one-man job. By the terms of the agreement between me and the AES, the latter appointed a review committee to pass upon each chapter, both as to general approach and as to accuracy of detail. This committee, whose personnel changed somewhat over the course of the project, was invaluable in catching occasional errors and offering suggestions for improvement. The book, however, is my responsibility; any remaining errors are mine.

It is hoped that the chapters in this book will speak for themselves; but several general statements belong here.

Metal finishing is a highly proprietary field; supply houses, large and small, offer patented or proprietary materials and processes for carrying out almost all the operations covered. The reader should be aware that publication of formulas or methods in this book carries with it no implication whether the material so discussed is the subject of a U.S. or foreign patent. Anyone planning to practice, on anything but an experimental and noncommercial scale, the processes discussed should take whatever steps are deemed appropriate to guard oneself against legal action.

Although the hazards involved in handling some of the chemicals mentioned are briefly discussed, this book makes no claim to thorough discussion of safety regulations in general. It is assumed that the reader is familiar with the elementary precautions required in dealing with potentially toxic or explosive materials, and in the hazards of mechanical and electrical equipment.

Some subjects are too active in the current literature to be adequately treated in a book such as this; this is especially true of governmental regulations concerning pollution abatement, labor conditions, and safety in the workplace.

At this writing many of these regulations are being modified and some are the subject of legal conflicts; there is simply no substitute for the latest periodical literature.

Finally, in recognition of the gradual (one hopes not too gradual) introduction of the International System (SI) of units in America, these units are used throughout the book. Conversion tables from "U.S. customary" to the newer units are offered in an appendix. The SI is explained in detail in several books; in particular ASTM E 380 offers a complete and authoritative guide to its use. All the principal societies serving the metal-finishing trade have endorsed, at least in principle, a schedule for gradual conversion to metrication. The inevitable temporary confusion resulting from simultaneous use of two systems is a small price to pay for the final change to universal use of the SI, now approved by almost all the industrial nations of the world.

FREDERICK A. LOWENHEIM

Acknowledgments

As previously stated, the chapters in this book were reviewed by a committee of the AES under the chairmanship of Professor H. B. Linford of Columbia University; my thanks to Dr. Linford and his committee for their helpful suggestions.

Rather obviously, several chapters in this book rely heavily on corresponding material in MODERN ELECTROPLATING, 3d edition, which I edited and which is published and copyrighted by John Wiley & Sons, Inc., New York. Some of the descriptive material on the individual elements also parallels similar paragraphs in my articles in the ENCYCLOPEDIA OF INDUSTRIAL CHEMICAL ANALYSIS, edited by F. D. Snell and L. S. Ettre, also published by John Wiley & Sons, Inc.

Although I have not consciously or deliberately resorted to direct quotation from these sources, there are similarities of organization of content, instances of paraphrasing, and even possibly inadvertent quotation of a sentence here and there. I thank the publishers, John Wiley & Sons, Inc., for permission to use this material.

I thank the American Society for Testing and Materials (ASTM) for permission to use material, either directly or by paraphrase, from Part 9 of their *Annual Book of Standards*. The author's many years of association with Committee B.08 of ASTM has been most helpful in the preparation of this book.

The American Electroplaters' Society has granted me unrestricted permission to quote or to use illustrations from any of their copyrighted material, a privilege that has been most valuable. In particular, I want to acknowledge the help of Mr. J. Howard Schumacher, Jr., General Manager of the Society; and that of his predecessor, Mr. Rodney Leeds, with whom I had the pleasure of about ten years of association as Technical Editor of the Society's journal, and who originally approached me about taking on this assignment.

ELECTROPLATING

PART ONE
FUNDAMENTALS

INTRODUCTION: THE WHAT AND WHY OF METAL FINISHING

A student entering upon a course of study should have a good idea of the scope of the subject: what is included, and, just as important, what is excluded.

Metals are "finished" for many reasons. A *finish* may be defined as any final operation applied to the surface of a metal article in order to lend it properties not possessed by the article in its "unfinished" form. Metals are shaped, formed, forged, drilled, turned, wrought, cast, etc., to give them a shape appropriate to their purpose, whether it be an automobile bumper, a bobby pin, a 1000-ton casting, a tiny electrical connector, or anything in between. Such operations are not "metal finishing"; they put the metal part into the desired shape, but may leave the surface unsuitable for the intended purpose. Hence, after these operations have been completed, the metal part may need to be finished.

Such finishing may consist merely of polishing the surface to make it smooth or to give it a bright appearance; another metal or metals may be applied to the surface to change its properties; the surface may be painted or lacquered by any one of a number of different techniques; a ceramic coating such as porcelain may be applied. In other processes the coating is derived from the substrate metal itself, usually to form an oxide or other compound of the substrate. These include such techniques as blackening of steel and anodizing and conversion coatings on aluminum.

Many metal parts are not finished, in the sense used here: rough castings may be suitable for their intended use just as they come out of the mold, except for removal of gates, risers, etc.; nails, nuts, bolts, wire, and many other metal products do not require any final operation to render them fit for use, although such articles may be finished if desired, and often are. But just as often, the metal does require some treatment before it can be considered suitable for the intended application; in other words, it must be finished.

Steel is the cheapest structural material available for countless uses; but iron and steel are not very resistant to corrosion, and for many applications a surface coating must be applied to separate the steel from the surrounding atmosphere or other ambient conditions, or to protect it electrochemically from corrosion. It may be electroplated, painted, porcelain enameled, or finished in other ways.

Many other metals in addition to steel must be protected from corrosion: aluminum, zinc in the form of die castings, and even copper, although copper does not corrode very rapidly in most environments. Therefore corrosion protection is one main reason for the existence of a metal-finishing industry.

In some cases basis metals need to be protected from their environments even though they do not corrode very much, because the particular environment concerned must itself be protected from the metal in question. For example, many chemical and pharmaceutical products are extremely sensitive to even traces of iron, so that the vessels used to contain them are often heavily plated with nickel or clad with precious metals such as platinum. The "tin can" is another example: many foods and beverages are not highly corrosive to iron, but even traces of iron would spoil their taste or acceptability; thus the steel which gives strength and durability to the container is plated with tin, and often further lacquered, to protect the contents as well as the can.

Another reason for finishing metals is to improve their appearance, either temporarily or for long periods. A steel automobile bumper, if unprotected from the atmosphere (which includes deicing salts and other road hazards), would quickly rust and become unsightly. Even a rusty bumper probably would continue to perform its function of protecting the car from damage in minor collisions for many years, for it would take a long time for the corrosion to degrade the strength of the steel to the point where it became useless. But consumers prefer that the bumper, and other metal parts of their cars collectively called "brightwork," remain shiny and relatively unrusted during the useful life of the vehicles. For this reason exterior hardware is almost universally "chromium-plated"—actually plated with copper and nickel, or nickel alone, followed by a very thin layer of chromium. In terms of commercial importance to the finishing industry, such finishes, classified as "decorative/protective," probably represent the most significant application of electroplating.

The list of articles which are finished for reasons of protection to the substrate and appearance is a long one: appliances of all types, golf clubs and other sporting equipment, hardware and tools—the list is almost endless.

Some finishes are purely decorative. Many objects meant to be used indoors, in a dry environment and where danger of corrosion is slight, are nevertheless finished with lacquers, paints, and electroplated coatings for purely aesthetic reasons. The very thin layer of gold applied to some articles of inexpensive jewelry has little or no protective value; it is there principally to attract a potential buyer.

There are many applications of metal finishing, some of them of increasing importance at present, in which neither corrosion prevention nor decorative appeal is the reason for using a finish. Copper is an excellent conductor of electricity and is therefore basic to such items as printed circuits and communications equipment. It does, however, quickly form tarnish films that interfere with joining operations such as soldering and that also render contact resistances unacceptably high in relays and switches. To make soldering easier, coatings of tin or tin-lead alloys are often applied to copper, and for better contacts overplates of gold are frequently required. Other surface properties may call for modification: if light reflection is important, a silver or rhodium plate may be necessary. In wave guides for radar, high electrical conductivity is the most important criterion, and silver is the preferred coating. Good bearing properties may require coatings of tin, lead, or indium. If a hard surface is required, chromium or nickel usually will serve. These few examples illustrate another use of metal finishing: to modify the surface properties, either physical or chemical, to render them suitable for the intended use.

Although the answer may be obvious, we may pause momentarily to consider this question: if these coating metals are necessary to provide the article with the desired properties, why go through the somewhat complicated process of metal finishing? Why not simply manufacture the article out of the desired metal in the first place? Usually the answer is cost or availability, and, in some cases, the properties of the metals concerned. An all-platinum chemical reaction vessel of practical production size would be prohibitively expensive; but a steel vessel, clad with a relatively thin layer of platinum, serves the purpose at far lower cost. An all-nickel automobile bumper would render the car a luxury for the rich, aside from the fact that the required amount of nickel would probably be unobtainable. A tin can made entirely of tin would not only be more expensive than the food inside, but would also have no physical strength: tin is a very soft and weak metal. Chromium in massive form is almost impossible to work into useful shapes. In sum, metal finishing allows the use of relatively inexpensive metals like steel and zinc for the bulk of the article, while affording to the exterior the selected properties of the coating chosen. It allows us to make use of the structural strength of steel without suffering its tendency to corrode quickly in use, to enjoy the easy castability of zinc alloys while rendering them attractive and corrosion-resistant, and to benefit from the good conductivity of copper while preserving its ability to be joined by soldering.

Metal finishing is a very broad subject, and not all aspects of it can be considered within the confines of this book. What is included and what is omitted are largely matters of arbitrary selection; we are guided by a consideration of those operations that might ordinarily be carried out in a metal-finishing shop. Such a shop would probably engage in electroplating as well as such related operations as anodizing, "electroless" plating, and the application of conversion coatings such as chromates, black oxides, and the like. It might include various barrel-finishing operations, such as tumbling and vibratory finishing, and what is called "mechanical plating."

It would probably not, except in unusual circumstances, have a section devoted to ceramic coatings or porcelain enameling. These operations are, true enough, metal finishing; but the high-temperature science and technology involved are quite unrelated to our principal theme and call for entirely different background and training. Nor will we consider vacuum metalizing and metal spraying. Various surface operations such as chromizing and siliconizing are not truly surface treatments, since they affect the bulk of the basis metal as well as the surface. Other high-temperature processes such as hard facing are also excluded.

On the other hand, we shall include a few subjects that by a strict definition are not subsumed in the term *metal finishing*. Of increasing commercial importance is the operation of plating on nonconducting substrates, especially plastics such as acrylonitrile-butadiene-styrene (ABS) and polypropylene. Since these are not metals, finishing them is strictly outside our scope; but by universal agreement the technique of plating on plastics is a branch of metal finishing. In addition, the actual manufacture of metal articles by the process of electroforming is not strictly within our province, but it is so closely related to electroplating, and its fundamentals are so nearly identical, that it will be considered briefly.

The principal emphasis in this book will be on electroplating, including the operations required before an article can be satisfactorily plated, the plating process itself, and what must be done after plating to render the article suitable for its final purpose. Also considered, but more briefly, will be such related operations as anodizing, processes not requiring the use of electric current such as conversion coatings, electroless or autocatalytic plating, and immersion coatings.

1 Background: The Place of Electrochemistry in Technology

We have defined metal finishing: it is a treatment that alters the surface of a metal, or occasionally a nonmetal, without significantly altering its bulk properties, at least intentionally. (In some cases, unfortunately, when we operate on the surface of a metal, there is a chance that the effect of this operation may extend to its bulk, usually to the detriment of its properties. The principal example of this effect is the "hydrogen embrittlement" of some steels caused by preplating operations or electroplating itself.)

Metal finishing as such is not a recognized scientific discipline like, say, organic chemistry: it borrows its theory and practice from many sciences and technologies. Its two main branches are organic coatings and electroplating; because the technology of organic coatings is almost entirely unrelated to our main subject, we shall not consider it. Our concern is with that method of applying a finish known as electroplating and methods related to it; this process in turn involves many sciences and technologies, but basically it is a branch of electrochemistry.

Electrochemistry may be defined as the science that deals with the interconversions between electrical and chemical energy. This interaction may involve the use of an electric current to bring about a chemical reaction or a chemical reaction to generate an electric current. It is an extremely important branch of chemical science, and technology. Investigators since Faraday (1791–1867) have been concerned with the nature of electrolytic solutions, together with corrosion phenomena and many other natural or artificial processes that involve the transfer of electrons from one substance to another. Technologically, many of the world's most important materials are produced by electrochemical means or depend upon electrochemical knowledge for their existence or preservation. The various applications of electrochemical technology are summarized in the next few paragraphs.

So far as appropriate we shall use a simple solution of copper sulfate, $CuSO_4 \cdot 5H_2O$, in water for illustration. If a direct electric current (unidirectional, or dc) is passed through such a solution, copper will be deposited on the negative electrode (cathode). If the deposit is fairly thin and in suitable form—smooth, perhaps bright, and adherent to the cathode—we have produced an electrodeposited coating, which is the province of the electroplater and the main topic of this book. In a slight variation, the deposit may have all these properties except adherence to the cathode: the latter is thus removable, leaving the deposit as an entity in itself: it has been *electroformed*. Electroforming is a branch of electroplating technology.

But the reason for carrying out the operation could be to deposit a thick coating, without too much regard for its appearance or smoothness, using as the positive electrode (anode) relatively impure copper and relying on the action of the current to deposit only the copper and leave the impurities behind. This is *electrorefining,* the basic theory of which is not much different from that of electroplating; but since its aims are not the same, the economics and engineering aspects differ considerably. A large proportion of the world's copper is thus electrorefined, as well as nickel, silver, gold, and some tin and other metals.

The solution of copper sulfate may be electrolyzed for still other reasons: we may employ as anode a material that does not dissolve in the solution as a result of the passage of the current—an "inert" anode. In such a case, the anodic reaction will be decomposition of the solvent—water—to yield oxygen, which is normally wasted; and the cathode deposit of massive copper has thus been recovered from solution in a form that can be further processed for the market. The copper sulfate solution was perhaps prepared by leaching an ore or other low-grade source with sulfuric acid, and the copper has been recovered by *electrowinning*. In addition to some copper, a fair proportion of the zinc produced results from this type of operation; much secondary tin is recovered by detinning plants using electrowinning as the final step.

Finally, our copper sulfate solution may be a laboratory sample whose copper content must be determined. Using an inert anode again, this time of platinum, we arrange conditions so that *all* the copper is deposited from the solution, and we obtain the desired result by determining the gain in weight of the cathode. This operation is called *electroanalysis*. Although limited in application, such methods retain their value in analytical chemistry.

The four processes described thus far are subsumed under the heading of *electrodeposition,* which is our main concern. But electrochemistry is much broader than that, and to orient the reader we shall mention some of its other aspects.

The solution in question need not be aqueous: if aluminum oxide is dissolved in molten cryolite (sodium aluminum fluoride) and electric current is passed between a carbon anode and a molten aluminum cathode, aluminum metal forms at the cathode. All the world's aluminum is produced in this way. Electrolysis of molten salts also produces practically all our magnesium and sodium. Thus these metals also are produced by (nonaqueous) electrowinning processes.

If an electric current is passed through a solution in water of common salt—sodium chloride—and if steps are taken to prevent mixing of the solution near the anode (anolyte) with that near the cathode (catholyte), then the product at the anode will be chlorine gas and that at the cathode will be sodium hydroxide, known in the trade as caustic soda. Both chlorine and "caustic" are basic tonnage chemical products without which the modern chemical industry could not exist. With very minor exception, all chlorine and caustic are produced electrochemically. If potassium chloride is used in place of sodium chloride, potassium hydroxide and chlorine will be the products.

Many other chemical reactions can be made to take place by the passage of an electric current through a conducting solution. Among inorganic chemicals so produced are sodium chlorate, sodium hypochlorite, potassium permanganate, and some hydrogen peroxide. Most organic electrochemical reactions are of academic interest, but the production of an important intermediate for nylon—adiponitrile—is carried out electrochemically.

Most metals, except the "noble" ones, tend to corrode and thus sooner or later become useless. Corrosion processes have been shown to be basically electrochemical, and thus an important branch of the study of electrochemistry is the investigation of corrosion phenomena: why metals corrode, the mechanism of the corrosion process, and—of most practical concern—methods for preventing or slowing down the process of corrosion.

When we turn our attention to the reverse of the situation, the generation of an electric current by chemical reactions, we are in the realm of galvanic or "electrogenetic" cells: primary cells, such as the familiar dry cell; secondary or rechargeable cells, such as those that make up the lead-acid battery that starts your car or the nickel-cadmium battery that operates many "cordless" appliances; and others, many of them in early stages of development, such as the "fuel cell."

Finally, electrochemists want to know the mechanisms—the kinetics and thermodynamics—of the phenomena we have so briefly touched upon: the interactions between ions in solution, the mathematical formulation of the laws that govern electrolytic conduction, anode and cathode reactions, and the like. So the final branch of electrochemistry, the most fundamental and yet the least understood, is *theoretical electrochemistry*.

Thus electrochemistry is a very broad and basic science. Electrodeposition is a branch of it, and electroplating is one branch of the technology of electrodeposition. To sum up, we may define electrodeposition as the application of electrochemistry to the deposition of a coating on an electrode by means of electrolysis; and electroplating as electrodeposition in which the deposited coating, usually a metal, has useful and desirable properties, as these properties are defined by the user of the product.

2

Some Definitions and the Basics of Electrodeposition

CURRENT RELATIONSHIPS

We introduce here some of the fundamental concepts necessary for an understanding of the process of electrodeposition. A few of these subjects will be dealt with in more detail in later chapters.

The passage of a unidirectional current (direct current) through a solution is associated with the movement through it of charged particles called *ions*. The terminals leading the current into the solution are *electrodes;* the pole at which the chemical reactions that take place are oxidations is the *anode;* the pole at which the reactions are reductions is the *cathode*. In an electrolytic cell, which draws its current from an outside source such as a battery or rectifier, the anode is the positive electrode and the cathode the negative electrode, but ''positive'' and ''negative'' are not basic to the definitions. The ions that move, or *migrate,* toward the anode are *anions* and have a negative charge; the ions that migrate toward the cathode are *cations,* and they have a positive charge. The solution itself is called the *electrolyte;** that part of the electrolyte immediately surrounding the anode is the *anolyte,* and that part immediately surrounding the cathode is the *catholyte*. The total process of decomposition due to the passage of a current is called *electrolysis*.

The definitions offered in the preceding paragraph refer, of course, to electrolytic solutions (electrolytes), i.e., those which are at least partly composed of ions rather than of uncharged molecules, and which therefore conduct the current. Many solutes (nonelectrolytes) when dissolved in water or other solvents do not ionize but remain in molecular, uncharged form—e.g., sucrose, benzene, and many other organic compounds. Except as minor additives to some

*The term *electrolyte* has two meanings; the other meaning refers to the substance which when dissolved in a solvent gives rise to a conducting solution—an electrolyte in the first meaning. The context will almost always tell which is meant.

plating solutions, the study of electrodeposition is hardly concerned with such nonelectrolytes.

Conductors

The electric current may be conducted through a medium in two principal ways: metallic or electronic, and electrolytic, conduction. When the conduction is metallic or electronic, it is not accompanied by any movement of material through the conductor or by any chemical reaction (unless the heating effect of the current causes the metal to oxidize, etc.). It is dependent on the fact that metals, and a few other substances having some metallic properties, possess in their structure electrons not tightly bound to any particular atom and so free to move under the influence of an applied potential. In the class known as electrolytic conductors, conduction is accompanied by movement of matter through the medium and by chemical reaction at the electrodes. The conducting species are the charged ions that exist in these solutions; each charge is associated with a particular ion, which moves with the charge under the applied potential. Electrolytic conductors include solutions of acids, bases, and salts; fused salts; a few solid substances; and hot gases. (Semiconductors represent a third class, somewhat intermediate between the two.)

For the most part, in our study of electrodeposition we are interested in electrolytic conductors only, and usually, though not exclusively, the solvent will be water. Metals and most alloys are much better conductors than electrolytes; so, for the most part, the resistance to the current due to metallic conduction can be neglected, and we will be concerned only with the conductance characteristics of electrolytes. (There are exceptions: when plating long lengths of wire or strip at high currents, the resistance of the metal itself must be taken into account.)

The passage of electricity from one type of conductor to the other is always accompanied by chemical reaction. This means that when electrons leave a metallic electrode and enter an electrolyte, or vice versa, a chemical reaction must take place at the interface between the two.

Electrolytic Conductance

There is (with minor exceptions) no good evidence for the existence of molecules of true salts, either in the solid state or in solution; their behavior can be explained entirely on the basis of their complete dissociation into ions. Deviations from such behavior are, at least potentially, explicable by the electrical effects between particles of unlike charge. In the crystal of, say, sodium chloride, which is a typical salt, each sodium ion is surrounded equidistantly by six chloride ions, and each chloride ion by six sodium ions. Thus it is meaningless to choose arbitrarily one of the six chloride ions to pair with a sodium ion and to call the pair a "molecule" of sodium chloride. In aqueous solution also, although there are interionic forces that may momentarily form "ion pairs," it is not productive to think of molecules of sodium chloride as existing in solution. The solution consists of sodium ions, Na^+, and chloride ions, Cl^-, often closely associated with solvent molecules, in a sea of the solvent water.

Many substances which do in fact exist as true molecules in the pure state form ions in water solution or when dissolved in other solvents. For example, hydrogen chloride, HCl, in the pure state exists as discrete molecules; but when it is dissolved in water, a reaction takes place:

$$HCl + H_2O \rightarrow H_3O^+ + Cl^-$$

Sulfuric acid ionizes in the pure state according to

$$H_2SO_4 + H_2SO_4 \rightleftharpoons H_3SO_4^+ + HSO_4^-$$

in addition to several other more complex reactions; but when sulfuric acid is dissolved in water, the principal reaction is

$$H_2SO_4 + H_2O \rightarrow H_3O^+ + HSO_4^-$$

and to a lesser extent

$$HSO_4^- + H_2O \rightleftharpoons H_3O^+ + SO_4^{--}$$

Strong bases such as sodium hydroxide similarly ionize to $Na^+ + OH^-$.

Many substances do not ionize completely when dissolved in water; most such substances are acids and bases, but some salts having partially covalent character are also in this class. For example, acetic acid reacts with water according to

$$CH_3COOH + H_2O \rightleftharpoons H_3O^+ + CH_3COO^-$$

and ammonia reacts according to

$$NH_3 + H_2O \rightleftharpoons NH_4^+ + OH^-$$

In the cases of acetic acid and ammonia, therefore, ionization is not complete and an equilibrium is set up, as shown by the two-way arrows.

Whether the solutes are "strong" (completely ionized) or "weak" (partially ionized) electrolytes, however, at least some ions must be formed in order for the solution to be a conducting medium. A solution of sucrose in water does not conduct an electric current appreciably and is of little interest in connection with the study of electrolysis.

It will be noted that the hydrogen ion has been written as H_3O^+ rather than the possibly more familiar H^+. The latter notation would indicate a hydrogen atom that has lost one electron, which would be equivalent to a bare proton. It is extremely unlikely that a species with such a large ratio of charge to size could exist in such a polar medium as water. It must be hydrated; hence the notation H_3O^+, which is equivalent to $H(H_2O)^+$. There is no assurance that even this is necessarily correct: it might be $H(H_2O)_x^+$ where x might be any fairly small number. But it is convenient to write H_3O^+ as an admission that H^+ cannot be right.

Similarly, most other ions in water solution must be hydrated; i.e., in their immediate vicinity water dipoles must be closely associated with them. Certainly when many salts crystallize from water solution, a definite number of molecules of water accompany them: $CuSO_4 \cdot 5H_2O$, $ZnSO_4 \cdot 7H_2O$, etc. The number of water molecules in the crystal may or may not be the same as the degree of hydration in the solution. These molecules of water of hydration are important in kinetic studies of anode and cathode processes, but in writing chemical reactions it is customary to omit them, since they take no direct part in the total stoichiometry of a reaction. For the same reason, it is usually convenient and justifiable to use the term *hydrogen ion* and the symbol H^+ as a shorthand for the whole group of hydrates.

Under the influence of an applied voltage, the ions move toward the electrodes: the cations toward the cathode and the anions toward the anode. Each ion moves at a particular rate characteristic of that ion; this rate, under conditions such that the applied voltage is unity, is called its *mobility,* or its individual ionic conductivity. The total conductivity of a given solution would be the sum of the mobilities of the ions it contains, except that electrical forces between the ions interfere with this simple assumption unless the solution is extremely dilute. The mobilities of individual ions differ considerably, so that in any solution it is likely that more current is carried by the cation than by the anion, or vice versa. The proportion of the total current carried by a given ion is its *transference,* or *transport, number.*

Generally, ions move through solutions as units, impeded by the drag of the surrounding envelope of solvent molecules and also by the retarding effect of ions of opposite charge. In water, however, the hydrogen and hydroxyl ions are transported by a different mechanism which involves the transfer down the chain of water molecules of a positively charged hydrogen atom, H^+, which is the hydrogen nucleus or a bare proton. The original configuration

$$H_3O^+ : H_2O$$

by the transfer of a proton H^+ from left to right becomes

$$H_2O : H_3O^+$$

And similarly the configuration

$$H_2O : OH^-$$

by transfer of a proton from left to right becomes

$$OH^- : H_2O$$

This is a much more rapid process than actual movement of a hydronium ion H_3O^+ through the solution, as would be required if the ion were, say, $Cu(H_2O)_x^{++}$. Consequently the mobilities of the hydrogen (hydronium) ion H_3O^+ and of the hydroxyl ion OH^- in water are much higher than those of any other ions: about 5 times for H_3O^+ and 2.5 times for OH^-, on the average. Therefore solutions of strong acids and bases are much better conductors than any other aqueous solutions, and in electrolytic processes free acid or free alkali is often used to improve the conductivity of solutions.

Electrical Relationships

Michael Faraday, perhaps the greatest experimental scientist in history, enunciated his laws of electrolysis in 1833, and these laws have remained unchallenged ever since. They are basic to both the understanding and the practical use of electrolytic processes. They may be stated as follows.

1. The amount of chemical change produced by an electric current is proportional to the quantity of electricity that passes.
2. The amounts of different substances liberated by a given quantity of electricity are proportional to their chemical equivalent weights.

Stated another way, these laws correctly predict that: (1) by measuring the quantity of electricity that passes, one has a measure of the amount of chemical change that will thereby be produced; and (2) once knowing the chemical equivalent weight of a substance, one can predict the amount of that substance that will be liberated by a given quantity of electricity.

No true exceptions to these laws have ever been confirmed. Apparent exceptions can always be explained by the failure to take into account *all* the chemical reactions involved, or, occasionally, by the partially nonelectrolytic nature of the reaction. Two examples may make this clear; practical applications of similar examples will be encountered throughout our study.

It might be expected that a given quantity of electricity would deposit, say, 1 g of zinc on a cathode; yet we find that only 0.85 g is actually deposited. But this apparent failure of the law is in reality due to the fact that, in addition to the zinc deposited, some hydrogen is evolved. If this were measured, it would be found that Faraday's law has been followed exactly.

In another case, a given quantity of electricity should have dissolved 1 g of zinc from a zinc anode; in fact, 1.1 g of zinc dissolved. But since zinc dissolves in many solutions without the aid of an externally applied electric current, some of the weight loss was due to corrosion—an electrochemical reaction that requires no external application of current.

The chemical equivalent of an element is its atomic weight divided by the valence change involved in the reaction. Thus although an element has only one atomic weight, it may have several equivalent weights. If iron is deposited from the iron(II) or ferrous state, $Fe^{++} \rightarrow Fe^0$, the valence change is 2, and the chemical equivalent weight is its atomic weight, 55.85, divided by 2, or 27.925. If the reaction were to start from the iron(III) or ferric state and result in metallic iron, $Fe^{3+} \rightarrow Fe^0$, the valence change would be 3, and the chemical equivalent weight would be 55.85/3, or 18.617; finally, if the reaction is merely the reduction of iron(III) to iron(II), $Fe^{3+} \rightarrow Fe^{++}$, the valence change is 1 and the chemical equivalent weight is equal to the atomic weight.

By definition, the coulomb (C), the unit quantity of electricity, is one ampere flowing for one second, 1 A·s. A practical definition is the amount of electricity that will deposit 0.001118 g of silver. The quantity of electricity required to deposit 1 gram-equivalent weight of silver, and therefore by Faraday's law 1 gram-equivalent weight of any element, is 107.868/0.001118 = 96 483 C; this quantity is called the *faraday* ($\mathfrak{F}$). The figure 96 500 is sufficiently accurate for all practical purposes. Faraday's laws may now be stated in the form of an equation:

$$g = Iet / 96\,500 \tag{2-1}$$

where g = grams of substance reacting, I = current in amperes, e = chemical equivalent weight, and t = time in seconds.

Obviously, the electrochemical equivalent of an element or compound can be stated in various units, and several are useful in practice. For example, in Eq. (2-1) if I = 1 A and t = 1 s, then 1000 g = mg/C (milligrams per coulomb); if t = 3600 s, g = g/A·h (grams per ampere-hour). Other units may be similarly calculated such as kg/A·day by use of the proper factors. See Appendix A-1.

If the voltage of the cell reaction is known, the energy consumed in an electrochemical reaction also may be easily computed: volts × ampere-hours = watt-hours (W·h); V × A·s = W·s or joules (J), etc.

To determine the quantity of electricity flowing in a cell, several methods can be used. For rough work, it is sufficient to read an ammeter several times during a reaction, calculate an average current, and use a timing device such as a stopwatch to determine time; multiplying the two yields A·h. Ampere-hour meters are commercially available and yield somewhat more accurate results if the amperage fluctuates significantly during the run. For most accurate results, an apparatus called a *coulometer** must be used. This is an electrochemical cell, used in series with the cell under study, in which care has been taken to ensure that the electrode reaction takes place with no side reactions. Best known are the silver, copper, and iodine coulometers. In the first two, the weight of silver or copper deposited is determined; in the third the amount of iodine set free by the reaction is titrated. From these figures, the quantity of electricity passing in the coulometer is calculated from Faraday's law, and since the cell was in series with the unknown cell, this must be identical to the quantity in the unknown.

Several electromechanical and other fairly sophisticated coulometers have been described,† but for routine use in electroplating experiments the classical copper coulometer, though not the most accurate available, is adequate.

In the copper coulometer, the electrodes are sheets of pure copper; the electrolyte is 150 g copper sulfate pentahydrate, 50 g concentrated sulfuric acid, and 50 ml of ethanol, in 1 L of distilled water. (The addition of ethanol is said to minimize the side reaction $Cu^{+} \rightarrow Cu^{++}$ at the surface of the electrolyte because of absorption of oxygen from the air; the ethanol oxidizes slowly to acetone and acetic acid, but it is often omitted.) The cathode current density should be between 20 and 200 A/m². The electrolyte may be stirred by a stream of small

*This instrument was formerly called the voltameter, for Alessandro Volta.

†See, for example, A. Weissberger and B. W. Rossiter (eds.), "Techniques of Chemistry," vol. 1, Part IIA, Electrochemical Methods, Wiley-Interscience, New York, 1971, pp. 235, 689.

bubbles of carbon dioxide. The size of the electrodes is determined by the current being measured, in such a way that the current density will fall within the range cited. Since the electrochemical equivalent of copper is 31.77 g/𝔉, the deposition of 1 g of copper corresponds to the passage of 3037 C of electricity.

Current Efficiency

It has been stated that the total amount of chemical change at an electrode is exactly proportional to the quantity of electricity passing. Often, however, we are interested in only one of the several chemical changes taking place, and any current used up in causing other changes is considered "wasted." In the usual electroplating situation, our interest focuses on the quantity of metal deposited at the cathode or dissolved at the anode, and any hydrogen evolved at the cathode or oxygen at the anode represents a waste of electricity and a reduction in the efficiency of the process. Thus we speak of *current efficiency* (CE) as the ratio of the desired chemical change to the total chemical change, or

$$\mathrm{CE} = 100 \times \mathrm{Act/Theo} \tag{2-2}$$

where CE is current efficiency in percent, Act is the weight of metal deposited or dissolved, and Theo is the corresponding weight to be expected from Faraday's laws if there were no side reactions.

Cathode efficiency is current efficiency as applied to the cathode reaction, and anode efficiency is current efficiency as applied to the anode reaction.

Some examples of the use of Faraday's laws in practical situations are appended to this chapter.

Current Density

The total current flowing through the cell permits calculation of gross product; but in electroplating we are interested not so much in the total weight of metal deposited as in its thickness—both average thickness and the distribution of the deposit on the cathode. The average thickness will depend on both the total amount of metal deposited *and* the area over which the deposit is spread; its distribution will depend upon how evenly the deposit covers the cathode. It is important in electroplating, therefore, to consider the variable called *current density,* defined as current in amperes per unit area of electrode. Usual units for this factor are amperes per square meter, A/m^2; other units found in the literature include milliamperes per square centimeter, mA/cm^2; amperes per square decimeter, A/dm^2; and, in English units, amperes per square foot, A/ft^2 (sometimes written ASF, but this usage is not recommended), and amperes per square inch, $A/in.^2$. Conversion tables will be found in Appendix A-2.

Current density is a very important variable in all electroplating operations; the character of the deposit, its distribution, the current efficiency, and perhaps even whether a deposit forms at all may depend on the current density employed. Anode current density is also important. Illustrations of these statements will be found in subsequent chapters.

Although current density has been defined as current per unit area, we usually do not know the actual area of an electrode unless it is perfectly smooth and can be measured geometrically. For most situations, however, it is sufficient to equate apparent area to actual area; exceptions include some theoretical considerations and situations where the electrode is so rough that the difference is highly significant.

Current Distribution

The current divided by the apparent area yields an average figure. Except for the simplest geometries of a cell, such as when the anode and cathode are concentric, the current is not uniform over the surface of an electrode. In fact, the manner in which the current distributes itself over an electrode surface in any practical case is quite complicated, usually far too much so to be simply calculated from geometry. As indicated in Fig. 2-1, current will tend to concentrate at edges and points, and unless the resistance of the solution is extremely low (lower than in any practical case), it will flow more readily to parts near the opposite electrode than to more distant parts. Thus, except for the simplest parts subject to electroplating, the thickness of deposit, which depends on the current density, will not be uniform over the surface. Since one usually desires to deposit

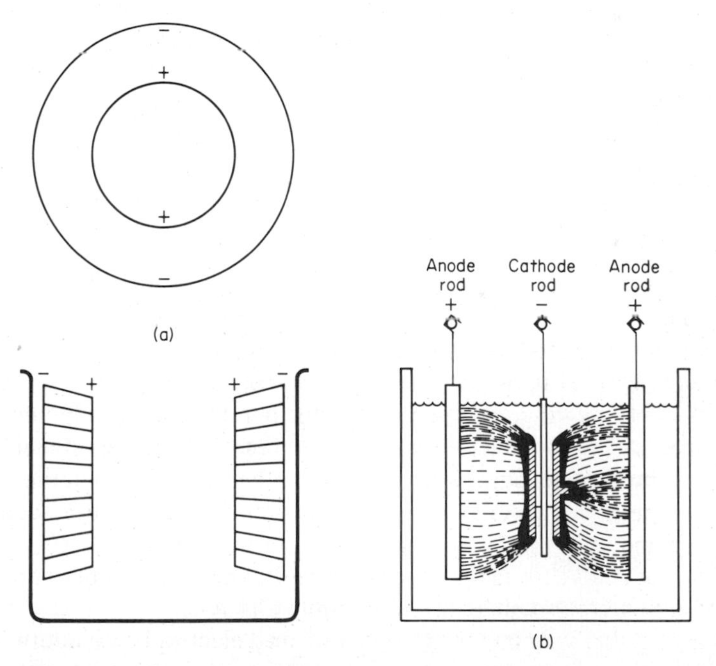

Fig. 2-1 Current distribution with (*a*) anode and cathode concentric, (*b*) shaped cathode. Closeness of lines indicates relative distribution of current. *(From ASTM B 507, by permission.)*

a certain minimum amount of metal on particular surfaces of the part, any excess metal deposited on other areas of the surface is a waste of metal, which may be very significant if the metal is expensive, as most are. Therefore the electroplater takes pains to position the cathodes in the plating tank so as to decrease the difference between the high- and the low-current-density areas; the designer can help by designing parts more amenable to uniform plating.

Some solutions also tend to decrease the difference between the thickest and the thinnest deposits on a given part; such solutions are said to have good *throwing power*. Throwing power may be defined as the improvement in metal distribution on a part over that which would be expected from purely geometric considerations. It is a very important property of electroplating solutions which will be discussed in more detail later.

POTENTIAL RELATIONSHIPS

The previous section, concerning current relationships, dealt with fairly straightforward concepts: a given quantity of electricity produces a definite and at least theoretically predictable amount of chemical change. This current, in turn, is driven through the conducting medium, either electrolytic or metallic, by a potential difference, or voltage; and the voltage necessary to force a given current through a conductor is given by Ohm's law:

$$E = IR$$

where E = volts, I = amperes, and R = ohms. Since electrolytic resistances are generally much greater than metallic resistances—the resistance of a 1 percent solution of sodium chloride is about 50 million times that of an equal section of copper—in considering electrolytic processes, we can generally neglect the resistance of the metallic portions of the circuit and focus our attention on only the electrolyte. In other words, given copper or similar conductors (bus bars) of sufficient capacity for the external portions of the circuit, our only concern will be the voltage-current relationships in the electrolyte itself and at the interfaces between it and the two electrodes.

The potential relationships within the electrolyte and at the electrolyte-electrode interfaces are not, however, as straightforward as the current relationships, and some of them, such as overvoltages, are not yet thoroughly understood. Here we introduce some of the more important concepts; further discussion will be left for a later chapter.

Electrode Potentials

When a metallic electrode dips into a solution containing ions of that metal, an equilibrium is set up between the tendency of the metal to enter solution as ions

$$\mathrm{M} \rightarrow \mathrm{M}^{n+} + ne^- \qquad e^- \text{ represents one electron}$$

and the opposing tendency for the ions to lose their charge and deposit on, or in, the metal

$$M^{n+} + ne^- \rightarrow M$$

In the absence of an external voltage, this exchange is a dynamic one. Initially, one of these reactions may occur more rapidly than the other, but this will increase the rate of the reverse reaction and an equilibrium will result: metal ions will go into solution at the same rate as ions are discharged. This will occur long before any concentration change can be detected by analysis. But before this equilibrium is set up, a charge separation will occur: if the ionization is initially faster, the metal will become negatively charged relative to the solution; if the deposition is the faster, the metal will become positively charged relative to the solution. The resultant potential between the metal and the solution is called the *electrode potential.*

Such a potential can be set up in systems other than those involving metals. For example, if hydrogen gas is bubbled over a platinum (inert) electrode in a solution of hydrogen ions, the potential depends on the equilibrium

$$H_2 \leftrightharpoons 2H^+ + 2e^-$$

This potential (the hydrogen electrode) is the reference point from which others are measured. A similar potential is set up between oxygen gas and a solution of hydroxyl ions:

$$O_2 + 2H_2O + 4e^- \rightleftharpoons 4OH^-$$

And if a metal can exist in two different ionic forms, such as iron(III) and iron(II), the equilibrium is

$$Fe^{3+} + e^- \leftrightharpoons Fe^{++}$$

The preceding equation describes what is known as a *redox* (oxidation-reduction or ORP) *potential,* although there is no fundamental difference between it and the others: the reaction

$$M^{2+} + 2e^- \rightleftharpoons M$$

is also an oxidation-reduction.

The magnitude of the potential between a metal and a solution of its ions is given by the *Nernst equation**

$$E = E^0 + \frac{RT}{n\mathfrak{F}} \ln a \qquad (2\text{-}3)$$

where E^0 is a constant characteristic of the material of the electrode; R = the gas

*More precisely,

$$E = E^0 + \frac{RT}{n\mathfrak{F}} \ln \frac{a}{a_M}$$

where a_M is the activity of the metal itself. With pure metals this is normally unity, and so it can be neglected; but it is a factor in considering the electrode potentials of alloys. In general, the activity of the oxidized state of the reactant is in the numerator of the logarithmic term and that of the reduced state in the denominator.

constant, 8.3143 $J \cdot K^{-1} \cdot mol^{-1}$; T = absolute temperature in kelvins; $\mathfrak{F}$ = the faraday; n = valence change; and a = activity of the metal ion. If approximate treatment is sufficient, we may use instead of a the activity c = the concentration of metal ions, where $c = C\alpha$, C = the stoichiometric concentration, and α is the "apparent degree of ionization" as given by the conductance ratio.*

When a (or c) = 1, ln a = 0, so that $E = E^0$. E^0 is called the *standard electrode potential;* it is the potential of an electrode in contact with a solution of its ions of unit activity. By definition, the standard potential of the hydrogen electrode is zero at all temperatures.

Single electrode potentials cannot be measured directly. It is always necessary to construct a cell in which the test electrode is coupled to a "reference" electrode. In theory, though seldom in practice, it is possible to determine the standard electrode potential of a metal by measuring the potential of a cell in which one electrode is the metal in contact with its solution at unit activity and the other electrode is the hydrogen electrode or some other electrode whose potential versus the hydrogen electrode is known, such as the saturated calomel or the silver–silver chloride electrode. Often such direct measurements are not feasible, and the standard electrode potentials are calculated from heats of reaction or by other thermodynamic methods.

A tabulation of E^0 for the elements results in the familiar *electromotive (emf) series,* as shown in Table 2-1. In this table, the elements having the greatest negative potentials have the greatest tendency to pass into solution; they are the *base* metals. Those having the greatest positive potentials tend to be inert; they are the *noble* metals. This convention as to signs of the potentials is not universally accepted, and in some texts the signs are reversed. This unfortunate confusion is often avoided by using the unambiguous terms *base* and *noble* metals and potentials.

The electromotive series is familiar to most chemists, and it is sometimes stated that "any metal higher (more negative) in the series will displace from solution any metal below it in the series." But this is not strictly true under all conditions, and the series should be used with care in practical cases. First, it is calculated for metal ions at unit activity, a condition very seldom encountered. Second, kinetic factors are often just as important as theoretical equilibria. In other words, just because a reaction *should* take place from thermodynamic considerations is no guarantee that it *will* do so.

There are other limits to the practical use of the emf series, which is calculated for metals in their so-called standard states. The previous history of an electrode may have an effect on its potential: a metal under strain is negative (active) to the same metal unstrained; large crystals may have a different potential from small

*It is instructive to follow the dimensions of this equation to verify that it does indeed result in a true potential. The dimensions of voltage in SI units are $kg \cdot m^2 \cdot s^{-3} \cdot A^{-1}$; those of R, the gas constant, are $J \cdot K^{-1} \cdot mol^{-1}$; in turn, joules, J, are $kg \cdot m^2 \cdot s^{-2}$; and the faraday is $A \cdot s \cdot mol^{-1}$. One may wish to factor out the fraction and determine for oneself that the equation does yield volts equals constant voltage plus an additional voltage, depending on the molarity or activity.

Table 2-1 The Electromotive Force Series

Electrode	*Potential, V*	*Electrode*	*Potential, V*
$Li \rightleftharpoons Li^+$	−3.045	$Co \rightleftharpoons Co^{++}$	−0.277
$Rb \rightleftharpoons Rb^+$	−2.93	$Ni \rightleftharpoons Ni^{++}$	−0.250
$K \rightleftharpoons K^+$	−2.924	$Sn \rightleftharpoons Sn^{++}$	−0.136
$Ba \rightleftharpoons Ba^{++}$	−2.90	$Pb \rightleftharpoons Pb^{++}$	−0.126
$Sr \rightleftharpoons Sr^{++}$	−2.90	$Fe \rightleftharpoons Fe^{3+}$	−0.04
$Ca \rightleftharpoons Ca^{++}$	−2.87	$Pt/H_2 \rightleftharpoons H^+$	0.0000
$Na \rightleftharpoons Na^+$	−2.715	$Sb \rightleftharpoons Sb^{3+}$	+0.15
$Mg \rightleftharpoons Mg^{++}$	−2.37	$Bi \rightleftharpoons Bi^{3+}$	+0.2
$Al \rightleftharpoons Al^{3+}$	−1.67	$As \rightleftharpoons As^{3+}$	+0.3
$Mn \rightleftharpoons Mn^{++}$	−1.18	$Cu \rightleftharpoons Cu^{++}$	+0.34
$Zn \rightleftharpoons Zn^{++}$	−0.762	$Pt/OH^- \rightleftharpoons O_2$	+0.40
$Cr \rightleftharpoons Cr^{3+}$	−0.74	$Cu \rightleftharpoons Cu^+$	+0.52
$Cr \rightleftharpoons Cr^{++}$	−0.56	$Hg \rightleftharpoons Hg_2^{++}$	+0.789
$Fe \rightleftharpoons Fe^{++}$	−0.441	$Ag \rightleftharpoons Ag^+$	+0.799
$Cd \rightleftharpoons Cd^{++}$	−0.402	$Pd \rightleftharpoons Pd^{++}$	+0.987
$In \rightleftharpoons In^{3+}$	−0.34	$Au \rightleftharpoons Au^{3+}$	+1.50
$Tl \rightleftharpoons Tl^+$	−0.336	$Au \rightleftharpoons Au^+$	+1.68

ones; and even the nature of the metallographic crystal face exposed to the solution may determine its potential.

In Eq. (2-3, if numerical values are substituted for R and $\mathfrak{F}$, if T is taken as 25°C (298 K), and if logarithms are taken to the base 10 instead of base e, we obtain

$$E - E^0 + (0.059/n) \log a \text{ (or } \log c \text{ approx.)} \tag{2-4}$$

Now E^0 and n are constants for any given electrode reaction; therefore the potential difference between two identical metallic electrodes dipping into solutions that differ only in the concentrations of their ions is given by this equation and depends only on these concentrations. For a tenfold change in the concentration of a univalent ion, the potential will change by 59 mV; for a bivalent ion, it will change by 29.5 mV, etc. As a corollary, it will be seen that a potential difference will be set up between an anode and a cathode in an electroplating cell when the current is shut off. Momentarily the solution around the anode will be more concentrated, and the solution around the cathode more dilute, than the bulk of the solution. Until convection or agitation has equalized matters, therefore, the situation is that of a metal dipping into solutions of different concentrations of its ions. Such a situation is known as a *concentration cell*. Such cells are also important for corrosion problems, as shown in Chap. 3.

OTHER POTENTIAL EFFECTS

The discussion up to this point has been concerned with electrode systems at equilibrium—in other words, with reversible processes. The equilibrium is dynamic, with metal ions being discharged and metal atoms being ionized, but these two effects cancel each other and there is no net change in the system. The small currents involved are called *exchange currents*. For useful reactions to occur—metal deposition at the cathode, metal dissolution at the anode, for example—the system must be moved away from this equilibrium condition: an external source of current must be supplied (or current must be generated internally, as by contact of dissimilar metals—the effect is the same), and this introduces an element of irreversibility into the system. We are no longer able to discuss strictly equilibrium conditions, and the situation becomes more complicated.

This external potential, required to make it possible for useful electrode reactions to take place at practical rates, is made up of many elements.

Perhaps most obvious is the potential required to overcome the resistance of the electrolyte. Electrolytes obey Ohm's law, $E = IR$. If the resistance of the bulk of the electrolyte remains constant (as it will if the heating effect of the current is neglected), then of course the higher the current density, the higher the potential needed to drive the current through the solution.

In addition to this obvious requirement for an additional external potential, more subtle changes take place at the electrodes, leading to what is variously termed *overvoltage, polarization,* and *overpotential*. Unfortunately, electrochemists do not agree on the application of the words *polarization* and *overvoltage*. The following discussion is much simplified.

The equilibrium established at the interface of an electrode and the solution has been discussed above; the equilibrium is dynamic, in which charge carriers pass through the interface in both directions at equal rates (the exchange currents). The exchange current density i_0 is the same in both directions. This means that if the anodic and cathodic partial current densities are denoted by i_+ and i_-, then at equilibrium

$$i_+ = |i_-| = i_0$$

($|\;|$ means the absolute value of, neglecting sign).

When an external source of current is imposed on the system, so that electrode reactions instead of balancing each other in dynamic equilibrium take place in a net direction—that is, $i_+ \neq i_-$—considerations of electrode kinetics are superimposed on the thermodynamic equilibria. Kinetics is concerned with the mechanisms of reactions rather than with their total result—with the individual steps whereby, for example, a metal ion in solution, surrounded by its coordinated shell of water molecules or other ligands, finally comes to rest as part of a metallic deposit on a cathode. The total reaction may be

$$M^{++} + 2e^- \rightarrow M$$

but this tells us nothing about the path from left to right in the equation. Each step in that path may be affected by kinetic factors which add to the potential necessary to drive the reaction. These "resistances" to the progress of the reaction make themselves felt in various increases in the voltage required to make it occur; these are subsumed under the heading "overvoltage." Although a detailed consideration of electrode kinetics is beyond our scope (see texts on electrochemistry such as those cited in the Bibliography), we mention some of the factors involved.

In short, in an operating electrolytic cell through which an appreciable current density is passing, the potential will differ from the equilibrium potential:

$$\eta = E_i - E_{eq}$$

where E_i is the potential when current is flowing and E_{eq} is the equilibrium potential. η is the overpotential or overvoltage. This overpotential may have several causes.

During the passage of current the activities (related, usually directly, to the concentrations) of the reactants near the electrodes change. This causes *concentration overpotential.* When deposition at the cathode has proceeded for a finite time, the concentration of metal ions near the cathode decreases, because the ions are not replenished as fast as they are deposited. This change in concentration lowers the equilibrium potential according to the Nernst equation:

$$\eta_{conc} = \frac{RT}{n\mathfrak{F}} \ln \frac{a_E}{a_0} \tag{2-5}$$

where a_E is the activity of the depositing ion next to the electrode, a_0 is the activity in the bulk of the solution, and n is the number of electrons involved in the discharge of the ion. Within a few milliseconds after current is interrupted, η_{act} (activation overvoltage, below) disappears, and η_{conc} can be measured if the electrode is reversible. In other words, a potential-measuring device across the cell will continue to register a voltage for a short time even after the current is interrupted. This portion of the overvoltage is concentration overvoltage. It disappears gradually, as convection restores the original concentration throughout the solution and at the face of the electrode. For cathodic processes, a_E is less than a_0; therefore η_{conc} is negative. At soluble anodes the changes are comparable but in the opposite direction.

Overpotential may be required to overcome various kinetic barriers to the reaction; the most commonly encountered is hydrogen overvoltage. This is an example of *activation overvoltage*. For any reaction to take place, a minimum energy exists which the reactants must possess; this may be thought of as an energy or potential barrier that must be overcome for the reaction to proceed. This barrier is illustrated in Fig. 2-2. Cathodic overpotential shifts the energy level of ions (in the layer next to the cathode) nearer to the potential barrier, so that it becomes easier to surmount: more ions can cross it in a given time, to form a deposit on the cathode. Effects at the anode are in the opposite direction but comparable.

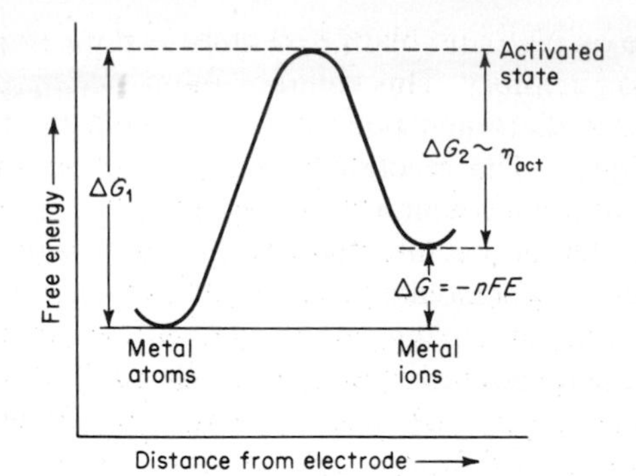

Fig. 2-2 Energy barrier in metal deposition.

That portion of the total electrode overvoltage which produces these effects is called activation overvoltage, η_{act}. Part of this is accounted for by the fact that most metal ions are hydrated in solution, and the coordination sphere of the ion must be distorted and the water molecules freed from their association with the metal ion in order for deposition to take place.

If η_{act} is more than about 50 mV, it can be described, for the cathodic process, by the *Tafel equation:*

$$\eta_{act} = a + b \log i \tag{2-6}$$

where a and b are constants that depend on the mechanism of the reaction, the activities of the substances involved, and the temperature; i is the current density. This equation shows that over a considerable range the activation overvoltage is a logarithmic function of the current density.

For most metals depositing on a cathode, η_{act} is fairly small and can usually be neglected in practical cases unless i, the current density, is very large. However, certain metals—notably those of the iron-cobalt-nickel triad—and all gases evidence considerable activation overpotential. Hydrogen overvoltage, in particular, is of great importance in electroplating processes; it accounts for the fact that some metals, such as zinc, can be deposited at all from aqueous solutions.

From the relation $E = E^0 + 0.059 \log a$, it would be expected that hydrogen ($E^0 = 0$) would be discharged from an HCl solution of unit activity at a cathode potential of 0 V. Actually, hydrogen bubbles do not form until the potential becomes considerably more negative:

$$E = E^0 + (0.059/n) \log a - \eta \tag{2-7}$$

at 25°C, where η = overvoltage.

Hydrogen overvoltage in particular has been extensively studied, and many values have been reported. It depends not only on the current density and the other factors mentioned above, but also on the physical and chemical nature of the cathode. It is higher on smooth surfaces than on rough ones, and its value varies from negligible on platinum black and quite low on graphite to perhaps as much as a volt or so on mercury. This value on mercury explains why extremely active metals such as sodium and potassium can be deposited into a mercury cathode to form amalgams. This reaction is of great importance in the production of sodium and potassium hydroxide in mercury cells.

A resistance overpotential occurs when an electrode acquires a surface film that in itself possesses a substantial resistance. In some instances, one electrode or both is covered with a film having a resistance different from that of the bath; this contributes to the total potential, sometimes markedly. In the anodization of aluminum, for example, it is the major factor. And many other anodes in practical electroplating baths form resistant films, especially in cyanide and stannate electrolytes.

When the anode and cathode reactions are not equal and opposite, another cause of additional potential enters; this involves the energy required to decompose a reactant in the system. In the case of aqueous solutions, it involves evolving hydrogen at the cathode and oxygen at the anode, which together require that sufficient energy be furnished to decompose water. This may, following Potter (see Bibliography), be called the *decomposition potential.*

In practical electroplating situations, both electrode reactions are often irreversible. Oxygen may be evolved at the anode and hydrogen at the cathode by the decomposition of water. In any working electrolytic cell, the rate of the cell reaction is determined by the current: in general, the greater the externally applied potential, the greater the current and therefore the faster the reaction. The relationship between the applied potential and the current flowing often takes the form shown in Fig. 2-3. The current first increases slowly (AB) and then suddenly, and further increases in applied potential cause relatively large increases in current. The onset of this sudden increase represents the beginning of an appreciable decomposition rate of the reactants. The portion AB of the curve is ascribed to extraneous reactions, and it is usual to extrapolate the upper portion CD back to zero current at E. The value of the potential at point E is known as the decomposition potential or voltage of the compound being decomposed in the cell. It has little or no theoretical importance, but is relevant to the

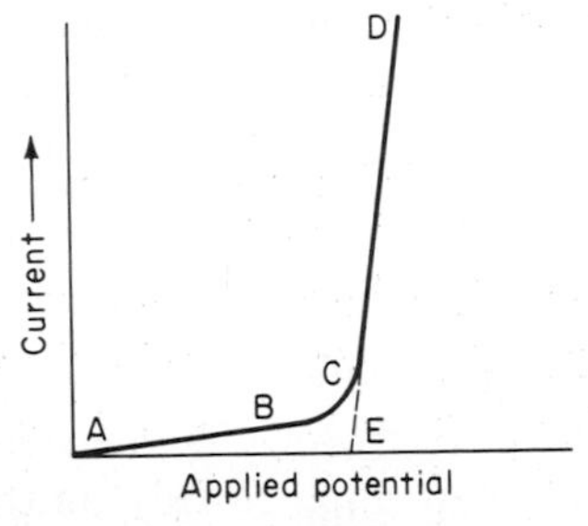

Fig. 2-3 Generalized decomposition curve.

practical operation and economics of electrolytic processes. It may be called the *practical* decomposition potential.

In most electroplating situations, the compound being decomposed at the electrodes is water (the potentials connected with the reactions ions → metal and metal → ions cancel out). We may consider the decomposition of water in the electrolysis of dilute sulfuric acid at inert electrodes, e.g., platinum. At the cathode the formation of hydrogen gas by the electrode reaction

$$2H^+ + 2e^- \rightarrow H_2$$

cannot proceed until the reversible hydrogen electrode potential is exceeded (in the negative direction). Similarly, the formation of oxygen at the anode

$$2H_2O \rightarrow O_2 + 4H^+ + 4e^-$$

takes place only when the reversible oxygen potential is exceeded (in the positive direction) at the anode. Therefore the potential difference between cathode and anode that must be exceeded if the decomposition of water is to take place is given by the difference between the appropriate reversible potentials for the hydrogen and oxygen electrodes in the system.

The hydrogen electrode potential, for convenience, may be put at zero (i.e., the activity of the hydrogen ion is 1). Then the necessary minimum potential difference that allows water to be electrolyzed (at 25°C) in air is equal to the reversible oxygen potential under the conditions $a_{OH^-} = K_w/a_{H^+} = 10^{-14}$ (see page 514, Chap. 26), and the pressure of O_2 = 0.2 atm (air contains about 20 percent oxygen). This potential is about 1.23 V.

However, the hydrogen and oxygen potentials are not reversible at most electrodes, and so to this figure must be added the overvoltages already discussed.

Theoretically, the decomposition potential, or the potential required to decompose electrolytically any compound, may be calculated from its free energy ΔG of formation:

$$-\Delta G = n\mathfrak{F}E$$

Examples of the Practical Use of Faraday's Laws

2.1. A current of 15 A was passed through a copper sulfate solution for 10 min. What weight of copper is deposited on the work if cathode efficiency is 100 percent?

Solution: 15A × 10 min × 60 s/min = 15 A × 600 s = 9000 C = 9000 C ÷ 96 500 C/$\mathfrak{F}$ = 0.0933 $\mathfrak{F}$. But 1 $\mathfrak{F}$ deposits 63.54 ÷ 2 = 31.77 g Cu; therefore 0.0933 $\mathfrak{F}$ deposits 0.0933 × 31.77 = 2.964 g Cu.

2.2. In Example 2.1 the area of work in the tank was 1500 cm^2. What is the average thickness of the copper deposit? The density of Cu is 8.93 g/cm^3.

Solution: 2.964 g (total Cu) ÷ 1500 cm^2 = 0.00198 g/cm^2 of Cu. If we consider 1 cm^2 of surface, a film 1 μm thick would have a volume of 10^{-4} cm^3 (10^{-4} cm × 1 cm^2) and would weigh 10^{-4} cm^3 × 9.93 g/cm^3 or 8.93 × 10^{-4} g. Actually, 1 cm^2 of the film weighs 19.8 × 10^{-4} g. Therefore, the film thickness is 19.8 ÷ 8.93 or 2.2 μm.

2.3. At a current of 15 A, how long must work be plated in a copper sulfate solution to achieve an average thickness of 5 μm of Cu if the cathode efficiency is 100 percent and the area of the work is 1000 cm^2?

Solution: Total volume of Cu is 5 μm × 1000 cm^2 = 5 × 10^{-4} cm × 10^3 cm^2 = 5 × 10^{-1} cm^3. The weight of Cu = 5 × 10^{-1} cm^3 × 8.93 g/cm^3 = 4.465 g. Plating 4.465 g of copper requires 4.465 ÷ 31.77 = 0.141$\mathfrak{F}$ = 0.141 × 96 500 = 13 562 C (A·s) or 13 562 ÷ 60 = 226 A·min. At 15 A this requires 226 ÷ 15 = 15 min.

2.4. Total area of work in the tank is 1.5 m^2; at an average current of 1000 A, the average thickness of zinc deposited was 25 μm in 15 min. What was the current efficiency?

Solution: Total volume of zinc deposited is 1.5 m^2 × 25 × 10^{-6} m = 37.5 × 10^{-6} m^3, or 37.5 cm^3. The density of zinc is 7.14 g/cm^3; so the weight of zinc is 7.14 × 37.5 = 268 g. 1500 A × 15 min × 60 s/min = 1500 × 900 A·s = 1.35 × 10^6 C; divided by 96 500, this equals 14$\mathfrak{F}$. This should deposit 14 × (65.38/2) = 458 g of zinc. Actual weight of Zn was 268 g; efficiency is 100 × (268/458) = 59 percent.

2.5. In plating a continuous strip 0.9 m wide with tin from an acid solution (valence change = 2), the desired strip speed is 500 m/min, current efficiency is 100 percent, and the thickness of tin is to be 0.4 μm (each side) at a current density of 5000 A/m^2. How long must the plating tank be, assuming that the strip passes straight through the tank with no return bends?

Solution: Density of tin = 7.31 g/cm^3; atomic weight = 118.7.

(*a*) 0.4 μm tin both sides = 0.8 μm total; on each square meter of strip the total volume of tin = 1 m^2 × 0.8 × 10^{-6} m = 0.8 × 10^{-6} m^3 = 0.8 cm^3; the weight of tin per m^2 = 0.8 × 7.31 = 5.85 g/m^2. Since the strip is 0.9 m wide, this requires 5.85 × 0.9 = 5.27 g tin per linear meter of strip.

(*b*) To deposit 5.27 g tin requires (5.27/118.7) × 2 = 0.0888$\mathfrak{F}$ or 0.0888 × 96 500 = 8569 A·s = 143 A·min on each linear meter of strip.

(*c*) Let the length of the tank in meters be L. Then at 500 m/min, each linear meter of strip is in the tank L/500 min, and in this time it must receive 143 A·min of electricity, at 5000 A/m^2.

(*d*) Each linear meter of strip has an area (both sides) of $2 \times 0.9 \times 1 = 1.8$ m^2. At 5000 A/m^2, total current to this area $= 5000 \times 1.8 = 9000$ A. Since the required quantity of electricity is 143 A·min, time available is 143 A·min $\div$ 9000 A $= 0.016$ min, which is the time the strip must be in the tank. This is also (as shown above) $L/500$ min; so $L/500 = 0.016$; $L = 8$ m.

The foregoing calculation is presented stepwise for instructional purposes. A convenient formula for strip plating is

$$t = \frac{KIEL}{S}$$

where t = coating weight in g/m^2 (each side of strip)
K = a constant depending on the coating metal, expressed in g/A·min
I = current density in A/m^2
E = current efficiency expressed as a fraction (100 percent = 1)
L = length of pass, in m
S = speed of strip, in m/min

2.6. In chromium plating, insoluble anodes are used and all metal deposited must be replaced chemically by adding chromic acid, CrO_3. Assuming cathode efficiency of 18 percent and disregarding drag-out, how much chromic acid must be added to the tank for each 1000 A·h of current passed?

Solution: Electrochemical equivalent of Cr $= 52 \div 6 = 8.67$ g/$\mathfrak{F}$ or 0.323 g/A·h; for 1000 A·h = 323 g Cr at 100 percent efficiency or 58.1 g Cr at 18 percent efficiency. $CrO_3/Cr = 100/52 = 1.923$; therefore 1.923×58.1 g CrO_3 = 112 g CrO_3 added per 1000 A·h of electricity.

Note: In any practical case, drag-out and other losses cannot be neglected as they are in this example.

2.7. In a laboratory experiment to determine the cathode efficiency of a nickel-plating bath, the cathode of a copper coulometer in series with the experimental cell was found to have gained 1.025 g. In the experimental cell, the weight of nickel deposited was 0.905 g. What is the current efficiency of the nickel-plating bath?

Solution: Total current (C) was $1.025 \div (63.54/2) = 0.03226\mathfrak{F}$. At 100 percent efficiency, this should deposit $58.71/2 = 29.355 \times 0.03226 = 0.947$ g nickel. Since weight of Ni deposited was 0.905 g, efficiency is $100 \times 0.905/0.947 = 95.6$ percent.

3 Corrosion

INTRODUCTION AND DEFINITION

Few metals are found "native"—i.e., in their uncombined state—in nature. Among those few are copper, silver, gold, the platinum metals, and a few rare finds of meteoric iron. It will be noted that all these except iron are the noble metals of the emf series, p. 19. In order to produce the other metals that are used in today's technology, it is necessary to expend relatively large amounts of energy to decompose their compounds, or ores, which are usually oxides or sulfides. It is reasonable to suppose that most materials in the lithosphere have found their most stable configurations over the millennia. The effort and energy required to recover a metal from its ores produces a material that is "unstable" with respect to its preferred condition. And even for those metals that do occur in the elemental state, as mentioned above, such occurrences are not the rule: the occasional finds of native copper, for example, would fall ridiculously short of satisfying the demand for the metal, and almost all of it must be recovered from its ores just as iron must be.

A corollary to these facts is that most metals, if given the chance, will revert to their combined state. A more "scientific" way of expressing this idea is to say that the free-energy change of the reaction $M + X \rightarrow MX$ (where M is a metal and X is any anion with which the metal is associated in the natural state) is negative; i.e., the reaction tends to proceed to the right. To reverse the reaction, energy must be supplied.

When metals thus revert to their combined state, we say that they *corrode*. The copper alloy bronze has relatively little tendency to corrode in most atmospheres, and during the Bronze Age the problem of corrosion was probably not too serious. But when bronze weapons and other artifacts gave way to iron, when the Iron Age began, the problem presented itself and has been with us ever since. The authors of the Bible were familiar with it.*

*"Lay not up for yourselves treasures upon earth, where moth and rust doth corrupt." (*Matt.* **6**, 19)

Corrosion is the result of the reaction of metals with nonmetallic elements of their environments. The compounds so formed, called *corrosion products,* are usually either oxides or salts; their nature often influences the course of the reaction and may determine whether additional protection is required. The most widely used metal, iron, unfortunately is little protected by its natural corrosion product, rust, and so usually requires some form of corrosion protection.

Corrosion control is based on preventing or slowing down the reaction of a metal with its environment. This aim may be approached through controlling the environment, the nature of the metal, or the nature of the interface between the two.

Controlling the environment may entail elimination of such corrosive agents as oxygen, humidity, dust, and sulfur gases. An example is the deaeration of boiler waters.

The Nature of the Metal

Controlling the nature of the metal means choosing for any construction a metal that is resistant to the corrosive agents in the environment where it will be used. Cathodic protection of pipelines and similar structures might be considered a subclass of this method, since it alters the nature of the metal by deliberately giving it a potential less anodic than its normal one. Careful design of the structure can often lessen corrosion damage: avoidance of dissimilar metal couples, and of crevices and blind recesses, is important, as we will show.

Protective Coatings

The third means of reducing corrosion hazards—controlling the nature of the interface between the metal and its environment—includes protective coatings, both metallic and nonmetallic, and it is the one considered in this book.

Occasionally the protective coating is natural: if the initial corrosion product is highly insoluble and is formed in intimate contact with the metal, thus providing continuous coverage, corrosion will start but not proceed beyond the initial phase. Thus the resistance of lead to sulfuric acid solutions is due to a thick protective film of insoluble lead sulfate on the surface. The passivity of certain metals such as the stainless steels is due to a surface film, often of monomolecular thickness. Passivity may sometimes be induced by adding a passivating agent to the environment, as by adding soluble chromates to water in recirculating water systems and air-conditioning equipment. Other means of control include adding inhibitors to the environment; these generally act by forming a protective adsorbed film on the surface of the metal. But the most common form of corrosion control is the addition of a coating to separate the metal from its environment.

The Importance of Corrosion

The more or less gradual degradation of the countless metallic structures on which the modern world depends—from bridges to nuts and bolts, from automobiles to boiler tubes and nuclear reactors—is a practical problem of enormous consequence. The cost of corrosion, including both the steps taken to prevent or reduce it and the materials rendered useless by it, is almost impossible to estimate; one authority has put it at about $6 billion annually for the United States alone.*

TYPES OF CORROSION

Corrosion reactions may be classified in many ways; a convenient one divides them into two classes: direct combination of metals (or metal ions) with nonmetallic elements, and those reactions in which the metal dissolves, frequently in an aqueous environment, and later combines with nonmetallic constituents to form corrosion products. In so dissolving, the metal replaces hydrogen or some other metal. Direct combination is sometimes referred to as "dry corrosion"; examples of this are oxidation and sulfidation reactions. The replacement reactions are called "wet corrosion." When metals corrode in the atmosphere either or both of these processes may take place: high temperatures and dry air favor the former, moist air favors the latter. Both types are electrochemical, depending on the operation of electrochemical cells on the surface of the metal.

Direct Reaction, or Dry Corrosion

Direct reaction includes oxidation in air; reaction with sulfur vapors, hydrogen sulfide, or other constituents of a dry atmosphere; and reaction with liquid metals such as sodium. Most such reactions occur at relatively high temperatures; steel, which is stable in dry air, rapidly forms "mill scale" at the temperature of the steel furnace and during hot-rolling and forging. Some metals such as molybdenum are resistant to oxidation at room temperature but form oxides when heated; worse yet, molybdenum oxide, unlike iron oxide, is not solid at these temperatures but vaporizes, and the whole mass of metal simply "evaporates"—a case of so-called catastrophic corrosion. Such properties severely limit the utility of some of the "refractory" metals and offer one reason for the interest in protecting them with a coating of other metals.

The reaction $M + \frac{1}{2}O_2 \rightarrow MO$ does not appear to involve electrochemical mechanisms, but on closer study it turns out that even this form of corrosion basically depends on an electronic exchange mechanism involving a flow of current. High-temperature corrosion of this type is not of prime concern here, and we must be satisfied with a simplistic statement: molecular oxygen is initially

*This estimate was published in 1971 and does not include any correction for inflation.

absorbed on the metal surface. Here it decomposes into atoms, which ionize according to $\frac{1}{2}O_2 + 2e^- \rightarrow O^{--}$. The other half of the circuit is provided by the ionization of the metal: $M \rightarrow M^{++} + 2e^-$. The oxide and metal ions combine, forming the initial layer of the oxide film. Metal ions continue to be formed at the surface, and electrons diffuse through the oxide layer and ionize the oxygen at the surface. The oxide ions diffuse into the oxide layer and react with the metal ions; thus the oxide layer gradually increases in thickness. In some cases it may be the metal that ionizes, and the metal ions that diffuse to the surface, with the same result. Whether this type of corrosion continues to an unacceptable level or ceases without doing much damage depends on the properties of the metal oxide, how permeable it is, and how adherent it is to the metal surface.

Wet or Electrolytic Corrosion

The solution of a metal in a liquid occurs at discrete sites, anodes; the half-reaction may be represented, for a divalent metal, as

$$M \rightarrow M^{++} + 2e^- \tag{3-1}$$

This represents the formation of a metal ion, M^{++}, in solution, leaving two electrons in the metal. To maintain electrical neutrality, a corresponding cathodic reaction that consumes electrons must occur. In acidic solution the cathodic reaction may be

$$2H^+ + 2e^- \rightarrow 2H \qquad 2H \rightarrow H_2 \tag{3-2}$$

In nearly neutral solutions, more commonly found in natural environments, such as many atmospheres, soil, or seawater, the cathodic reaction, instead of being the formation of hydrogen, is

$$\tfrac{1}{2}O_2 + H_2O + 2e^- \rightarrow 2OH^- \tag{3-3}$$

Here the oxygen dissolved in the electrolyte combines with the water and electrons to form hydroxyl ions.

A corrosion cell, then, consists of an anode and a cathode in electrical contact with each other and with an electrolyte. The driving force for the reaction is determined by the difference in potential or emf between the anode and the cathode.

We have already seen that the potential between a metal and a solution of its ions is given by the Nernst equation:

$$E = E^0 + (0.059/n)\log(M^{n+}) \tag{3-4}$$

where n is the number of electrons transferred in the reaction. When gaseous products are involved, they enter the Nernst equation as pressure (in atmospheres). Thus the reversible potential of a cathodic reaction involving oxygen, as in Eq. (3-3), is given by

$$E = E^0 + (0.059/2)\log[(pO_2)^{1/2}/(OH^-)^2] \tag{3-5}$$

The logarithmic term contains the oxidized species in the numerator and the reduced species in the denominator; and the effective concentrations or pressures are raised to the power corresponding to their coefficients in the equation of the reaction. Putting the two half-reactions together, we obtain for the complete corrosion reaction:

$$M + \tfrac{1}{2}O_2 + H_2O \rightarrow M^{++} + 2OH^- \qquad (3\text{-}6)$$

The emf of the corrosion cell is the difference between the two half-cell potentials:

$$\Delta E_{\text{cell}} = E_{\text{cathode}} - E_{\text{anode}} = E_c^0 - E_a^0 + (0.059/n)\log\frac{(pO_2)^{1/2}}{(OH^-)^2(M^{++})} \qquad (3\text{-}7)$$

The E^0 in this equation are the standard potentials, already defined in Chap. 2 and listed in Table 2-1.

The change in free energy, ΔG, accompanying a cell reaction is given by

$$\Delta G = -n\mathfrak{F}\Delta E_{\text{cell}} \qquad (3\text{-}8)$$

in which n and E_{cell} have the same meanings as before; $\mathfrak{F}$ is the faraday, 96.5 kJ (23 050 cal) per volt. If the reaction is to occur spontaneously, ΔG must be negative; that is, ΔE_{cell} must be positive. The higher the free-energy change, the greater the tendency for the reaction to occur. This tendency is based on thermodynamics, but kinetic factors often play a significant part, especially in determining the rate of the reactions. Although thermodynamic tendencies indicate whether the reaction can take place, these kinetic factors determine the rate. Obviously in corrosion reactions, reaction rates are of utmost practical importance. The rate is fixed by the amount of current that flows—amperes per unit area per unit of time. The kinetic factors are complex, but something can be said about the relative importance of those that increase and those that decrease the rate of corrosion. But before considering corrosion rates, we must illustrate some typical corrosion cells and how they operate.

Examples of Electrolytic Corrosion

Example A. The container may be either an H tube or an ordinary beaker with a porous diaphragm, as shown in Fig. 3-1(a) and (b). In either case the purpose of the horizontal leg of the H, or of the diaphragm, is to separate the solution into two parts, so that gross mixing is prevented but ionic migration from one part to the other is not hindered. We now place a solution of sodium chloride (common salt) in the container, and in each half we place a piece of zinc metal that has been weighed previously. Each piece of zinc is attached by means of a clamp to a copper wire, and the external circuit is completed through an ammeter.

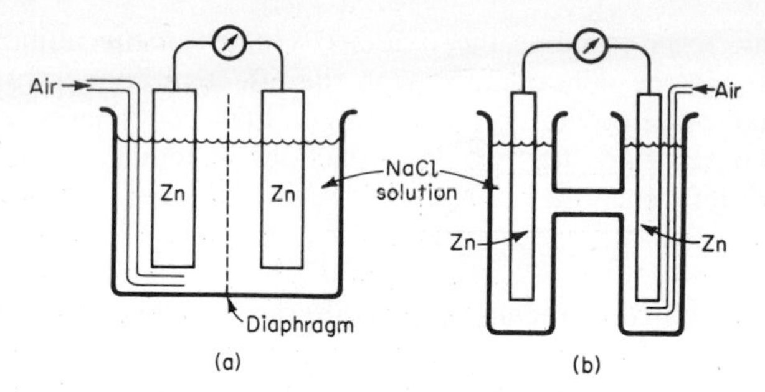

Fig. 3-1 Example A.

As we have seen (p. 16), each piece of zinc in contact with a solution will tend to dissolve until this tendency is exactly balanced by the opposite tendency for zinc ions to be discharged, and a potential will be set up between the electrode and the solution, as given by the Nernst equation. Since the two electrodes are identical, dipping into identical solutions, the potentials will also be identical, and no current will flow: the ammeter needle will remain at zero.

If we now bubble air or oxygen over one of the electrodes, a current will register on the ammeter. If the flow of oxygen is stopped, the current will be maintained for a few moments and then gradually fall to zero. On weighing the two pieces of zinc, it will be found that the electrode in the still compartment—that over which no oxygen passed—has lost weight, while that over which oxygen was bubbled remained the same. Moreover, if a record is kept of the current flowing times the total time during which it flowed, the loss in weight of the zinc will be seen to be exactly that calculated from Faraday's laws.

This example seems somewhat artificial, not corresponding to real situations. But real situations are very similar, as may be seen from the following examples.

Example B. If a piece of unrusted steel be inserted into a beaker of salt solution or seawater and left there for some time, rust will form as expected; see Fig. 3-2. And, at least in the early stages of corrosion, the rust will form near the bottom of the steel panel, far removed from the source of oxygen, which in this case is the

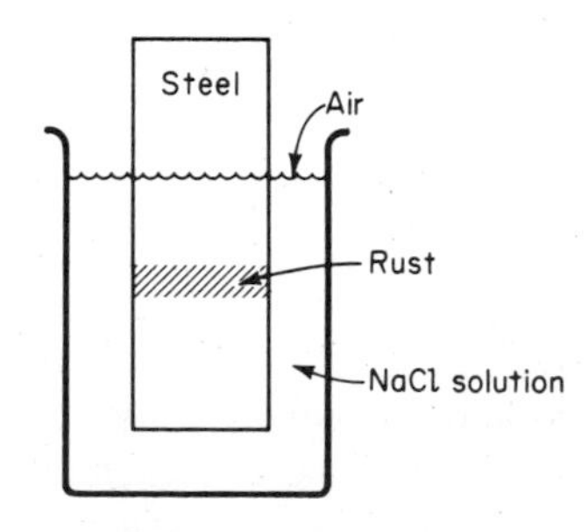

Fig. 3-2 Example B.

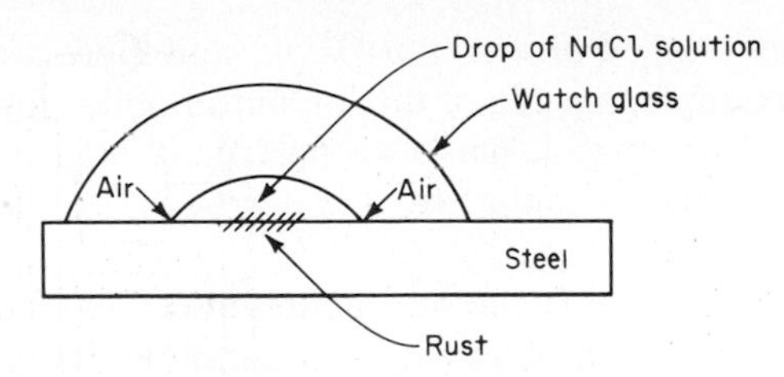

Fig. 3-3 Example C.

air above the solution. As before, oxygen is necessary for corrosion to occur, but it occurs on those areas of the metal where oxygen is least available. With a single piece of metal, as contrasted with the two separate pieces in Example A, we have no simple way of observing the flow of current. But it is not difficult to imagine that such flow takes place through the metal itself instead of through an outside conductor—in other words, that the circuit is completed internally instead of externally.

Other simple and related examples serve to reinforce the point.

Example C. Place a piece of steel on a horizontal plane and place a drop of electrolyte—salt water or even rainwater—in the middle of the panel; see Fig. 3-3. Making sure that the water does not evaporate, perhaps by placing an inverted watch glass over it, observe the formation of rust. It will be found that the initial rusting takes place under the center of the drop, once again as far as possible from the oxygen source.

Example D. Immerse a piece of steel in a neutral salt solution, and place on the panel a curved nonmetallic object such as a chemical watch glass as shown in Fig. 3-4. Rusting takes place in an annular ring around the area of contact between the glass and the steel. The cause is the same as before: the portions of the steel to which oxygen has the most ready access are those marked *a* in the figure, while the portion adjacent to the contact point receives less oxygen and corrodes.

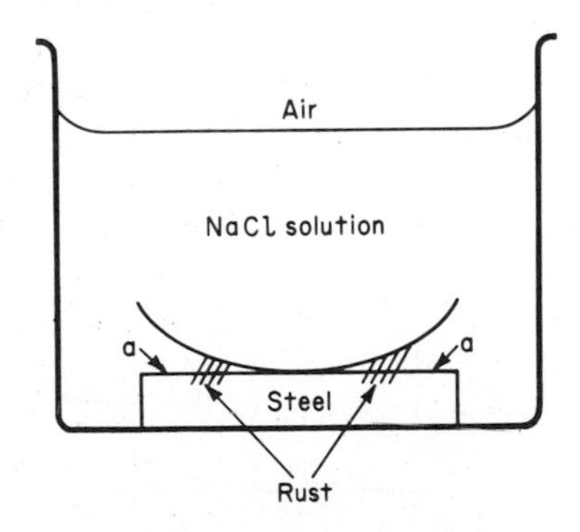

Fig. 3-4 Example D.

This phenomenon is called *crevice corrosion,* and there are many examples of it in practice. For example, a piece of dirt on automobile trim protects the metal directly beneath it from oxygen, while air has ready access to the surrounding metal; the metal immediately under the dirt corrodes.

Example E. Now return to the H cell or diaphragm cell of Example A. This time, replace one of the pieces of zinc with a copper panel. Perhaps a slight current will flow through the external circuit, but it will be of short duration. If we now bubble air over the zinc panel, we find little or no change; but if oxygen is bubbled over the copper panel, a significant current will flow, and we will find that the zinc again has lost weight. The same effect will occur if we add to the solution surrounding the copper an oxidizing agent such as potassium permanganate. Finally, we may carefully replace the sodium chloride solution surrounding the copper panel with a solution of copper sulfate: once again a current will flow and will be sustained.

Example F. To illustrate the effect of strain in a metal, the following experiment will serve (see Fig. 3-5). Immerse an iron nail in a jelled neutral salt solution to which some potassium ferrocyanide has been added, as an indicator for iron ions, as well as a little phenolphthalein, which turns red in the presence of alkali. After some time, the solution adjacent to the head and point of the nail will turn blue, while that near the middle turns red. Iron has dissolved at the ends of the nail, and alkali has been produced around the middle. During manufacture of the nail, the head and the point were formed by cold-working, which placed these parts in a strained condition. Obviously this has made them more active than the relatively unstrained body of the nail. The only reaction that could have caused the formation of hydroxyl ions near the main portion is

$$\tfrac{1}{2}O_2 + H_2O + 2e^- \rightarrow 2OH^- \tag{3-9}$$

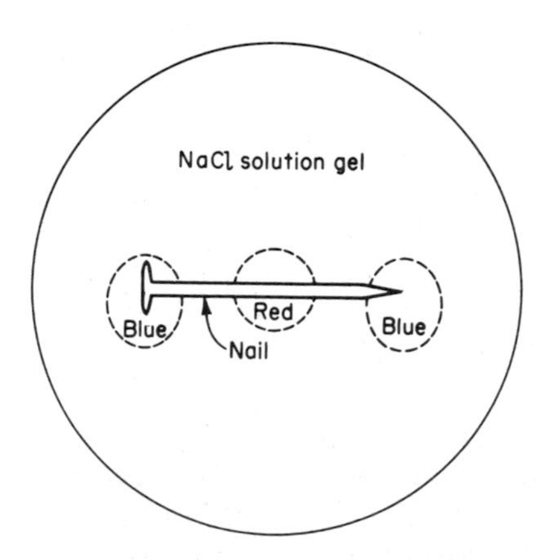

Fig. 3-5 Example F.

We must now attempt to explain these observations and to relate them to conditions in the real world outside the laboratory.

Explanations. A. When no air was available to either electrode—or, more accurately, since the solutions surrounding both electrodes contained equal amounts of dissolved oxygen—the electrodes were identical. No potential existed between them, and there was, therefore, no driving force for a chemical reaction. When excess oxygen was made available to one electrode, the other corroded. The potentials must have been different, in order for a current to be driven through the circuit. The potential at the still electrode depended on the Nernst equation:

$$E = E^0 + (0.059/n) \log a \qquad (3\text{-}10)$$

or, in this case since zinc is divalent,

$$E = E^0 + 0.0295 \log a \qquad (3\text{-}11)$$

For zinc E^0 is known to be -0.76 V (see Table 2-1); and, since the concentration of zinc ions in the solution was originally zero and even after some corrosion has occurred is very small, log a will be a large negative number.

At the oxygen electrode, the potential is as given in Eq. (3-5). The cell potential is shown in Eq. (3-7).

An electrolytic reaction requires both an anode and a cathode; we now have an anodic half-reaction, $Zn \rightarrow Zn^{++} + 2e^-$, and a cathode reaction, as shown in Eq. (3-9). The potentials are different, current can flow, and a reaction can take place.

Also shown by this example is that oxygen, or some similar oxidizing agent (see Example E), is necessary for corrosion to occur. The possibility of an anodic reaction, in this case the oxidation of zinc to zinc ions, is always present; but to complete the cell, a corresponding reduction or cathodic reaction must be available.

B, C, D. These examples depend on "differential aeration" effects. At the portions of the metal closest to the supply of oxygen, Eq. (3-9) represents the cathodic half of the electrolytic cell and the anodic half-reaction is solution of the metal, in this case iron. Since in these nearly neutral solutions iron ions are not soluble, iron oxide (hydrated) is formed and appears as the familiar rust. In each case, oxygen was more available to one part of the metal than to another. The part distant from the oxygen source became the anode and corroded; the part where oxygen was most easily available was the site of the cathodic reduction of oxygen to hydroxyl ions.

This type of corrosion is of great practical importance. Usually it will be found that, in metals partly immersed in seawater, for example, corrosion occurs not at, but just below, the water line. As already mentioned, particles of dirt on automobile trim shield the portion immediately below them from air, and corrosion occurs beneath the dirt particle. Another type of corrosion due to this effect is so widespread that it has been given a name of its own: crevice corrosion. Blind holes, crevices in shaped parts, and the like are the practical counterparts of the setup illustrated in Fig. 3-4. Those portions within the crevice or blind hole will corrode, and the portion just outside the crevice will act as a cathode. A typical form of crevice corrosion occurs in ordinary chicken wire. At each point where the wires cross, a crevice is formed; corrosion takes place at these crossings—the points marked with arrows in Fig. 3-6.

E. When a zinc panel and a copper panel are placed in electrical contact, we know from their electrode potentials that a potential difference exists between them, such that zinc should corrode. However, although zinc can corrode in a salt solution, as shown by Example A, there is no possibility of a corresponding cathodic reaction on the copper: the only species present are sodium ions, which cannot be reduced at these potentials; chloride ions, already fully reduced; and a little oxygen dissolved in the solution, which is soon exhausted but accounts for the short-lived current observed. If now we bubble air over the copper, the situation is the same as in Example A. And when we replace the sodium chloride solution with one of copper sulfate, we allow a cathodic reduction reaction to take place and we furnish the necessary complement, the other half-reaction, $Cu^{++} + 2e^{-} \rightarrow Cu$. In fact, in this case we have nearly reproduced an early form of battery, the Daniell cell. The example illustrates another very general cause of corrosion, contact between dissimilar metals, often called *galvanic corrosion*.

F. This example shows that metals are not necessarily homogeneous; in fact, they seldom are. Thus although the whole nail is made of iron, the head and the point have undergone strains during the manufacturing process, and these strained parts have potentials different from the relatively unstrained body of the nail. Since the head and the point corroded, they must have been anodic sites, where the reaction was simply solution of the metal. The body of the nail was the

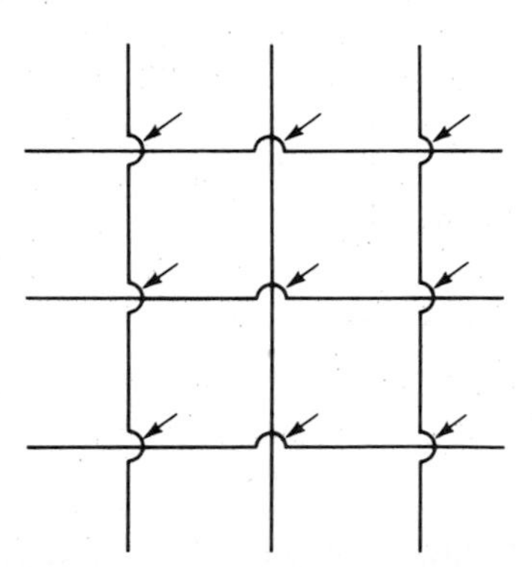

Fig. 3-6 Crevices in chicken wire (marked by arrows).

site of the cathodic reduction of oxygen to form hydroxyl ions, indicated by the red color of the phenolphthalein.

The two different physical states of the iron acted as if they were dissimilar metals. Analogous cases of corrosion in an apparently homogeneous metal arise from inclusions, segregation at grain boundaries, and different phases in alloys.

TYPES OF CORROSION DAMAGE

In general, corrosion can take two forms, one of which is usually far more serious than the other. We have seen that, for corrosion to occur, there must be both anodic and cathodic sites; these may be microscopically close together, as on the surface of an apparently homogeneous metal, or quite far apart, as with some types of pipeline corrosion. In either case they must exist and be electrically connected. The total amount of metal corroded depends entirely on Faraday's laws: the corrosion current flowing and the time during which it flows determine the weight of metal dissolved.

If the anodic area is large relative to the cathodic area, the anodic current density will be small and the total weight of metal dissolved will be distributed over a large area; if the anodic area is small relative to the cathodic area, the same amount of metal will be dissolved from a much smaller area. In the first case, the result is generalized corrosion; in the second, pitting or localized corrosion occurs. The second is usually the far more serious of the two.

This is not difficult to see. If a tank made of steel is filled with a mildly corrosive liquid and it corrodes at a fairly slow rate over its entire inside area, then the thickness of the steel will be eroded only gradually, and the tank may have a useful life of many years (assuming that the slight iron contamination of its contents is not harmful). But if the same amount of corrosion is concentrated at a few spots, the anodic areas, then it will not be long before the tank will be perforated, and it is no comfort to know that most of the tank is still in fine shape—it still will not hold a liquid!

In controlling corrosion, therefore, it is important to guard against situations where anodic sites are small in area relative to cathodic sites; however, generalized corrosion often can be tolerated. A good example of the practical application of this principle is given by the use of microdiscontinuous chromium plating, as discussed later and in Chap. 16.

CORROSION RATES

As the corrosion current flows and anodic and cathodic reactions occur, as already discussed, M^{++} ions accumulate about the anode, and near the cathode OH^- ions accumulate and dissolved oxygen is consumed. These reactions occur rapidly in comparison with the processes of diffusion and convection which carry away the M^{++} ions and replenish the dissolved oxygen. The changes that result change the reversible electrode potentials; and the greater the current density, the greater the change in the local environments, therefore the greater the change in potential. This change in potential with current density is an aspect of polarization (see Chap. 2). Polarization shifts the cathode potential in the anodic direction and the anode potential in the cathodic direction; thus it acts as a counter-emf which reduces the effective cell potential, or ΔE_{cell}.

Polarization also occurs if an energy of activation is required for the occurrence of the electrode reaction. It is observed that hydrogen evolution does not take place even though thermodynamic conditions are favorable: the potential at which hydrogen is evolved from the cathode in a corrosion cell is more negative than the reversible value predicted from the Nernst equation. The difference between the observed and the theoretical values is due to kinetic factors and is called the *hydrogen overvoltage*. The lower the hydrogen overvoltage, the more easily hydrogen is evolved and the greater the magnitude of the corrosion reaction for cells in which hydrogen evolution is the main cathodic reaction.

Activation energy, and thus hydrogen overvoltage, increases with current density (in the negative direction) according to the Tafel equation (see p. 22):

$$\eta = a + b \log i \tag{3-12}$$

In addition to the polarization caused by concentration effects and by overvoltage, polarization is caused by the resistance of the electrolyte as well as the resistances of any films formed on the corroding metal by the formation of slightly soluble salts or hydroxides. Polarizing films of this type may retard or even prevent further corrosion.

Corrosion rates are determined largely by the polarization behavior of the cell. The effective corrosion emf may be reduced by polarization practically to zero; in this case, the corrosion rate is limited to that necessary to maintain the polarization. Thus the progress of corrosion may be controlled by the magnitude of either anodic or cathodic polarization, or by both.

Figure 3-7 is typical of graphs that often are used to illustrate corrosion rates and their dependence on anodic and cathodic potentials. Potential difference is plotted against corrosion current density which determines the corrosion rate. At the start, the difference in potential between the anode and the cathode is at a maximum; this is the driving force available for the corrosion process. Some of this driving force is used to overcome reactions and processes occurring at the electrode surfaces: this is the polarization, already discussed. The polarization always tends to decrease the driving force, making the anode more cathodic and the cathode more anodic, and it increases with current density. Therefore the two lines eventually meet; where they meet determines the maximum possible corrosion rate.

Many factors tend to prevent the corrosion rate from ever reaching this theoretical maximum. One, illustrated in Fig. 3-7, is the resistance of the electrolyte. If this is high (line *a*), much of the potential difference will be used up in overcoming this resistance, and the corrosion current will be only that shown

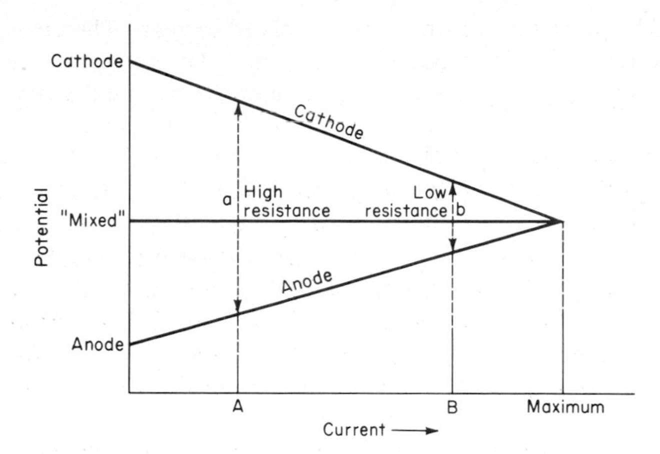

Fig. 3-7 Polarization curve in corrosion.

at *A*. If the resistance of the electrolyte is low, the potential required to overcome it will be less, as in line *b*, and the corrosion rate can be much higher (point *B*). Intermediate cases, of course, will yield intermediate rates.

It is not always true that polarization at the anode and polarization at the cathode are about equally effective, as in Fig. 3-7. If the cathode area is large and the anode small, most of the polarization will occur at the anode, and the cathodic polarization curve will be rather flat. Similar considerations govern the reverse situation.

Practical Consequences

Now we must examine briefly some of the practical aspects of these corrosion mechanisms as they are likely to affect metal finishers, and in particular electroplaters.

Dry corrosion, the direct reaction of a metal with its environment as in oxidation, is rarely a concern in metal finishing. Rarely, of course, does not mean never; some metals such as molybdenum which are subject to high-temperature oxidation require coating with a more resistant metal in order for their other properties to be used. Much more frequent, however, are cases of wet corrosion.

Differential aeration cells are met with in such phenomena as crevice corrosion; here the designer can be of more direct assistance than the metal finisher, by designing parts so that crevices, blind holes, and other similar areas are avoided in the original blueprint of the part. Once designed into the part, they are very difficult to guard against, and the finisher should, as much as possible, warn the designer of the hazards involved.

But the principal cause of corrosion in connection with metal finishing is the dissimilar metal contact. It is well known that a shipbuilder should not fasten a copper ship bottom with steel rivets: the large cathodic area (copper) and the small anodic area (steel), combined with the high-conductance electrolyte (sea-

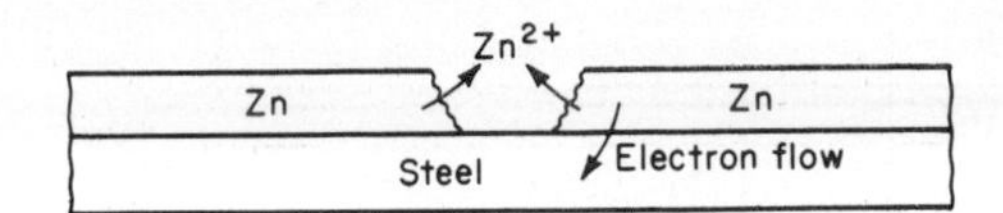

Fig. 3-8 Corrosion at a break in zinc coating on steel.

water), provide ideal conditions for corrosion of the steel. In fact, the rivets will soon corrode away, and the whole ship may be lost.

However, by their very nature electroplated coatings on metals do involve dissimilar metal contact, and the necessary electrolyte to complete the circuit is usually available in the form of contaminated moisture if of nothing more conductive. Such dissimilar metal contacts may be protective or not, depending on the nature of the metals.

If steel is coated with zinc (or cadmium), the coating metal is more active (anodic) than the basis steel. In the presence of corrosive conditions it is the zinc that corrodes anodically, and the steel is cathodic and is protected. Even if small pores or discontinuities develop in the coating, the steel does not corrode for some time, until practically all the zinc has disappeared from the immediate neighborhood of the pore. The situation is illustrated in Fig. 3-8. Because the zinc is ''sacrificed'' by itself corroding as it protects the steel, such coatings are often termed *sacrificial coatings*. Zinc is the most widely used coating metal for protecting steel from corrosion, although it has little purely decorative value.

The opposite condition obtains when a zinc die casting is coated with another metal for decorative purposes. We have seen that zinc is among the more active metals in the emf series, and zinc die castings are often electroplated with copper, nickel, and chromium for appearance. Now the anodic metal is the substrate, and the cathodic metals constitute the coating. When a pore or other discontinuity develops in the coating, the substrate becomes the anode, and since the area of exposed zinc is very small compared to that of the cathodic coating, corrosion can be very rapid. The condition is illustrated in Fig. 3-9. A similar condition exists when electroplating on aluminum, which is even more active than zinc. When electroplated zinc or aluminum corrodes, the failure may take place in several stages: first, a pore in the coating either is present initially or is formed in some way, as by corrosion of the coating or by mechanical damage. When the pore reaches the substrate, the latter corrodes, at first merely at the site of the pore. But as corrosion proceeds, it progresses laterally under the

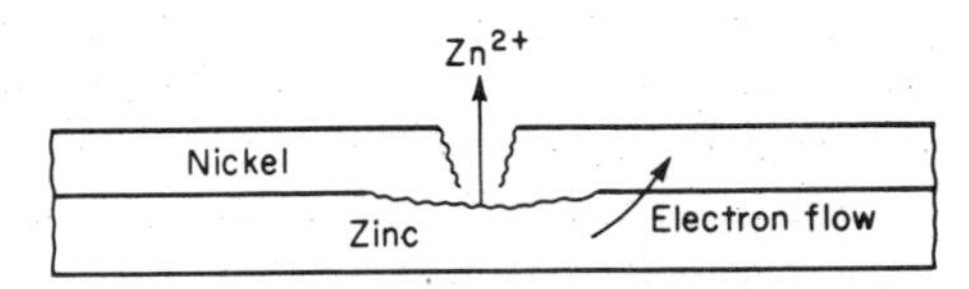

Fig. 3-9 Corrosion at a break in nickel coating on zinc.

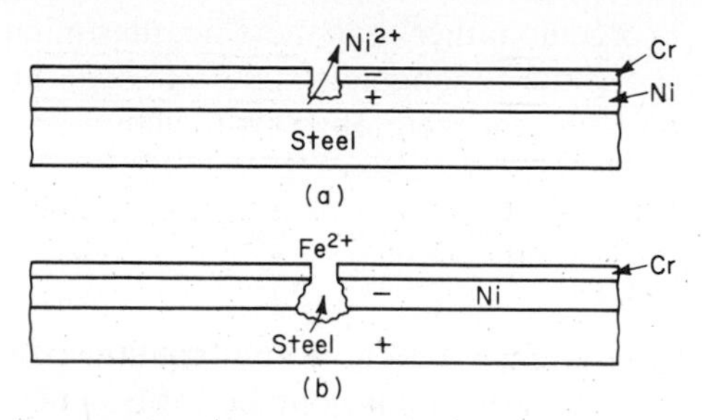

Fig. 3-10 (*a*) Initial stage of corrosion, Ni/Cr on steel; (*b*) later stage of corrosion, Ni/Cr on steel (thicknesses not to scale).

coating, forming a small blister. Unless the corrosion products are insoluble and fill up the pore to prevent further damage, the lateral corrosion finally extends far enough to cause enlargement of the blister and even exfoliation of the coating, which in extreme cases simply peels off.

Thus when a metallic coating is applied to an active metal like zinc or aluminum, the only way in which serious corrosion damage can be prevented is to ensure that the coating is free of discontinuities, in other words, to isolate the active metal from its environment.

A different condition is encountered when the substrate metal is steel and it is to be coated with decorative/protective electroplates such as nickel and chromium. Both nickel and chromium tend to become passive in most environments, and thus to behave cathodically; steel is the anode in the resulting corrosion cell, so that the situation thus far is analogous to coated zinc or aluminum. But it is more complicated, in that the usual protective system consists not of one metal, but usually of at least two, nickel and chromium; often of three, copper, nickel, and chromium; and sometimes effectively of even more than three, since nickel can be deposited in several different conditions, some of which are more active than others. Figure 3-10 illustrates the simple system nickel/chromium on steel; here chromium is cathodic and steel anodic, with nickel intermediate. Chromium, however, can be deposited in a form known as microdiscontinuous, which has many microscopic cracks or pores. Now the nickel, anodic to the chromium, corrodes laterally, and since much more of it is available, eventual penetration to the basis steel is much delayed (Fig. 3-11). See also Chap. 12.

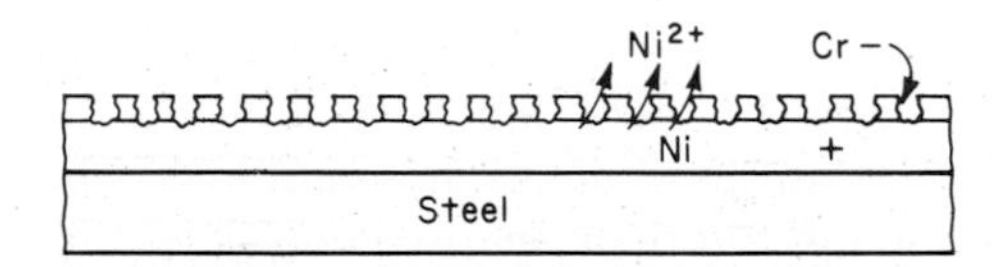

Fig. 3-11 Microdiscontinuous chromium; corrosion is distributed over a large area of nickel.

When the coating is organic rather than metallic, dissimilar metal contact is no longer of concern; the organic coating acts merely to separate the substrate from a corrosive environment, and the main object is to achieve a coating free of pores and scratches.

TESTING CORROSION

In order to determine whether a metal, a combination of metals, or a coating system is suitable for use in a given environment, it is necessary to test it in that environment. It is well to repeat here that corrosion is not an action but an interaction, which involves both the metal and its environment. It is pointless to test a panel in dilute sulfuric acid if the contemplated use involves exposure to seawater.

As nearly as possible, corrosion testing should be carried out in an environment that simulates that in which the part will be used. For most finished metal articles, this requirement is not at all easy to meet. Because corrosion is usually a fairly gradual process, the user or manufacturer cannot afford to wait for the results of a test carried out in strictly natural environments: by the time results are available, the purpose of the test—to predict the performance of the item—would be negated. Therefore many accelerated tests have been devised, with the aim of speeding up the process of corrosion while at the same time simulating its effects. But because the nature of the environment is so important to corrosion processes, it is difficult to speed up the process without changing it, and the best that can be hoped is that accelerated tests will be fairly accurate in predicting the actual service performance of an electroplated or otherwise finished metal article.

Natural environments are of many kinds and contain many types of corrosive chemicals. If a metal part is to be finished for a specific service, then it should be tested for corrosion resistance in as near a replication of that service environment as possible. Perhaps corrosion may be accelerated by heat; for example, tinplate is often tested in contact with the foods it will contain, but instead of the "room" temperature at which most canned foods are stored, higher temperatures are used in the test room so that any failures will be noted more quickly. Even such acceleration is not entirely free of doubt, since it is possible that the corrosion mechanism can change with temperature, and not all corrosion reactions are accelerated by heat.

Many electroplated items, however, are destined for service in various environments whose nature is not known at the time of manufacture. The best example of this type of service is provided by the bright work on automobiles: bumper bars, grilles, hubcaps, door handles, and other decorative trim. The product may be used in any one or a combination of several types of atmospheres, which differ greatly in their aggressiveness toward electroplated coatings and substrates.

Many industrial organizations and testing societies maintain outdoor exposure stations where metals and other materials can be exposed to the elements continuously, and note can be taken of their behavior by inspecting them at suitable time intervals. This sort of testing is slightly accelerated, and for most purposes except long-range research not sufficiently so to obtain speedy results.

A closer approach to actual service conditions is obtained when specimens are mounted on "mobile" exposure. Thus panels may be placed on trucks driven normally on highways during a northern winter, subjected to deicing salts as well as the usual dirt and debris characteristic of such use, or mounted on tugboats plying, say, New York Harbor. Such testing offers the advantage of severe exposure under real-life conditions; it suffers from the drawback that specimens may be mechanically damaged, lost, or impossible to trace.

Truly accelerated laboratory tests—whose aim is to provide answers to the question "how well does the finishing system resist corrosion?"—have been developed and are in general use. However, how well such accelerated tests predict actual behavior under service conditions is an unanswered question. Such tests include the following:

Neutral Salt Spray (Fog). Widely used and specified, the test consists essentially of spraying the specimens with a mist of 5% or 20% sodium chloride solution in a closed cabinet, under strictly specified conditions of temperature and spray rate. Many standardizing bodies, including ASTM, offer detailed directions for running the test: ASTM B 117.

Acetic Acid Salt Spray (Fog) Test. Similar to the above, with the mist solution acidified with acetic acid for faster action: ASTM B 287.

Copper-accelerated Acetic Acid Salt Spray (CASS) Test. Similar, but with the addition of copper salts to the mist solution: ASTM B 368.

Corrodkote Test. The specimen is coated with a slurry of kaolin containing copper nitrate, ferric chloride, and ammonium chloride. The slurry is allowed to dry, and the coated specimen is placed in a humidity chamber; it is removed, cleaned, and examined after stated intervals of time: ASTM B 380.

Electrochemical Corrosion Test (ECT). The specimen is made anodic in a specified electrolyte; under carefully controlled conditions corrosion is stated to occur within a few minutes.

Sulfur Dioxide Test. The specimen is suspended in a closed chamber in the presence of sulfur dioxide gas.

Humidity tests of various kinds.

None of these tests has been shown to accurately predict service behavior under all conditions. The neutral salt spray, though still specified, has lost favor for research purposes, for it often gives the wrong answer in comparing the relative merits of different coatings. For example, it predicts that, thickness for thickness, cadmium plate is preferable to zinc; but this is true only at the seashore. For some coatings it may serve as a measure of porosity, but little else.

The modified salt spray tests are somewhat better, and at this writing the CASS test is in regular use for predicting the service behavior of automobile hardware, as is the Corrodkote test.

The ECT test has been in use for a relatively short time and by few investigators, so that its possibilities are not known.

The sulfur dioxide test has found little favor in the United States but has been used in Europe.

In addition, various specialized tests have been devised which are appropriate to particular metals, such as the "FACT" test for anodized aluminum (ASTM B 538).

In the use of any accelerated test, the purpose of the test must be clearly understood. Two principal uses of corrosion tests may be recognized: (1) as a research tool, to determine whether a coating system under evaluation is satisfactory or to compare two different coating systems; and (2) as an inspection device, to monitor the quality of the product of a manufacturing operation. For the first application, accelerated tests are of limited value: their results must be evaluated with great care and with caution against jumping to conclusions. It must be remembered that most tests were developed for a given coating system, and, after some years of experience, it has been concluded that the test fairly accurately predicts service performance for that system. But the corrosion process depends on the chemistry of the metals involved; if these change, the artificial environment of the accelerated test may or may not be appropriate for the new system.

As inspection tools, however, accelerated tests can function very well. Once a manufacturing quality level has been established, the test merely indicates whether quality control is doing its job.

Evaluating the Results

The corrosion engineer's usual method of expressing corrosion rates is in weight loss per unit area per unit time, such as milligrams per square decimeter per day (mdd). Measurements of this type normally are not practical for the metal finisher. Evaluation of corrosion depends on the use for which the coating was applied. If the purpose of the coating is to provide good electric contact, testing after exposure must involve measuring the potential drop across the relevant contact faces; if it is to provide solderability, some form of solderability test must be applied. A large proportion of electroplated coatings is applied for decorative purposes, and since the decorative merit of a specimen is basically a subjective judgment, inspection criteria are also somewhat subjective. Present trends in evaluating parts and specimens subjected to corrosion tests, whether by outdoor exposure or accelerated, involve a count of the defects observed, their nature, and their total effect on the appearance of the item.

Coatings anodic to the basis metal—primarily zinc and cadmium—are applied mainly to prevent or delay the onset of substrate corrosion products. When the substrate is steel, this is ordinary rust. Therefore the result of the inspection is an estimate of "percentage area rusted"; the lower, the better the rating of the specimen.

When the electroplated coatings add decorative appearance to the objective of rust prevention, the types of defects encountered are more numerous and the inspection criteria correspondingly more complicated. Some defects, such as pores and pits extending to the basis metal, cause actual rusting; others, such as surface pits, stains, and discolorations, detract from the appearance but do not involve corrosion of the substrate. Inspection guides are available that indicate preferred methods of evaluating the results of such exposure tests; they ascribe different values to the different types of defect, and the result is a set of numerical ratings which are susceptible to averaging and to statistical treatment: ASTM B 537.

It cannot be overstressed that actual service is the final test of corrosion resistance of any material, whether finished or not. The various corrosion tests available can give at best approximations, even though in many cases fairly close approximations, of how the part will fare in service. The nearer the test conditions approach actual service conditions, the better will be the approximation to actual experience. On the other hand, sometimes it is necessary to balance speed in obtaining results against the accuracy of slower but more reliable testing. This trade-off should be recognized by the user, who should realize that there is no point in a speedy result if it is not at least reasonably accurate.

Finally, we should point out that the subject of corrosion has been by no means thoroughly discussed in this chapter. Many important aspects—stress corrosion cracking, corrosion fatigue, and others—have not been mentioned. Excellent texts are available, and this chapter has considered only those aspects of corrosion that are of principal interest to the metal finisher.

4 Metallurgy

It is not possible in a short chapter to present even a summary of the metallurgical knowledge that might be needed in metal-finishing situations. Here we shall attempt only a review of some rudiments and an outline of the types of test procedures that a metal finisher may be called upon to use.

Since metallurgy is the science of metals, one would expect to begin by defining a metal. It is difficult to formulate a definition of the word *metal* in such a way that it will include all those substances that are metals and exclude all those that are not—which is the purpose of a definition. A *metal* is "an aggregate of atoms in which the bonding is such that electrons can move in response to an applied electric field, even when that field is quite small, and in which the number of electrons that can be so moved is comparable with the number of atoms."* This definition has the virtue that it avoids the enumeration of such properties as ductility, luster, high specific gravity, and thermal and electrical conductivity that are usually included in attempts at defining this elusive concept. It also emphasizes the most important characteristic of metals: that their mode of conducting electricity is electronic, and that their atomic structure is such that the valence electrons are more or less free of direct association with any particular atom and thus are free to move under the influence of an applied electric field.

The elements considered to be metals are those to the left of the heavy line in the periodic table of Fig. 4-1 (except hydrogen). There are some borderline cases—a few elements adjacent to the line. Both silicon and phosphorus possess allotropic modifications that are metallic; arsenic, selenium, and tellurium are often included, and antimony is sometimes excluded; germanium is borderline; and tin has a "nonmetallic" modification that is stable at low temperatures. For the rest, there is little doubt that the elements to the left of the line are metals and those to the right are nonmetals.

*B. Chalmers, "Physical Metallurgy," John Wiley & Sons, Inc., New York, 1959, p. 15.

THE SCOPE OF METALLURGY

Metallurgical science is divided into two main branches, which are fairly distinct from each other. *Chemical metallurgy* is concerned with the processes of recovering metals from their ores; *physical metallurgy* involves the mechanical and physical properties of metals, the methods for treating the products of the chemical metallurgist's art so that they are suitable for practical applications, and the examination of metallic structures and how these structures correlate with their properties.

Chemical metallurgy is, of course, of great importance; without the processes of extraction and refining, the physical metallurgist would have nothing to work with. The subject of extractive, or chemical, metallurgy is outside our scope for present purposes.

Physical metallurgy may be further subdivided: its principal divisions are metallography, property studies, and mechanical studies.

Metallography is concerned with determination of the internal structure of metals; it uses for this purpose any available technique. Formerly these were mostly confined to the use of the light microscope, but advances in instrumentation have offered this branch of the science such tools as the electron microscope, the scanning electron microscope (SEM), and x-ray and electron diffraction. Metallography is concerned with the techniques for preparing a metal for examination, with the use of any appropriate tool for looking into its structure, and with interpretations of the results.

Studying the properties of metals means determining their alloying behavior and such properties as their ductility, strength, and hardness—in fact, any and all properties except the strictly chemical ones such as corrosion resistance and reactivity.

METALS — NONMETALS — INERT GASES

																	INERT GASES
H																	He
Li	Be											B	C	N	O	F	Ne
Na	Mg											Al	Si	P	S	Cl	Ar
K	Ca	Sc	Ti	V	Cr	Mn	Fe	Co	Ni	Cu	Zn	Ga	Ge	As	Se	Br	Kr
Rb	Sr	Y	Zr	Nb	Mo	Tc	Ru	Rh	Pd	Ag	Cd	In	Sn	Sb	Te	I	Xe
Cs	Ba	La	Hf	Ta	W	Re	Os	Ir	Pt	Au	Hg	Tl	Pb	Bi	Po	At	Rn
Fr	Ra	Ac															

Lanthanides

Ce	Pr	Nd	Pm	Sm	Eu	Gd	Tb	Dy	Ho	Er	Tm	Yb	Lu

Actinides

Th	Pa	U	Np	Pu	Am	Cm	Bk	Cf	Es	Fm	Md	No	Lr

Fig. 4-1 The metals in the periodic system.

Finally, mechanical metallurgy studies the means of testing such properties and the methods of fabricating metals into useful structures: rolling, forging, shaping, annealing, and hardening.

For metallographic examination, samples are cut from the area of interest, polished to a flat, mirrorlike surface, and chemically etched to reveal their structure. Etchants are chosen so that they react at different rates with the various parts of the specimen; thus on microscopic examination grain boundaries, inclusions, different phases,* and the like will show up as lighter or darker areas. Much skill and practice are required both in preparation of the specimen and in interpretation of the results.

Property studies include the study of alloys and their behavior under various kinds of heat treatment. By studying the constitutional diagrams of alloys (considered briefly below), it is often possible to predict how they will behave when heated to specified temperatures and then either slowly or quickly cooled to room temperature.

Mechanical metallurgy uses the tests available to measure such properties as tensile strength, hardness, creep, and others that determine the practical utility of metals and alloys for their intended applications. Although electrodeposited metals are still metals, usually they are so thin that many of the common tests applied to metals in bulk are not applicable to them, and either entirely new tests, or modifications of standard tests, have been devised to determine their physical and mechanical properties. For example, the tensile strength of ordinary metals is commonly tested in a powerful machine that pulls a specimen apart until it breaks under the applied load; the load per unit area at failure tells us the tensile strength, and the amount of stretching exhibited by the specimen at the breaking point is a measure of its ductility. But for thin electrodeposits, such tests normally are not practical; applicable tests are mentioned later.

METALLIC STRUCTURES

Examination of the internal atomic arrangement in metals has shown that most fall into one of three structures, shown in Fig. 4-2. Other more complicated structures are not uncommon, however. The three structures most usually encountered are face-centered cubic (abbreviated fcc), body-centered cubic (bcc), and hexagonal close-packed (hcp). The unit cell of each arrangement contains the smallest number of atoms in the array that, infinitely repeated, makes up the structure of the massive metal. A grain of metal is made up of unit cells repeated in three directions, such that in a grain all the atoms are arranged in a lattice that is based on the unit cell. In most metal lattices there are many planes parallel to one another, and each plane is more or less densely populated with atoms. Planes which are densely enough packed with atoms to behave like solid sheets of material slip past one another when a load is applied to the solid (like a pack of playing cards); this produces what is known as *plastic deformation.* Metallic structures are considered further below.

*See the glossary, p. 61, for some terms not defined in the text.

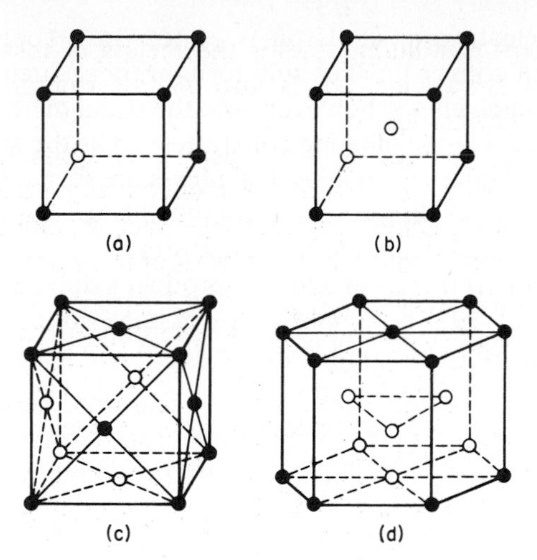

Fig. 4-2 Metallic structures: (*a*) simple cubic; (*b*) body-centered cubic, bcc; (*c*) face-centered cubic, fcc; (*d*) hexagonal close-packed, hcp.

ALLOYS

An alloy is a mixture of at least two elements, at least one of which is a metal. This definition may be modified for practical purposes, because metals encountered in practice are seldom pure. And so, strictly speaking, all "practical" metals are alloys, and the definition becomes so all-inclusive as not to be very helpful. Usually, the criterion is whether the second component is deliberately added or whether it is an adventitious impurity: if the former is true, the result is an alloy; if the latter holds, then an impure metal results. Many of the metals used in practice in their massive form are, in fact, alloys. Steel is the best example, but aluminum alloys, nickel alloys, and many others are also useful. The properties of most metals are improved by judicious alloying. Sterling silver is only 92.5 percent silver, with the remainder being copper; little pure aluminum is used. Alloying enables the metallurgist to modify the properties of the major constituent by the addition of large or small amounts of other elements, to improve workability, corrosion resistance, hardness, strength, and many other properties. Of the metals in large-scale use, only copper is used in relatively pure, unalloyed form.

On the other hand, most electrodeposited metals are not alloys, in the sense in which we have used the term. The major electroplating metals—copper, nickel, chromium, tin, silver, zinc, and cadmium—are usually plated as pure metals. Among the more common electroplated alloys are bronze (copper-tin), tin-lead, brass (copper-zinc), and various gold alloys. The last commonly contain only small amounts (1 percent or so) of the alloying metal.

Metals can alloy with one another in the solid state; that is, if a piece of copper and a piece of zinc are placed in intimate contact for some time, an intermediate layer of copper-zinc alloy will form at the interface. The same thing can happen in the case of an electrodeposit. If tin is plated on a copper basis metal, an intermediate layer of copper-tin alloy will form in time. Such slow processes are not practical for forming alloys, however, and the usual method is to melt at least one of the metals and add the alloying constituent to it; the second metal may be added as a solid or a liquid. Many practical alloys are formed by making use of a "master" alloy, that is, an alloy containing a much larger proportion of the added metal than is to be contained in the final mixture.

The combining of two metals in alloying produces different behaviors according to the metals involved. The simplest type of behavior occurs either when the two metals are completely immiscible in the liquid and solid states or when they are completely miscible in both states.

Two metals that are completely immiscible with each other in the liquid state—just as oil and water are—form two layers in the liquid. Even when the liquid metals are soluble in each other to a limited extent, they still form two layers when this mutual solubility is exceeded. Such alloys also remain as two layers when they solidify.

Two metals that have complete mutual solubility in the liquid state behave much like water and ethanol; that is, the liquid alloy cannot be distinguished by appearance from a one-substance liquid. This resemblance may carry over to the solid state also. Such a system is known as a *solid solution* because the two metals are mutually soluble in each other. For example, silver and gold are completely soluble in each other, in both the liquid and the solid states. Silver may be added to gold, or gold to silver, in any proportion, with little effect on the microscopic appearance of the mixture, except for its color. Copper and nickel behave in the same way.

The lack of change in microscopic appearance does not mean that no changes take place in properties, such as electrical conductivity and mechanical properties. The hardening and strengthening of gold by the addition of silver is familiar. Copper and nickel produce alloys that are much stronger than either metal alone, as shown by the familiar Monel metal.

Constitutional Diagrams

A *constitutional diagram* (or equilibrium diagram) is a graph that shows the temperatures at which some change takes place in an alloy. When only two metals are involved, such a diagram is called a *binary* constitutional diagram. Figure 4-3 is such a diagram for the copper-nickel system. The temperature is plotted along the ordinate and the percentage of one of the constituents along the abscissa; the latter may be expressed in either weight percent or atomic percent. The upper line gives the temperature of beginning of solidification of an alloy containing the percentage by weight of nickel indicated along the base line. The lower line specifies the temperature at which solidification ends in an alloy containing the indicated percentage of nickel. The upper line is known as the liquidus line (or simply liquidus) and the lower line as the solidus line. Thus an alloy freezes over a range of temperatures corresponding to the vertical distance

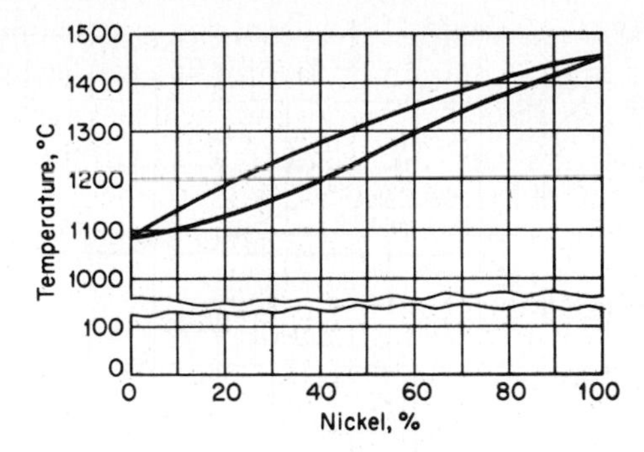

Fig. 4-3 Nickel-copper system of alloys.

between the two curves, rather than at a fixed temperature as a pure metal does. For example, an alloy containing 50 percent nickel and 50 percent copper begins to solidify at about 1310°C and is completely solid when it has cooled to about 1240°C.

Diagrams such as Fig. 4-3 also give information concerning the process of melting. Its lower line indicates for each composition the temperature at which melting begins, and its upper line tells where melting is complete.

So far we have considered only the solid solution type of alloy. Another important type is the so-called eutectic alloy, an idealized diagram of which is shown in Fig. 4-4. No practical example of this type of alloy behavior is observed, but a close approximation is offered by the system tin-lead, the equilibrium diagram of which is shown in Fig. 4-5. This system is of great practical importance, since it includes the soft solders and terne plate. The figure shows that the addition of lead to pure tin lowers the melting point, as does the addition of tin to pure lead; and at point *B*, the eutectic composition, the melting point is a minimum. But tin and lead have little mutual solubility, and the solid alloy consists of two phases, easily distinguishable under the microscope after appropriate preparation of the sample, as shown in Fig. 4-6.

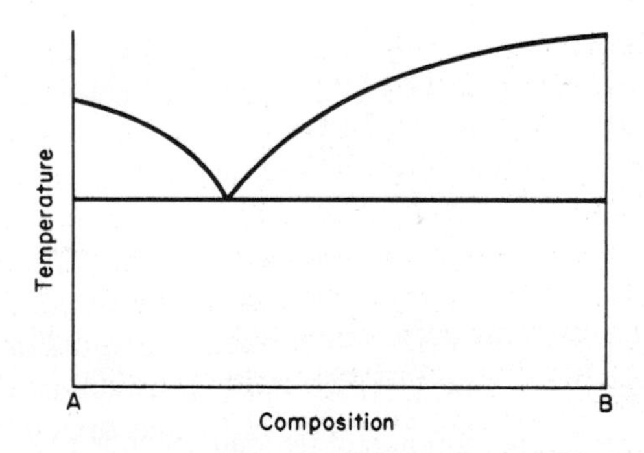

Fig. 4-4 Simple eutectic system with no mutual solubility.

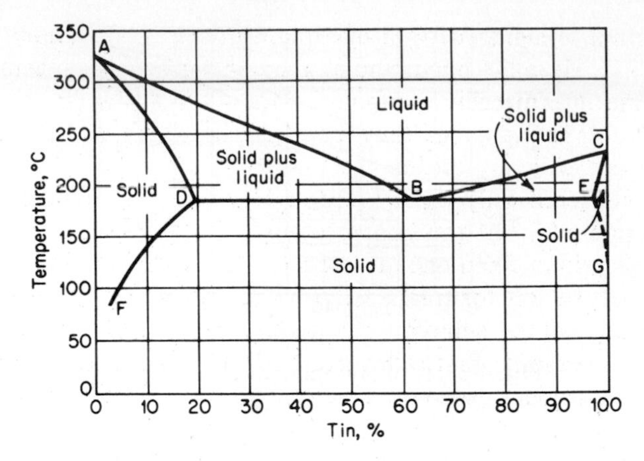

Fig. 4-5 Tin-lead system.

In Fig. 4-5, the upper line *ABC* represents the temperature at which the molten alloys begin to solidify; the lower line *ADBEC* represents the temperatures at which they are solid. For example, the alloy with 10 percent tin begins to solidify at about 300°C and is completely solid at about 270°C. At any composition except that of the eutectic, *B,* there exists a range over which the alloy is partly solid and partly liquid. In other words it is "pasty," a very important property in the practical use of solders.

The third important type of alloy is the intermetallic compound; formation of

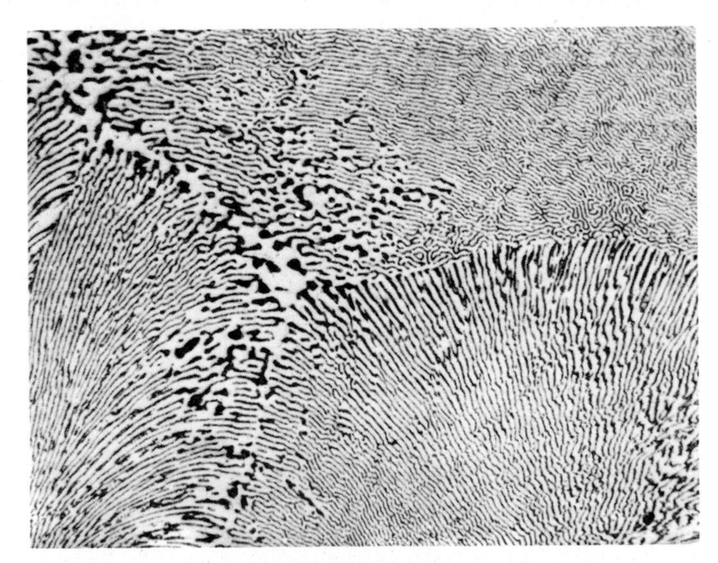

Fig. 4-6 Photomicrograph of tin-lead eutectic (63% tin, 37% lead). Not chill-cast; etched 2% nital; original magnification × 200. *(Tin Research Institute.)*

such a compound is signaled by a maximum in the constitutional diagram, as shown in Fig. 4-7. Here the maximum at about 29 percent magnesium signals the formation of the intermetallic compound Mg_2Sn. Intermetallic compounds are true chemical compounds, but they often do not obey the standard rules of valence. They are typically brittle.

We have considered only binary alloys. Many useful alloys consist of more than two components, but the construction of equilibrium diagrams naturally presents complications when one attempts to represent them in two dimensions. This subject must be left for treatises on metallurgy.

Some knowledge of the behavior of various types of alloys is valuable to metal finishers for two reasons: first, alloys often form the substrate which they are called on to electroplate; second, electrodeposition of alloys is of increasing importance.

When the substrate is an alloy, it is often helpful to know what type of alloy it is. A solid solution alloy, like a liquid solution, is homogeneous, and complications seldom arise from the presence of different phases which may react differently to the plating sequence. But an eutectic alloy, such as soft solder, is composed of two or more different materials, and the problem of plating on such an alloy is the same as trying to plate on a complex workpiece made of different metals.

THE STRUCTURE OF METALS—ONCE MORE

Of the various devices that have been used in studying the structure of metals, the greatest amount of information has been yielded by the microscope. But

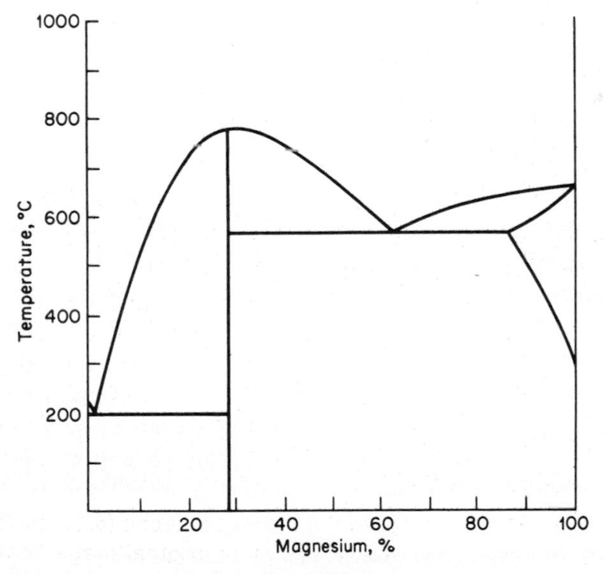

Fig. 4-7 Tin-magnesium system.

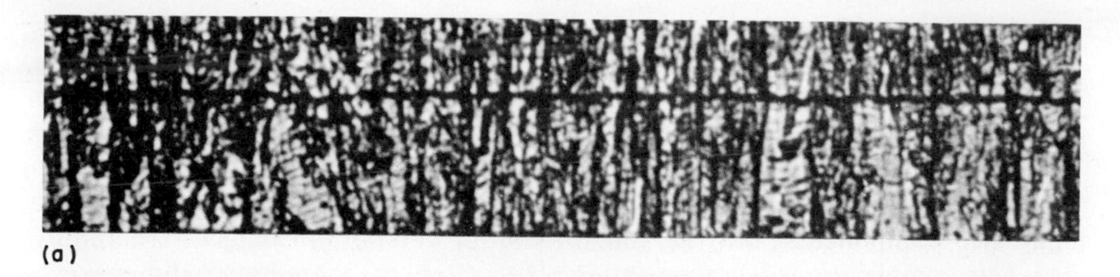

(a)

(b)

Fig. 4-8 Grains (photomicrograph). Top: Nickel, Watts bath, pH 5.3; magnified 200×; shows columnar structure. Bottom: Nickel, Watts bath plus organic brighteners; magnified 200×; shows laminar structure. *(Courtesy J. D. Thomas, General Motors Research Laboratory.)*

because even the thinnest sections of metals are opaque, the metallurgical microscopist cannot examine specimens in the way that a biologist can, by cutting thin sections and viewing them by transmitted light. The metallurgist first polishes a specimen until it is very smooth and flat; this is necessary so that the whole surface to be examined will be in focus at the same time. Then some method for bringing out details is required; in its absence the surface might merely act as a mirror. This operation is known as *etching*. The result might be as shown in Fig. 4-8, which is a typical photomicrograph (or micrograph).*

Etching reagents, or etchants, differ, and it is important to choose the right

**Not* microphotograph, which would connote simply a small photograph.

one. The basis of their action is that they react faster with some features of the surface than with others, so that structure is made evident on visual examination.

The ordinary or optical microscope is limited to magnifications of perhaps 2000 or 3000 diameters. The magnification available to the metallurgist (and to other workers) was vastly increased by the development in the 1930s of the electron microscope; this uses a beam of electrons, instead of light rays, and multiplies the magnification that can be obtained by at least one order of magnitude. Because metals are opaque, most early work with the electron microscope was done with "replicas"; i.e., the surface of the etched metal specimen was exactly reproduced by means of a plastic material molded to its contours, but which was transparent to electrons. During the late 1950s new etching and preparative techniques made it possible to take electron microscope pictures in the transmission mode direct from the metal specimen.

The Arrangement of Atoms in Metals

A few metals reveal their crystalline nature when they are frozen from the melt; bismuth, for example, freezes in well-defined cubic crystals that can only result from some regular pattern in the solid. For most purposes, however, in order to investigate the arrangement of the atoms in metals, it is necessary to use x-ray diffraction techniques.

One of the simplest of these is the x-ray powder diffraction pattern, an example of which is shown in Fig. 4-9. This is obtained by means of an apparatus shown schematically in Fig. 4-10.

Planes in a crystal do not reflect x-rays as a mirror reflects light; x-rays are reflected by the cooperation of many parallel planes containing atoms, as shown diagrammatically in Fig. 4-11. To produce a line in a pattern such as Fig. 4-9, three factors must be related:

The distance between planes, normally called a

The wavelength of the x-rays, λ

The angle at which the x-ray beam strikes the planes, θ

Refer to Fig. 4-11, which shows a group of parallel planes perpendicular to the plane of the paper: the distance between them is d. Two rays of an x-ray beam of

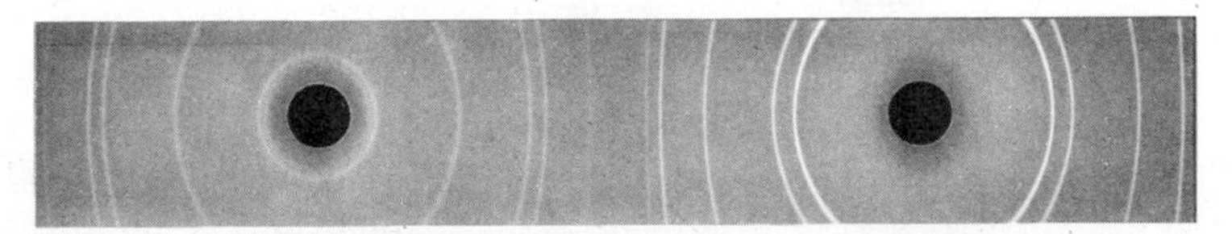

Fig. 4-9 Typical x-ray powder pattern. *(Courtesy Professor Rolf Weil, Stevens Institute of Technology.)*

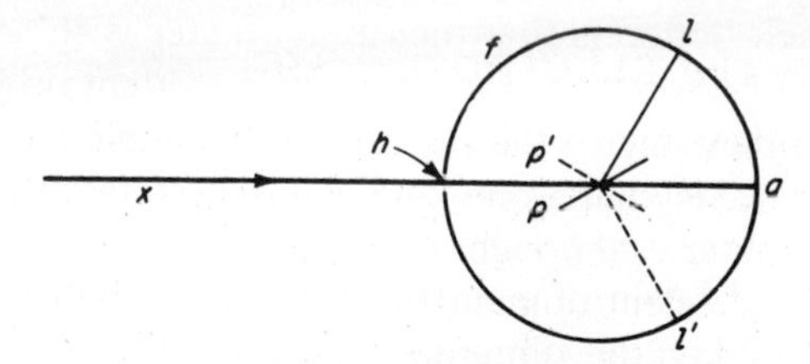

Fig. 4-10 Reflection of x-rays by planes of atoms. Key: *f*, photographic film bent in the form of a circle; *p–p'*, plane of atoms reflecting x-rays; *l*, line made on film by reflected ray; *x*, beam of x-rays; *h*, hole through which beam enters camera; *a*, nonreflected rays strike film here. *(From B. A. Rogers, "The Nature of Metals," copyright 1964 by American Society for Metals; reproduced by permission.)*

wavelength λ strike the two top planes of atoms at angle θ and are reflected at the same angle. To reach point *f*, comparable to point *c* reached by the ray *abc*, ray *def* must travel an additional distance *xey*. Therefore when the two rays reach the positions *c* and *f*, they will not be in phase unless *xey* is exactly equal to a wavelength λ. It is therefore necessary to adjust the angle of the incoming rays so that *xey* will equal λ; under these conditions the rays will reinforce one another rather than cancel and will produce a line on the picture. The distance *xey* may also equal 2λ, in which case it gives rise to a second-order reflection.

From a knowledge of the x-rays used and the angle through which the rays are deflected, as well as other background information, the x-ray crystallographer can deduce much about the internal arrangement of atoms in a crystal, including the crystal of a metal. These deductions and conclusions have already been summarized on p. 48.

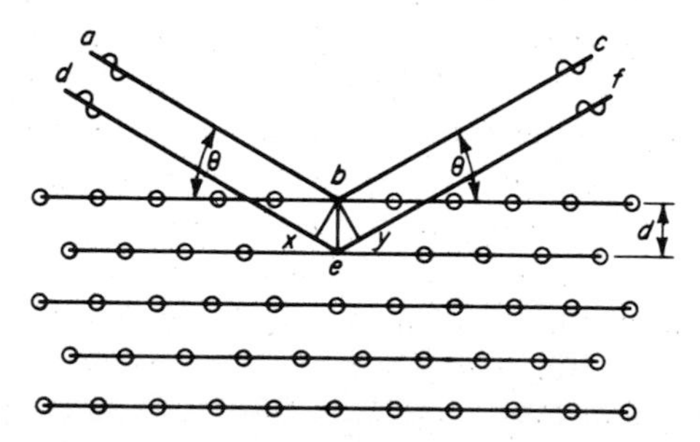

Fig. 4-11 Reflection of an x-ray beam from a set of parallel planes (see text). *(From B. A. Rogers, "The Nature of Metals," copyright 1964 by American Society for Metals; reproduced by permission.)*

METALLURGY FOR THE ELECTROPLATER

The electroplater is interested in the properties of metals for two reasons: they are (usually) the substrates which one is called upon to finish, and they are also the finish which one applies. Although the properties of the substrate are often of little direct interest to the electroplater, there are nevertheless many instances in which they are important. If the substrate is an alloy consisting of more than one phase, the cleaning and finishing operations may have to be modified to take account of any disparity in the acceptance of a finish by the two or more phases. The finishing operations may call for heat treatments of various types; in fact, even the operations of polishing and buffing generate considerable heat. The finisher must be concerned about whether the application of heat will change the properties of the substrate in an undesirable manner. The danger of "hydrogen embrittlement" may have to be kept in mind in finishing substrates particularly susceptible to this kind of damage. In some cases the possibility that the substrate may alloy with the finish, with results either favorable or unfavorable, must be considered.

When it comes to the metallurgical properties of the finishes themselves, the metal finisher may or may not be concerned, depending on whether the physical or mechanical properties of the finish are important to the use of the final product. Often these properties are of little moment: in most purely decorative finishing, where the appearance of the product and the retention of this appearance for a reasonable time are the only considerations, such properties as hardness, tensile strength, and the like are usually not important. Similarly, when the plated metal is soft, ductile, and weak, like tin or lead, its physical properties usually are not of great significance.

On the other hand, many metallic electrodeposits are used in applications where their physical properties are of great importance. This is almost always true in electroforming, electrotyping, and in many other applications in which the deposit must evidence, e.g., high hardness and wear resistance, such as in the chromium plating of gun barrels and forming dies and tools, or softness and ductility, such as the copper plating of plastics as an undercoat for further finishing.

In general, electrodeposited metals have the same structure as metals produced thermally; that is, examination of the atomic or crystal structure of the massive metal will be valid for electrodeposits also. There are, however, some exceptions to this generalization: for at least the first few layers of deposited atoms, the electrodeposit often tends to follow the structure of the substrate rather than its own preferred structure. Some metals can be deposited in a state that would be metastable or even unstable in the massive condition; and in a few cases, electrodeposited alloys exhibit phases not encountered in the constitutional diagram of the binary system. Illustrations may be found in the deposition of γ-manganese, which is unstable at room temperature; this form, however, soon reverts to the stable form on standing. Again, although the stable form of tin at very low temperatures is α or gray tin, when tin is plated at low temperature, it is always in the ordinary metallic or β form. The tin-nickel alloy NiSn, containing 65 percent tin and 35 percent nickel, is an intermetallic compound which does not appear on the constitutional diagram of the tin-nickel system; it has been produced only by electrodeposition.

Deposits from practical plating solutions are crystalline in the same sense that thermally prepared metals are. The metal is composed of building blocks, each of which is a fairly simple arrangement of a few atoms, repeated in three mutually perpendicular directions. Crystals that touch each other continuously to make up a metallic body are called *grains*. Although the size and shape of a grain may bear no relationship to the crystal habit of the metal, these grains may be as important as crystal habit in determining the properties of the metal.

Most of the common metals have crystal habits in one of the three systems already considered: face-centered cubic (fcc), body-centered cubic (bcc), or hexagonal close-packed (hcp). For thermally prepared metals, ductility is usually best for fcc and worst for hcp, with bcc intermediate; this is related to the ease with which planes of atoms in the metal can slip past one another.

Grains usually behave like crystals, except that other factors such as grain boundaries may complicate their response to applied stresses. Approximately, as the grain size of a metal is decreased, the metal behaves less and less as its atomic structure would predict; this is because the smaller the grain size, the greater the ratio of grain boundary material to the total volume of metal. Since many practical electrodeposits are fine-grained, it follows that the metallurgical properties of deposits often are not determined by the crystal habit of the massive metal. The metallurgical behavior will be influenced greatly by such factors as grain size, amount of codeposited foreign material, internal stress, and preferred orientation.

Besides the departure from the regular crystalline arrangement of the atoms that occurs at grain boundaries, there are other defects within the grain in electrodeposited metals; these include dislocations, twins, and codeposited foreign matter. Dislocations are important in determining mechanical properties. Since plastic deformation occurs by movement of dislocations through a crystal, high strength (or resistance to plastic flow) results from obstructions to their movement by other dislocations, foreign matter, and twins and grain boundaries. Such defects may also affect electric resistance, magnetic properties, and corrosion resistance.

(The strengthening of deposited metals by incorporation of foreign material has been given practical application by the deliberate codeposition of powders, fibers, or filaments in the deposit; such "composite" materials have been used in applications requiring exceptionally high strength.)

Epitaxy

The structure of the basis metal often has an effect on that of the electrodeposit, depending on a variety of factors. If the interatomic distances in a lattice plane of the substrate match those of the lattice plane of the deposit, the structure of the substrate may be continued into that of the deposit. This is called *epitaxy,* or *epitaxial growth.* Plating conditions often determine whether epitaxy will occur; if these conditions result in high overvoltages, produced, for example, by high current densities or some bath additives, three-dimensional nuclei may be formed which tend to overcome the relationship between the substrate and the deposit. Conversely, elevated bath temperatures and low current densities, which permit the migration of atoms to sites where they can be incorporated into the existing structure, favor epitaxy.

Grain size and grain shape can easily be observed by the ordinary metallurgical microscope (but see the following paragraph). In the absence of other factors, simple plating solutions such as the copper sulfate bath produce large-grained columnar deposits with considerable anisotropy; solutions based on complex compounds such as copper cyanide, or solutions containing active addition agents, tend to yield fine-grained deposits with no obvious preferred direction of growth. Colloidal material and suspended particles in the plating bath tend to yield fine-grained deposits; both colloids and particles may be trapped in the deposit. In general, large-grained deposits are soft, weak, and ductile, while fine-grained deposits are hard, strong, and brittle; but this is a general guideline rather than an invariable rule.

Microscopic examination of deposits, for determination of grain size and preferred orientation, may be misleading. X-ray or electron diffraction is the only reliable way to determine these properties; what appear to be grains under the microscope sometimes turn out to be colonies of many small crystals not resolvable by the microscope. Use of the electron microscope often can resolve such questions.

Bright electrodeposits are usually fine-grained but not invariably so.

The electron microscope and the scanning electron microscope have become increasingly important tools for the examination of the structural features of electrodeposited metals. The latter, in particular, enables the visualization of surface features, including codeposited materials, nucleating centers, and much else.

HEAT TREATMENTS

Unlike thermally prepared and worked metals, heat treatment of electrodeposits is not often practiced, although it can be important in individual cases. Almost all electrolytic tinplate is "reflowed"—i.e., the tin deposit is melted and resolidified—but tin is a low-melting metal and this is an exceptional case. Low-temperature heat treatments are fairly common, especially as a means of relieving hydrogen embrittlement of the substrate metal. It is necessary to exercise care, employing only times and temperatures that will accomplish the release of the hydrogen without altering the desirable properties of the substrate. Internal stress can be relieved by heat treatments at temperatures much below those required for annealing.

The possibilities of heat treatment may be modified by the presence of addition agents in the bath from which the deposit was plated. Some addition agents, for example, used in bright nickel plating cause codeposition of sulfur; and such deposits, when heated even moderately, may develop brittleness owing to the formation of nickel sulfide which segregates at grain boundaries. Another example is the tin deposits from some "bright acid" tin-plating solutions; satisfactory as plated, heating may cause bubbles or blisters arising from the decomposition of the codeposited organic addition agents.

TESTING THE PHYSICAL PROPERTIES OF ELECTRODEPOSITS

Tensile Strength and Ductility

Tensile strength and ductility of electrodeposits may be of great importance to the performance of the plated article. They are important factors in determining the degree to which the plated article will resist stress corrosion cracking; this term includes any combined action of static tensile stress and corrosion which leads to failure by cracking, and it takes many forms. These properties also are important in preplated strip and wire; if the coating is lacking in strength and ductility, it will fail during fabrication. The same is true, of course, of any finished article which must undergo significant deformation after plating.

There are no widely accepted specifications for determining the tensile strength of electrodeposited metals. The standard technique, using specially shaped "tensile specimens" and pulling them apart in a machine used for massive metal, is difficult to adapt to the very thin (compared to bulk metal) deposits. Two ductility tests for deposited metals are specified by ASTM. One (ASTM B 489) involves bending the plated specimen over mandrels successively smaller in diameter until the coating cracks. In the other (ASTM B 490) the deposit, in the form of a foil detached from the substrate, is held between the jaws of a micrometer; the jaws are closed until the foil cracks. Neither test gives results which necessarily correlate with values obtained by conventional methods, but such results are internally consistent and at least of comparative value.

Both tensile strength and ductility (percent elongation) of electrodeposits can be evaluated by the bulge test proposed by Read and his coworkers. The sample is deformed by blowing a bubble with oil under pressure; from the size of the bubble when it breaks and the oil pressure needed, the strength and ductility can be calculated (see Chap. 23).

Internal Stress

Many electrodeposits are in a state of stress; this is called *internal stress*. One type, macrostress, is made evident when a substrate plated on only one side is forced by the deposit into the shape of a letter C. If the deposit is on the concave side (inside) of the C, the macrostress is tensile; if it is on the convex side (outside), it is compressive. Such stresses can cause distortion of the substrate, cracking of the deposit, loss of adhesion, and other adverse effects. Several methods of evaluating internal stress are available, as discussed in Chap. 23.

Microstresses manifest themselves primarily in an increase in hardness. They can be determined only by line broadening in x-ray diffraction.

Hardness

Hardness is an easily understood concept, yet it is not a measure of any single property of a metal. It can be fairly easily measured and is often related to properties such as tensile strength and ductility. The usual qualitative relationships between hardness, tensile strength, and ductility do not always hold for electrodeposits. The concept, therefore, must be used with caution; in particular, hardness and wear resistance do not necessarily go together.

The measurement of hardness may be based on several experimental techniques, including scratch width, abrasion resistance, and resistance to penetration. Only the last has been generally recognized and used.

Indentation methods are of two main kinds. In one, an indentor is forced into the metal under specified conditions, and the depth of penetration is a measure of hardness. The Rockwell machine is the best known of this type, but it is seldom applicable to electrodeposits, because the depth of penetration is greater than the thickness of most such deposits. The method can be used for thick, "hard" chromium deposits. Because electrodeposits are normally thin, ordinary hardness tests would be greatly influenced by the hardness of the substrate, and it is necessary to use special techniques, most of which depend on the determination of the size of an indentation produced by a hard indentor under specified loads. The size of the impression produced by the special indentor is measured by a microscope and expressed in various ways according to the method used in measuring it. The most common methods of this second type are Brinell, Vickers, and Knoop, all named after the worker who originated the test and its equipment.

The standard Vickers machine uses loads usually too high for testing electroplates. To overcome this, either the shape of the indentor may be changed, as in the Knoop method, or the load can be decreased. Equipment is available that uses loads in the gram range rather than kilograms as used in the original Vickers method. See Chap. 23.

Glossary

Allotropy. The existence (particularly in the solid state) of two or more crystalline or molecular structural forms of an element.

Anisotropic. (n., **anisotropy**). Having properties that differ according to the direction in which they are measured.

Anvil effect. An error in hardness measurements caused by testing a specimen in such a way that the hardness of the specimen support (the anvil) affects the result.

Creep. Deformation (see) in which irreversible strain occurs while the stress is fixed.

Crystal. A body of material with an orderly repeating arrangement of atoms.

Crystal habit. The type of atomic arrangement in a crystal.

Deformation, elastic. Deformation in which the strain appears and disappears simultaneously with the application and removal of the stress.

Deformation, plastic. Deformation in which the strain occurs at the same time as the stress, but does not vanish when the stress is removed.

Dislocation. A disturbed region between two essentially perfect parts of a crystal that are incompatible with each other.

Ductility. The amount of strain that a body can withstand without breaking.

Eutectic (adj.). Of the lowest possible melting point; said of an alloy or mixture whose melting point is lower than that of any other alloy or mixture of the same ingredients. (n.) A substance having these properties.

Grain. An individual crystal in a polycrystalline body.

Hydrogen embrittlement. Loss of ductility and strength caused by the presence of interstitial hydrogen in a metal.

Phase. A mechanically separable, homogeneous part of a system.

Preferred orientation. An arrangement of grains in a material wherein all or most of the grains are oriented so that in them some crystallographic feature bears a constant directional relation to some structural reference feature.

Strain. Deformation caused by the application of a stress.

Stress. Load per unit area.

Tensile strength. The maximum load per unit area that a body can support without breaking.

Twin. Portion of a grain wherein the crystal lattice is the mirror image of that elsewhere.

PART TWO

ELECTROPLATING: PRINCIPLES AND PRACTICE

Section A

PREPARATION FOR FINISHING

5 Chemical Preparation for Finishing

With negligible exceptions, parts as received in the finishing department simply cannot be introduced into an electroplating solution without pretreatment of some kind. The electroplate (or other finish) is expected to be adherent to the substrate, and such adhesion will not be attained unless the parts are reasonably clean before entering the electroplating bath. Therefore the parts must be pretreated in order to render their surfaces amenable to accepting a satisfactory deposit.

It is necessary to define the word *clean* for our purposes. The ideally clean surface would consist of atoms of the substrate metal, uncontaminated by any foreign material at all. Such a surface is extremely difficult to obtain, even in the laboratory, and is practically impossible under shop conditions. A practical definition of the word *clean* is "containing no contaminants that would interfere with satisfactory deposition of an adherent finish." It follows that a clean surface will mean different things, depending on the nature of the finish to be applied and the nature of the finishing operation itself, since some finishing operations are much more sensitive to contaminants than others. For the same reason, it is not possible to specify in detail in this chapter exactly what steps must be taken to render an article suitable for finishing—in other words, to spell out the exact cycles required in the prefinishing sequence. We enumerate some general principles applicable to the prefinishing cycle—usually called, inclusively, cleaning—and indicate some of the variations necessitated by variations in the nature of the substrate, the nature of the finish to be applied, or both. Many guides are available (see the Bibliography) that offer details of individual cycles appropriate for various substrates.

The necessary steps in the cleaning cycle, then, depend on two factors: (1) the nature and quantity of the soil, and (2) the nature of the finish to be applied. The first is usually more important than the second, since sources of variation are more numerous.

Before being received in the finishing department or in the plating shop (the two are equivalent for present purposes), parts have been subjected to one or all of many operations, all of which can leave residues of foreign material on them. Fabrication, stamping, grinding, polishing, buffing, handling, and shipping each contribute to the soil that must be removed before the parts can enter a finishing cycle. In addition, by merely being exposed to normal shop atmosphere any object receives deposits of dust and airborne contaminants, and most metals acquire films of oxides, sulfides, or other incipient corrosion products.

These prefinishing operations will leave on the parts, in addition to normal shop soil and oxide films, heat scale, quenching oils, rust-proofing oils, drawing oils, stamping and die lubricants, flushing oils, and even on occasion zinc phosphate coatings. Polishing and buffing steps leave residues of the fats and waxes used in buffing compounds, as well as the abrasives from the compounds themselves or from grinding. It follows that the soils to be removed before electroplating are complex. Furthermore, they will change character on standing, usually in the direction of becoming more difficult to remove. Such residues can also attack the metal surface, leading to later troubles. Therefore, in addition to all the other variables, time of standing must be considered.

HOW CLEAN IS CLEAN?

It is little wonder that frequently the complexity of the problem of removing all the contaminants mentioned causes trouble, and that this trouble is made manifest in a poor finishing job: the electrodeposit is nonadherent, discolored, or unsatisfactory in some other respect. When troubles arise in a plating operation, the experienced finisher tends to suspect that the root of the problem is in the preplating cycle. In other words, the cleaning operation is suspect unless vindicated. It is expensive and wasteful to find out, only after the whole plating sequence has been completed, that something was wrong at the very beginning of it. Therefore it is helpful to have a test for cleanliness which can be applied before the part enters the plating sequence. Such a test will tell the operator whether the part is satisfactory for further processing and whether the cleaning cycle is operating properly.

Many such tests have been devised, of greater or lesser sensitivity and ease of application. Gravimetric tests involving actual weighing of the soil left on a part are sometimes useful. Many organic soils fluoresce under ultraviolet light, and this property has been used as a measure of "dirtiness." For research purposes many of these tests have value, but for practical shop use the so-called water-break test is almost universally accepted.

This test depends on the fact that clean metal surfaces are hydrophilic and will shed water in an unbroken sheet, whereas if traces of soil remain on the surface, the water will run off, leaving unwetted areas which are easily observed. Simple visual examination thus suffices to give an idea of whether the part is suitable for further processing.

The test must be applied carefully: if the part retains a film of alkali or alkaline cleaner from the last cleaning operation, the test can be easily masked to show a clean surface when that is not the fact. Therefore, before the test is applied, the part should be rinsed thoroughly after the final alkaline cleaning, immersed briefly in dilute acid, rinsed in cool, clean water, and then observed.

Although the water-break test is not especially sensitive to very small amounts of soil, it appears to be sufficiently so for most practical purposes. A more sensitive test, though one not so easily applied, is the atomizer test devised by Linford and Saubestre. Here the suspected surface is sprayed with water from an atomizer. On a clean surface the water droplets will agglomerate into an unbroken sheet, whereas on a soiled one the droplets will remain separate. In spite of its probable advantages, this test is not widely used, presumably because the familiar water-break test is sufficiently sensitive in practice.

These tests reveal organic or hydrophobic soils only; oxide films and other inorganic soils will still pass.

The final test, of course, is the end result: is the finish satisfactory? When trouble is suspected in the cleaning line and other tests do not show it up, a good test is to clean a sample part thoroughly by hand, using any method that is known to yield a clean surface under the circumstances. Then this sample part is plated along with regular production; if it is satisfactory while regular production is not, the trouble is in the cleaning, and if both are unsatisfactory, the trouble is elsewhere.

TYPES OF CLEANING

Although there are many subdivisions, the preplating operations included under the general heading of cleaning are of three general types: organic solvent degreasing, alkaline cleaning, and acid pickling. Usually they are applied in that order, although for some extremely heavily soiled work some rough cleaning of a mechanical nature may be necessary preceding all of them.

Organic Solvents

Many soils consist of oils and greases of various types, waxes, and miscellaneous organic materials. These can be removed by appropriate organic solvents, either by dipping the part in the solvent or, more frequently and preferably, by the process known as *vapor degreasing*. In this process the organic solvent is vaporized in a boiler beneath the work; the vapors condense on the work, forming a liquid which does the cleaning and drips back into a solvent sump. The process has the outstanding advantage over merely dipping the part in the solvent that the solvent is being continuously distilled; consequently, the liquid in contact with the work is always pure, uncontaminated solvent.

Most vapor-degreasing solvents are chlorinated hydrocarbons, usually with an inhibitor added to minimize hydrolysis by the moisture in the atmosphere producing hydrogen chloride, which is corrosive to many metal substrates:

$$\underset{\text{chlorinated hydrocarbon}}{RCl} + H_2O \rightleftharpoons HCl + \underset{\text{alcohol}}{ROH}$$

The commonly used solvents are trichloroethylene and perchloroethylene, 1,1,1-trichloroethane (methyl chloroform), and chlorofluorocarbons such as trichlorotrifluoroethane (sold by one supplier under the inclusive tradename Freon). The properties of these solvents are listed in Table 5-1.

Vapor degreasers are of many general designs, but they all consist at least of (1) a boiler, above which is (2) a free space where the work is placed, and (3) a set of cooling coils to prevent the vapors from escaping into the air of the shop. In some designs the work is first rinsed in the liquid solvent, then raised into the vapor space. Other designs include means for constantly distilling the solvent and removing the accumulated sludge, as well as other refinements; see Fig, 5-1.

For some types of soil, vapor degreasing can do more harm than good: when the soil consists of inorganic (solvent-insoluble) residues adhering by means of organic waxes and the like (as in some buffing compounds), the removal of the organic portion may simply leave the inorganics more firmly adherent to the substrate than before.

No organic solvent is completely nontoxic; in addition, an atmosphere saturated with the solvent vapor is dangerous simply because it contains insufficient oxygen. Therefore precautions are mandatory in cleaning out or otherwise repairing a vapor degreaser. No one should enter such an apparatus without a breathing mask and a partner outside to observe.

Alkaline Cleaners

Preliminary vapor degreasing may or may not be required for a given type of work entering the finishing department, but some type of alkaline cleaning is almost universally necessary. Alkaline cleaners may be "heavy-duty" or "light-duty"; they may be "soak" or electrolytic; and they may be applied in various ways: by dipping, spraying, or even by hand scrubbing. They must be formulated specifically for specific jobs, depending on many factors: the nature of the

Table 5-1 Physical Properties of Selected Organic Solvents

Solvent		*Boiling point at 1 atm (100 kPa), °C*	*Vapor pressure at 20°C, torr (kPa)*	*Specific gravity*	*Threshold limit value (TLV), ppm**
Name	*Formula*				
Perchloroethylene	$Cl_2C{=}CCl_2$	121	14 (1.8)	1.625	100
Trichloroethylene	$HClC{=}CCl_2$	87	58 (7.7)	1.456	100
1,1,1-Trichloroethane (Methylchloroform)	$H_3C{-}CCl_3$	74	105 (14.0)	1.325	350
Methylene chloride	CH_2Cl_2	40	352 (46.8)	1.335	500
Trichlorotrifluoroethane	$CCl_2F\ CClF_2$	48	273 (36.3)	1.42	1000

*Permissible concentration in atmosphere of work place; the higher this number, the less toxic the material.

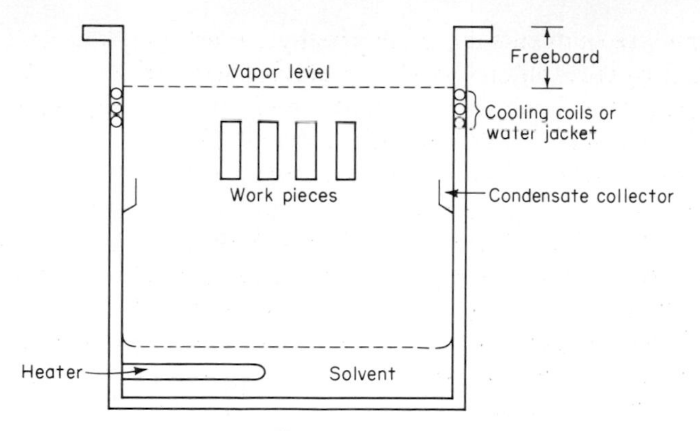

Fig. 5-1 Essentials of a vapor degreaser.

surface to be cleaned (the basis metal); the nature of the dirt or soil to be removed; the sensitivity of the finishing process to be applied, i.e., the degree of cleanliness required; the method of application of the cleaner—soak or electrolytic; and pollution abatement requirements.

With regard to the last, since by their very action cleaners get dirty, they must be disposed of at intervals, since no good way of rejuvenating a spent cleaner is known. Such disposal should be accomplished in accordance with applicable government regulations, and this may influence the type of ingredients of the cleaner.

The surface to be cleaned must be considered in choosing the ingredients of the cleaner; compounds that have little effect on some metals would be far too corrosive for others. Ferrous metals can withstand highly alkaline solutions; copper and its alloys must be cleaned in milder solutions. Table 5-2 lists some typical cleaner compositions and their operating conditions.

Proprietaries

Inspection of Table 5-2 brings up a point about cleaning that will arise many times in considering plating solutions: although the compositions of typical cleaners are set forth, enabling finishers to formulate their own materials, it is often better to make use of proprietary materials offered by the supply houses. In the present study of "fundamentals" it is obviously necessary to speak in terms of definite chemical compounds rather than proprietary formulations, so that the reader will gain some insight into the workings of the chemicals and their mixtures. In the practical carrying out of the processes, however, the experience and background of the specialists who offer proprietary materials are often of sufficient value to balance the obviously higher cost. The user also gains the (usually) expert guidance in solving problems offered by these suppliers. Their formulations generally are based on the same principles as are set forth here; the only difference is that their exact nature is maintained as a trade secret.

Table 5-2 Cleaner Formulations

Ingredient / *Basis Metal:* / *Method of application:*	*Steel* *Electro*	*Copper* *Electro*	*Zinc* *Electro*	*Brass* *Electro*	*Aluminum* *Soak*
Sodium hydroxide, NaOH, %	50	25	20	10	—
Sodium metasilicate, Na_2SiO_3, %	40	40	40	40	40
Sodium tripolyphosphate, $Na_5P_3O_{10}$, %*	5	10	10	10	40
Sodium carbonate, Na_2CO_3, %	4	23	28	38	10
Sodium bicarbonate, $NaHCO_3$, %	—	—	—	—	5
Surfactant: 40% sodium linear alkylate sulfonate, %	1	2	2	2	5
Cleaner concentration, g/L	60–120	30–60	30–45	25–45	30–60
Current density, A/m^2	500–1000	200–500	200–500	150–500	—
Temperature, °C	80–boil	70–82	65–75	60–70	70–82

*Or chelating agent.

Nature of the Soil

Soils are of two general types: organic and inorganic. The organic soils consist of mineral, animal, and vegetable oils and residues from prior cleaning and pickling operations: slushing, quenching, and cutting oils and buffing and drawing compounds used in fabrication of the part. They are not pure compounds but complex mixtures.

Although animal and vegetable (though not mineral) oils can be saponified by the alkaline ingredients of cleaners, saponification actually plays a very minor role in the cleaning process. *Saponification* is the reaction of an ester to form an alcohol (usually glycerol) and the alkali metal salt of the acid corresponding to the ester; this alkali (usually sodium) salt is called a *soap*. The general reaction is

$$\underset{\text{ester}}{(RO)_3C_3H_5} + 3NaOH \rightarrow \underset{\text{soap}}{3RONa} + \underset{\text{glycerol}}{C_3H_5(OH)_3}$$

where R is an organic radical. The saponification reaction is so slow that it plays little part in actual cleaning of metal surfaces. Both types of oil, saponifiable and nonsaponifiable, are removed rather by processes of emulsification, solubilization, and preferential wetting.

Ease of removal of the soil depends on its composition. Soils containing polar groups will be adsorbed on the surface of the metal; and if they contain large amounts of free fatty acids, metal soaps will form that are strongly attached to the surface. (In a metal soap a heavy metal ion substitutes for Na in RONa, above.) This attachment grows stronger with time, and especially at elevated temperatures. Such soaps are difficult to remove; hence their formation should be avoided.

Pickle and cleaner residues are either inhibitors used to control the pickling action of acidic solutions or metallic soaps formed as mentioned.

Inorganic soils include rust and tarnish, solid dirt and dust, scale, smut, and cleaning residues. These are usually removed by alkaline cleaning or acid pickling. Rust and tarnish are insoluble in water but usually can be removed either in acidic solutions or by alkaline chelating solutions. Insoluble solid particles often adhere to the oil on the surface and are removed along with the oil. Pickling smuts are residues of carbide or graphitic material remaining after acid treatment; they are difficult to remove. Inorganic cleaning residues may be oxide films or films of phosphates, silicates, and the like which remain on the surface after alkaline cleaning and are not removed by rinses and acid dips. Better rinsing is the preferred way to eliminate this sort of trouble.

The cleaning process to be used also depends on the finishing process to follow; not all plating sequences require the same degree of cleanliness. Alkaline cyanide and alkaline stannate plating solutions tolerate a degree of "dirtiness" that would be deleterious to some acidic plating solutions such as bright nickel. Therefore cleaning cycles may be shorter or less severe preceding such alkaline solutions than preceding a bright-nickel bath. This tolerance should not be overdone, however; it is poor practice to rely on a plating bath to act as a combined cleaning-plating solution (as was sometimes done in the early years of the plating industry).

The method of applying the cleaner is important in its formulation. Soak cleaners can tolerate the addition of foaming agents to a much greater degree than spray or electrolytic cleaners; for the latter, low-foaming surface-active agents must be used. A foam blanket is helpful, however, in suppressing alkali-containing spray and mist.

The quality of the water supply influences the type of cleaner. Where soft water is available, cleaners containing soaps may be used; but in the more common case of hard waters (unless they are treated) soaps must be avoided or else the cleaner should contain sequestering or chelating agents to overcome the formation of calcium and magnesium soaps, which are insoluble.

TYPES OF CLEANERS

Solvents

Solvent cleaners are usually limited to precleaning for the removal of large excesses of oils and greases. Vapor degreasing may remove the oil from solid dirt particles without removing the solids; these are now dried out and more firmly adherent to the work than ever, and they will be very difficult to remove in subsequent alkaline cleaning.

The solvent used in a vapor degreaser should possess to the greatest extent possible the following desirable properties. It should have good solvent ability for the soils to be removed, should present no explosion or fire hazard, and should be inert to the metals being cleaned. It should be stable and nontoxic. Preferably it should boil below 125°C and possess low specific heat and latent heat of vaporization, and its vapors should be heavier than air. Finally, it must be acceptable under governmental regulations such as pollution control requirements and OSHA and other safety regulations.

Chlorinated hydrocarbons such as tri- and tetrachloroethylene (perchloroethylene), properly inhibited, come close to meeting most of these requirements; they are, however, somewhat toxic. 1,1,1-Trichloroethane is somewhat less toxic, having a tolerance about three times that of "trichlor" and "perchlor," but it is significantly more expensive. The chlorofluorocarbons are less toxic, and almost completely inert.*

Solvent Emulsions

Solvent emulsion cleaners are produced by the addition of oil-soluble emulsifiers to solvents. They afford simultaneous contact of the surface with water and a

*At this writing the safety of all chlorinated hydrocarbons is in question, from several aspects. Chlorofluorocarbons are suspected of depleting the ozone layer of the stratosphere, thus perhaps promoting skin cancers; and many chlorinated organics are possible carcinogens. These matters are controversial and by no means settled; current literature should be consulted.

solvent that preferentially wets the metal. The system is especially useful in removing dust and solid particles mixed with oils and greases or pigmented drawing and buffing compounds. There is still an oil film on the metal, however, so that alkaline cleaning is still required. Preliminary cleaning in a solvent emulsion can greatly reduce the load on the final alkaline cleaner. Solvent cleaners must be entirely removed before plating. Formulations of these types of cleaners are almost entirely proprietary.

Alkaline Cleaners

Alkaline cleaners must be water-soluble; their water solutions must have these properties: they should wet the surface being cleaned, and preferentially wet and penetrate the soil to be removed; they must be able to temporarily emulsify these soils including oils, greases, and solid dirt particles. If necessary, they should soften hard water. They should rinse freely, and not attack or tarnish the metal surface. In addition, they should have high buffer capacity, i.e., tolerate drag-in of acidic substances introduced with the metal without change of pH. Finally, they should not form excessive foam or suds during cleaning or rinsing.

Given these properties, the cleaner will remove dirt effectively in a reasonable time and produce a metal surface that will accept a satisfactory electroplate.

Ingredients of Cleaners

No single alkali makes a good cleaner; a combination of alkalis with appropriate soaps or surfactants, chelating agents, and other ingredients is necessary for maximum effectiveness. The more common ingredients of alkaline cleaners are the following.

Soda ash (sodium carbonate, Na_2CO_3) is highly alkaline, and, depending on market conditions, may be the cheapest alkali available. It provides good buffering and water softening, and has advantages over caustic soda for formulation of dry-mix compounds.

Caustic soda (sodium hydroxide, NaOH) is a most important ingredient in cleaners. It is often a cheaper source of alkali than soda ash, and is also more alkaline; thus it more effectively saponifies fats and oils (even though this action is usually not important, as previously discussed); it reacts with amphoteric metals to form soluble salts, splits esters, and attacks many organic compounds. It has the highest conductivity of the sodium-based alkalis (potassium hydroxide is more conductive but usually is ruled out on the basis of price). It is used in all heavy-duty cleaners and in alkaline derusters and alkaline permanganate scale conditioners.

Phosphates are useful mainly for their water-softening properties; they also contribute to dispersion of soils (or peptization), efficient rinsing, and control of scale. There are many phosphates, including the ortho, pyro, meta, and poly. Trisodium phosphate (TSP, $Na_3PO_4 \cdot 12H_2O$) is used to some extent alone as a soak or spray cleaner. Disodium phosphate (Na_2HPO_4) is used for buffering in the pH range of 8 to 10. Monosodium phosphate (NaH_2PO_4) is used in iron phosphating compounds but seldom in cleaners. Tetrasodium pyrophosphate (TSPP, $Na_4P_2O_7$) is a good sequestrant for zinc, copper, and magnesium, but less good for calcium; it is not very soluble. Of the various phosphates, sodium tripolyphosphate ("tripoly," $Na_5P_3O_{10}$) is the most useful in metal cleaners and in fact in the whole detergent industry. It is an excellent water softener, it sequesters or chelates many metal ions, and it stores and compounds well (i.e., is compatible with other dry ingredients of mixtures). In addition it contributes to good rinsing.

Silicates are good buffers, and when compounded with surfactants, they are good wetting, emulsifying, and deflocculating agents. They are good inhibitors and have high pH and conductivity. They are usually present in heavy-duty cleaners. Of the many silicates, those most used are sodium metasilicate, Na_2SiO_3, which is used both in the anhydrous form and as the pentahydrate, and sodium orthosilicate, Na_4SiO_4, which is widely used in cleaners for steel. Silicates are often specified by the ratio of SiO_2 to Na_2O in the formula; the metasilicate would be a 1:1 compound ($SiO_2 \cdot Na_2O$); the orthosilicate is a 1:2 compound ($SiO_2 \cdot 2Na_2O$). Many compounds and mixtures with both lower and higher ratios are produced. Silicates as ingredients of metal cleaners are a somewhat controversial subject, and they can cause trouble in subsequent plating operations because they are difficult to rinse. Some process specifications call for nonsilicated cleaners.

Chelating agents have become important in compounding of cleaners, especially with the demand for formulations containing little or no phosphate. Phosphates are suspected of causing eutrophication of lakes and streams and are banned in some jurisdictions; chelating agents can replace phosphates for many purposes, although at higher cost. The most widely used chelating agents in metal cleaners are sodium gluconate, sodium citrate, trisodium nitrilotriacetate (NTA), tetrasodium ethylenediamine tetraacetate (EDTA), and triethanolamine; see Fig. 5-2. These compounds can soften water and tie up many metal ions. They are used in many nonphosphated cleaners, etchants for aluminum, alkaline derusting and descaling agents, and in electrocleaners.

Soaps or synthetic detergents (syndets) or both are added to cleaners to decrease surface and interfacial tensions, emulsify oils, and suspend solid particles. Syndets can be high-foaming, low-foaming, or nonfoaming, and the proper selection is important in electrolytic and spray cleaning. Soaps are used in heavy-duty soak cleaners where soft water is available; syndets are effective in both hard and soft water, and have replaced soaps for most uses, but there are a few occasions where soap works better.

About 700 surface-active agents (surfactants) are commercially available; they are broadly classified as cationic, anionic, or nonionic. This nomenclature describes the nature of the hydrophobic (water-insoluble, organic-soluble) group. A typical anionic surfactant is sodium dodecyl sulfonate; the anionic (active) group is $C_{12}H_{25}O^-$, balanced by SO_3Na^+. The cationic type is represented by trimethyl dodecyl ammonium chloride, in which the active cationic group is $C_{12}H_{25}N(CH_3)_3^+$, balanced by the simple chloride ion Cl^-. Nonionics are of many

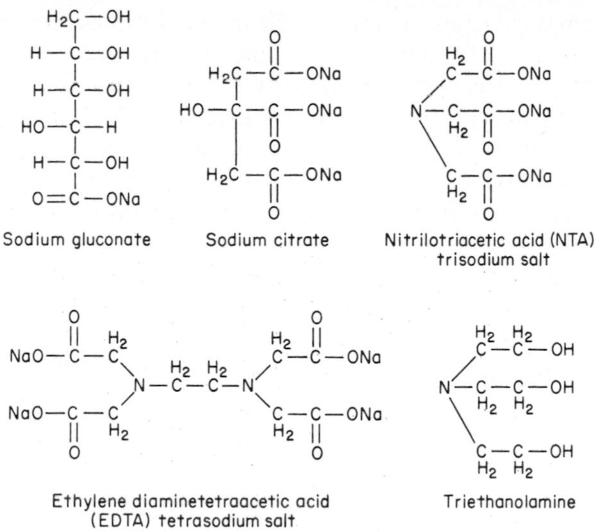

Sodium gluconate

Sodium citrate

Nitrilotriacetic acid (NTA) trisodium salt

Ethylene diaminetetraacetic acid (EDTA) tetrasodium salt

Triethanolamine

Fig. 5-2 Structures of typical chelating agents.

types, having the common property, as the name implies, that they do not ionize; a typical one is polyethylene oxide, $(CH_2{-}CH_2O)_nH$, where n is about 10. The anionics are used in electrocleaners; combinations of nonionics and anionics are useful in soak and spray cleaners. Proper compounding enables the regulation of foam.

With increasing governmental regulation of effluents, it is almost universally required that surfactants be biodegradable; i.e., they can be destroyed by the bacteria present in sewage and waste treatment plants.

Most widely used surfactants in metal cleaning are the sodium linear alkyl sulfonates and the fatty alcohol sulfates for electrocleaners; the sodium linear alcohol sulfonates and oxyethylated alcohol nonionics in soak cleaners; and low-foaming nonionics in spray cleaners.

HOW CLEANERS ARE APPLIED

Hand Cleaning

When the volume of work is small or pieces are too large to fit into available equipment, hand cleaning is feasible. There is no set procedure; the cleaner is applied by any convenient method, including the use of brushes, swabs, or cloths. Operators must be protected: suitable goggles, gloves, and protective clothing must be used. Toxic or inflammable solvents should not be permitted. Hand cleaning is a useful way of checking the operation of a cleaning sequence, as previously described.

Alkaline Soak Cleaning

In this method parts are immersed in tanks of hot alkaline cleaning solution. Concentration and temperature of the cleaner should be as high as possible considering the nature of the work, in order to minimize the time for satisfactory cleaning. Agitation of the solution is helpful: the cleaning action depends on wetting and gradual emulsification of oils and greases, and occasionally saponification of these is also involved. Dirt and solid soil are bound to the surface by the oils and greases; they are removed along with the binders when the latter are dislodged. Heat and agitation speed up the wetting, emulsification, and saponification. The force of the moving solution aids in dislodging the soil and preventing it from redepositing on the work.

Ultrasonic Cleaning

Application of ultrasonic energy, high-frequency sound waves above 20,000 Hz, is helpful in improving the action of many cleaners. It aids removal of soil in blind holes, crevices, and other hard-to-reach areas, by cavitation of the solution. The ultrasonic energy is introduced by means of transducers strategically placed in the cleaning tank. This type of cleaning may be too costly in installation and operation for average types of work not requiring the utmost in cleanliness; but it has proved very useful in processing pieces that are inherently valuable and small enough to justify the installation—expensive jewelry, semiconductors, and other precision parts.

Machine Cleaning

Rapid alkaline degreasing usually is done by spray washing in automatic or semiautomatic machines. In spray washing the force of the spray adds mechanical action to the chemical action of the cleaner. Large excesses of dirt and soil are well removed in precleaning operations by the machine method, when the volume of work justifies the cost of the installation.

Electrocleaning

Soak cleaning can be effective if long enough times are allowed for the cleaning action, but electrocleaning is usually preferred. In electrocleaning the work is made either the cathode (called *direct cleaning*) or the anode *(reverse cleaning)*; the other electrode is inert, usually steel. Electrocleaning adds to the chemical action of the cleaner the mechanical action caused by copious gas evolution at the surface of the work; the gas helps in dislodging the soil and simultaneously brings up fresh solution to the surface.

In *cathodic or direct cleaning* hydrogen gas is evolved at the work surface. At a given current density, twice as much hydrogen is evolved at the cathode as oxygen at the anode:

$$2H_2O \rightarrow 2H_2 + O_2$$

The negatively charged work also repels negatively charged particles of dirt; in highly alkaline solutions most colloidal particles are negatively charged. On the other hand, the negatively charged work will attract positively charged metal ions, forming a metal smut. For this reason separate tanks should be used when nonferrous metals are being cleaned by both cathodic and anodic cleaning. The hydrogen evolved at the work surface may penetrate the metal, causing hydrogen embrittlement of steel. Nickel and nickel alloys must be cleaned cathodically; anodic cleaning would cause passivity, and passive nickel will not plate properly.

In *anodic or reverse cleaning* the gas evolved on the work is oxygen; only half as much oxygen is evolved as hydrogen at the cathode so that the mechanical action is not as effective. However, the positively charged work will repel metal ions; smuts formed during cathodic cycles are thus dissolved or dislodged. Copper and copper-base alloys should be cleaned for only short periods anodically unless the cleaner is specially inhibited to prevent formation of tarnish films.

Because cathodic and anodic cleaning have individual advantages and disadvantages, as described, the usual procedure is to use both, either in separate tanks or by reversing the current in the same tank. For similar reasons, periodic reverse (PR) current is widely used, especially with alkaline derusting compounds.

ELECTROPOLISHING

Electropolishing produces an excellent surface for subsequent electroplating; in addition, it is used as the final finishing method for some metals, notably stainless steels. It provides a chemically and physically clean surface and removes mechanical surface damage as well. Many metal surfaces that have undergone mechanical preparation may have surface asperities which may be detrimental to the production of uniform and pit-free electroplates; electropolishing can remove these defects.

Although electropolishing can remove light tarnish films and light oil and grease, the latter should preferably be removed before electropolishing. If this is not done, the oil and grease or their degradation products accumulate on the surface of the electropolishing solution and will deposit on the work as it is withdrawn. Heavy rust and heat scale also should be removed before electropolishing.

Work for electropolishing is racked as for plating; but because current densities are usually higher by a factor of 5 to 10, contacts and splines must be heavier. The same racks can be used for electropolishing and for subsequent plating.

For electropolishing the work is made anodic in the solution; several solutions may be used. Most common ones are based on mixtures of sulfuric and phosphoric acids in various proportions. For ferrous alloys the proportion is about 50/

50 (by weight), at 55 to 105°C; for copper about 85/15 phosphoric/sulfuric at 45 to 60°C; and for brass, phosphoric acid with 5–7 percent chromic anhydride (CrO_3, commonly called chromic acid), at about 60°C. Current densities range from 500 to 4000 A/m^2; time ranges from 2 to 7 min.

Mixtures of perchloric and acetic acids were among the earliest of the electropolishing solutions developed, originally for very small-scale metallographic purposes. Following a disastrous explosion and fire in Los Angeles, probably caused by carelessness in observing precautions, such solutions have been completely out of favor at least for shop use.

RINSING

Complete rinsing is an integral part of every cleaning operation (and of every plating operation as well). All the partially dissolved, partially loosened, emulsified, or suspended soils must be removed from the surface of the work. In addition, the film of cleaning solution adhering to the surface should be removed. The subject of rinsing is considered in Chap. 7.

PICKLING: REMOVAL OF SCALE AND OXIDES

Heavy rust and scale should be removed before the metal enters the cleaning cycle; methods for removing them will depend on their nature. They can be minimized by careful control in heat treating or any other treatments required.

After alkaline cleaning, a thin film of oxide or tarnish is usually present on the cleaned surfaces; this must be removed before electroplating. Furthermore the alkaline film left by the alkaline cleaners, even after good rinsing, should be neutralized before the metal enters any acidic plating bath.

Removal of heavy scale, heat-treat scale, oxide, and the like requires more stringent treatment than removal of light tarnish and oxide films, but the chemical action is similar in both. Acids, primarily sulfuric and hydrochloric, are generally used; some dry acid salts having similar action are also employed, depending on the nature of the metal and the scale.

Acids include almost all the common ones: sulfuric, hydrochloric, nitric, phosphoric, hydrofluoric, fluosilicic, fluoboric, chromic, and sulfamic. Acid salts used are sodium bisulfate, ferric chloride, ammonium persulfate, and the acid salts of the other acids mentioned.

Sulfuric acid is the most widely used, because it is the cheapest acid, and it gives off no noxious fumes as does hydrochloric. It is used for pickling steel, copper, and brass; with chromic acid or dichromates for desmutting and deoxidizing aluminum; and with hydrofluoric or nitric acid for descaling stainless steel. Anodic sulfuric acid etching is used for removing the last traces of smut and scale from steel; cathodic pickling may be used when dimensional tolerances are important and no metal removal can be tolerated.

Hydrochloric acid is somewhat more vigorous in action than sulfuric and can be used at room temperature, but the resulting fumes may be a problem.

Nitric acid is an ingredient of several bright dips. In combination with hydrofluoric acid it is used for removing heat scale from aluminum, stainless steel, nickel and iron alloys, titanium and zirconium, and some cobalt alloys.

Phosphoric acid is useful for removing rust from steel and, in combinations, for stainless steel, aluminum, brass, and copper. Fluoboric acid is an effective pickle for lead alloys or soldered copper and brass parts before plating.

Sodium bisulfate, $NaHSO_4$, is in effect a convenient way of handling sulfuric acid in dry form, thus eliminating the hazards of handling this corrosive and dangerous material. It is the basis of many proprietary pickling compounds.

Pretreatments

In some cases heat scales do not dissolve readily in acids, or the acid treatment may need to be so vigorous that the substrate metal is attacked to an unacceptable extent. The nature of the scale can be changed to make it more readily soluble in dilute acids; occasionally alkaline cleaning, using a chelated cleaner, is beneficial. Potassium permanganate plus caustic soda as a pretreatment is helpful in many such cases.

Molten salt pretreatments can remove difficult scales. Molten sodium hydroxide plus sodium hydride (NaH, formed *in situ* by the use of sodium metal and hydrogen) reduces scale and renders it removable by a simple water quench followed by attack with dilute acids. Some molten sodium hydroxide treatments contain oxidizing agents such as sodium nitrate. Most of these treatments are proprietary.

Pickling, like cleaning, can be made more effective by the employment of current. Cathodic pickling is used on stainless steels; anodic pickling is effective for removing the last traces of smut and scale from high-carbon steels.

Corrosion inhibitors are widely used in descaling and pickling, but in pickling before plating they can cause trouble and are generally avoided.

For specific formulas for removing scale and for pickling various substrates, see Table 5-3.

Specific Prefinishing Cycles

Having explained some fundamentals of preparing metals for electroplating or other finishing, we must now consider briefly some specific pretreatment cycles for the more commonly encountered substrates. Many of these are set forth in various ASTM "Recommended Practices" to which reference should be made for details; some are summarized below.

Table 5-3 Solutions for Scale Removal, Pickling, and Bright Dips (% by volume)

Basis metal	*Scale removal*	*Acid pickle before plating*	*Bright dip*
Low-carbon steel	15–25% H_2SO_4, 50–82°C 25–85% HCl, rt* das† 120 g/L, 25–60°C	4–10% H_2SO_4, rt 5–15% HCl, rt das 60–120 g/L rt	25–30 g/L oxalic acid + 10–15 g/L 30% H_2O_2 + 0.08 g/L H_2SO_4 (low-carbon and high-carbon steel)
High-carbon steel	Mechanical	10% HCl, rt	
Stainless steel	10–20% HNO_3 + ½% HF, 50–60°C, das 120 g/L, cathodic, 70°C H_2SO_4 + CrO_3 + HF, each 60 g/L, rt	20–50% H_2SO_4, 65–82°C then 1% H_2SO_4 + 0.1% HCl, rt 5–50% HCl or H_2SO_4, cathodic, rt das 120 g/L, cathodic, rt	25% H_2SO_4 + 4% HCl + 3% HNO_3
Copper alloys	10–40% H_2SO_4, rt 10–15% H_2SO_4 + 15–30 g/L $Na_2Cr_2O_7$, rt 25% H_2SO_4 + 12.5% HNO_3, rt	4–10% H_2SO_4, rt	40–45% H_2SO_4 + 20–25% HNO_3 + 0.1% HCl, rt CrO_3 270 g/L, rt 55% H_3PO_4 + 20% HNO_3 + 25% CH_3COOH, 55–80°C
Zinc		0.25–1% H_2SO_4, rt	
Nickel		5% H_2SO_4, rt 20% HCl, rt das 120–180 g/L, rt	60% H_3PO_4 + 20% H_2SO_4 + 20% HNO_3, rt

*rt = room temperature.

†das = dry acid salts.

CLEANING METALS PRIOR TO ELECTROPLATING (ASTM B 322)

Several stages are required for adequate cleaning:

Stage 1. Precleaning—solvent, emulsion, or alkaline spray, to remove the bulk of the soil

Stage 2. Intermediate alkaline cleaning

Stage 3. Final electrocleaning, to remove traces of solids and other adherent contaminants

One or more of these stages may be eliminated in some cases, depending on the type and amount of soil on the pieces as received; but it is usually preferable to maintain a multistage operation, because it improves the life and efficiency of the cleaning solutions.

Before Cleaning

Cleaning can be improved and simplified if consideration is given to controlling the soil initially. Overbuffing leaves excess compound on the work; long aging between buffing and cleaning hardens the compound and makes it more difficult to remove, Liquid buffing compounds are easier to remove than solid ones and should be considered when applicable. Drawing compounds with polymerizing oils, or white-lead pigments, should be avoided because they are difficult to clean off. Proper choice of lubricating oils in machining, with an eye to ease of removal, pays off in better cleaning. Prolonged storage, or drying, of emulsion drawing compounds after metal working should be avoided. Fingerprints are troublesome and can be avoided by handling work with clean gloves after buffing or polishing. Cleaning should be done as soon as possible after the last metal-working operation, since soils are removed more easily when fresh than after they have hardened. In short, every precaution should be taken to make the job of cleaning as easy as possible, by appropriate precautions during the metal-working operations.

The Metal

The properties of the metal being finished affect the cleaning sequence. Hardness and ductility are factors in selecting buffing or polishing methods. The chemical activity of the metal plays an obvious role in the selection of the cleaner. Aluminum is subject to overetching in alkalis; aluminum and zinc are sensitive to pitting attack; zinc and brass tarnish easily. Zinc die castings are particularly sensitive to attack by cleaners. Design of parts is important; where possible, small indentations that can trap solid particles and buffing compounds should be avoided. The "skin" of die castings should not be cut through during buffing, since the subsurface is more sensitive than the skin. As with the remainder of the cleaning cycle, some precleaning defects do not show up until after the complete plating cycle.

Stage 1: Precleaning

Precleaning is designed to remove large excesses of soil, reducing the load on the later stages and enabling the use of milder solutions later on. Several methods of precleaning are used: cold solvent, vapor degreasing, emulsifiable solvent, solvent emulsion spray, invert emulsion cleaners, and hot alkaline spray with or without solvent emulsion.

Cold cleaning involves a single dip in one of various solvents, usually chosen from mineral spirits (naphtha, not often used because of fire hazard), trichloroethylene, perchloroethylene, 1,1,1-trichloroethane (methyl chloroform), methylene chloride, and trichlorotrifluoroethane. The chlorinated solvents are effective for many types of soil, but not for those based on soaps. Cold cleaning is rarely a complete preparation for plating and must be followed by other steps.

Vapor degreasing uses the same solvents as listed above except for mineral spirits. Usually the work is first sprayed with clean solvent, or immersed in warm or boiling solvent; this removes much of the soil mechanically. After this the work is placed in the condensing part of the degreaser, where hot, clean solvent vapors condense on the work. Vapor degreasing can be used on all types of metal. It simplifies the cleaning of parts made up of several metals, because the action is solvent rather than chemical. And there is no danger of overcleaning caused by any difference of chemical reactivity among the metals concerned. Because vapors penetrate well, this method is effective on blind holes, recesses, perforations, crevices, and welded seams. It can be supplemented by ultrasonic cleaning in the solvent rinse chamber.

Vapor degreasing is effective for solvent-soluble soil and chemically active lubricants. Some insoluble soils such as buffing grits, metal chips, and dust are flushed away at the same time as the soluble soils—grease and oil—dissolve. It is not effective on metal salts, scale, carbon deposits, some inorganic welding and soldering fluxes, and nongreasy fingerprints. In many instances degreased work can proceed directly to mild electrolytic cleaning, skipping the intermediate stage.

Emulsion Cleaners

Oils and high-boiling hydrocarbons such as kerosene can dissolve most greases, especially at high temperatures. Addition of emulsifiers, soaps, and surfactants enhances the penetrating power of the organic solvent and allows removal of the solvent along with the soil by power flushing. In addition, contact of the metal with the aqueous phase favors removal of soils not soluble in the organic phase.

This principle is applied in several ways: use of straight emulsifiable solvents, unstable emulsions or diphase cleaners, invert types of emulsions, and stable emulsions. (An emulsion is a more or less stable mixture of two or more immiscible liquids held in suspension by a substance called an emulsifier. An invert emulsion is water in oil, as contrasted to the usual oil-in-water type.) Rust inhibitors or alkali cleaners may be added to the water phase. Agitation is important.

Emulsion cleaners are used at temperatures up to 80°C. High temperatures favor good cleaning, but care must be taken not to exceed the flash point of the solvents.

Soils accumulate in solvents, so that they must be either discarded or purified. In vapor degreasing equipment, the solvent is recovered by distillation and the soil discarded. Emulsifiable solvents must be discarded occasionally. Their disposal may be troublesome.

Stage 2: Intermediate Alkaline Cleaning

Intermediate alkaline cleaning removes solvent residues and soil which has been softened or conditioned by the precleaning cycle. If the solvent cleaning step is omitted, this stage serves as a precleaning stage followed by final alkaline cleaning. Metals sensitive to alkalis, such as zinc, should not be so treated, since time in the alkali should be minimal. Sometimes *precleaning* is used as a term for alkaline cleaning before electrocleaning, especially when solvent cleaning is omitted or carried out in a different department. As stated, the intermediate alkaline cleaning stage can sometimes be omitted and the work transferred directly from vapor degreasing to final electrolytic cleaning. More often the intermediate stage is required.

This intermediate, or heavy-duty, cleaning can take several forms:

1. Soak cleaning, at a concentration of 30 to 120 g/L of alkaline cleaner at 80°C to boiling, for 3 to 15 min. Ultrasonic energy may be added, at temperatures of 70°C to boiling. The cleaners usually contain surface-active soap-like compounds, which foam on vigorous agitation.
2. Spray cleaning, at 4 to 15 g/L of an appropriate cleaner at 50 to 80°C for 1 to 3 min, with spray pressures of 70 to 350 kPa (10 to 50 psi). Foaming can be a problem. Even when the cleaner is formulated so as to be low-foaming or nonfoaming, foam can result from the formation of soaps in the cleaning action itself. Barrel alkaline cleaning is usually carried out at 7 to 45 g/L, at somewhat lower temperatures than for soak cleaning.

Good alkaline cleaning is affected by many factors. Concentration of the cleaner must be a compromise among many variables, and is usually best set by tests. The highest practicable temperature should be used; a rolling boil is helpful in producing agitation, but other factors sometimes limit the feasible temperature. Alkaline cleaners operate by lifting of the oil film, and this takes time, which must be sufficient to do the job. Agitation is accomplished by spray cleaning; in soak cleaning by maintaining a rolling boil, by mechanical agitation, movement of the work, or pumping of the solution; or by ultrasonics. In any case, good agitation is paramount.

Stage 3: Final Electrocleaning

As has been stated, an absolutely clean surface is not only unnecessary for satisfactory electroplating, but is also probably unobtainable. The object of the final electrocleaning step is to render the metal surface free of any *objectionable* surface films and to replace them with films that are suitable for further processing in the electroplating bath. If the parts have been subjected to stage 1 and stage 2 cleaning, such objectionable films that must now be gotten rid of are precleaner residues, minor amounts of soils and oils, and solid particles not completely removed by prior treatment; in other words, stage 3 cleaning is primarily insurance against later trouble in the plating steps. This is usually accomplished by electrocleaning, which has been already described.

PREPARATION OF INDIVIDUAL SUBSTRATES

Since this text considers the fundamentals of metal finishing, it is beyond our scope to go into minute detail regarding the preparation of each individual metal surface that the electroplater may be called upon to finish. It is strongly recommended that the student refer to relevant texts that offer detailed directions, among which, in addition to those cited in the general Bibliography, are the various ASTM "Recommended Practices," a few of which are summarized below.

Low-carbon Steel (ASTM B 183)

The principal steps for cleaning low-carbon steels have already been outlined in the main body of this chapter. They consist of removal of oils, greases, and caked-on dirt by cleaning; removal of scale and oxide films by pickling; and removal of smut left by the preceding step and activation of the surface. Removal of bulk soil, fabricating lubricants, and finishing compounds is often done by "precleaning" before the parts are received by the finisher.

Precleaning, if not already done, is accomplished in alkaline soak cleaners, spray cleaners, or solvents, as already described. As always, this step should be carried out as soon as possible after mechanical finishing to avoid aging of the soil.

Electrocleaning for low-carbon steel should be anodic (reverse) in a suitable electrocleaner at about 60 to 120 g/L. Current density is 500 to 1500 A/m^2 at about 6 V; temperature is 75 to 100°C, for 1 to 4 min.

Acid pickling follows (rinsing understood) in 15 to 50% (vol) hydrochloric acid or 10% (vol) sulfuric acid, or proprietary salts, at room temperature for times sufficient to remove oxides and scale.

In place of acid pickling, alkaline derusters may be used. A solution containing 180 g/L of sodium hydroxide, 120 g/L of sodium cyanide, and 80 g/L of a chelating agent of the EDTA, NTA, or gluconate type, at about 40°C and using a current density 200 to 500 A/m^2 is used.

Very heavy scale may be better removed by electrolytic pickling, either anodic or cathodic, in 5 to 10% sulfuric acid at 50 to 65°C and about 400 A/m^2. Anodic pickling avoids hydrogen embrittlement; cathodic pickling provides a somewhat brighter surface. Mechanical treatments such as shot blasting, tumbling, or sandblasting avoid hydrogen embrittlement but may work-harden the surface.

High-carbon Steel (ASTM B 242)

There is no definite cut-off point between "low-carbon" and "high-carbon" steels, although a carbon content of 0.35 percent often is considered the dividing line. For electroplating purposes case-hardened steels are included but alloy steels are not. Electroplating on high-carbon steels presents somewhat more difficult problems than on low-carbon steels; they have a greater tendency to become embrittled during the cleaning and plating cycles, and maximum adhesion of the electroplate is somewhat more difficult to obtain. High hardness is a major cause of cracking of steel during or after plating. Depending on the nature of the parts, steels harder than Rockwell C45 to C62* should not be electroplated. Problems with hydrogen embrittlement increase with hardness. A 30-min bake at about 205°C will provide relief of internal stress before plating. After plating, relief of embrittlement usually requires 3 to 24 h at about 190°C.

Many treatments are of assistance in maintaining a satisfactory surface during fabrication and mechanical finishing; these are beyond our scope; however, the ASTM practice or equivalent source should be consulted for practical details.

Electropolishing is especially adaptable to high-carbon steel; it removes highly stressed metal and metal debris from the surface without forming smut. Proprietary mixtures of sulfuric and phosphoric acids are used.

Iron Castings (ASTM B 320)

The principal difficulties encountered in plating on cast irons derive from the casting process and the carbon or graphite inclusions in the surface. Precleaning is largely mechanical, to remove gross defects; cleaning is conventional. Because of the low hydrogen overvoltage on graphite, cast iron is often difficult to plate with zinc; in this case a strike in a cadmium, tin, or tin-zinc alloy plating bath renders the surface receptive to a zinc plate.

*In some cases the upper limit is as low as C35. Individual specifications should be consulted.

Stainless Steel (ASTM B 254)

Stainless steel is an inclusive term for a number of iron alloys with chromium, nickel, or both, often containing additional elements including titanium, niobium, tantalum, molybdenum, and tungsten. The principal problem in plating on stainless steels derives from their passivity; this is, of course, not accidental, for the "stainless" property is a function of this very passivity. Because stainless steels are corrosion-resistant, usually they are not finished further by electroplating, but occasions do arise for plating on them: for color matching (especially with chromium-plated parts), lubrication during cold heading, and for a few other specialized purposes.

The corrosion-resistant properties of the stainless steels are due to an adherent thin transparent film of oxides. This film re-forms quickly if removed, and the principal aim of preplating operations is to remove the film while preventing it from re-forming. This is called *activation* and is the critical step in electroplating on stainless steel.

Precleaning and alkaline cleaning sequences are conventional, as for carbon steels. After cleaning, activation must be accomplished before plating: the thin film of oxides must be removed, and the parts not allowed to dry but transferred quickly into the plating cycle. Cathodic activation in various acids is used; mixtures of sulfuric and hydrochloric acids without current are also satisfactory in special cases.

Copper and Copper-base Alloys (ASTM B 281)

The preplating cycles for copper and copper alloys are standard, as already described. If the plating solution is acidic, the cycle includes vapor or solvent degrease, alkaline electroclean, acid dip, and plate (rinse between all steps). If the plating solution is alkaline, a cyanide dip is added before plating. Electrolytic cleaning is usually anodic (reverse), or if both polarities are used, the anodic cycle is the last.

Copper and its alloys may be bright-dipped either before the cleaning cycle or as a part of it. A typical bright-dipping bath is included in Table 5-3. The bright dip gives off toxic fumes (nitrogen oxides) and must be adequately ventilated.

Zinc Alloy Die Castings (ASTM B 252)

Zinc alloy die castings normally contain, in addition to the zinc, about 4 percent aluminum, 0.04 percent magnesium, and either 0.25 or 1 percent copper; impurities such as lead, cadmium, tin, and iron are limited by specification. Zinc alloy die castings are chemically active and are dissolved or etched by prolonged contact with acid or strongly alkaline solutions; thus immersion times should be short.

The preparative steps include: (1) smoothing of parting lines; (2) smoothing of rough or defective surfaces, if required; (3) buffing, if required; (4) precleaning and rinsing; (5) alkaline electrocleaning and rinsing; (6) acid dipping and rinsing; and (7) copper plating.

Parting lines are smoothed by mechanical polishing with abrasive-coated wheels or belts, tumbling with abrasive media, or vibration with abrasives. Step 2 is accomplished similarly. Die castings are buffed to produce a mirror finish suitable for plating with conventional plating solutions, If solutions with good leveling power are available for plating copper and nickel, the buffing step may be omitted. (Leveling is discussed in Chap. 8).

Preliminary removal of the bulk of the buffing compounds and other soil should be done as soon as possible after buffing and polishing. Solvent cleaning and vapor degreasing, emulsion cleaning, and various aqueous detergents are satisfactory, as are power spray alkaline cleaners. Typical solutions for the last-named consist of 10 g/L of mixed alkalis such as trisodium phosphate, sodium metasilicate, and sodium bicarbonate, along with a small amount (about 1 g/L) of sodium hydroxide; temperature is about 80°C at a pressure of 170 to 245 kPa (25 to 35 psi).

Alkaline electrocleaning as a final step is required to remove last traces of oils, greases, and other soils. Anodic electrocleaning is preferred to cathodic for zinc die castings, at current density (cd)* from 150 to 300 A/m², for 30 to 60 s. Typical solutions contain 30 to 40 g/L of mixed alkalis including sodium metasilicate and trisodium phosphate, 0.5 g/L of surfactant, and no more than 0.5 g/L of sodium hydroxide, at 70–80°C. Thorough rinsing is understood between all steps.

Acid dipping follows the last alkaline cleaning step. Sulfuric acid at a concentration of about 0.25 to 0.75 percent for about 30 to 60 s is normal.

After thorough rinsing, a copper film is applied in order to prepare the zinc casting for further plating. This strike is applied in a copper cyanide solution; its thickness should be at least 1 μm if the following plate is to be a bright copper in high-temperature cyanide solutions, and at least 5 μm for parts that will be nickel-plated directly without a further copper undercoat. If a bright leveling copper is to follow, an intermediate thickness of 3 to 4 μm is recommended. The copper strike solution contains 20–45 g/L copper cyanide, 10–20 g/L free sodium cyanide, and 15–75 g/L sodium carbonate. Potassium cyanide may be substituted for the sodium salt, and other additives are sometimes used. Cathode cd ranges from 250 to 650 A/m² at temperatures from 50 to 55°C. Copper plating is considered in more detail in Chap. 12.

Aluminum and Its Alloys (ASTM B 253)

Aluminum alloys pose special problems for electroplating; aluminum is a very active metal (negative electrode potential) that forms a natural, impervious oxide film which must be removed before plating. This film forms rapidly, and it is not sufficient merely to remove it; it must be replaced by a more receptive film for satisfactory plating. Chemically pure aluminum is seldom used, and the various

*The common abbreviation for current density, cd, is used frequently throughout the text.

aluminum alloys have various microstructures which cannot all be treated alike. Common alloying elements include silicon, copper, manganese, and magnesium, and many others may be encountered. As has been mentioned (Chap. 3), any metal which can be plated on aluminum will be lower (more noble) in the electromotive series than the substrate; consequently, corrosion problems will be aggravated unless the electroplate is practically free of pores.

In spite of these difficulties, aluminum is widely finished with electroplates for many purposes. The standard copper/nickel/chromium sequence is used for decorative purposes, as are brass, silver, and gold. Silver may be applied to improve surface conductivity; brass for adhesion to rubber; copper or tin for assembly by soft soldering; chromium to reduce friction or wear; zinc to lubricate threaded parts; and tin to reduce friction or to provide better electrical contact in bolted electrical conductors such as bus bars.

Many methods have been proposed for plating on aluminum; most are now obsolete. Early methods relied on some kind of roughening of the surface to provide mechanical adhesion; most surviving methods rely on replacing the natural oxide film with one which is more receptive to plating. An exception is the use of anodizing, to yield an oxide film of controlled properties; this is still used in special cases. The two most commonly used methods are the zinc-immersion and the tin-immersion processes.

Cleaning and conditioning pretreatments are necessary, as for all metals; vapor degreasing or other solvent treatment, as usual, removes the bulk of oils and greases. This is followed by mild etching in an alkaline cleaner, containing typically 25 g/L of sodium carbonate and 25 g/L of trisodium phosphate at 70–85°C for 1 to 3 min.

Following cleaning, a conditioning treatment usually is required. For pure aluminum and Al-Mn alloys, this may be a dip in 50 percent (vol.) nitric acid. For Al-Mg and Al-Mg-Si alloys this nitric acid dip is preceded by a 2- to 5-min etch in 25 percent (wt.) sulfuric acid at 85°C to remove undesirable microconstituents. For casting alloys containing large amounts of silicon, a 3- to 5-s dip in 3 parts nitric acid plus one part 48% hydrofluoric acid may be used before the zinc-immersion treatment.

Another highly useful pretreatment is the so-called double zinc immersion; in this process, the first zinc-immersion dip removes the oxide layer and replaces it with a zinc film. This zinc film is then dissolved in 50% nitric acid; the surface is then in good condition for receiving the second zinc-immersion film.

Deposition of the final zinc-immersion film is carried out in a highly alkaline solution of sodium zincate, prepared by dissolving zinc oxide in sodium hydroxide solution. Concentrations vary widely, and proprietary processes are also available. For best performance the zinc film should be as thin as possible consistent with satisfactory subsequent electroplating. The weight of zinc should be about 150 to 470 mg/m^2, and preferably not over 300 mg/m^2.

A typical zinc-immersion bath consists of 400 to 500 g/L sodium hydroxide and 80 to 100 g/L zinc oxide, used at about 25°C. A modified zinc-immersion bath containing Rochelle salts and ferric chloride possesses some advantages over this basic formula; it gives more uniform coverage in the ensuing electroplating operations, has a greater operating range when the double zincate process is used, and yields somewhat better corrosion resistance of the total finishing system. This modified formula is: NaOH, 525 g/L; ZnO, 100 g/L; $KNaC_4H_4O_6 \cdot 4H_2O$ (sodium potassium tartrate, Rochelle salt), 10 g/L; $FeCl_3 \cdot 6H_2O$, 1 g/L. It is operated at room temperature at immersion times of 30 to 60 s.

Other modified zinc-immersion processes have been proposed, including several proprietaries; one containing nickel has been used widely.

A disadvantage of these zinc-immersion solutions is high viscosity, which renders rinsing and drag-out somewhat troublesome. Several more dilute baths have been proposed, of similar composition but much lower concentrations; these are more free-rinsing, but on the other hand have less reserve of zinc metal and therefore require more frequent replacement. In using these dilute solutions immersion time must be short—not over 30 s.

Although the zinc-immersion pretreatment may be regarded as the standard preparatory method for plating on aluminum, it is not entirely satisfactory, and many attempts have been made to find improvements. Among the more promising is a tin-immersion treatment from a stannate solution followed by a bronze (copper-tin) strike. This process is proprietary and cannot be discussed here, but it is finding wide acceptance in industry.

After the application of the zinc film, the part can be further processed by any method suitable for plating on zinc, provided that the process does not penetrate the extremely thin zinc film and attack the underlying aluminum. Although silver, brass, nickel, zinc, and chromium can be plated directly on the zinc film, it is usual to apply a copper strike before proceeding further. The copper strike solution has the following composition: copper cyanide, CuCN, 40 g/L; total sodium cyanide, NaCN, 50 g/L; sodium carbonate, Na_2CO_3, 30 g/L; Rochelle salt, $KNaC_4H_4O_6 \cdot 4H_2O$, 60 g/L; free sodium cyanide, 4 g/L. Temperature is 35–45°C, and pH 10.2 to 10.5; electrical contact should be made before immersion in the bath ("hot contact"), and the initial current density should be high, about 250 A/m^2. After about 2 min the current density may be halved and deposition continued for another 3 to 5 min. After this initial copper strike any desired electroplate may be applied.

As mentioned, aluminum is more active (less noble) than any metal which can be plated on it; for this reason, once a break develops in the coating, corrosion may be very rapid. In addition, the zinc film is, in some environments, actually anodic to aluminum and thus tends to protect both the substrate and the electrodeposit. The zinc film may be undermined and lead to lifting or peeling of the whole deposit. This is one reason why the zinc film should be as thin as possible.

Magnesium (ASTM B 480)

The problems in plating on magnesium are similar to those encountered with aluminum, but perhaps even more pronounced. The preparative procedures are similar: clean, pickle, zinc-immersion dip, copper strike. The pickle is a solution containing chromic acid, ferric nitrate, and potassium fluoride; this is followed by activation in a phosphoric acid–ammonium bifluoride solution. The zinc-immersion bath differs in composition from that for aluminum but serves a similar purpose.

Other Metals

The usual substrates that the metal finisher is called upon to electroplate are iron and steel; copper, bronze, and brass; zinc die castings; and aluminum. Accordingly we have gone into some detail regarding the preparation of these metals, even though not enough to obviate the need to refer to specific directions. Less often and for special purposes almost any metal may have to be electroplated, and most present specific problems to the finisher. Among them are lead and lead alloys and the so-called refractory metals—titanium, zirconium, niobium, tantalum, tungsten, molybdenum, and uranium. For some of these refractory metals no entirely satisfactory method has been published, but at least passable results can be obtained by careful adherence to prescribed methods.

Problems arise because of the very reactive nature of most of these metals; although they seem stable in air, this stability is due to an adherent film of oxide. During the course of the preparative cycle, it is necessary to remove this oxide film in order to render the surface receptive to an adherent electrodeposit. As with aluminum, it is not sufficient merely to remove the oxide layer, however, since it would simply re-form on exposure to air. Therefore some other film must be substituted. The nature of this replacement film differs in each specific case, so that a knowledge of the chemistry of the substrate metal is required in order to make an intelligent choice of a processing cycle. Details will be found in various ASTM "Recommended Practices" for individual substrates, as well as in standard texts as listed in the Bibliography.

Nonconductors

The technology of plating on plastics and other nonconductors advanced rapidly in the decade beginning about 1965, and is considered in a separate chapter.

6 Mechanical Preparation for Finishing

A crude ingot or casting of metal may undergo many mechanical and thermal treatments as it is being transformed into its final shape as a useful article. Most of these treatments are of little concern to the metal finisher: casting, rolling, shaping, cutting, stamping, forging, grinding, and the like are carried out before the parts are received in the finishing shop. Two mechanical processes, however, are normally the province of the finisher: polishing and buffing, as well as some variations of these having similar purposes.

Polishing involves the removal of small amounts of metal by means of abrasives. It produces a surface that is free of the larger imperfections left by grinding, and is a preliminary to buffing, which is similar in outline to polishing but because it employs finer abrasives removes very little metal and can produce an extremely smooth, even mirror, surface.

Unlike the electrochemical aspects of metal finishing, the operations of polishing and buffing have been the subjects of comparatively little fundamental study and have been brought to their present state of development principally by experience. They are, therefore, of minor concern in a study of the fundamentals of metal finishing, but since they form an integral part of most such processes, we must consider them briefly. Since this text is not generally concerned with the engineering aspects of metal finishing (which does not by any means imply that these aspects are unimportant), only the bare outlines of mechanical prefinishing can be offered here.

The need for polishing and buffing will differ depending on the surface finish—the finish of the part as received and the finish desired in the final article. Except where automatic or semiautomatic processes can be used, polishing and buffing are labor-intensive processes and are therefore avoided where possible. When "mill finishes" are adequate, polishing, buffing, or both may be dispensed with entirely, and many articles can proceed directly from fabrication to finishing. Just as often, however, some smoothing of the surface is required before finishing.

Often buffing (but not polishing) is also a part of the finishing process itself:

Where the electroplating operation yields a mat finish and a mirror finish is desired, the electroplate itself may be buffed. Sometimes an intermediate electrodeposit is buffed in preparation for further plating; for example, a steel part may be copper-plated and the copper then is buffed before subsequent nickel and chromium plating. This is done because copper is more easily buffed to a high luster than the basis steel; scratch marks and other imperfections in the substrate are masked by this intermediate step. The need for intermediate buffing has decreased markedly since the introduction of plating processes that produce mirror-bright finishes without the need for a bright substrate, and of other processes that have good leveling power (discussed in Chap. 8). Nevertheless, the necessity for in-process buffing has not entirely disappeared.

POLISHING AND BUFFING

Polishing differs from grinding in that grinding ordinarily involves the use of a solid wheel, normally composed almost entirely of the bonded grit (abrasive). Grinding typically leaves a fairly rough surface. Polishing follows grinding and precedes buffing. Its purpose is to remove a significant amount of metal and to smooth the surface in a preliminary way, preparatory to the more refined smoothing offered by buffing. Buffing usually follows polishing and smooths the metal surface and improves its appearance, but removes little metal.

Both polishing and buffing are carried out by bringing the work into contact with wheels revolving at fairly high speeds. The main difference between the two operations lies in the construction of the wheels and the abrasives used in

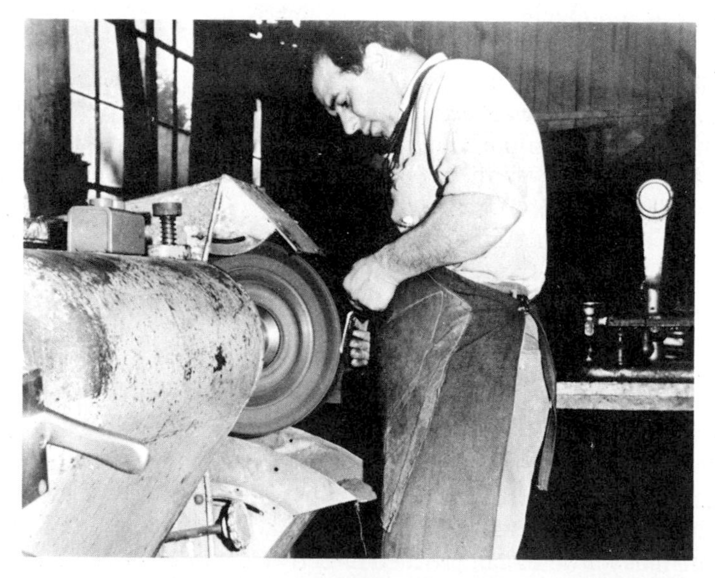

Fig. 6-1 Buffing wheel in use. *(Lea Manufacturing Co., Waterbury, Conn.)*

connection with them. Polishing may also be carried on by means of endless belts coated with abrasives; the choice between belts and wheels is determined by the configuration of the parts and the degree of automation appropriate to the operation—in effect, the total costs. Both polishing and buffing, as already stated, are relatively labor-intensive, and skilled polishers are becoming scarce. Belt polishing is somewhat less demanding of skilled labor than wheel polishing.

POLISHING

Polishing wheels are usually of muslin, canvas, felt, or leather. Construction of the wheels can be varied to make available wheels of varying flexibility, best adapted to the shape and surface condition of the object to be finished. The hardest wheels are made of individual disks of canvas cemented together; the softest are composed of muslin disks sewn together. Intermediate grades consist of wheels formed from sewed sections of muslin disks either glued or cemented together.

The abrasives required for polishing are fastened to the surface of the wheel by an adhesive, usually hide glue or silicate cements. The glue must be of high grade, and strict precautions are required to prevent bacterial degradation of the adhesive. Wheels are "set up" in a series of individual operations including melting the glue, adding the abrasive, coating the wheel, and setting the glue. Silicate cements have the advantage of requiring fewer precautions and of being applicable at room temperature, without the need for special equipment.

Abrasives

Natural abrasives such as emery, or corundum (names for a naturally occurring aluminum oxide), as well as artificial ones such as synthetic aluminum oxide (alumina) and silicon carbide are used. The most widely used abrasive in wheel polishing is an artificial aluminum oxide grain.

Other abrasives are used for special purposes. Tripoli is a natural amorphous silica containing some iron oxide which colors it. It is too soft and fragile for use on ferrous metals, but is suitable for buffing aluminum, copper, brass, and zinc die castings. Soft or amorphous silica also occurs naturally but, unlike tripoli, it is free from iron and is white. It is used alone for coloring nonferrous metals or mixed with tripoli to give added coloring ability to so-called cut-and-color compounds; see Buffing, below.

Chromium oxide is used for highest color buffing of stainless steels, chromium, and nickel plate. Silicon carbide fractures readily and is used in polishing operations where sharp cutting edges are desired. Rouge (iron oxide) is preferred for high coloring on copper, brass, gold, and silver. Crocus is a less refined form of rouge; it is used mainly for coloring in the cutlery trade. Lime compounds give good mirror finishes on nickel plate, as well as on brass, copper, and aluminum. Pumice (an igneous rock, impure aluminum silicate) is used with Tampico brushes for scratch finishes, in compounds for buffing plastics, and with water for buffing plastics and glass that cannot tolerate the heat generated by normal buffing. Emery, mixed with grease binders to form a paste or cake, is used for lubricating and refining the cut of polishing wheels and on Tampico brushes for deburring.

Speeds

Efficient operation of polishing wheels depends largely on the speed of rotation, usually expressed in surface feet per minute (sfpm). (This is easily calculated: surface speed = $\pi d \times$ rpm; 1 ft/min = 0.00508 m/s and 1 m/s = 197 sfpm.) Usual speeds are in the range of 30 to 40 m/s (6000 to 8000 sfpm) when glue is used as the adhesive. At higher speeds the heat evolved tends to break down the glue, but with use of silicate adhesives wheels may operate up to 45 m/s (9000 sfpm). At too low speeds the abrading action is retarded, and there is a tendency for the abrasive to be torn from the wheel. Certain metals having a low tolerance to the heat generated must be polished at low speeds.

Abrasive belts are similar in principle to wheels; they are endless belts coated with abrasives and glue, by either the user or the manufacturer of the belt. Belts of cloth or paper, precoated by the manufacturer, are available in a complete range of grit sizes, ready for use without the necessity for setting up that is required for polishing wheels. Cloth-backed belts are best adapted to manual operation and polishing of irregularly shaped articles; paper-backed belts are more efficient for automatic polishing of flat sheets. They may be resurfaced several times, provided that they are not torn in use.

BUFFING

Buffing may be used for the production of several types of finishes: satin (such as "brushed" and "butler" finishes), preliminary smoothing, cut-and-color for smoothness and some luster, and color buffing for producing high gloss or mirror finishes.

Buffing wheels are usually of muslin, of various constructions: sewed, pocketed, or folded, etc. Buffing wheels in general are more flexible than polishing wheels. Grades of muslin are specified by thread count: the denser the thread (higher the count), the less flexible the wheel and the harder the surface to be buffed.

Abrasives for buffing (see above) are commonly applied in the form of a greaseless compound formed into a bar which combines the binder (glue) and the abrasive. Many grades are available, depending on the metal to be buffed, the amount of metal to be removed, and the finish desired. Wheel speeds in general are somewhat lower than for polishing, in the range of 25 to 30 m/s (5000 to 6000 sfpm). The abrasive compound may also be applied to the wheel in the form of a liquid which is sprayed on. Liquid compounds have several advantages over solid bars: Since the liquid may be applied continuously rather than intermittently, the proper amount of compound can always be present on the wheel, and

Fig. 6-2 Abrasive belt polisher and sander. *(Lea Manufacturing Co., Waterbury, Conn.)*

less compound is wasted by overapplication; buffing wheel wear is lessened for the same reason; less compound is packed into crevices in the work, which is therefore easier to clean; production is increased because of less down-time for application of compound; no unusable nubbins of bar compound are left over; and there is less fire hazard than with bar compounds, since insufficient compound on the wheel causes excess frictional heat which may cause the wheel to catch fire.

Bulk Finishing

When the nature of the parts allows, several methods of bulk finishing are available. In some of them, rotating barrels are used. These methods minimize hand labor and are therefore more economical when they can be used. Barrel finishing consists of placing the parts, either alone or with suitable media, in a closed horizontal or open oblique cylinder or container; this barrel is rotated at a speed that allows the parts and the medium to make continuous contact. Operations may be conducted either dry, with materials such as leather scrap or wooden pegs plus dry-burnishing compounds including lime, alumina, rouge, and sawdust; or wet, in which a liquid medium (which may be water, to act as a lubricant) is added.

Another bulk finishing technique is *vibratory finishing,* which is especially adapted to zinc die castings. In this process the parts are deburred, buffed, or both in so-called vibratory machines; these are equipped with eccentric weights that move the die castings and the abrasive media in a circular path while the units oscillate at 1200 to 2000 cycles/min, with an amplitude of 2.5 to 5.5 mm. The abrasive media are of various shapes and materials, depending on the shape of the castings and the amount of metal removal required. The media come in the shape of cones, cubes, triangles, and cylinders; they are made of abrasive-loaded plastics or ceramics. They are mixed with the parts in a volume ratio of about 5 :

1. The media have a bulk density of 800 to 1250 kg/m^3 (0.8–1.25 g/cm^3) or more for some ceramic materials. A liquid medium, usually a soap solution, is mixed with the media and parts to prevent direct impingement of the media upon the zinc surfaces. Vibratory finishing is said to yield both better surfaces for subsequent plating or organic finishing and lower costs since the degree of automation is high and much handling is avoided.

In some cases a two-step process is preferred, using coarse abrasive media for the rapid removal of surface defects and finer media for final buffing or surface smoothing. Use of chemical "accelerators" such as sodium bisulfate or sodium dichromate is claimed to speed up the process by yielding higher rates of metal removal.

7
Rinsing and Water*

When a part is transferred from one treating solution to another, or when it leaves the final treating solution, it carries with it some of the solution in which it has been immersed. In most cases, this adhering solution must be eliminated before the part enters the next step in the sequence; and in all cases, the final processing solution must be eliminated before the part is dried and readied for shipment. To accomplish this, the parts being finished must be rinsed free of this adhering solution.

Perhaps at one time in the past, water to accomplish this necessary rinsing was cheap, endlessly available, and of good quality; and the disposal of the rinsewater, now contaminated with the solutions that it carried off, was of little concern: it could merely be run to a sewer system or a nearby stream. Whatever may have been true then, these conditions no longer hold; good-quality water is often expensive, and pollution abatement requirements forbid the easy methods of disposal once practiced. As a consequence, rinsing techniques have assumed increasing importance and have become the subject of considerable study, both theoretical and practical.

When rinsing is an in-process step, its function is to prevent or decrease the contamination of the following solution in a finishing sequence by the contents of the one preceding it. The importance of this step differs according to the nature of the particular sequence; if the preceding step is a dip in a cyanide solution (e.g., for tarnish removal) and the subsequent step is plating in a cyanide bath, the rinse step is of little consequence and can even be omitted. But if the preceding step was copper cyanide plating, from which the parts enter a bright-nickel-plating solution, contamination of the nickel bath by the contents of the copper bath can be seriously deleterious to the operation of the nickel-plating process. And if rinsing is the final step, preceding drying and packing of the parts

*The publications of J. B. Kushner (see "Metal Finishing Handbook," Bibliography) are acknowledged for some of the content, though not the actual text, of this chapter.

for shipment, any solution remaining on the parts must be eliminated to prevent discoloration, staining, or corrosion of the parts. So final rinsing is without exception an important step in the total finishing process.

When parts, and the wires or racks required to position them in the plating tank, are removed from the solution, the solution that adheres to them as they are removed is called *drag-out*. When the parts and associated racks enter a plating or other solution from a preceding process, the solution adhering to them as they enter is called *drag-in*. The significance of drag-out and drag-in varies with the nature of the process. Drag-out has both technical and economic aspects. From a technical point of view it is often actually beneficial: without it, impurities and degradation products might build up in the solution to unacceptable levels, and drag-out helps in minimizing them. Economically, drag-out may or may not be highly significant. If the solution being dragged out is a copper sulfate plating bath, the loss of a little copper and sulfuric acid is probably negligible (so far as the cost of the solution is concerned; it may add appreciably to the costs of waste treatment). But if it is a gold-plating solution, the finisher must, to stay in business, recover as much as possible of its valuable constituents. Thus rinsing fulfills another function: that of recovering valuable materials and reducing the cost of waste treatment.

Drag-in also may be important or not, depending on the susceptibility to contamination of the solution into which the contaminants are "dragged"; this aspect has already been mentioned. The composition of both drag-in and drag-out is, of course, the same as the solution which is its source, unless suspended particles are included, in which case their composition is to some extent unpredictable. A principal function of rinsing is to reduce to a minimum the quantity of drag-in and drag-out, by rendering these solutions as dilute as possible.

Soon we shall consider methods of making rinsing as efficient as possible. Independently of rinsing itself, however, there are several means available for lessening the magnitude of the job it has to do.

Design of parts being finished, and their positioning in the processing tank, have a great influence on the amount of solution that must be rinsed off. If the parts have blind holes or concavities, it is obvious that much more solution leaves with the parts when the concavity faces up than when it faces down (see Fig. 7-1). When possible, such parts should be provided with drain holes so that they do not act as containers to carry solution out of the first tank and into the second. *Time of drainage* over the treating solution also has a significant effect; if time is allowed for solution to drain back into the treating tank, less is dragged over into the rinse. This expedient must not be overdone; if drainage time is too long, the solution may evaporate on the surfaces of the work, leaving dry salts which are more difficult to rinse off. This effect will be more serious if the temperature of the solution is high. Another effective method is to direct a fine *spray* of water upon the work as it leaves the tank; this washes a proportion of the processing solution back into the tank, and that much less is dragged out. The use of this method is limited by the amount of water that can be tolerated by the solution in the processing tank without unduly diluting it.

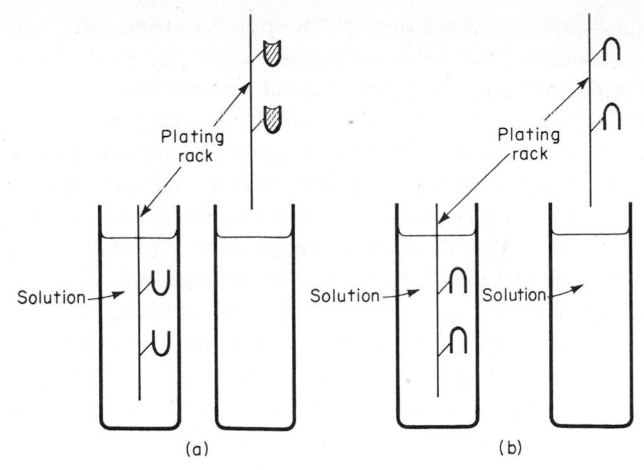

Fig. 7-1 Positioning of cup-shaped parts in plating tank to minimize drag-out: (*a*) wrong, (*b*) right.

In spite of any or all of these precautions, some solution will adhere to the work leaving a plating (or other treating) tank, and it is the function of rinsing to replace this film of adhering solution with a film of water (or, very rarely, some other solvent). In the case of in-process rinsing, the adhering film of water may be allowed to carry some salts, the amount depending on the nature of the process coming next in the sequence. In the case of final rinsing, this film of water should almost always be as pure as possible, since the next step will be drying and delivery either to shipping or to some other department of the plant. Any salts dissolved or suspended in the water film would then cause staining of the work.

In the now rare case where water is cheap and abundant, the route to good rinsing is merely to use so much water that the adhering film is rendered practically "infinitely dilute" by brute force. This expedient is almost always ruled out by resultant high water bills and waste-treatment costs. A better approach is to use as little water as will do the job, and this means getting the most effective use out of every drop. An approximate mathematical treatment of what happens when a part is removed from a plating tank and immersed in a rinsing tank can demonstrate how this can be done. The treatment is not completely rigorous because certain assumptions and approximations are involved, but it yields results that are borne out in practice.

THE RINSING EQUATION

The following symbols are used.

C_0 = the concentration of salts in a processing tank, g/L.

θ = the volume of processing solution adhering to a piece or group of pieces removed from a processing tank at any one time (drag-out per load of work), L. This is assumed constant for each such load.

W = the rate of flow of water into and out of a rinse tank, L/min.

m = interval between loads thus removed and dipped into the rinse tank, min.

C = the maximum concentration known to be safe for the film adhering to the work when it is removed from the rinse tank or the rinsing sequence, g/L. (Ideally $C = 0$, but this is unattainable in practice, so C is set by past experience, taking into account such factors as staining of the work, waste-treatment costs, tolerance for impurities of the next process in the sequence, and value of the salts lost in the adhering film. It is also assumed that the entering water is pure; this is seldom the case, but it is obviously impossible to attain a surface any purer than the entering water itself.)

When the system reaches an equilibrium state, the following equation holds:

$$C = C_0 \frac{\theta}{mW + \theta} \quad (7\text{-}1)$$

But since θ is very small compared to mW, this approximates to

$$C = C_0(\theta/mW) \quad (7\text{-}2)$$

Or, in words, the concentration of the salts clinging to the work when it is removed from the rinse tank equals the concentration in the processing tank, multiplied by the volume of drag-out, and divided by the amount of water flowing into the rinse tank in the interval between loads of work processed. C may thus be minimized by (1) maximizing the interval between loads (but this slows down production) and (2) maximizing the amount of water flowing into (and out of) the rinse tank (but this has already been shown to be far too costly). How, then, can we make use of this knowledge?

Let the first rinse tank take the place of the processing tank, and transfer the work from this tank into a second rinse tank. The same equation holds; the drag-out from the first rinse tank has the concentration, as shown, $C_0(\theta/mW)$. Set this equal to C_1, and let C_2 be the equilibrium concentration of salts adhering to the work when it is removed from the second tank. Then

$$C_2 = C_1(\theta/mW)$$

but since $C_1 = C_0(\theta/mW)$, $C_2 = [C_0\theta/(mW)](\theta/mW)$, or

$$C_2 = C_0(\theta/mW)^2 \quad (7\text{-}3)$$

Following the same line of argument, if we now add a third rinse tank, the concentration of salts adhering to the work when it leaves the third rinse will be

$$C_3 = C_0(\theta/mW)^3$$

Or, in the general case, with n rinses the final film adhering to the work after the nth rinse will be

$$C = C_0(\theta/mW)^n \quad (7\text{-}4)$$

Recall that this equation involves the approximation of neglecting the second term in Eq. (7-1); it also assumes perfect mixing, as discussed below.

This is known as the *rinsing equation,* and it is the basis for the well-established fact that good rinsing combined with economical water use is attained by countercurrent rinses. Fresh water enters the last rinse and moves counter to the movement of the work, as shown in Fig. 7-2.

The importance of the rinsing equation is perhaps best appreciated by a practical example. Assume that the work is being withdrawn from a chromium-plating solution having a concentration of chromic acid (CrO_3) of 375 g/L. (This is admittedly an extreme case, since most plating solutions are not as concentrated as this; but it is perfectly practical and for other cases the calculations are analogous.) Further assume that in order that the work dry satisfactorily, free of stains, the final concentration out of the last rinse should be no more than 0.007 g/L. Also assume that each load of parts drags out with it about 0.004 L of liquid, and that a rack or load of work is processed, i.e., arrives in the rinse, every 2 min. Therefore

$$C = 0.007$$
$$C_0 = 375$$
$$m = 2$$
$$\theta = 0.004$$

The unknown is W, the quantity of water that must flow through the rinse tank, in liters per minute. With one rinse, we solve the rinsing equation for W:

$$C = C_0(\theta/mW) \qquad \text{or} \qquad W = (C_0\theta)/mC \tag{7-5}$$

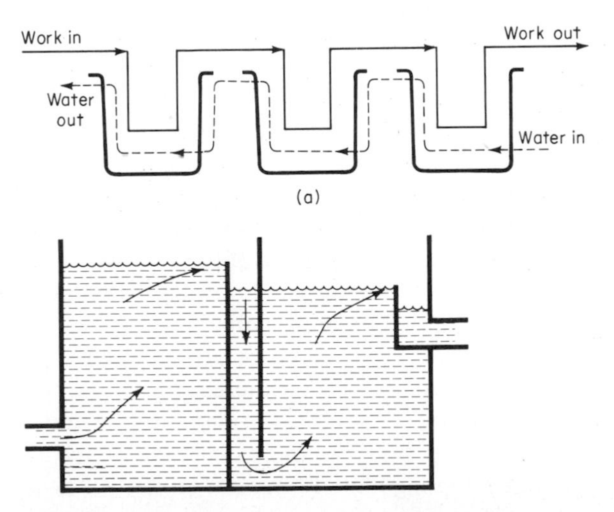

Fig. 7-2 (*a*) Countercurrent (series) rinsing. Key: solid line, work flow; dashed line, water flow. (*b*) Double-wall countercurrent rinse.

$W = (375 \times 0.004)/(0.007 \times 2) = 107$ L/min. At this rate of flow, assuming an 8-h day, the water use will be $107 \times 60 \times 8 = 51{,}400$ L/day; in a 300-day year, water use for rinsing will be $51{,}400 \times 300 = 15{,}400{,}000$ L/yr (4 million gal), just for one rinse tank.

If now we substitute two countercurrent rinse tanks for the single rinse, we similarly solve the general rinsing equation, Eq. (7-4), $C = C_0(\theta/mW)^n$, where $n = 2$, for W in the same way:

$$W = (\theta/m)(C_0/C)^{1/2} = 0.46 \text{ L/min}$$

The water consumption for two rinse tanks is reduced to 223 L/day or 67,000 L/yr, a reduction of 230 times. Repeating the calculation for three rinse tanks, we obtain $W = 0.076$ L/min, or 11,000 L/yr, a further reduction of 6 times. A fourth rinse tank would reduce water consumption to 4400 L/yr, a factor of only an additional 2.5 times.

We may conclude that the addition of a second rinse tank is enormously effective in reducing water consumption; addition of a third tank is usually worthwhile; and further rinse stations, while effective, probably do not warrant the expense of installation and maintenance. In practice, it is customary to have at least two countercurrent rinse tanks, and usually three are preferred; few installations use more than three.

In deriving the equations presented above, a tacit assumption is made that there is perfect mixing between the adhering film of processing solution (θ) and the water flowing through the rinse tank (mW), yielding the concentration of salts in the rinse tank, $\theta/(mW + \theta)$. To the extent that perfect mixing is not attained, these equations do not accurately reflect the facts, and rinsing will not be as efficient as predicted. To approach this perfect mixing, several expedients are available.

MIXING

When the work is processed by hand through the various stages of the finishing procedure, operators frequently swish the racks up and down and back and forth in the rinse tank; when this motion is sufficiently vigorous, it promotes good mixing between the film and the water in the tank. When automatic machines are used, it is more common to rely on agitation of the rinsewater itself. This may be provided by a copious flow of water into and out of the tank, but this defeats the purpose of water conservation. Therefore some form of mechanical agitation is required: by propeller, pump, ultrasonics, or air agitation. Good agitation in a rinse tank may also be provided, with no increase in water usage, by adding a reducer or spider (similar to that for air agitation) onto the end of the water inlet pipe. This increases the water velocity. In any case, some form of agitation must be provided so that the adhering film of drag-out mixes well with the rinsewater. The importance of agitation increases as multiple rinse tanks are employed, since the flow of water itself is less and less effective as water usage is reduced.

Use of sprays to rinse off the work is another expedient that can be and is used to replace, or more often to supplement, simple immersion in the rinse tank. Applicability of sprays is limited by the design of the parts: parts having hidden pockets or deeply recessed areas may simply trap the spray, which may not run off freely. Spray nozzles tend to clog, and maintenance problems may be increased by their use.

Good mixing can also be promoted by adding a wetting agent (surfactant) to the final rinse. Viscous solutions are more difficult to rinse off than less viscous ones. Rinsing is favored by diffusion as well as by the forced convection provided by agitation; and diffusion is favored by using warm or hot water in the rinses. This also favors evaporation and makes it more practical to use the first (most concentrated) rinse as a recovery rinse, feeding the drag-out back into the plating tank and thus reducing both waste-treatment problems and loss of valuable solution.

RINSE TANKS

Typically rinse tanks are constructed so that fresh water enters from the bottom and leaves from the top. When two or more rinse tanks are cascaded (in series), the work flows counter to the flow of water, with baffles placed so as to provide agitation of the water and flow from bottom to top (see Fig. 7-3).

Naturally, rinse tanks must be large enough to accommodate the largest work piece or rack to be handled, but tanks should not be overlarge, since agitation would thereby be decreased. Time of contact between the work and the rinse must be adequate to ensure good mixing. With satisfactory agitation, this time is found to be about 30 to 60 s in most cases.

Monitoring the Rinse

Since the final rinsewater constitutes the film which adheres to the work as it leaves the processing sequence, the concentration of this rinse must be monitored. If the water is unnecessarily pure, it is a sign that too much water is being used; and if it is too concentrated in salts, drying will leave stains on the work. Although chemical analysis could be used for this control, a simpler and more convenient method is employment of a conductivity meter, which can be set up

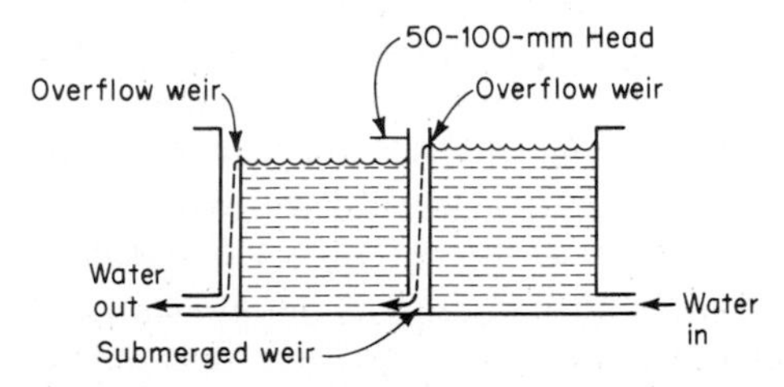

Fig. 7-3 Preferred design of multiple-rinse tank.

to control a solenoid valve to regulate the flow of water to the rinse tanks. Since conductivity is a rough measure of the salt concentration of a solution, such a meter can ensure that water flow is sufficient to yield good rinsing without excess use of water. Simple conductivity bridges and associated controls are commercially available.

Drying

Unless the work is to proceed to further processing solutions, the final step in finishing is drying. If the last rinse is a hot-water immersion, evaporation of the water is easier and quicker; and if the work is massive, its retained heat may evaporate the water film without further effort. Otherwise, various methods are used: the work may be dried in an absorbent medium such as sawdust; by centrifuging; by heat such as infrared radiation, warm air, or ovens; or by employing a dip in a rapidly evaporating organic solvent. Solvent-based water-displacing liquids are also available. Which to use depends on the nature of the work—its size, whether it is fragile—and to a great extent on mere preference. When hot water is used, the final rinse must often be of deionized or distilled water to prevent stains caused by calcium and magnesium salts on the work.

WATER

In what we have had to say about rinsing, it is clear that water is the most important factor in this process. In practically all the other operations of electroplating and metal finishing, water is also a leading component; yet most of us take an adequate supply of good-quality water for granted. If this was ever the case, it is no longer so. Many sections of the country and the world are short of water, and even when water is abundant, its quality is often questionable. In most countries of the world and in most states of the United States, water that has been used in an industrial process and thereby contaminated cannot simply be disposed of by running it to a municipal sewer system or to a convenient stream; it must be treated to conform to local government standards of purity.

Waste treatment will be considered briefly in a later chapter; here we must mention the subject of water supply. Because this subject is worthy of complete texts in its own right, and because it is by no means peculiar to the principal subject of this text but of central importance to almost every chemical and processing industry, a complete discussion will not be attempted. Our main reason for introducing the subject is to make students aware of its importance and to deter them from simply taking water for granted.

Water is the basic solvent in practically all metal-finishing operations that employ any liquid at all; the only exceptions of importance are solvent and vapor degreasing and organic finishing.

The two criteria of a water supply are its quantity and its quality. Little need be said about the first; enough water must be available to make up the various solutions that will be required for cleaning, plating, rinsing, and other finishing operations considered in this book, as well as cooling water and water for such auxiliary requirements as sanitary facilities. We have already discussed means of minimizing the quantity of water necessary for efficient rinsing. Since it is likely that all process water must be treated before it is disposed of, economy in its use is advisable in all the other operations also.

Water may be safe to drink and still be entirely unsuitable for many metal-finishing purposes. Tap water, as delivered by local municipalities, varies widely in purity; even more variation is encountered in privately operated supplies such as river water and well water.

Impurities present in most water supplies may include any or all of the following: dissolved mineral salts; dissolved organic compounds; dissolved gases; suspended solids; microorganisms such as yeasts, molds, and bacteria; and even macroorganisms such as small fishes and crustaceans, insects, and the like. Simple filtration suffices to rid the water of the last named and of suspended solids. Dissolved gases are seldom significant to metal-finishing processes, and the quantity of dissolved organic matter is usually so minor as to be of little concern, although there are important exceptions to this statement. For the metal finisher, the principal source of concern is dissolved mineral salts. The most common are calcium and magnesium salts, found in *hard water.*

Hard water is water containing an amount of calcium and magnesium ions sufficient to render washing in it with ordinary soap difficult or *hard*. This is because soap is a sodium (or potassium) salt of a fatty acid (typically stearic) which, on contact with calcium ion in solution, forms the calcium salt of this acid. This calcium salt is insoluble, thereby causing curds. (For calcium read calcium plus magnesium in this discussion.) *Soft* water may contain many impurities, but does not contain sufficient calcium ions to form troublesome amounts of this calcium soap.

Water hardness is traditionally measured in terms of grains per gallon; a better measure, more in conformance with the metric system, is parts per million (ppm). Both are expressed as calcium carbonate, $CaCO_3$; the formula weight of $CaCO_3$ is 100, which simplifies calculations. One ppm = 0.058 gr/gal, and 1 gr/gal = 17 ppm. The dividing line between hard and soft water is not a definite number, but any water with a hardness greater than 120 ppm is considered hard; with less than 120-ppm hardness, the water is considered soft.

Note that although water quality is often expressed in terms of hardness, many ions do not contribute to hardness but may nevertheless be detrimental to some finishing operations. Ions such as sodium, potassium, and ammonium do not precipitate soaps and therefore do not count as hardness; yet they may produce undesirable effects in some metal-finishing processes. In many cases total dissolved solids (TDS) is a more informative measure of water quality.

Suggested water-quality standards for various industries are listed in Table 7-1. Standards for drinking water as given by the U.S. Public Health Service are listed in Table 7-2. Note that industrial water must often be lower in at least some impurities than drinking water.

Table 7-1 Suggested Water-Quality Tolerances for Various Industries*

	Tolerance, ppm										
Industry	*Turbidity*	*Color*	*Taste, odor*	*Hardness as $CaCO_3$*	*Alkalinity as $CaCO_3$*	*Iron*	*Manganese*	*Organics as O_2 consumed*	*Chlorides as Cl*	*Total solids*	*pH*
Electroplating	1	5	Low	0	1	0.1	0.1	1	1	2	7.0
Carbonated beverage	0–2	10	None	250	50	0.2	0.2	10	—	850	—
Brewing	5–10	10	Low	100	75–100	0.1	0.1	10	100	500–1000	7.0
Canning	5	5–10	None	25–50	25–50	0.2	0.2	—	—	850	7–8
Laundries	1	2	Low	50	60	0.1	0.1	10	—	—	6–6.8
Kraft paper	40	25	—	100	75	0.2	0.1	—	200	300	6.8–7.3
Textile, cotton	1.5	5	—	10	75–100	0.05	0.05	—	—	200	—

Quality Tolerances for Ultrapure Water

	Turbidity	*Color*	*Taste, odor*	*Hardness as $CaCO_3$*	*Alkalinity as $CaCO_3$*	*Iron*	*Copper*	*Organics as O_2 consumed*	*Chlorides as Cl*	*Resistivity, MΩ*	*pH*
Semiconductor	0	0	—	0	0	0.005	0.005	0.1	0	18	7.0

*From D. M. Considine (ed.), "Chemical and Process Technology Encyclopedia," copyright © McGraw-Hill, New York, 1974, 1160.

Used with permission of McGraw-Hill Book Co.

Table 7-2 U.S. Public Health Standards for Drinking Water* (Partial list)

Constituent	*Maximum concentration, ppm*
Total solids	500
Chloride, as Cl^-	250
Sulfate, as SO_4^{--}	250
Nitrate, as NO_3^-	45
Copper (Cu)	1.0
Lead (Pb)	0.05
Iron (Fe)	0.3
Manganese (Mn)	0.01
Cadmium (Cd)	0.01
Zinc (Zn)	5.0
Chromium (hexavalent, Cr^{6+})	0.05
Cyanide, CN^-	0.2

*From Considine, op. cit., p. 1159. By permission.

WATER TREATMENT

Hard water may be treated to render it merely soft, that is, to eliminate only the calcium and magnesium ions; to eliminate or reduce all ionic impurities; or to eliminate (or reduce) all impurities, both mineral and organic. Which treatment is chosen depends on the eventual use of the water and the requirements of the process in which it will be used, as well as upon costs.

Water softening is carried out by passing the water through a bed of an ion-exchange material called a *zeolite,* which collects calcium ions and substitutes sodium ions for them:

$$Na_2Z + Ca^{++} \rightarrow 2Na^+ + CaZ$$

where Z symbolizes the zeolite anion. Zeolites are either natural or, more usually today, artificial hydrated sodium aluminum silicates of the general formula $Na_2O \cdot Al_2O_3 \cdot nSiO_2 \cdot xH_2O$, which have the property of readily exchanging their cations for those present in the solution. When their capacity for such exchange is exhausted, they can be regenerated by passing salt brine through them, when this reaction takes place:

$$\underbrace{2Na^+ + 2Cl^-}_{\text{brine}} + CaZ \rightarrow Na_2Z + \underbrace{Ca^{++} + 2Cl^-}_{\text{calcium chloride}}$$

The calcium chloride is sent to waste (or treatment) and the sodium zeolite is ready for reuse. This type of water softening is common in home water softeners and is useful industrially in cases where the resulting sodium content of the water is not detrimental. Ion-exchange resins (see below) may be used similarly to zeolites.

Water may also be softened by the lime-soda process, which depends on the precipitation of calcium carbonate by sodium carbonate, but this process is not commonly used in the finishing industry.

If more complete purification is required, water may be *deionized* by the process known as *ion exchange* (often abbreviated IX). The softening process described above is a special case of ion exchange.

Ion Exchange

Ion-exchange resins are insoluble acids or bases which have the property of exchanging ions from solutions. During the ion-exchange reaction, the ion-exchange resins are converted into insoluble acids, bases, or salts. Cation-exchange resins contain fixed electronegative charges which interact with mobile counter-ions having a positive charge. Anion-exchange resins have fixed electropositive charges and exchange negatively charged anions.

A classification of ion-exchange resins by type, active ion-exchange group, and configuration of the active group on the polymer is given in Table 7-3.

Ion-exchange resins are usually used in fixed-bed processing equipment for softening and deionizing water. Equations for the removal of sodium chloride, a typical mineral impurity, are as follows. For the service (or exhaustion) step:

$$Na^+ + Cl^- + RSO_3H \rightarrow H^+ + Cl^- + RSO_3Na \qquad (7\text{-}6)$$

For the regeneration step:

$$2RSO_3Na + H_2SO_4 \rightarrow 2Na^+ + SO_4^{--} + 2RSO_3H \qquad (7\text{-}7)$$

These equations represent the reversible exchange of sodium ions for the hydrogen ion from the sulfonic cation-exchange resin. When the resin is depleted of hydrogen ions, it is regenerated with a dilute (5%) solution of sulfuric acid [Eq. (7-7)]. The service step usually lasts for several hours, the regeneration step for about 30 min.

The removal of NaCl from water (deionization) is completed by passage of the cation-exchange effluent through a bed of anion-exchange resin. For the service step:

$$H^+ + Cl^- + ROH \rightarrow H^+ + OH^- + RCl \qquad (7\text{-}8)$$

For the regeneration step:

$$RCl + Na^+ + OH^- \rightarrow ROH + Na^+ + Cl^- \qquad (7\text{-}9)$$

The resin is regenerated by treatment with a 5% sodium hydroxide solution [Eq. (7-9)].

The sum of the two ion-exchange reactions is that hydrogen ions take the place of the cations in the water, and hydroxyl ions take the place of the anions;

Table 7-3 Classification of Ion-exchange Resins*

Type	*Active group*	*Typical configuration*
	Cation exchange	
Strong acid	Sulfonic acid	⬡— SO_3H
Weak acid	Carboxylic acid	~ CH_2CHCH_2 ~ (with COOH on the CH)
Weak acid	Phosphonic acid	⬡— $PO(OH)_2$
	Anion exchange	
Strong base	Quaternary ammonium	⬡— $CH_2N(CH_3)_3Cl$
Weak base	Secondary amine	⬡— CH_2NHR
Weak base	Tertiary amine (aromatic)	⬡— CH_2NR_2
Weak base	Tertiary amine (aliphatic)	—$CHCH_2$ NCH_2 (OH on CH; CH_2 on N)

*From Considine, op. cit., p. 620. By permission.

hydrogen ions plus hydroxyl ions combine to produce water. As a result, the water is essentially freed of all ionic impurities, to produce "deionized" or "DI" water.

Distillation

Although deionized water may be very low in mineral salts, the ion-exchange procedure has little or no effect on organic impurities in the water. Frequently these are of no importance to metal-finishing processes, but in some bright-nickel-plating baths, for example, they can be sources of trouble. Even the resins themselves can introduce small amounts of organic materials into the water. If process water of exceptional purity is required, distillation is the best method of producing it. But this process, even with efficient triple-effect evaporators, tends to be quite expensive owing to its high requirement for heat energy. Because of this expense, use of distilled water for metal finishing is usually confined to the production of inherently valuable parts where the extra cost of the water is of little consequence or where very high purity is absolutely essential, as in some areas of the semiconductor industry.

In some plants, boiler condensate is available and, assuming no organic inhibitors have been added to prevent boiler-tube corrosion, can be a source of pure water of distilled grade.

Reverse Osmosis

Another technique, reverse osmosis (RO), is finding considerable application in treatment of wastewater, and it could serve for purifying process water as well. In this technique raw water under pressure (1.4 to 4.2 MPa or 200 to 600 psi) is forced through a semipermeable membrane. The membranes used have a controlled porosity which allows rejection of dissolved salts, organic matter, and particulate matter, while allowing water to permeate through the membrane.

When pure water and a saline solution are on opposite sides of a semipermeable membrane, the pure water diffuses through the membrane and dilutes the saline water; this is called *osmosis,* and the effective driving force of any saline solution is called its *osmotic pressure.* By exerting pressure on the saline solution, the osmosis process can be reversed. In the reverse osmosis process, pressure in excess of the osmotic pressure is applied to the saline solution. Fresh water permeates the membrane and collects on the opposite side, where it is drawn off as product.

Although other schemes for treating wastewaters have been developed, those mentioned are the most important for producing process water of high quality. Ion-exchange is probably the most widely used.

Section B

FUNDAMENTALS OF ELECTROPLATING

8 Electrochemistry Applied to Electroplating

FUNDAMENTALS

Electroplating is defined as the "electrodeposition of an adherent metallic coating upon an electrode for the purpose of securing a surface with properties or dimensions different from those of the basis metal."* For present purposes the definition must be broadened slightly, to include plating on nonconductors such as plastics.

As we have seen, the process of electrodeposition has many branches (Chap. 1). The keyword in our definition of electroplating is *adherent*. In the other two principal applications of electrodeposition—electrorefining and electrowinning—the deposit need be only sufficiently adherent to the cathode so that it does not fall off during the operation, whereas in electroplating the deposit becomes an integral part of the work being electroplated, and it is expected to adhere to the basis metal during the useful life of the object.† (It is true that even in refining the form of the cathode deposit is of some concern, but by and large the comparison is valid.)

There are some important processes in which the electrode of interest is the anode; anodizing of aluminum is the principal one. For the most part, however, the word *electrode* in the definition refers to the cathode. Thus the physical embodiment of an electroplating process consists of four parts: (1) the external circuit, consisting of a source of direct current (dc), means of conveying this current to the plating tank, and associated instruments such as ammeters, voltmeters, and means of regulating the voltage and current at their appropriate values; (2) the negative electrodes or cathodes, which are the material to be

*ASTM B 374.

†In electroforming and electrotyping, adhesion to the cathode is avoided by special techniques, but the processes are so similar in other respects that they are generally considered a branch of electroplating.

plated, called the *work,* along with means of positioning the work in the plating solution so that contact is made with the current source; (3) the plating solution itself, almost always aqueous, called by platers the "bath"; (4) the positive electrodes, the anodes, usually of the metal being plated but sometimes of a conducting material which serves merely to complete the circuit, called *inert* or *insoluble* anodes. See Fig. 8-1.

The plating solution, of course, is contained in a tank, which must be of a material appropriate to the solution it contains: often plain mild steel for alkaline solutions, and of steel lined with resistant material for acid solutions. Such linings may be of rubber, various plastics, or even glass or lead.

The typical plating tank will have three bare copper conductors running down its length: these are called bus bars, and they are insulated from the tank itself by various means such as ceramic insulators. The two outside bars are connected to the positive side of the dc source, and on them are hung the anodes, usually by means of hooks (Fig. 8-2). The central bus bar is connected to the negative side of the dc source and holds the work, usually held on *racks* which are similarly hung on the cathode bar by hooks. The racks themselves are so constructed as to hold one or many parts, depending on their size and shape, and are often custom-made for the particular work being processed. The racks usually are covered with insulating material except where they make contact with the work and the cathode bus bar. When the work consists of many small parts (screws, nuts, small electric connectors, and the like) which do not lend themselves to being hung individually on plating racks, they may be placed in bulk in a barrel, which takes the place of the cathode and is rotated in the plating bath so that all parts at some time come into contact with a cathode placed inside the barrel. The barrel

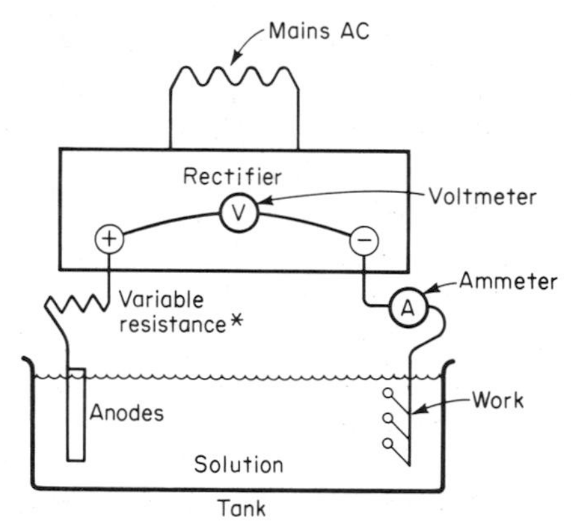

Fig. 8-1 Schematic of electroplating layout. The variable resistance is required only if the rectifier serves two or more tanks in parallel; otherwise voltage is regulated at the rectifier.

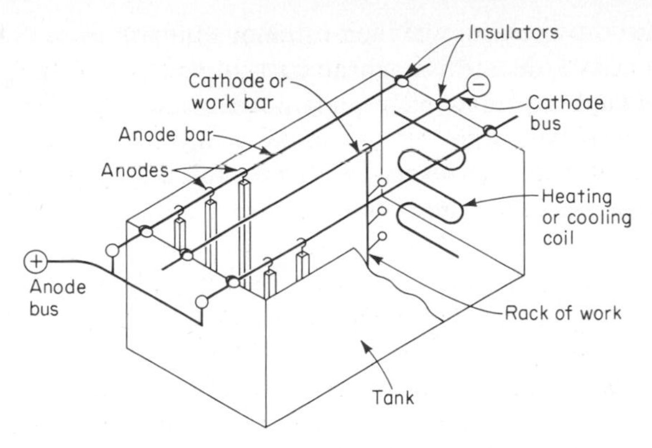

Fig. 8-2 Cutaway view of typical plating tank.

has holes, too small to permit the parts to fall out but large enough to permit fairly good circulation of the solution and passage of the electrolytic current. Barrels are of many types; some are self-contained (oblique barrels) and hold the solution, the anode, and the cathode contact, thus dispensing with a plating tank altogether. Barrels meant to be inserted into a plating tank may be of many shapes and of many materials, and the cathode contacts may be so-called danglers, buttons, or of other forms. *Barrel plating* is no different *in principle* from plating on racks, though it has its own problems of design and plate distribution.

PURPOSE

The purposes for which articles are electroplated have already been mentioned in the Introduction, but they are worth repeating: electrodeposited coatings may be applied for (1) appearance, (2) protection, (3) special surface properties, or (4) engineering or mechanical properties. The distinctions between these aims are not, of course, clear-cut, and there are many overlapping categories. A deposit applied purely for appearance must be, at least to some extent, protective as well. But the classification is convenient.

Decorative Plating

Many metals do not possess much ''eye appeal'' or lose such pleasing appearance as they do possess rather quickly on exposure to ordinary conditions: zinc die castings and ordinary steels, which are the least expensive metals available for most articles, are dull and not very attractive. A thin coating of a metal such as chromium enhances their appearance and adds to their sales value. Chromium can be plated in a bright condition, and when properly applied, it maintains its brightness over long periods. For this reason chromium is the most common electroplate for decorative purposes. But since coatings of chromium thick enough to retain some measure of their original appearance are difficult to apply and very expensive, decorative chromium deposits are almost universally applied over undercoats of copper and nickel or nickel alone, and the final chromium coating is very thin.*

Other electroplated metals used for decorative effects include gold, silver, brass, bronze, nickel, copper, and rhodium. Lead and tin are sometimes used for special effects.

Protective Plating

The function of protecting the basis metal from degradation overlaps that of improving its appearance. The common copper/nickel/chromium composite applied to automotive hardware and numerous other items not only imparts a pleasing appearance, but also protects the substrate from corrosion. However, when corrosion prevention is the only, or principal, aim of the coating, zinc is the most economical and most effective metal available. Although zinc can be plated in a bright condition, or brightened after plating by so-called conversion coatings, this coating does not retain its brightness very long in service and would not be chosen where decorative appeal is the main consideration. Thickness for thickness, however, zinc is the most economical coating for preventing steel from rusting. Cadmium is far more expensive than zinc, but is superior to it for some environments, especially marine.

Tin is not normally protective to steel, but in the particular conditions obtaining inside the sanitary can (sanitary meaning containing foods or beverages) the usual potentials of tin and iron are reversed and tin becomes protective. Tin plating of steel for use in the "tin can" is the largest single use of electroplating, in terms of tonnage of product.

Special Surface Properties

This category cannot be characterized by generalities: each use has its own particular reason for being. Soft solder, a tin-lead alloy sometimes containing other minor constituents, is used widely in the communications and electronics industry for making electrical connections, and for this purpose the parts to be joined must be easily and quickly wetted by the molten solder, often without the use of any corrosive fluxes. Although copper, almost universally used as the electrical conductor, is wetted by solder when its surface is fresh and untarnished, it quickly tarnishes and becomes difficult to solder. A coating of tin or tin-lead alloy renders the surface far more solderable.

Light reflection is another surface property that can be modified by coating; both silver and rhodium are used for this application. Mating surfaces in electri-

*Chromium plate is often called *chrome plate,* and this usage is now legitimized by some dictionaries. The present author will continue to wage a (solitary?) battle: "chrome" is chromium ore, the metal is chromium.

cal and electronic assemblies often require that contact resistance between the parts be held to a minimum, and that this resistance remain low during hundreds or thousands of "makes" and "breaks." Gold plating is used for such applications, as is palladium.

Engineering or Mechanical Properties

This category might be considered a subclass of the previous one, in that in both cases modification of physical properties is the aim of the coating process. They are separated because in most (not all) cases coatings for so-called engineering uses are thicker than the former, often quoted in millimeters (mm) rather than micrometers (μm). Usually the physical properties of most interest here are hardness and wear resistance. Sometimes plating is used to build up the dimensions of a part that has been worn down or mismachined.

Chromium, in thicknesses far greater than those used for decorative purposes, is used to face gun barrels and to form dies, rolls for paper-making machinery, Diesel engine cylinders, and other parts requiring high hardness and resistance to wear. Appearance is of little moment. Nickel is used for similar reasons.

Bearing properties often can be significantly improved by a coating; chromium can be applied for this purpose in a condition known as "porous," which retains an oil film better than smooth metal. Aluminum alloy pistons for internal-combustion engines are tin-plated to prevent scoring of the cylinder walls by the very abrasive aluminum oxide during the running-in period. Lead and its alloys, indium, and silver are also used for similar purposes.

Occasionally the chemical rather than the physical properties of the deposit are of engineering interest. In the case-hardening of steel by carburizing, often it is necessary to "stop off" some areas which are not to receive the hardening treatment, and copper is plated on such areas. In the less common process of nitriding, tin can serve as the nitrogen barrier, but since tin would melt at the process temperature, the tin-copper alloy bronze is chosen.

Similar purposes are served by several processes which by a strict definition do not qualify as electroplating, but are nevertheless considered as part of the electroplater's repertory: autocatalytic or electroless plating; immersion plating, or deposition by replacement; conversion coatings; and anodizing. These will be considered in later chapters.

THE PLATING BATH

Of the four essential parts of an electroplating system—the work, the solution, the anodes, and the external circuit—this chapter is concerned with all but the last. The external circuit, consisting of the current source, conductors, and associated instrumentation, is a subject for the electrical engineer and will be considered in this book only to the extent of mentioning the influence of the type of current on the results of the electroplating operation.

The plating bath is practically always an aqueous solution containing a compound of the metal to be deposited. Nonaqueous solutions, in which the solvent may be an organic or inorganic liquid or a fused salt, are of great theoretical interest, but there is hardly any commercial use of such solutions; the only present exception is the plating of aluminum from an organic electrolyte, practiced by only a few highly specialized shops. A few of the so-called refractory metals, such as tantalum, niobium, zirconium, and tungsten, have been plated in adherent and coherent form from fused electrolytes on a large scale, but again only by a few specialists. Otherwise all electroplating baths are aqueous.

Ingredients of a Plating Bath

Every plating bath contains ingredients which serve one or more of the following functions.

1. To provide a source of the metal or metals being deposited
2. To form complexes with ions of the depositing metal
3. To provide conductivity
4. To stabilize the solution, e.g., against hydrolysis
5. To act as a buffer, i.e., to stabilize the pH (See Chap. 24)
6. To modify or regulate the physical form of the deposit
7. To aid in dissolving the anodes
8. To modify other properties, either of the solution or of the deposit, peculiar to the specific case

This does not mean that all plating baths contain eight ingredients, since some compounds perform more than one of these functions, and in some instances not all of the listed functions are necessary.

Why are these functions required? This question will be addressed in the ensuing discussion. In summary,

1. Obviously the bath must contain the metal to be deposited.
2. Complex formation is not always required, and some metals are plated from simple salt solutions. In many cases, however, it is found that the deposits obtained from complex ions are superior to those from simple ions.
3. Any ionic solution conducts electricity, but many metal salts are rather poor conductors (their ions have low mobilities), and to avoid the necessity for the employment of high voltages, "conducting salts" are often added.

4. Most metal salts are subject to hydrolysis, since most metal hydroxides are insoluble:

$$MX + H_2O \rightarrow M(OH)\downarrow + HX$$

In some alkaline baths, absorption of carbon dioxide from the air would precipitate metal compounds unless an acceptor for carbon dioxide were present.

5. Many plating solutions are highly acid or highly alkaline, and for these pH control is a minor concern. For that class of solutions known as "neutral," i.e., with pH between about 5 and 8, control of pH within prescribed limits is important, and buffering is usually necessary.

6. When direct current is passed from an anode to a cathode through a solution containing a depositable metal ion, the metal will deposit on the cathode; but frequently the deposit will be useless, consisting of trees and nodules, either nonadherent or noncoherent, unless additives are present to control its physical form.

7. Unless anodes are deliberately inert—i.e., act merely to introduce current into the solution—it is desired that they replenish the metal deposited at the cathode so that the composition of the bath remains relatively stable. Anodes of some metals tend to become "passive," i.e., to act as inert anodes, unless specific ions are present that tend to break down this passivity.

8. Finally, some baths require specific additives for specific purposes. An example is the addition to most zinc cyanide baths of sulfur compounds such as sodium polysulfide to precipitate impurities that would interfere with satisfactory plating.

These several functions will become clearer in the discussions of individual plating metals. Some considerations common to all or most plating solutions are discussed below; it is further advisable at this point to review Chap. 2 on the fundamental aspects of electrochemistry.

ASPECTS OF THE DEPOSITION PROCESS

Electrode Potentials.

The emf series (Table 2-1) shows that copper is 0.345 V more positive than hydrogen in a solution of copper(II) ions (at unit activity). This potential, as we have seen, is a result of the opposing forces in the dynamic equilibrium:

$$Cu^{++} + 2e^- \rightleftharpoons Cu^0$$

and represents the potential at which the rates are equal under standard conditions. If by applying an external potential we lower the copper potential to, say, 0.340 V, the rate of the anodic or dissolution reaction is decreased by the additional supply of electrons which bind the copper more firmly to the metal lattice; simultaneously the added electrons in the metal increase the attraction for hydrated copper(II) ions from the solution, which become attached to the metal and are deposited. In short, the deposition reaction is faster than the dissolution reaction, and copper plates out.

If the potential be raised externally to 0.350 V, the effects are reversed. Copper dissolution is faster than its deposition, or on balance copper dissolves in the solution.

In practical plating situations, such small deviations from equilibrium potentials as 0.005 V (5 mV) are not sufficient to drive the deposition process at significant rates, and a greater potential difference is required.

Hydrogen ions are always present in water solutions, so that evolution of hydrogen is always a possibility. To evolve hydrogen from a copper sulfate solution at pH 4 would require a cathode potential more negative than about −0.24 V, or less than 0 V if the activity of the acid is 1. But if copper(II) ions are present at unit activity, the potential cannot become much more negative than +0.345 V, or somewhat less if the activity of the copper ions is less than 1, because as soon as the potential drops, copper ions are plated out and thereby consume all the available electrons. Thus the potential cannot drop to that required for hydrogen evolution until nearly all the copper has been deposited, at least from the solution immediately surrounding the cathode; hydrogen is not evolved and the current efficiency remains 100 percent.*

On the other hand, in a nickel bath, with a standard potential of −0.23 V, hydrogen can be produced along with the nickel, and the current efficiency in nickel plating thus may approach, but seldom reaches, 100 percent.

To generalize: at the cathode, the reaction requiring the least negative potential will take place, exclusively if the deposition potential of any alternative process is much more negative, and almost exclusively if the alternative potential is only a little more negative. If the potential is made sufficiently negative to produce a second reaction, that reaction may occur simultaneously. Thus some metals cannot be deposited at all from aqueous solutions; the potential necessary to deposit aluminum is more than 0.8 V negative to the potential required to

*To find the copper ion activity in the cathode film that would allow deposition of hydrogen, solve the Nernst equation:

$$E = E^0 + (0.059/n) \log a_{Cu^{++}}$$

for $a_{Cu^{++}}$:

$$0 = 0.345 + (0.0295 \log a) \qquad \log a = -(0.345/0.0295) = -11.695$$

or, roughly, $a_{Cu^{++}} = 10^{-11}$. At such a low copper ion activity a and c (concentration) may be considered equal. Further, this calculation does not allow for hydrogen overvoltage, which would make hydrogen evolution even more difficult; it depends on the nature of the cathode.

deposit hydrogen, even from alkaline solutions. Hence in aqueous solutions of aluminum compounds, only hydrogen is produced at the cathode (some claims in the older literature to the contrary are baseless). This argument holds *a fortiori* for the deposition of sodium from aqueous solutions. There are exceptions when the hydrogen overvoltage on the cathode is extraordinarily high, and in addition the cathode material can form alloys with the depositing metal—both conditions hold for mercury cathodes. Polarography depends on the possibility of the deposition of many otherwise nondepositable metals into a mercury cathode, and the alkali metals can be deposited into a mercury cathode to form amalgams, as in the mercury cell for the production of caustic soda.

Reactions at the anode may be considered in the same fashion: the reaction requiring the least positive potential occurs, unless the potential becomes so positive that a second reaction is possible. The usual anode reaction is dissolution of the metal, but other reactions are possible. If the potential becomes sufficiently positive, water may be oxidized and oxygen may be evolved: $2H_2O = O_2 + 4H^+ + 4e^-$; or the anode metal may be oxidized to some compound of the metal. In the case of copper, oxygen evolution would require a potential about 2 V more positive than that required to dissolve copper, so that copper anodes ordinarily dissolve at 100 percent efficiency in a simple acidic copper bath.

In such a simple copper bath, containing a copper(II) salt and an acid, the anode and cathode have the same potential in the absence of an external potential—assuming that the cathode is also copper, as would be the case as soon as a minute amount of copper had been deposited on it. A small externally supplied potential is sufficient to start the current flowing. But if the anode is of lead, platinum, or some other metal insoluble in the bath, no current can flow until the applied potential becomes sufficient to evolve oxygen.*

Polarized Anodes

Under some conditions, an anode that would ordinarily be expected to dissolve in the bath becomes coated with a film of oxide or other compound of the metal; this film may be invisible, but it prevents solution of the metal in the bath. The anode ceases to dissolve and is said to be polarized. (Sometimes such anodes are referred to as "passive," but this terminology is not technically correct for all such cases.) Such a film may be an insulator, as is the case with anodizing of aluminum, where higher and higher potentials are required to drive the current as the film thickens. Other anodic films are conductive; the anode remains inert but reaction continues with evolution of oxygen, as with platinum. The potential may become high enough to cause transport of ions through the film; anodic dissolution may continue, usually with the production of ions in some higher valence state. This happens, for example, with tin anodes in an alkaline stannate solution, which at low potentials dissolve in the divalent state, at higher potentials as quadrivalent tin; at still higher potentials only oxygen is produced.

*If we neglect an initial transient current which charges the capacitance of the cell as a condenser.

COMPLEX IONS

Most metal ions in aqueous solution are associated with water molecules; the ionic species of copper(II) in solution, rather than being Cu^{++}, are in fact $Cu(H_2O)_n^{++}$, where n is a small number, probably 4. Other metal ions are similarly associated with a small number of water molecules.

However, this association is fairly loose and has little or no effect on the potential of the metal electrode as given by the Nernst equation:

$$E = E_{M^{n+}}{}^0 + (RT/n\mathfrak{F})\,(\ln a_{M^{n+}}) \qquad (8\text{-}1)$$

Other species, however, can form much more stable bonds to metal ions; the result is what are known as *complex ions*. In such ions the complexing agent *(ligand)* is more or less firmly bound to the metal ion, and the effect on the electrode potential may be very great. In simple (noncomplex) solutions and for rough calculations, it can be assumed that the activity of a metal ion is about equal to its concentration. This assumption introduces an error, perhaps of an order of magnitude or so, but the approximation affects a logarithmic term, and in any case other assumptions limit the accuracy of the calculations. But if complexing takes place, the activity of the metal ion may be decreased by many orders of magnitude, depending on the strength of the bonding in the complex.

For example, in a simple aqueous solution containing 1 mol/L of copper(II) ion, the electrode potential of copper (E^0 = +0.34 V) is

$$E = 0.34 + 0.059/2 \log a \qquad (8\text{-}2)$$

a is about 0.5 when c (concentration) = 1, so

$$E = 0.34 + (0.0295)(-0.3) = 0.257\ \text{V} \qquad (8\text{-}3)$$

This is still a fairly large positive potential.

If now sufficient cyanide ion, in the form of sodium or potassium cyanide, is added to the solution to form the complex ion $Cu(CN)_3^{--}$ [the copper will be simultaneously reduced to the copper(I) oxidation state and E^0 will be changed accordingly], although the concentration of copper in the solution remains the same, its effective concentration, or activity, is reduced to about 10^{-18} because the copper-cyanide complex ion is so stable. E^0 for copper(I) is about 0.55 V, so the electrode potential of copper in a solution containing cyanide ions is

$$E = 0.55 + (0.059)(\log 10^{-18}) \quad \text{or} \quad 0.55 - 1.04 = -0.5\ \text{V} \qquad (8\text{-}4)$$

Copper has been changed from an electropositive (noble) metal to one having an electronegative (active) potential about the same as that of iron.

Many ions as well as cyanide form stable complexes with metal ions; and conversely, metal ions differ widely in their tendency to form complexes with various ligands. As a general rule, the so-called transition metals tend to form the most stable complexes, while the nontransition metals form fewer and less stable ones. Transition metals are those with unfilled d or f electron shells in their commonly occurring oxidation states, and they include many of the metals important to the electroplater: copper, silver, and gold; chromium, iron, nickel, and the platinum metals. Some metals, such as tin, lead, and zinc, are not transition metals, and they follow the general rule in that they form fewer and less stable complex ions than the former group.

From the electroplater's standpoint, the most important anion that readily forms stable complexes with metals of interest is cyanide. Cyanide forms complexes of greater or lesser stability with copper, silver, gold, the platinum group, the iron-cobalt-nickel triad, and zinc and cadmium. The last two are much less stable than the others, and accordingly cyanide ion has less effect on the activity of zinc ion than on copper ion. On the other hand, the cyanide complexes of iron and nickel are so stable that these metals cannot be usefully plated from cyanide solutions (although this may also be due to kinetic factors). Tin and lead do not form complexes with cyanide; this fact enables the formulation of baths for plating copper-tin alloys, as will be shown.

Other ligands, of which there are many, include the ethylenediaminetetraacetate (EDTA*) ion, commonly used in many analytical procedures; it is finding some application in plating-bath formulations as the search for substitutes for cyanide progresses. Although cyanide-containing baths are eminently satisfactory as far as plating results go, pollution control poses a severe problem economically because of the toxicity of cyanide, and efforts to replace it are continuing. EDTA has several analogs, also of interest in this connection.

Pyrophosphate forms soluble complexes with several metal ions, and some pyrophosphate baths are commercially important. Amphoteric metals owe their solubility in alkalis to hydroxyl complexes, such as those of zinc and tin: $Zn(OH)_4^{--}$ and $SN(OH)_6^{--}$. The halide ions form complexes, usually not as stable as those of cyanide, with many metals; in fact, it is probable that most solutions of metal chlorides containing excess chloride ion contain at least small amounts of chloro-complexes of the type $M^nCl_{n+x}^{x-}$, for example, $SnCl_6^{--}$ or SnF_6^{--}. Complexing ability in the halogen series declines in the order F > Cl > Br > I.

Complexing behavior is important in many aspects of electroplating. The fact that complexing agents have differing effects upon different metals makes possible the electrodeposition of alloys of metals that are so far apart in the emf series that codeposition from simple baths is all but impossible.

For example, in simple salt solutions the standard potential of zinc is about 1.1 V more negative than that of copper. But if enough cyanide ion be added to form the complexes $Cu(CN)_3^{--}$ and $Cu(CN)_4^{3-}$, the electrode potential of copper is rendered much more negative, as shown above, while that of zinc is not changed very much, since the zinc cyanide complex $Zn(CN)_4^{--}$ is relatively weak. Consequently the potentials of the two metals can be made approximately equal, and the copper-zinc alloy brass can be deposited.

In the case of the copper-tin alloy bronze, the situation is somewhat simpler: cyanide ion forms the cyanocuprate(I) ion $Cu(CN)_3^{--}$ as well as ions containing more cyanide; and tin forms no cyanide complex. Hydroxyl ions complex with

*EDTA also is used as an acronym for ethylenediaminetetraacetic acid.

VARYING THE CN^- CONC. IN THE PLATE BATH WILL VARY THE ELECTRONEGATIVITY OF TRANSITION METALS, MAKING THE METAL LESS ELECTRONEGATIVE (MORE STABLE) AS MORE CN^- IS ADDED.

tin to form the stannate complex $Sn(OH)_6^{--}$ [hexahydroxystannate(IV)], and copper forms no hydroxyl complex ion. Therefore the potential of copper is regulated by the cyanide concentration and that of tin by the hydroxyl concentration; the two potentials can be regulated independently of each other.

These effects are useful in other ways. If zinc is immersed in a copper sulfate solution, copper is deposited rapidly on the zinc, because the potential of zinc is so negative that it transfers electrons readily to copper ions, converting them to copper metal, while the zinc dissolves:

$$Zn + Cu^{++} \rightarrow Cu + Zn^{++}$$

This process is called *immersion* or *displacement deposition*. Such deposits are often spongy and nonadherent; deposition may be so rapid that the deposit is essentially "burned," as discussed later (p. 144). Therefore it is not possible to plate copper satisfactorily on a zinc surface from simple copper salt solutions. But in a cyanide-copper bath, the copper potential is sufficiently negative so that there is no tendency for the immersion or displacement reaction to occur. An external source of potential is required, and copper can be readily electroplated on zinc from cyanide baths.

The situation when iron is immersed in a simple copper sulfate bath is similar—an immersion deposit is formed:

$$Cu^{++} + Fe \rightarrow Cu + Fe^{++}$$

and the deposit is, again, unsatisfactory. Accordingly, copper is plated on steel from cyanide baths, at least to start with (after a film of copper sufficient to cover the steel is built up, the parts are often transferred to an acid copper solution for further plating). The explanation, however, for the success of cyanide copper solutions for plating on steel is not the same, since cyanide also complexes iron so strongly that from potential considerations alone the immersion deposition should still take place. In this case it is hypothesized that the reaction between iron and cyanide ion is so slow that the calculated potential is not reached.

The sometimes observed superior properties of deposits from complex ion baths over those from simple ion baths were at one time attributed to the low concentration of free ions in the complex solutions. Later work has indicated that the fine-grained deposits from cyanide baths are due rather to the effects of adsorbed cyanide ions; many complex baths yield unsatisfactory deposits.

Older theories formulated the deposition from, for example, a cyanide copper bath in two steps:

$$Cu(CN)_4^{3-} \rightleftharpoons Cu^+ + 4CN^-$$
$$Cu^+ + e^- \rightarrow Cu$$

At least in many instances that have been investigated, no evidence for this mechanism has been observed, and deposition is regarded as taking place directly from the complex:

$$Cu(CN)_4^{3-} + e^- \rightleftharpoons Cu + 4CN^-$$

In the case of zinc cyanide complex baths, at least two types of complex ions are present: $Zn(CN)_4^{--}$ and $Zn(OH)_4^{--}$. Deposition is believed to take place from the hydroxo-complex, or from $Zn(OH)_2$ in the cathode film.

FREE CYANIDE IS THE PLATING BATH EQUIVALENT OF EXCESS IN BOILER FLUE GAS

"Free" Cyanide

In the formulations of many cyanide plating solutions, both "total" and "free" cyanide are specified, and analytical procedures are presented for determining them. The total cyanide is unambiguous: it is the amount of cyanide ion, expressed either as CN^- or as sodium or potassium cyanide, present in the solution, whether or not tied up in a complex ion, and easily determined by standard analytical procedures. The concept of "free" cyanide, however, is not so clear. It is intended to represent the concentration of cyanide ion beyond that required to form the metal-cyanide complex in question. Therefore its determination requires a knowledge of the formula of the metal-cyanide complex, and this is not always known with any certainty. In the case of copper, the complexes possible include $Cu(CN)_3^{--}$, $Cu(CN)_4^{3-}$, and perhaps complexes containing even more cyanide. Similar situations obtain for the cyanide complexes of other metals. Therefore, since we do not know how much of the total cyanide is tied up with the metal, we cannot know how much of it is free. Hence where possible it is preferable to specify, and determine, total cyanide; if free cyanide must be known, the formula of the complex is chosen arbitrarily, as the most probable of the several possibilities.

The Nernst Equation for Complex Ions

As we have seen (Chap. 2), the potential of a metal electrode in equilibrium with its ions is given by the Nernst equation:

$$E = E_{M^{n+}}{}^0 + (RT/n\,\mathfrak{F})(\ln a_{M^{n+}}) \tag{8-5}$$

where E is the electrode potential; $E_{M^{n+}}{}^0$ is the standard potential for the reduction of the metal ion M^{n+} to the metal M^0; R is the gas constant, T the absolute temperature, $\mathfrak{F}$ the Faraday equivalent, and n the formal charge on the ion (its oxidation number); and $a_{M^{n+}}$ is the activity of the metal ion in the solution. R and $\mathfrak{F}$ are constants; room temperature is taken as 25°C, or $T = 298$; and to convert from Naperian to common (base 10) logarithms, the factor is 2.3. Therefore the equation reduces to

$$E = E_{M^{n+}}{}^0 + (0.059/n)(\log a) \tag{8-6}$$

When complex ions are involved, in order to determine the activity a, it is necessary to consider the stability constant of the complex (the inverse of its dissociation constant). The formation of the complex ion may be represented by

$$M^{n+} + qX^{p-} \rightarrow MX_q^{n-pq} \tag{8-7}$$

where X is the complexing ion of charge p and q is the coordination number of the metal, i.e., the number of coordinate bonds to the complexing ion X. The Nernst equation may now be written in the form

$$E = E_{M^{n+}}{}^{0} - (RT/n\mathfrak{F})\ln K_f + (RT/n\mathfrak{F})[\ln(a_{MX_q}^{n-pq} / a_{X^{p-}}{}^{q})] \tag{8-8}$$

where K_f is the stability constant of the complex and a is the activity of the various species denoted by subscripts. Further discussion of this subject is beyond our scope; tables of stability constants for most common metal complexes can be found in standard handbooks.

We have seen above how complexing action can prevent the formation of immersion deposits, for example of copper on zinc, by complexing the copper with cyanide. This phenomenon can also be used in a reverse sense: to enable the deposition by immersion of a metal higher (less noble) in the emf series upon a more noble metal. A practical example of such a reaction is the deposition by displacement of tin on copper. The reaction

$$Cu + Sn^{++} \rightarrow Cu^{++} + Sn$$

will not normally occur, since tin is electronegative to copper; but if thiourea be added to the tin salt solution, the potential of copper is rendered sufficiently negative so that tin deposits on it, in this case in a useful form. Other ligands than thiourea will also serve. This reaction is used in practice to deposit thin coatings of tin on copper for some purposes, such as for coloring—to distinguish the two wires in a copper conductor, for example. (Its value in printed circuits, for soldering purposes, is controversial.) Such deposits are invariably thin, because action ceases as soon as the substrate is completely covered.

THE PROCESS OF ELECTRODEPOSITION

Before considering the details of plating individual metals, we must discuss some factors that are common to all electroplating processes. These include both very practical matters such as current efficiencies and deposit distribution, as well as some theoretical considerations important to an understanding of the phenomena of electrodeposition.

Current Efficiency

This subject was mentioned in Chap. 2. The ratio of the weight* of metal actually deposited to that which would have been deposited if all the current had been used in its deposition, multiplied by 100, is called the *cathode efficiency*. If all the cathode reactions are taken into account, the cathode efficiency is always 100 percent. Similar considerations apply at the anode, to yield the anode efficiency. The side reaction that takes place at the cathode is usually the evolution of hydrogen, and that at the anode the evolution of oxygen, both resulting from the

*Strictly speaking, mass not weight, but we accept the less formal nomenclature in this text.

decomposition of the solvent water; but there are exceptions. If ferric ion (Fe^{3+}) is present, it may be reduced at the cathode to ferrous (Fe^{++}), thus using some of the current without visible gas evolution or metal deposition; and the reverse reaction may take place at the anode. In fact, a solution containing both Fe^{3+} and Fe^{++} may be electrolyzed indefinitely without any change in its composition; the reactions $Fe^{3+} + e^- = Fe^{++}$ and $Fe^{++} = Fe^{3+} + e^-$ simply balance each other. Other reducible ions may also cause low cathode efficiencies without hydrogen evolution. If nitrate ion is present in the solution, it is easily reduced to lower valencies, such as nitrite, nitrogen, or ammonia, at the expense of metal deposition.

The total weight of metal deposited is not the only consideration in practical plating: deposits should also be acceptably distributed upon the substrate. Usually it is desired that the distribution be as uniform as possible. Most specifications require that deposits on so-called significant surfaces* equal or exceed a specified minimum thickness, and if, in order to achieve this, excess metal must be applied to other surfaces, metal is wasted and plating times are unnecessarily prolonged. Such uniformity of distribution is approximated by proper positioning of the parts in the plating bath, by use of conforming anodes, or of "bipolar" anodes, or (to a very limited extent) by placing the anodes as far as practical from the work, and by employment of solutions with good "throwing power." These subjects are discussed in more detail later.

Deposit Growth

The usual formulation of a metal deposition process, $M^{n+} + ne^- \rightarrow M$, represents only the total stoichiometry of the reaction and does not take into account any intermediate steps between the beginning and the end of the process. Metals do not deposit as continuous sheets from one part of the cathode to the others; metal ions, carrying with them their accompanying ligands (water molecules or complexing ions), attach themselves at certain preferred sites, losing in the process some of the water or other ligands, forming bonds with the cathode surface while their charges are partially neutralized. These *adions* diffuse over the surface to various irregularities in the surface such as kinks, edges, or steps, where they are now incorporated into the metal lattice. As these *growth sites* travel across the face of the crystal, monatomic growth layers are produced; they grow until they encounter adsorbed impurities, where they agglomerate to form growth stacks consisting of several layers. This lateral growth proceeds until several neighboring lattices meet to form a boundary at the contact lines; the individual structures thus formed are called *grains*. Further growth now proceeds outward, and the thickness of the deposit is thus built up.

*Those normally visible, directly or by reflection, which are essential to the appearance or serviceability of the article when assembled in normal position; or which can be the source of corrosion products that deface visible surfaces on the assembled article. (ASTM)

Epitaxy

(See also Chap. 4). When metal atoms arrive at the surface of the substrate during deposition, they tend to occupy positions that continue the grain structure of the substrate, even though this structure is not typical of the normal lattice structure of the depositing metal. If the lattices of the substrate and of the coating metal are similar, geometrically and dimensionally, this basis-metal structure can be continued almost indefinitely. This type of growth is called *epitaxial.*

If the lattice structures differ substantially, the growth pattern usually shifts gradually toward that of the deposit. Patterns of growth may also be radically altered by adsorbed material in the growing surface; such material may be incorporated in the deposit and may prevent the formation of the normal lattice and inhibit the formation of large grains. Certain crystal faces may grow more rapidly than others, causing orientation of the grains. At high rates of growth (high current densities) there may not be time for incoming atoms to find positions of greatest stability. All these factors tend to prevent epitaxial growth.

Stress

Many deposits are formed in a stressed condition, either in tension (more common) or in compression. These stresses are often of little importance, especially in deposits of weak and ductile metals like tin, lead, and cadmium; but they can also be very significant, in extreme cases causing cracking or peeling of the deposit. For many engineering applications stress is among the properties that must be controlled. Stresses may arise from many causes: mismatch of lattice parameters, or incorporation of foreign material such as oxides or hydrated oxides, water, sulfur and carbon (usually arising from addition agents), hydrogen, or metallic impurities. These obstruct the formation of normal lattice structures, or they form brittle intergranular deposits. If such interferences are absent, the mechanical properties of the electrodeposit resemble those of the thermally prepared metal.

Adhesion

Except for the special case of electroforming, in which it is desired to strip the electrodeposit from the substrate, electrodeposits are applied with the expectation that they will adhere to the substrate during the useful life of the article; and if proper operating practice is followed, this will be true. The first layer of deposited atoms engages the lattice forces of the substrate; the strength of the bond is approximately the same as that of the basis metal unless there is a major mismatch between the two lattices. The strength of the adhesion of the electroplate is very nearly the tensile strength of the substrate. When peeling occurs, it is a sign of poor substrate preparation, or occasionally of other faulty practices. Tests for adhesion are considered in Chap. 23.

FORMULATION OF PLATING BATHS

Plating baths are of two general types: acid and alkaline. Acid baths usually are solutions of relatively simple salts with little complex formation; alkaline baths are by their very nature complex, since the metal is contained in the anion. "Neutral" baths are sometimes added as a third type. They have a pH range of about 5 to 8, and they may be solutions of simple salts, as with nickel, or of complex ions as in the various pyrophosphate solutions.

Plating baths are usually fairly concentrated solutions, since if they were dilute, the metal ion would be rapidly depleted from the catholyte (the solution near the surface of the cathode). Since electrolytic theory has been studied principally from the standpoint of dilute solutions, much of the knowledge of plating baths is perforce empirical.

Choice of Salts

To achieve the high concentrations of metal required to form practical plating baths, soluble salts must be used, so that the field is somewhat limited. Most nitrates are soluble, but the nitrate ion is reduced at potentials too positive to allow most metals to be deposited. Another anion most of whose salts are quite soluble is perchlorate; and although the dangers of handling perchlorates in the concentrations required for plating baths are probably exaggerated, prejudice founded on a few disasters (where concentrated perchloric acid was involved) has militated against consideration of these salts. Fluoborates share most of the advantages of perchlorates without their possible hazards.

Anion effects must also be considered. Although the anion does not enter directly into the plating process in the case of acid baths, the nature of the anion nevertheless may have profound effects on the deposition process and the nature of the deposit. These effects may be due to adsorption on the cathode; the nature of the anion also affects the activity of the metal ion.

Most acid baths are formulated with sulfates: nickel, cobalt, zinc, tin, and copper are the main examples. Since lead sulfate is insoluble, some other anion must be used, and the choice usually falls upon fluoborate, although fluosilicate and sulfamate baths are also used.

Many metal chlorides are sufficiently soluble, and baths formulated with chloride include those for plating nickel, iron, and zinc. Chloride is also used in small amounts in the nickel sulfate bath, to prevent passivity of the anodes.

Bath formulations must take account of cost as well as technical factors. Fluoborates of most metals are extremely soluble, and the activities of the metal ion in fluoborate solutions are higher than those in sulfate baths. Both the higher solubility and the higher activity allow the use of current densities above those possible with sulfate baths. Fluoborates are more expensive than sulfates and are available only in solution form, making them more expensive to ship; cost must be balanced against speed of production. Many metals are plated from fluoborate solutions: nickel, copper, lead and tin-lead alloys. Many other metal fluoborates are commercially available though not extensively used.

Sulfamates are, in general, similar in their solubilities to fluoborates; nickel sulfamate baths are used principally for engineering applications, and lead and indium sulfamate baths have been recommended. Under some conditions the sulfamate radical can hydrolyze to sulfate:

$$H_2O + SO_3NH_2^- \rightarrow NH_4^+ + SO_4^{--}$$

which may be disadvantageous.

Little use has been made of organic anions. Tin can be plated from phenol- or cresolsulfonic acid solutions, and sulfonic acids have been recommended for other metals, but little commercial use has resulted except for tin. Most organic acids are weak, i.e., only partially ionized, such as acetic; and many others, though strong acids, are too expensive for practical consideration. In many cases, the choice of anion is a matter of availability of the metal salts: nickel and copper sulfates, for example, are readily available commercial chemicals, whereas metal salts of many other acids would be specialties, commanding high prices.

Although acid baths are regarded as solutions of simple metal ions (hydrated), some complexing action undoubtedly does take place. Weak complexes are formed with sulfate, and chloride ion forms many complexes with metals, as has been mentioned. Almost the only anions having practically no tendency to form metal complexes, i.e., to act as ligands, are fluoborate and perchlorate.

Alkaline or Complex Ion Baths

The most important complexing agent in alkaline baths is cyanide ion: cyano-complexes are used for plating copper, cadmium, gold, silver, zinc, and indium. Most cyano-complexes are decomposed by acids, with evolution of the poisonous gas hydrogen cyanide; the baths must therefore be definitely alkaline. An exception is the cyano-gold [cyanoaurate(I)] complex; it is unusual in that it is stable in solutions with a pH as low as 3. Cyanide forms complexes with many other metals—iron, nickel, cobalt, and the platinum metals—but these solutions do not yield useful deposits.

The number of cyanide ions coordinated with the central metal atom in the complex is not known with certainty, and actually several complexes with increasing numbers of cyanide ligands may exist, depending on the total cyanide concentration of the solution. See above, p. 127, where it was shown that the calculation of free cyanide, although perhaps useful for control purposes, has no foundation in solution chemistry.

Cyanide baths in general have better macrothrowing power (see p. 148) than acid baths. The metal is contained in the anion, which is actually moving away from the cathode; i.e., it has a negative transport number. This, combined with the liberation of cyanide ions at the cathode, leads to high concentration polarizations, which increase with increasing current density. Consequently cathode efficiency of metal deposition decreases strongly with an increase in current density, meaning that at high-current-density points (nearest to the anode) current efficiency is lower than at low-current-density areas, and the metal distribution tends to be more even than the current distribution. On the other hand, high concentration polarization leads to poor microthrowing power (see p. 150).

Cyanide ion performs an important function at the anode also. Unless sufficient excess cyanide is present, the copper(I) ions which form as the result of anodic solution of the metal combine with cyanide to form insoluble copper(I) cyanide, CuCN, which coats the anode and prevents further current flow. Therefore enough cyanide ions must be present to dissolve the copper cyanide:

$$CuCN + xCN^- \rightarrow Cu(CN)_{x+1}{}^{x-} \quad (x = 1, 2, \text{ or } 3)$$

so that electrolysis may continue. Cyanide acts similarly in cadmium, silver, and zinc solutions.

At the cathode, excess cyanide has an opposite effect: it decreases current efficiency. Therefore the cyanide concentration must be a compromise between these two opposing effects. In copper plating, tartrates often are added to prevent the filming of the anode while having little effect on the cathode efficiency.

Sodium or potassium hydroxide is usually added to cyanide plating baths, for several reasons. Hydroxyl ion improves conductivity; a solution containing only the metal complex plus excess alkali cyanide conducts rather poorly. Hydroxyl ion also acts as an acceptor for carbon dioxide from the atmosphere; in its absence, cyanide ion would be decomposed with the liberation of hydrocyanic acid, since carbonic acid, although very weak, is stronger than hydrocyanic:

$$CO_2 + CN^- + H_2O \rightarrow HCN + HCO_3^-$$

or

$$CO_2 + H_2O \rightarrow H_2CO_3 \qquad H_2CO_3 + CN^- \rightarrow HCN + HCO_3^-$$

In the cyanide zinc bath, the situation is somewhat more complicated, since zinc forms both a cyanide and a hydroxo-complex, $Zn(CN)_4{}^{--}$ and $Zn(OH)_4{}^{--}$. Deposition from the latter is somewhat easier than from the cyano-complex, and it is believed that the species actually responsible for zinc deposition is $Zn(OH)_2$ present in the cathode film. In fact, zinc can be deposited from alkaline baths containing no cyanide and only the zincate ion, but the addition of cyanide appears to improve the deposit characteristics, for reasons not entirely understood.

Cyanide ions also hydrolyze slowly, yielding ammonia and formate ion:

$$CN^- + 2H_2O \rightarrow NH_3 + OOCH^-$$

Ammonia volatilizes from the alkaline solutions, and formate appears to be harmless. At anodes that are evolving oxygen, cyanide may also be oxidized to cyanate; in fact, this reaction has been proposed as a method of destroying cyanide in wastes:

$$2CN^- + O_2 \rightarrow 2CNO^-$$

Further oxidation to carbon dioxide is also possible:

$$2CNO^- + O_2 \rightarrow 2CO_2 + N_2 + e^-$$

Carbonate ion is thus invariably present in cyanide baths, from both atmospheric absorption and anodic oxidation. Unless it reaches very high concentrations, carbonate ion has little effect in most cyanide baths. Sodium carbonate can be reduced to low levels by "freezing out" because the decahydrate $Na_2CO_3 \cdot 10H_2O$ is much less soluble at low than at elevated temperatures. The much higher solubility of potassium carbonate rules out this possibility, but if necessary, carbonate can be precipitated chemically.

POTASSIUM PREFERRED OVER SODIUM FOR ALKALINE SOLUTIONS BECAUSE OF GREATER SOLUBILITY. LESS CHANCE OF DEVELOPING CRYSTALINE DEPOSITS ON PLATING SURFACE.

If carbonate builds up to excessive levels, there is a chance that sodium carbonate crystals may be included in deposits; because of the higher solubility of potassium carbonate, potassium cyanide and hydroxide are often preferred to the sodium salts in spite of their greater cost. Alkaline baths formulated with potassium salts have other advantages, including higher cathode efficiencies and better conductivity. The reason for the higher efficiency is not clear, though it has been suggested that it is a function of the higher mobility of the potassium ion. Positive ions other than sodium and potassium, to balance the anions, have not been considered for alkaline plating baths.

Carbonates tend to reduce anode polarization, perhaps by formation of the rather weak carbonato-complexes, or by their buffering action.

Although cyanide is the most common complexing agent, several others are in practical use. The alkaline stannate tin-plating bath contains the hydroxyl complex of tetravalent tin, stannate ion, or more formally hexahydroxostannate(IV), $Sn(OH)_6^{--}$.* At the cathode this ion may be reduced to the stannite ion [tetrahydroxostannate (II)] $Sn(OH)_4^{--}$, from which tin metal is deposited; but the accumulation of significant amounts of bivalent tin in the bath is severely deleterious to the character of the deposit. Alkali hydroxide is always added to stannate baths, to improve conductivity and to prevent hydrolysis of stannate to insoluble stannic hydrate (see Chap. 13 for details).

Other complex ions have uses for individual metals. Copper is plated from pyrophosphate baths, in which the complex ion is $Cu(P_2O_7)_2^{6-}$ or $CuP_2O_7^{--}$; the copper is in the 2+ oxidation state. Little is known of the mechanism of deposition from this complex.

The familiar chromium plating bath is formulated with chromic anhydride (commonly called chromic acid), CrO_3, which in water solution forms various complexes in which the chromium is present in the anion, such as dichromate $Cr_2O_7^{--}$. Chromium can exist in several oxidation states, including Cr(III) and Cr(II), but whether these species are intermediates in the deposition of chromium metal is still a matter of some speculation. Chromium deposition is in many ways

*This ion is often written SnO_3^{--}, and the alkaline salts $Na_2SnO_3 \cdot 3H_2O$ and $K_2SnO_3 \cdot 3H_2O$. It is almost certain that the proper formulas are $Na_2Sn(OH)_6$ and $K_2Sn(OH)_6$.

in a class by itself; in part the deposition process may be chemical, depending on reduction by hydrogen formed at the cathode.

Chloro-complexes are probably involved in the deposition of many metals from baths containing chloride ion, and gold can be plated from the tetrachloroaurate(III) complex, $AuCl_4^-$. Antimony deposited from chloride solutions always contains chlorine, yielding a form of metal known as *explosive* antimony. Several metals can be deposited from sulfur complexes: thiostannate, thiosulfate, or thiocyanate. Such deposits usually contain substantial quantities of sulfur. Complexes of copper with ammonia and various amines have been proposed, and to some extent have been used commercially. Iodide complexes for silver deposition have been investigated, as have tartrate and citrate baths for various metals including antimony. None of these baths has enjoyed much commercial application, probably because any advantages exhibited have not been convincing enough to outweigh their higher costs and the long experience of platers with the more familiar baths.

Wetting Agents

Wetting agents (surfactants) are used in bright-nickel plating to promote disengagement of hydrogen bubbles at the cathode. If these bubbles were allowed to adhere to the cathode until they grew large enough to detach of their own accord, they would prevent plating at the site of the bubble and pitting would result. Thus these additives, often proprietary (although their nature is well known), are usually marketed under the name of *anti-pits* or *depitters*.

In stannate tin-plating baths, which are highly alkaline and in which the electrode efficiencies are usually significantly less than 100 percent, gas evolution at the electrodes produces a spray containing caustic alkali. This spray is irritating to workers and can settle on shop equipment to form a film of white salts. Thus, in stannate baths, wetting agents are usually present to form a foam blanket which permits more orderly release of the gases. These wetting agents seldom need to be deliberately added since commercial stannates usually contain soapy material.

Chromium plating from chromic acid baths produces large quantities of hydrogen (the cathode efficiency at best is only about 20 percent), and insoluble anodes are used, so that oxygen is evolved at 100 percent efficiency. Most organic wetting agents are attacked by the strongly oxidizing medium, but fairly recent development of stable surfactants has permitted the use of foam producers that inhibit chromic acid spray and may render unnecessary the provision of strong ventilation over the tanks (although local regulations may still mandate it).

Conducting Salts and Buffers

The ions of most metal salts, whether simple as copper(II) in the copper sulfate bath or complex as cyanocuprate(I) in the cyanide copper bath, and their associated counter ions (sulfate, chloride, sodium, potassium, etc.) are fairly poor conductors of electricity, and in most baths either a strong acid or strong alkali, as appropriate, is added to increase conductivity. (We have already seen, p. 11, that hydrogen and hydroxyl ions are the best ionic conductors available, i.e., have the highest ionic mobilities.)

Thus sulfuric acid is added to copper sulfate and tin sulfate baths; sodium or potassium hydroxide is added to most cyanide and to stannate tin-plating baths. These additives have the additional function of suppressing hydrolysis:

$$\text{(acid)} \qquad Cu^{++} + 2H_2O \rightarrow Cu(OH)_2\downarrow + 2H^+$$

$$\text{(base)} \qquad Sn(OH)_6^{--} \rightarrow Sn(OH)_4\downarrow + 2OH^-$$

Carbonates form spontaneously in alkaline baths by absorption of CO_2 from the air; they are also deliberately added to cyanide copper, silver, and gold baths, although their function is not clear. It has been claimed that they increase conductivity, act as buffers, and increase cathode and decrease anode polarizations. Carbonates are deleterious to the operation of some baths, and their concentration is kept as low as possible.

In nickel plating, boric acid is added in concentrations sufficiently high that it exists in polymeric forms, which buffer in the 5 to 6 pH range; pH is an important variable in nickel plating from the sulfate-chloride (Watts) bath. Other salts that buffer in operating ranges include acetates, citrates, and tartrates, but they have found little use. Formates are used in some nickel, cobalt-nickel, and gold baths.

Additives for Anode Corrosion

Except when inert anodes are deliberately employed, as in plating of chromium, gold, and the platinum metals, the metal content of the bath should be replenished by solution of anode metal in amount equal to that plated at the cathode (or perhaps slightly more, to make up for drag-out). If this is the case, control of the bath is much simplified, and addition of metal salts is necessary only infrequently. This situation is desirable for at least two reasons: metal in the form of its salts is usually more expensive than the metal itself, and addition of a metal salt entails the addition of its associated counter-ion (sulfate, potassium, sodium, etc.), which is not consumed in the electrolysis and simply builds up in the solution, which sooner or later gets far out of balance.

In plating from a simple metal salt, the cathode reaction involves a temporary increase in the pH of the solution around the cathode, unless cathode efficiency is 100 percent:

$$H_2O + e^- \rightarrow \tfrac{1}{2}H_2 + OH^-$$

And unless the anode efficiency is also 100 percent, the anode reaction similarly involves a decrease in pH:

$$H_2O \rightarrow \tfrac{1}{2}O_2 + 2H^+ + 2e^-$$

If the electrode efficiencies are the same, these two effects cancel and the pH of the bath remains stable. But if anode efficiency in acid baths is much lower than that at the cathode, the pH of the bath decreases—hydrogen ion builds up—and control measures must be taken to balance it.

In the ideal case, electroplating involves, in effect, the mere mechanical transport of metal from the anode to the cathode: cathode and anode efficiencies exactly balance, and the metal content of the solution, as well as its pH, remains the same. It is for this reason that anode solution is an important factor in the satisfactory operation of most plating baths, and that substances are often added to the bath mainly to ensure an efficient anode solution. In a few cases, no attempt is made to use soluble anodes, and the maintenance of the bath is strictly chemical. In chromium plating from chromic acid baths, no satisfactory soluble anode has been developed, and chromic acid, CrO_3, is used for replenishment. This introduces no extraneous ions, however. In the plating of the precious metals, gold and the platinum group, economic factors are usually more important than ease of control; thus these are special cases. In tin plating from stannate baths, anode control can be troublesome, and it is thought by some operators that chemical replenishment is more satisfactory. In some cases of plating at extremely high current densities, it has been found that although cathode current densities can be raised to high levels, metal anodes simply cannot be made to dissolve at these same rates (current densities) without passivation, and chemical replenishment is resorted to. But in general the desirability of balanced anode and cathode reactions remains the rule. Sometimes it is possible to use a sufficiently high anode-to-cathode area ratio to provide both rapid deposition and effective anode solution.

Although nickel can be plated from sulfate baths, nickel anodes tend to become passive unless some ion is added to prevent it, and chloride is used. In cyanide baths, excess (“free”) cyanide is present to aid in anode solution; otherwise a film of the metal cyanide would form over the anode and inhibit further passage of current. Tartrates and carbonates also aid in preventing passive films from forming in cyanide baths. Hydroxide performs the same function in stannate tin baths.

MECHANISMS OF METAL DEPOSITION

A metal crystal or grain consists of ions of the metal located in regular lattice positions; the electrons required to render the mass of metal electrically neutral exist in a rather indefinite cloud of negative charge, so that the electrons are not specifically identified with any particular ion. The ionic charges are statistically neutralized by the electron cloud over a period of time and over a number of ions. The electronic cloud is free to move under the influence of an applied potential, giving rise to the well-known high conductivity of most metals.

An electrode normally carries an electric charge; when it is dipped into a solution, it attracts water molecules, which are dipoles. (A dipole is any assemblage of atoms, such as a molecule, having equal electric charges of opposite sign separated by a finite distance. Though the molecule as a whole is electrically neutral, it has a positive and a negative end, somewhat analogous to a magnet with its north and south poles. The water molecule is pictured as $\overset{\oplus}{H-H}$ with the oxygen below ($\overset{\ominus}{O}$) positive end near the hydrogen atoms and the negative end near the oxygen atom.) It also attracts ions carrying an opposite charge; these ions are held near the metal surface by forces of electrostatic attraction. Thus an electric double layer is formed (see Fig. 8-3) which has the characteristics of a condenser with a measurable capacitance. Metal ions from this double layer may reach the surface of the metal and find their way to stable positions in the metal lattice; in the process they release their ligands (water molecules or complexing ions), and the charges on the ions and the metal neutralize each other. The total process constitutes a spontaneous flow of cathodic current. At the same time, neighboring lattice ions may become loosened from lattice forces and coordinate with some of the adsorbed water molecules, finally moving to the ionic side of the double layer as hydrated ions, and then on into the solution. This process constitutes an anodic or dissolution current.

When the electrode potential is at equilibrium, these anodic and cathodic currents are equal; there is no net current or overall reaction. The current flowing in this manner is called the *exchange current*. Some metals are characterized by high exchange currents—tin and lead, for example. Others exhibit exchange currents as much as 10^5 times lower. The relative magnitudes of the exchange currents are important in the study of electrode reactions; high values usually mean that metal is deposited with little polarization, and vice versa.

The presence of adsorbed impurities or addition agents decreases the

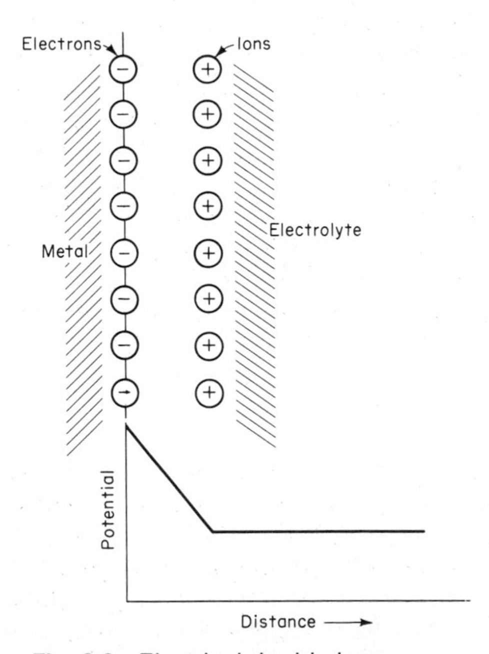

Fig. 8-3 Electrical double layer.

exchange currents, because these substances tend to form stronger bonds with the metal than the water ligands.

Polarization and the factors influencing it were discussed in Chap. 2, pp. 20-23. The total polarization is the applied potential required to sustain a steady current through a plating tank so that electrodeposition can proceed. It is the sum of the individual polarizations already discussed. After the first few seconds of plating with anodes of the metal being deposited, the tank voltage is the same as the total polarization. The voltage must be read between anodes and cathodes, and not at the rectifier; the latter would include irrelevant resistances of bus bars, etc.

Mass Transport

As metal is deposited upon a cathode, the solution in its immediate neighborhood is depleted in metal ions. If plating is to continue, these ions must be replenished. There are three ways in which this is accomplished. In reverse, a similar situation applies at the anode, and the same argument will apply.

Least important of these three ways is ionic migration, already discussed in Chap. 2. There are several reasons why ionic migration can supply few of the required ions at the cathode. Assuming that the metal is contained in a cation—i.e., the plating bath is acid and the metal is not tied up in an anionic complex—the rate of deposition is controlled by the total current, but the rate of in-migration of cations is only t times that current, t being the transport number of the ion. For most metal ions t is less than 0.5; if the bath contains other conducting salts or acids, the metal cation may actually conduct almost no current at all, and its migration approaches zero. Finally, in alkaline and cyanide baths, and some acid baths, the metal is tied up in an anionic complex and is actually migrating away from the cathode.

Ionic transport, then, is usually a negligible factor in replenishing the metal near the cathode. Much more important is *convection*—natural, artificial, or both. Convection involves the movement of substantial quantities of the solution relative to the electrodes. The electrodes may move, the solution may move, or both. Near the cathode, as metal is depleted, the solution becomes less dense and tends to rise along the face of the cathode. The opposite occurs at the face of the anode: solution tends to stream downward. These effects may tend to cause stratification in the plating bath, with the denser solution at the bottom and the lighter at the top.

Passage of current through a resistance causes heating. Thus the IR drop heats the solution by ohmic resistance. Often heating coils are used as well; sometimes cooling coils are required. In any case, temperature gradients are set up which oppose the stratification and aid in convective mass transport. If further convective movement is required, the cathodes may be agitated (by commercially available rod agitators), the solution may be stirred by propellers, or it may be pumped through heat exchangers for both temperature control and agitation. In automatic machines, often the cathodes are moved through the plating bath during their stay in the tank. In continuous plating, where the cathode is an unbroken length of metal such as strip steel or wire, the cathode is moved through the solution, sometimes at very high speeds.

Diffusion

However achieved, convection is very effective in bringing fresh solution, and thus fresh supplies of ions, to the vicinity of the electrodes. But at the surface of the electrode itself, and for a very short distance away from the surface, convection is negligible, and the final distance from the bulk of the solution to the actual face of the electrode must be negotiated through the forces of ionic migration (if they operate in the correct sign) and those of *diffusion*. Diffusion is the movement of a chemical species, either ionic or uncharged, through the solution as a result of a concentration gradient. It is a result of random motion of the ions or molecules; this motion tends to produce more uniform distribution of the various species throughout the solution. Thus the depletion of a species next to the cathode results in a movement of that species from the bulk of the solution toward the cathode.

This region next to the electrode, where the concentration of any chemical species differs from its concentration in the bulk of solution, is called the *diffusion layer*. The boundary between the diffusion layer and the bulk of solution is not a sharp line; it has been defined arbitrarily as the region where the concentration of any species differs from its concentration in the bulk of solution by 1 percent or more. In this region, as has been stated, convection is negligible.

The rate of diffusion, expressed in gram-ions or moles per square centimeter per second ($mol/cm^2/s$), is proportional to the concentration gradient at the electrode. The proportionality constant D is called the *diffusion constant*, expressed in square centimeters per second (cm^2/s). The rate may be expressed as

$$R = D(C_0 - C_E)/dN \tag{8-9}$$

where C_0 is the bulk concentration, C_E the concentration at the electrode surface, and dN is the effective thickness of the diffusion layer (sometimes called the *Nernst thickness*).

As dN decreases, diffusion rate increases, as shown by Eq. (8-9). Agitation decreases dN, thus increasing the diffusion rate. In the absence of agitation, dN is about 0.2 mm; and as agitation is increased, this value drops until, with the very severe agitation provided by a rotating-disk electrode or high-speed strip and wire plating, dN may be as little as 0.015 mm. Diffusion layers are, however, much thicker than the electric double layer, which is only about 10^{-6} mm (1 nm) thick.

Electrode surfaces are not usually perfectly smooth. If the profile of the electrode has dimensions about equal to dN, then the diffusion layer varies in thickness over the peaks and valleys, being thinner over the peaks. This fact is important in leveling; see the section Microthrowing Power and Leveling.

Hydrogen Overvoltage

If it were not for hydrogen overvoltage, many metals could not be deposited from aqueous solutions. If the pH of the bath is 5 (common for nickel plating, for example), the equilibrium potential for hydrogen evolution is about −0.295 V; in alkaline solutions with a pH of 10, it is −0.59 V. These potentials are significantly more positive than the reduction potentials of many common metals in these solutions; thus it would appear that only hydrogen would be deposited. Activation polarization is extraordinarily high in the evolution of hydrogen, and it is this that enables metal deposition or inhibits hydrogen evolution.

This hydrogen overvoltage depends on the nature of the cathode surface. At cathodes of tin, zinc, or lead, it can be more than 1 V, depending on the current density. It is lowest on platinum. On platinum black it is negligible, which is why platinum black is used in setting up the hydrogen electrode.

Impurities adsorbed on the cathode surface increase hydrogen overvoltage. Such impurities often include addition agents deliberately added for leveling or brightening the deposit.

In some cyanide solutions, the metal deposition potential and the potential for hydrogen evolution, including overvoltage, are close together, and hydrogen may codeposit with the metal; in other words, cathode efficiencies may be low. Hydrogen overvoltage is very low on graphite, which is a common inclusion in cast iron; thus it is difficult or impossible to plate zinc on cast iron, since all the current is used in depositing hydrogen. The practical way around this problem is to "flash" (flash means to make a very thin initial deposit) the cast iron with tin or cadmium, on which hydrogen overvoltage is high and which therefore can accept a zinc deposit.

Although we have discussed the deposition of metals and hydrogen in terms of electrode potentials, these are not always the determining factors. From the standpoint of their electrode potentials, it should be possible to electroplate such metals as tungsten and molybdenum from aqueous solutions with a pH of about 5. Nevertheless (in spite of claims in the literature), these metals cannot be deposited in pure form from aqueous solutions (except possibly at extremely low efficiencies, less than 1 percent). A similar situation exists with the cyanide complexes of nickel and iron. Although thermodynamically there is no reason for their failure to yield deposits, these metals cannot be deposited from cyanide solutions (again, except for possible extremely thin, almost monatomic layers). These failures have not been entirely explained; they have been attributed to kinetic factors, i.e., the extremely slow rates of reaction exhibited by the complexes, and possibly also by very low hydrogen overvoltages.

The case of chromium deposition from chromic acid baths, which takes place from the dichromate or similar anionic complexes, is anomalous. Since the chemistry of chromium is at least formally similar to that of molybdenum and tungsten, one might expect their electrodeposition behavior to be somewhat similar also. It may be noted, however, that even at best the current efficiency of chromium deposition is very low, about 20 percent or even less; and in spite of much effort the exact mechanisms of chromium deposition have not been

H																	He
Li	Be											B	C	N	O	F	Ne
Na	Mg											Al	Si	P	S	Cl	Ar
K	Ca	Sc	Ti	V	Cr	Mn	Fe	Co	Ni	Cu	Zn	Ga	Ge	As	Se	Br	Kr
Rb	Sr	Y	Zr	Nb	Mo	Tc	Ru	Rh	Pd	Ag	Cd	In	Sn	Sb	Te	I	Xe
Cs	Ba	La	Hf	Ta	W	Re	Os	Ir	Pt	Au	Hg	Tl	Pb	Bi	Pa	At	Rn
Fr	Ra	Ac															

Fig. 8-4 Periodic table of the elements: hydrogen and the metals depositable from aqueous solutions are within the heavy lines.

entirely elucidated. According to some workers, deposition is a "secondary" reaction involving reduction by hydrogen which is evolved in large quantities.

The metals that can be electrodeposited from aqueous solutions occupy a definite and compact region in the periodic table, as shown in Fig. 8-4. This suggests that the electronic structure, as well as hydrogen overvoltage and electrode potential, may have something to do with their ability to deposit; but if so, such a relationship has not been demonstrated.

Electrocrystallization

We have already seen that electrodeposition is the process of producing a deposit on an electrode by electrolysis; and electroplating connotes electrodeposition of a coating in a useful form. The term *electrocrystallization* is used to denote the mechanisms by which this deposit is formed; it attempts to look at the individual steps by which a metallic ion in solution is changed into a metal atom occupying a place in the regular lattice of the massive metal on the cathode.*

The deposition of copper from a copper sulfate bath, for example, is usually written

$$Cu^{++} + 2e^{-} \rightarrow Cu$$

But this equation merely summarizes the total stoichiometry of the reaction, and it neglects the many intermediate steps which undoubtedly take place during the electrocrystallization process. The deposition of copper from a cyanide solution, similarly, may be summarized:

*Deposits can also be produced on the anode, and not all deposits are metallic. For example, organic coatings may be deposited on either electrode, and a few metals or semimetals like selenium can be deposited on the anode. But for present purposes we are concerned, as electroplaters almost always are, with a cathodic deposit of a metal.

$$Cu(CN)_4^{3-} + e^- \rightarrow Cu + 4CN^-$$

But again, there must be many intermediate steps between the left-hand (reactants) and the right-hand (products) sides of the equation. These are what concern the study of electrocrystallization. This subject, sometimes called *electrodics,* is active in the current literature, and the answers are by no means all in. Here we can merely outline some areas of study.

The total energy required for the total process "metal ion in solution → metal atom in the lattice" is made up of several components. A metal ion probably reacts first with the cathode surface at some point on the flat area of a growing crystal; here it becomes adsorbed in a state intermediate between the ionic state in solution and the state existing in the metal lattice. It retains some of its ionic charge and some ligands, either water or other complexing agents. In this state it is called an *adion*. This adion moves over the surface by diffusion to a growth site, where it is incorporated into the growing lattice. The initial growth is thus lateral, producing a monolayer (a layer one atom thick); but at some point in the process this lateral growth stops, and a new layer is formed over the last one. Continuation of the process produces a crystallite or grain.

In the double layer next to the cathode, the metal ion is coordinated to several ligands, water molecules in the case of simple ions. These ligands may be released one by one, or the coordination number of the metal atom may temporarily increase to accommodate coordination to the metal in the depositing layer. In either case, the coordination sphere of the ion is distorted, and energy must be expended. In effect, the coordination to the ligand in solution is replaced by coordination to the lattice. More energy must be expended in coordinating to a kink or vacancy in the metal than in coordinating to a flat surface, which is why one supposes that the first step in deposition involves attachment to the flat, as already mentioned. The steps, in summary, appear to be:

1. Arrival at the double layer
2. Release of some ligands, one by one, to form an adion
3. Deposition at a flat area of the cathode
4. Lateral movement to a final place in the metal lattice

Electrocrystallization is no different in principle from other types of crystallization, such as growth from a melt, a vapor, or a saturated solution. The principal difference is that, as in other electrolytic processes, the potential available is under the control of the operator.

Diffusion Current

As increasing potential is applied across an electrolytic cell, the current increases, in obedience to Ohm's law, $E = IR$. Ohm's law is obeyed by electrolytes just as by metallic conductors, if electrode polarization is subtracted from the applied potential. If the concentration of depositing ions is relatively small, and that of nondepositing ions (the "supporting electrolyte," such as sulfuric acid in a copper sulfate bath) relatively large, a condition is reached at which further increases in potential cause only increased cathode polarization, and the current remains constant. (This is the situation, for example, in polarography.) This constant-current situation occurs when the depositing ions are plated out as rapidly as they can reach the surface of the cathode; the current then depends on the rate at which these ions can diffuse to the cathode, the diffusion rate already mentioned. This current is called the *diffusion current* or *limiting current;* the concentration of the depositing ion in the double layer is practically zero. Taking the electrode area into consideration, we have the *diffusion* or *limiting current density*. If the potential is made still more negative, some new process such as hydrogen evolution may take over, and the current again can rise.

Most electroplating baths do not quite reach this condition, because the concentration of depositing ions is usually larger than required for zero concentration in the double layer. But they approach it, and a *practical* definition of the limiting current density is the highest current density at which good deposits are obtained. Beyond this point, deposits are still produced, but they are generally dark, powdery, and noncoherent; they are said to be *burned*. Burned deposits are often observed on edges and points of a cathode which is otherwise covered with a satisfactory deposit; this is so because the current density at such areas is much higher than the average unless special precautions are taken.

The cause of burned deposits is not unequivocally explained. They have been attributed to high rates of hydrogen discharge, which causes a rise in pH at the cathode surface such that metal hydroxide or basic salts are precipitated and included in the deposit. If this explanation is valid, the utility of buffers, as in nickel plating, becomes clearer.

The mechanism does not explain all cases of burning; inclusions of water molecules as well as hydroxyl ions in the deposit have been found. Other inclusions such as strongly adsorbed anions and organic materials are also found. Regardless of mechanism, there is no doubt of the practical importance of the limiting current density as the highest obtainable under the conditions of the electrolysis for good deposits.

Burned deposits may be avoided by several expedients. Lowering the current density is the most obvious, but provision of higher mass transfer and diffusion rates is also effective; this is achieved by increasing agitation, higher temperatures, or higher concentrations of the depositing metal ion. In a few cases, as mentioned above, buffers can prevent the formation of basic salts or hydroxides in the cathode film.

Effect of pH

(See Chap. 24.) Depending on the particular plating bath, pH may or may not be a most important variable. Some highly acidic solutions, such as fluoborate baths and the chromic acid bath, are so concentrated in hydrogen ion that small changes in pH are unimportant. At the other extreme, the stannate tin-plating bath contains so much free alkali that minor variations are insignificant. For baths between these extremes, the pH of the solution influences many factors important to the satisfactory operation of the plating process.

pH influences the hydrogen discharge potential: it determines whether basic inclusions will be precipitated. It controls the composition of the complex containing the depositing metal, the adsorption of addition agents, and the cathode and anode efficiencies. Most of these factors are unpredictable, and the best pH range must be determined experimentally. pH is known to affect stress and hardness of some deposits; this effect is probably an indirect result of the inclusions in the deposit. In complex-ion baths, the pH can be a factor in the equilibria among the various possible complexes; for example, in zinc cyanide baths both hydroxozincate and cyanozincate ions are present, and the equilibrium between the two depends on pH.

pH both influences and is influenced by the efficiency of metal dissolution or deposition at both electrodes. At insoluble anodes, the bath is depleted in metal, which must be replaced chemically, and also becomes more acidic (lower pH) because the anode reaction is generally

$$2H_2O \rightarrow O_2 + 4H^+ + 4e^-$$

or in alkaline baths

$$4OH^- \rightarrow O_2 + 2H_2O + 4e^-$$

The resulting acid (hydrogen ion) may be neutralized and the bath simultaneously replenished if a metal compound is available which is soluble at the pH of the bath and does not at the same time add unwanted anions. In the zinc sulfate bath, for example, zinc metal (scrap) or zinc oxide (perhaps a roasted zinc ore as in the "Bethanizing" process for wire plating) will serve. Addition of zinc sulfate would replenish the zinc content but would not neutralize the acid formed, and the bath would build up in sulfate ion concentration.

The same situation exists if the anodes are dissolving at less than 100 percent efficiency, but of course to a lesser degree.

At the cathode, hydrogen evolution causes a rise in pH; the bath becomes more alkaline:

$$2H_2O + 2e^- = 2OH^- + H_2$$

If the anode and cathode efficiencies are equal, even if they are less than 100 percent, these two effects cancel out and the bath remains in balance (neglecting drag-out). If the current efficiency at the anode is higher than that at the cathode, the bath becomes more alkaline and metal ions accumulate; if the reverse is true, the bath becomes more acidic and metal ions are depleted. This is not true of stannate baths; release of hydroxyl ions from the $Sn(OH)_6^{--}$ complex, without corresponding replenishment of tin, renders the bath more rather than less alkaline.

Addition Agents

The character of an electrodeposited metal may be profoundly altered by the presence in the bath of very small amounts of certain materials, usually though not always organic or colloidal. Usually the effects are undesirable; but certain substances have the ability to produce smoother deposits, leveling deposits, or bright deposits, or to affect the internal stress in the deposit. Such materials are called *addition agents;* when they produce specific effects, they may be called levelers, brighteners, stress reducers, etc. Some solutions, e.g., simple acid solutions of stannous salts such as stannous chloride or sulfate, require addition agents to produce a useful deposit at all: in the absence of such an agent the cathode deposit is a loosely adherent collection of trees and large crystals of no utility even in electrorefining.

The amounts of addition agents required to produce these profound effects are generally very small; one molecule of addition agent must affect many thousands of metal ions. In order to be effective, the additive must in some way be adsorbed or otherwise included in the deposit. But the mechanism by which they exert their effects is not clear, and in spite of much fundamental research the development of an effective addition agent for a particular process remains, at least to some extent, a purely empirical search. Most effective additives increase the activation overpotential at the cathode, but there are exceptions.

Whatever the mechanism, most additives tend to favor the formation of new crystal nuclei on the metal surface and inhibit the growth of existing ones, giving rise to a deposit with finer crystal structure.

Metals differ in their susceptibility to the effect of additives, and the order of this susceptibility is roughly the same as the order of their melting points, hardness, and strength; it increases in the order Pb, Sn, Ag, Cd, Zn, Cu, Fe, Ni. Thousands of compounds are known that brighten nickel deposits from the sulfate-chloride bath, while it is only fairly recently that ways of brightening tin deposits from acid baths have been developed. The series also corresponds roughly to the increasing tendency of the ions to form complexes, to increasing activation polarization in deposition from simple aquated ions, and in reverse order to the hydrogen overvoltages on cathodes of the metal.

Addition agents are consumed during plating, and they must be replenished from time to time; decomposition products of the additive often are incorporated in the deposits, which therefore often contain traces of carbon, sulfur, or both. (Many effective additives are sulfur-containing organic compounds.) Although, as stated, thousands of compounds have been found to brighten nickel deposits, only a few are useful in a practical sense.

Most commercial additives for plating are proprietary and are covered by patents; the patent literature in this field is voluminous. "Practical" platers seldom know the nature of an additive they use, since it is sold under fanciful names having no relation to its chemical nature.

Deposit Distribution

It is not possible in practice to produce an electrodeposit of uniform thickness over the whole surface of a cathode, unless the latter is of extremely simple shape; even then the setup must be especially arranged. Edges and projections receive higher current densities than average, and recesses lower.

In the absence of polarization, the current distribution obtained is termed the *primary current distribution*. It depends only on Ohm's law; it can be calculated from potential theory (although calculations are complex and difficult) and can be measured by probes on a large scale, or up to 1000 mL by means of conducting paper. It depends only on the geometry of the system and is not affected by the properties of the electrolyte, including its conductivity.

When anodes and cathodes are very close together and either of them is of irregular profile (is rough), primary current distribution is far from uniform. As the electrodes are moved farther apart, the distribution becomes more uniform; but the greatest uniformity achievable by separating the electrodes is reached at fairly small distances, and further separation achieves little improvement.

As soon as cathodic polarization enters, i.e., as soon as metal deposition commences, the current distribution is changed to the actual or *secondary* distribution. Activation polarization enters immediately, and as soon as electrolysis has proceeded for any significant time, concentration polarization also becomes a factor and increases with time up to some steady value. Both types increase with increasing current density, and hence act to improve the current distribution; thus secondary current distribution is always more uniform than primary. The relative influence of ohmic resistances and of polarization depends on the scale of the system; the greater the roughness of the profile, or the greater the length of the Haring-Blum cell (see below), the greater the relative importance of ohmic resistance, and the closer the secondary to the primary distribution. This is one reason why the Haring-Blum cell, although a useful tool for determining relative throwing power, cannot yield quantitative results for actual plating situations.

Throwing Power

The distribution of metal deposit on a cathode is influenced, of course, by the actual current distribution, as described above. But it is also affected by the variations of cathode current efficiency with current density. In baths in which the cathode efficiency falls rapidly as cathode current density (cd) increases, the excess deposit on edges and projections, regions of high cd, will not be as great as would be expected from current distribution alone. Many cyanide baths and the alkaline stannate tin bath exhibit this tendency of rapid drop of efficiency with increasing cd, and accordingly exhibit high throwing power.

On the other hand, in chromium-plating baths the efficiency rises as the current density increases, at least within limits; and the nonuniformity of metal distribution is thereby exaggerated. Such a bath has ''negative'' throwing power.

Macrothrowing Power

The throwing power referred to just above is better termed *macrothrowing power,* since it refers to the distribution of deposit on cathodes having rather large-scale, or macroscopic, irregularities. It will be seen that the situation with regard to microscopic or minute irregularities in profile is quite different, and the term *microthrowing power,* discussed below, is used to describe this.

Macrothrowing power may be measured by the Haring-Blum cell (Fig. 8-5), by measuring the ratio of deposit thicknesses on two selected points of a bent cathode (Fig. 8-6) or in a Hull cell (Fig. 8-7), and by comparing this ratio with the ratios of the primary current distribution. As stated above, these measured quantities depend to some extent on the scale of the experiment and so are not absolute values; but in general a solution exhibiting good throwing power in such a test will exhibit it in practice also. The ratio of primary current densities is P, and that of actual metal distribution is M. Typical Haring-Blum throwing power cells (Fig. 8-5) have $P = 5$; i.e., the far cathode is 5 times as far from the anode as the near cathode. Also used are cells with $P = 2$. If, in an experiment in a cell with $P = 5$, the weight of deposit on the near cathode is 4 times that on the far cathode, the metal ratio is 4. The throwing power in the Haring-Blum formula is

$$T = \frac{P - M}{P} \times 100 \tag{8-10}$$

or, in this case, $(5 - 4)/5 \times 100 = 20$ percent. For ''perfect'' throwing power, that is, $M = 1$ (equal weights of deposit on both far and near cathodes) the formula yields a throwing power of 80 percent. Intuitively, one would prefer ''perfect''

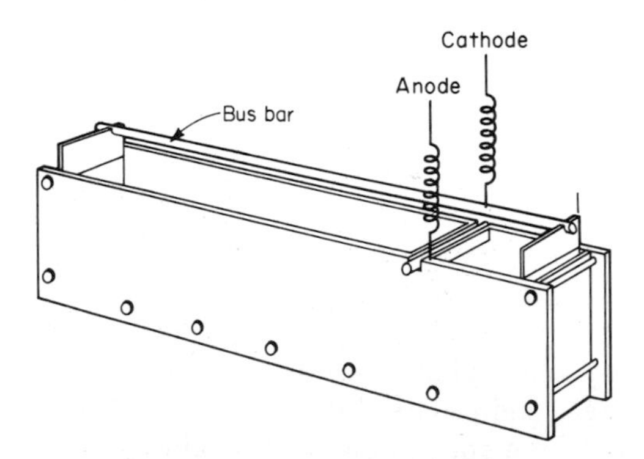

Fig. 8-5 Haring throwing-power box.

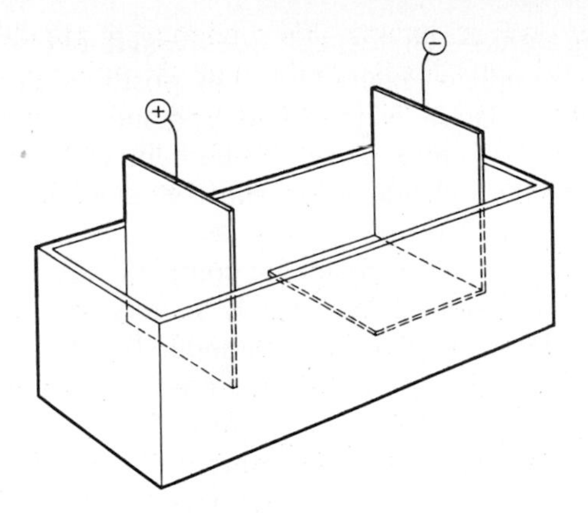

Fig. 8-6 Bent-cathode test.

throwing power to be called 100 percent. Further, for no deposit on the far cathode, T comes out to minus infinity; again, one intuitively dislikes a plating variable expressed as infinity. The result also depends on the value chosen for P (usually, as stated, either 2 or 5).

A formula for throwing power that avoids these objections is that of Field:

$$T = \frac{P - M}{P + M - 2} \times 100 \tag{8-11}$$

In this expression, when $M = 1$ (perfect throwing power), $T = 100$ percent regardless of the value of P; when $P = M$, throwing power is zero; and when M = infinity (no deposit on the far cathode), the throwing power is -100 percent instead of minus infinity.

Neither expression has any theoretical significance, but workers reporting experimental results should indicate which they are using.

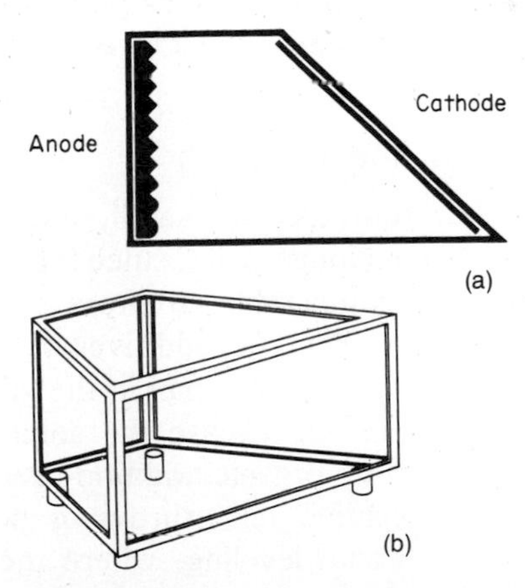

Fig. 8-7 Hull cell: (*a*) top view, (*b*) sketch of 267-mL cell.

Most acidic plating baths—sulfates, chlorides, and fluoborates of copper, zinc, nickel, and tin are examples—exhibit low but positive throwing powers; the nature of the metal is also a factor. The principal reason is that cathode efficiencies in these baths remain near 100 percent at high as well as at low current densities. Alkaline cyanide baths and the alkaline stannate tin bath in general have high positive throwing powers, as already mentioned. In all these baths the metal is incorporated in an anionic complex, and these generally show high concentration polarization, at least partly because the metal in the cathode film is actually migrating away from the cathode. In addition, these baths are characterized by a steep slope of the cathode efficiency–current density curve, again favoring good throwing power on macroprofiles. But the high concentration polarization tends to cause poor microthrowing power.

Since addition agents can profoundly affect polarization, they also have effects on macrothrowing power and microthrowing power. Thus some acidic plating baths may be exceptions to the stated generalities with certain additives.

Chromium plating is a special case: the bath is acidic, but the metal is in an anionic complex, and the current efficiency–current density curve has a positive slope. Its macrothrowing power is negative.

A different, but related, property of plating baths is *covering power;* this is the ability to produce any deposit at all at low current densities, e.g., in recesses. At very low current densities in some baths, the potential required for metal deposition may not be reached, and some other process may support the ability of the bath to conduct a current: hydrogen evolution, reduction of addition agents, or some reducible ion such as Cu(II) to Cu(I) or Fe(III) to Fe(II). Poor covering power is evidenced by failure to deposit metal in recesses or other areas of very low current density. The difficulty may often be avoided by temporarily using a high current density at the start of the electrolysis or by the use of a specially formulated bath for a preliminary *strike* deposit of the metal, to be followed by plating under standard conditions. Covering power can be estimated by determining the distance on a Hull cell cathode from the high-current-density edge to the point where there is no deposit (see Chap. 24).

Microthrowing Power and Leveling

When the depth of the profile is very small, the considerations that govern macrothrowing power may no longer hold, since the dimensions of the peaks and valleys of the profile are comparable to those of the diffusion layer itself. Depending on the solution used and the additives it contains, the deposit may be thicker over the peaks (or micropeaks) than in the valleys—poor microthrow—or it may be the reverse, in which case the solution is said to be leveling. Leveling usually is achieved by organic addition agents.

Figure 8-8 shows (1) negative microthrow or poor leveling, (2) so-called geometric leveling, and (3) true leveling, where the deposit is thicker in the valleys than on the peaks, and the plated article is smoother than the original work.

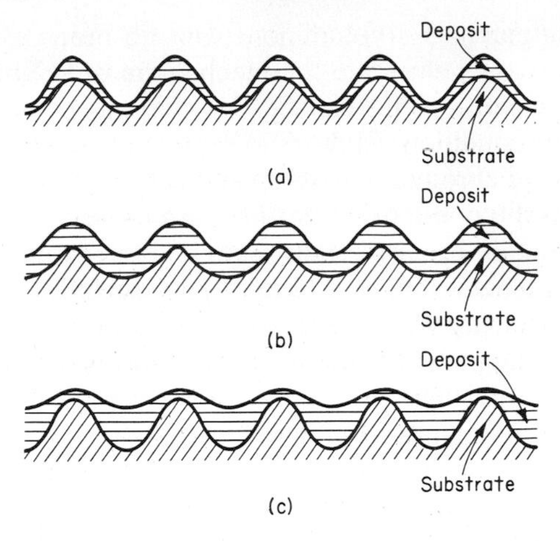

Fig. 8-8 Types of microthrowing power: (*a*) negative microthrow, (*b*) geometric leveling, (*c*) true leveling.

Leveling is practically important; use of leveling solutions may eliminate the requirement for buffing the substrate.

The theory of leveling, like that of brightening, is not fully developed; although there are many helpful hypotheses, it cannot be said that either phenomenon is understood. Both are dependent, in ways not entirely elucidated, on polarization and adsorption of addition agents. Current literature and more advanced texts should be consulted.

Brightening

Brightening and leveling are closely related, although many solutions capable of producing bright deposits have no leveling ability. A bright deposit is one having a high degree of specular reflection (a mirror), in the as-plated condition. If the substrate is bright to begin with, almost any deposit plated on it will be bright if it is thin enough; but a truly bright plate will be bright over a mat substrate, and it remains bright even when it is thick enough to hide the substrate completely. Baths without addition agents seldom or never produce bright deposits; the addition agents, usually organic compounds, which act to turn a dull or mat deposit into a bright one are called *brighteners*. These compounds are apparently adsorbed more strongly on the minute peaks in the substrate than on the valleys, and so tend to produce a more even deposit. But this explanation is far from explaining the mechanism of brightening, because a smooth deposit is not necessarily bright.

Because of its commercial importance, more work, both practical and theoretical, has been done on brighteners for nickel plating than for other metals. The literature, including patents, is voluminous, but no thoroughly satisfactory theory has been advanced to elucidate the mechanism of brightening.

Almost every plateable metal can be, by the use of appropriate additives, deposited in a bright condition. Chromium requires no addition agents for bright plating, but decorative chromium plate is usually so thin that the plate is bright if the substrate is and the operating conditions are correct.

Bright-plating baths in general require much closer control than baths formulated without brighteners; many additives are sensitive to impurities, their decomposition products may require removal by carbon treatment; excess brightener may embrittle the plate; and other complications may arise. Nevertheless, in many cases these added complications are worthwhile, because of the labor saved by the ability to eliminate buffing after plating.

ANODES

Anodes in a plating bath serve at least the first and often both of the following functions: they are *required* to complete the electric circuit and introduce current into the bath; i.e., they serve to remove the electrons introduced at the cathode. More briefly, they form the positive electrode of the electrolytic circuit. In most cases, but far from all, they also serve to replenish the metal content of the bath that is depleted by cathodic deposition.

Anodes may also be of two basic forms. They may be of massive metal, attached to the anode bar of the plating tank by means of hooks or other connectors. Or they may be small chips, broken pieces of an electrolytic cathode from a refinery, individual cast balls or other shapes, contained in an *anode basket* and connected to the electric circuit simply by random contacts between bits of metal and the basket containing them, or a contact rod inserted therein.

Both inert and active anodes have advantages and disadvantages, as have individual anodes and basket anodes.

Active anodes serve to replenish the metal content of the bath, and minimize the necessity for chemical additions, which are often more expensive than the metal itself; usually fewer analyses are necessary for control. Among the disadvantages of active anodes are that they require a substantial investment in inventory (with its accompanying interest charges), they can contribute impurities and insoluble matter to the bath, and they require monitoring to ensure that they remain active and do not form insoluble films which negate their function.

Advantages of inert anodes are many. They can be permanently installed in the tank through positive contacts requiring little or no maintenance; they do not require an inventory of metal; they do not change size or shape in use. Their principal disadvantage is that they contribute no metal to the bath to replace that plated out; this must therefore be accomplished by addition of chemicals, either controlled by analysis or added on a routine schedule. Chemical additions are usually more expensive than metal.

Individual anodes versus balls or chips in a basket also have their pros and cons. The individual anodes are not likely to be confused for some other metal by careless or inexperienced operators, a more frequent source of trouble than might be supposed. On the other hand, these anodes are not completely consumable; sooner or later a "sword" or "spear"—the remains of the anode after as much as possible has dissolved—must be sent back to a refiner or supplier for remelting. They are heavy and may be hard to handle. Basket-type anodes may be simply replenished by adding fresh balls or chips to the top of the basket, and are practically completely consumable, with no scrap problems. They are easier to handle. They do require somewhat more attention, and in some cases most of the current is carried by the basket instead of by the anode metal.

Soluble anodes usually dissolve in their lowest stable oxidation state. But if the metal can form an impermeable and insoluble film, consisting of its oxide or some other compound, this may happen, especially if the anode potential becomes high. Copper may form films of copper(I) cyanide; nickel may form films of oxide which conduct the current but cause oxidation of water rather than solution of metal. This passivity is usually avoided, in the sulfate-chloride (Watts) bath, by the inclusion of chlorides in the formulation of the bath.

The behavior of tin anodes in the stannate bath is more complicated, because tin may dissolve in the 2+ or 4+ states, or not at all, depending on the anode potential and the nature of the anodic film. This behavior is discussed in Chap. 13.

Lead or lead-alloy anodes are commonly used in chromium plating from the chromic acid bath; their inertness is due to an oxide film in these baths. Oxide films also account for the passivity of noble-metal anodes such as platinum and of active metals such as iron in alkaline baths. Titanium, commonly used in baskets for containing soluble anodes, owes its inertness to a highly resistant oxide film. This film can be broken down by some anions such as fluoride, and for this reason the use of titanium anode baskets is not recommended for some specific baths.

No commercially available metals are 100 percent pure, and the impurities in the anodes can be sources of troublesome contamination in some baths. Some grains or grain boundaries in the anode metal dissolve more rapidly than others; this may result in other grains being loosened before they dissolve. These loose grains may enter the bath and become attached to the cathode, causing roughness, nodules, or pits. More noble impurities in the anode metal may disengage from the anode and likewise form undissolved particles.* These undesirable effects are often avoided in practice by the use of anode bags or diaphragms, or by continuous filtration of the bath. Anode bags are, as the name implies, bags of material appropriate to the chemical nature of the bath which is finely woven enough to hold back anode sludges while permitting access of solution to the anodes. Diaphragms, less common, fulfill the same function but are installed in

*This type of reaction is useful in electrorefining: crude copper or nickel anodes dissolve, and purer metal is deposited on the cathode; the noble (platinum group, gold) metals contained in the ore do not dissolve but form anode "slimes" which are worked up for their content of precious metals. The value of these often pays for the refining.

the tank itself and serve to separate the anolyte from the catholyte. Continuous filtration, besides its obvious function, aids in agitation of the solution and may be used in conjunction with a heat exchanger for temperature control; if activated carbon is used as the filter medium, it also controls soluble organic impurities.

Some anodes in some solutions do not require the imposition of an cxternal potential to dissolve; they are chemically attacked by the solution even during idle periods. This is particularly notable with zinc anodes in the acid sulfate bath, but atmospheric oxygen can also cause copper anodes to dissolve in both acid and cyanide baths. This happens also with other anodes in cyanide baths when free cyanide is high. Current efficiencies may therefore appear to be higher at the anode than at the cathode. To some extent this is desirable, since it helps to replace drag-out; but if metal content continues to build up from this cause, insoluble anodes may be used to reduce the metal concentration.

However, there are limitations to the utility of insoluble anodes when they are used at the same potential as soluble anodes in most baths, including the copper and zinc sulfate and the cyanide baths. They will not correct high metal-ion concentration by evolving oxygen because the potential required for oxygen evolution is about 1 V higher than that required to dissolve the soluble anode. Therefore all the current will flow from the soluble anodes, and the insoluble anodes will carry no current. Anode metal will dissolve just as rapidly as if no insoluble anodes were present, unless the current density on them becomes so high that some other reaction takes over. Chemical attack is diminished because there is less surface area of soluble anode, but otherwise the insoluble-anode area is essentially useless. Insoluble anodes cannot carry current unless they are placed on a separate circuit with a higher potential.

However, this effect can be useful: lead-covered anode contacts can be used in sulfate baths; no current will flow from the contact as long as a soluble anode remains in the circuit.

These considerations apply to pure solutions and pure anode metal. In practice, some exceptions are noted, and sometimes inert anodes can be used to good

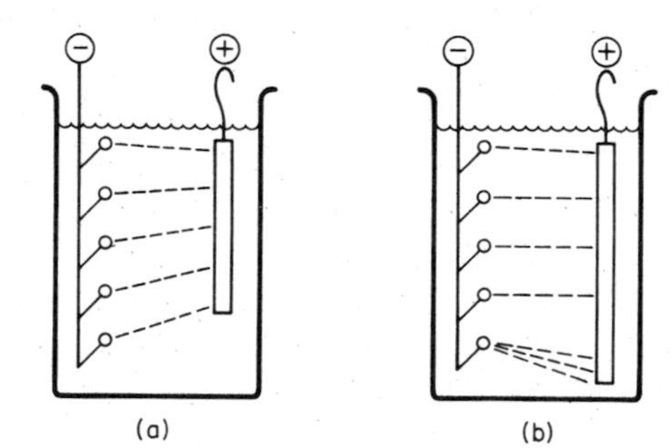

Fig. 8-9 Anode length: (*a*) anode shorter than rack of work, no excess current at bottom; (*b*) anode longer than rack of work, excess current at bottom of rack.

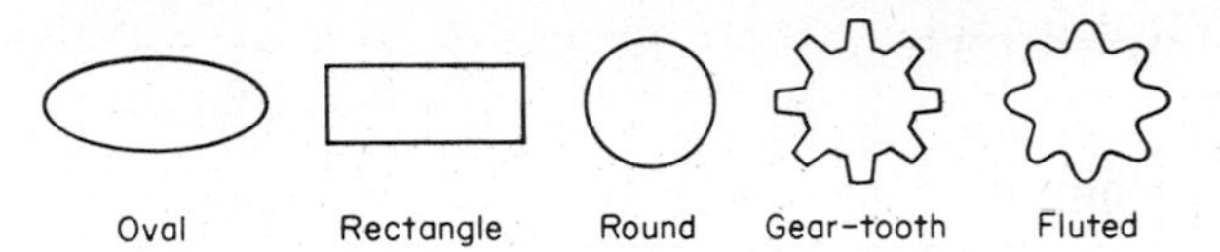

Fig. 8-10 Cross sections of anode shapes.

effect. Some bath constituents may be oxidized, such as ferrous ion in acid baths or ferrocyanide in cyanide baths, to permit current to flow from inert anodes without the necessity of oxygen evolution. In practice, therefore, steel anodes can sometimes be used along with soluble anodes in cyanide solutions and platinum anodes along with nickel in nickel-plating baths.

For most plating situations, the shape of the anodes is not especially important. Soluble anodes are made and sold in the shape of bars having round, oval, or rectangular cross sections. They are usually shorter than the cathode racks: this tends to decrease the excess current that would otherwise flow to the pieces of work at the bottom of the rack (see Fig. 8-9). Some plating solutions require that the anode current density be lower than cathode cd, in other words that anode area be greater than cathode area. In an effort to provide this additional area, some anodes are offered in cross sections resembling a gear-tooth (Fig. 8-10); the practical value of these shapes is controversial.

For solutions of good to moderate throwing power, the usual anode configurations are adequate. In plating complicated shapes having reentrant angles, or in plating the inside of pipe or other areas which are effectively shielded from the anodes, conforming and inside anodes are required.

Another means of supplying current to shielded sections of the cathode is the use of *bipolar* anodes. These are pieces of metal, usually inert in the particular solution, which are not connected to the electric circuit but are merely inserted between the active anodes and the cathode. Since their resistance is less than that of the electrolyte, the current takes the shortest electrical path, which is through the bipolar anode rather than through the solution. They thus become "bipolar"; where current leaves the solution and enters the metal, the metal becomes cathodic, and where it leaves the metal and enters the solution the metal serves as an anode. The use of a bipolar anode to direct current into a sharp angle is illustrated in Fig. 8-11.

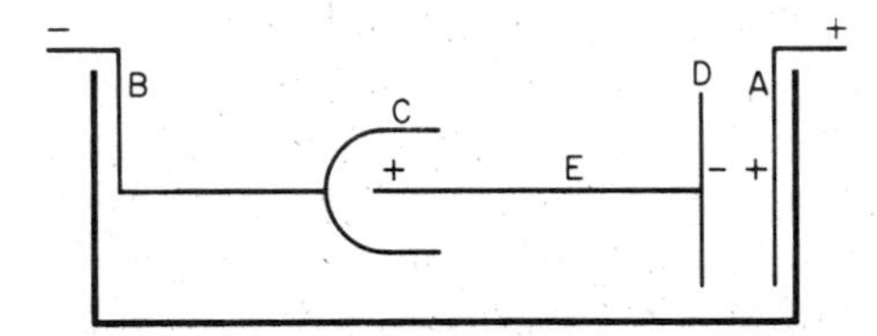

Fig. 8-11 Diagram of bipolar electrode. Key: A, anode; B, cathode connection; C, cathode; D, collector end; E, discharge end.

Effect of Temperature

Temperature is a critical variable in most practical plating baths, and for best results must usually be controlled within plus or minus about 2°C from the "optimum." An increase in temperature increases the rate of diffusion and increases ionic mobilities, and therefore the conductivity of the bath. It also increases the rate of evaporation, rate of hydrolysis of bath constituents, and the rate of decomposition of additives. Most compounds become more soluble with increasing temperature, but there are exceptions.

These opposing effects make it difficult or impossible to predict the best temperature of operation for any given bath, which therefore must be determined experimentally.

Thus heating or cooling coils are necessary for most plating setups. Passage of current through the resistance of the bath produces a heating effect, and unless this is balanced by evaporative cooling, temperature control is required.

Alloy plating is considered in Chap. 16.

9
Electricity and the Electroplater

In an electrical circuit, electric current is said to flow from the "positive" to the "negative" terminal: we connect the positive (plus) terminal of a battery, rectifier, or generator to the anodes or anode bar of the plating tank, and the negative (minus) terminal to the cathodes or cathode bus bar. This nomenclature was firmly set before it became clear that in metallic conductors the actual flow of electrons was in the opposite direction; the "negative" terminal provides electrons and the "positive" terminal collects them or acts as an "electron sink." It is too late to change this convention.

There are two basic types of electric current: unidirectional, direct current, or dc, in which the direction of flow is constant; and alternating current, or ac, in which the direction of flow reverses at definite, usually very short, intervals. For obvious reasons, the only type of current useful to the electroplater is direct current, or dc, and certain minor variations of it. Alternating current would have no net effect in either direction and would accordingly be useless for plating purposes.

Electricity as generated by power companies and transmitted over the country, however, is almost always in the form of alternating current, and for economic reasons it is transmitted at high voltages. These voltages may be extremely high in cross-country transmission lines; they are "stepped down" at local power stations. In most parts of the United States, the power actually supplied to residences is 220 V, 60-cycle ac, which is further reduced to 110 or 120 V at the house for lighting and driving the various appliances upon which modern living depends; some appliances work best on the original 220 V. (In England and many parts of Europe ordinary house current is delivered at 220 V.) This current is changing direction every 1/60 s and is known as 60-cycle, or in more modern terms, 60-Hz (1 Hz = 1 cycle/s).

Most electroplating processes require voltages much lower than the 110 or 120 V supplied by the power company, and at amperages much higher than the 15 or

30 A for which the usual house wiring is fused. These processes also require that the alternating current be changed into unidirectional or direct current (dc).*

The mathematical formulation of the behavior of alternating current is rather complicated and involves some fairly complex mathematics. Fortunately, the laws of direct current are much simpler and can be stated fairly straightforwardly.

Time-honored, but still useful instructionally, is the analogy between the flow of water and of dc electricity. In any hydraulic system, the rate of flow of water depends primarily on two factors: the pressure, or head that is produced either by a difference in elevation of the two parts of the system or by means of a pump, and the resistance of the pipes, which in turn depends upon their diameter and length, and to some extent on the smoothness or roughness of the internal surfaces of the pipes. These relationships may be expressed as

$$\text{Rate of flow} = K\,\frac{\text{pressure}}{\text{resistance}}$$

(K is a proportionality constant.) The rate of flow (liters per minute) is directly proportional to the pressure; it is inversely proportional to the resistance of the system. The resistance of a pipe can be increased by increasing its length or reducing its cross section, and to some extent by roughening its inside surface. The quantity of water, in liters (or some other unit), that flows will be equal to the product of the rate of flow and the time of flow.

In a direct-current (dc) circuit, the potential or electromotive force (emf, or voltage), expressed in volts (V), corresponds to the pressure in the hydraulic analogy. The obstruction that a wire or other conductor introduces to the flow of current is known as the *resistance* of that part of the circuit and is measured in ohms (Ω). The smoothness of the inside of the pipe in hydraulics can be compared (very roughly) to the individual characteristics of various conductors: some are "smoother"—oppose the flow less—than others of the same dimensions; copper is a better conductor than lead. The same laws hold: the resistance of any given conductor is directly proportional to its length and inversely proportional to its cross-sectional area.

The rate of flow of electricity is defined as the *current* and is expressed in amperes (A). The relationships are analogous:

$$\text{Current} = K\,\frac{\text{potential}}{\text{resistance}}$$

or

$$I(\text{current}) = E(\text{volts})/R(\text{ohms})$$

This relationship is known as Ohm's law and is the fundamental expression of current flow: in any circuit or part of a circuit the current is directly proportional to the potential and inversely proportional to the resistance. The law may be

*It is a common error to speak of dc current or ac current. This is redundant; dc *means* direct current, so dc current means direct current current.

expressed equally well in three identical forms, depending on which is most convenient:

$$E = IR \qquad I = E/R \qquad R = E/I$$

It is important to distinguish between the current I, the *rate* of flow of electricity, and the total *quantity* of electricity. The unit of current is the ampere, while the unit of quantity of electricity is the coulomb, the quantity of electricity delivered by one ampere flowing for one second.

The *power* delivered by a fall of water, or an electric current, depends on both the head or potential and the amount of water or current. The amount of work a current will do, or the heating effect it will produce by passing through a resistance, depends on both terms. To drop the hydraulic analogy at this point: The power depends on current multiplied by the voltage $E \times I$; this is expressed in watts (W): $W = EI$, or since $E = IR$, $W = I^2R$. The heat generated by a current flowing through a resistance depends on the resistance multiplied by the square of the current. And power, divided by the time during which it is exercised, yields the rate at which energy is delivered: energy, in joules, equals power, in watts, × time, in seconds; or J = W·s.

As stated, electric power as purchased from utility companies is almost always in the form of alternating current (ac), usually 60-Hz, which as the name implies reverses its direction 60 times each second, as diagramed in Fig. 9-1. Such power is delivered to the user at voltages nominally varying between 110 and 440 V or even higher, and may be 1-, 2-, or 3-phase. With the exception of a few rare cases, such power is not useful for electroplating purposes, and must be *rectified,* i.e., transformed into direct, unidirectional, current. Ideal unidirectional current, such as that provided by a battery, has the form shown in Fig. 9-2. Furthermore, the mains voltage is too high for most metal-finishing operations, and its amperage is too low, so that apparatus for transforming ac into dc also amplifies the amperage available.

Alternating current may be changed into direct current in several ways; formerly motor-generator sets were used for this purpose. In a motor-generator set, the mains ac drives an electric motor, which in turn drives a generator which generates dc at lower voltage and higher amperage than the primary ac supply. Such apparatus are rugged, possess long life if not abused, and withstand considerable overloads; they also provide, in general, purer dc than rectifiers. But owing to their high original cost, necessity for maintenance, and criticality in installation, they have been almost entirely superseded by rectifiers—at least in

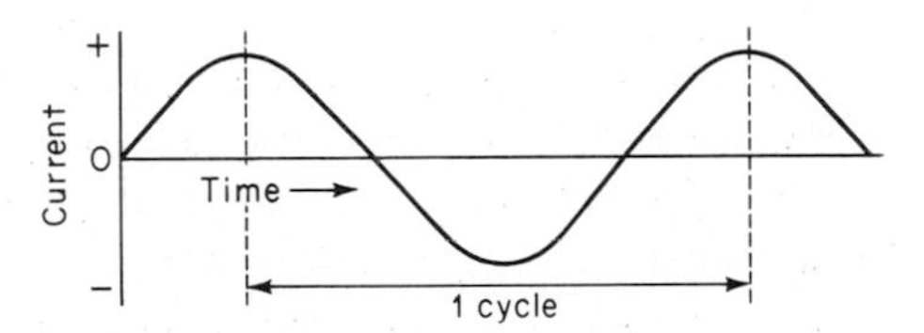

Fig. 9-1 Typical alternating current.

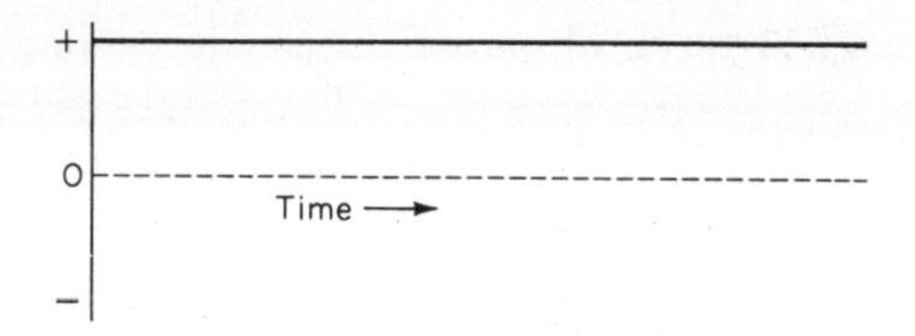

Fig. 9-2 Ideal direct current from a battery.

new installations; many motor-generator sets are still in operation, but few new ones are sold.

Rectifiers have the property of allowing current to flow freely in one direction but offering a very high resistance to current flow in the opposite direction, thus delivering dc at their terminals. Rectifiers also normally include step-down transformers, so that the voltage at their terminals is substantially reduced and the amperage correspondingly increased. Several materials, all classified as semiconductors, have the property that their resistance to current flow is much higher in one direction than in the other. One of the earliest to be commercialized was copper oxide; later copper sulfide, germanium, and selenium were used. At present the silicon rectifier dominates the field, and there are no indications that its position is being challenged; selenium rectifiers also enjoy some remaining popularity.

If only single-phase current is available at the rectifier, as in small bench-top rectifiers meant for use on ordinary 110 or 120-V lines, the current delivered has the waveform shown in Fig. 9-3. Such rectifiers are common for small bench-top applications, and the fluctuating-current supply is generally satisfactory for small-scale and experimental use. The rapid current fluctuations appear to have little or no effect in most electroplating operations, with the probable exception of chromium plating and a few other specialized applications. But for practical larger-scale shop use, the ac fed to the rectifier is 3-phase, and all three phases are rectified, leading to a dc output of much smoother form, as shown in Fig. 9-4. By proper internal wiring, rectifiers can be made to deliver dc almost as constant as that from a battery; the extent to which the dc deviates from the ideal is known as *ripple*. Ripple is normally of little importance in electroplating operations.

OTHER FORMS OF PLATING CURRENT

It is sometimes beneficial to utilize waveforms differing somewhat from straight dc. Much investigation has been carried out on the effect of special waveforms on the character of electrodeposits, usually in the hope of achieving improvements. As a rule, the theory behind these variations resides in an attempt to influence the double layer or diffusion layer next to the cathode, giving it time to re-form, or to influence in some other way the mechanism of electrocrystallization.

Several such waveforms are available to the metal finisher, provided by

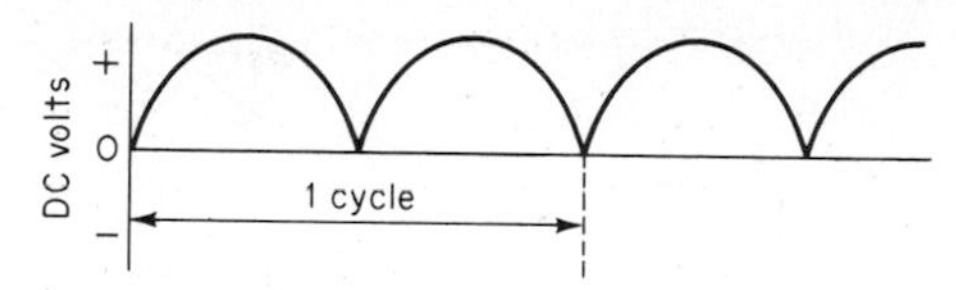

Fig. 9-3 Single-phase full-wave rectification.

electronic and electrical circuitry the details of which are beyond our scope. The presumed benefits of some of these special waveforms are still to a degree controversial; we can say only that they are for special and unusual circumstances, and for most electroplating applications straight dc as it comes from the rectifier is ample.

The simplest of these forms is "superimposed ac." In this mode, an alternating current is superimposed on the direct flow of the current, of amplitude insufficient to cause actual reversal of the current but enough to give it an undulating form diagramed in Fig. 9-5. Of doubtful utility in electroplating, it has been used to advantage in gold refining, apparently aiding in loosening slimes from the anode.

Of more general utility, and gaining in popularity, is the form known as *periodic reverse,* or PR. Here the direct current is made to change direction at preset intervals, which are much longer than the 60 cycles of common ac, so that for part of the cycle the current is flowing the "wrong way" for electrodeposition and is actually "deplating" the work. The "reverse" cycle, of course, must be of shorter duration or of less amperage, or both, than the "direct" cycle, or no net plating would take place (see Fig. 9-6). The more "deplating" takes place as compared with "plating," the more "sacrificial" the cycle is said to be. The loss in overall efficiency caused by these reverse cycles must be compensated by some improvement in the character of the deposit or of some other variable in the system; if this were not so, there would be no point in PR.

Periodic reverse may or may not be beneficial in electroplating processes; it appears to depend on the particular process under consideration. The current-reversal part of the cycle may act somewhat like electropolishing, in smoothing down incipient nodules or high points in the deposit, resulting in a smoother plate. The loss in cathode efficiency may not be as great as would be calculated simply from the times of forward and reverse cycles, since the relative anode and

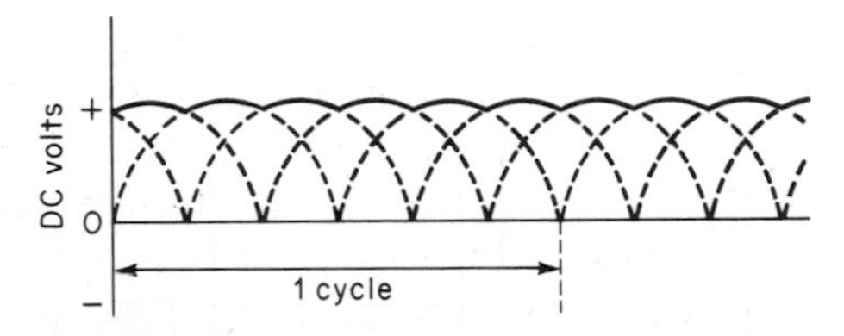

Fig. 9-4 Three-phase full-wave rectification.

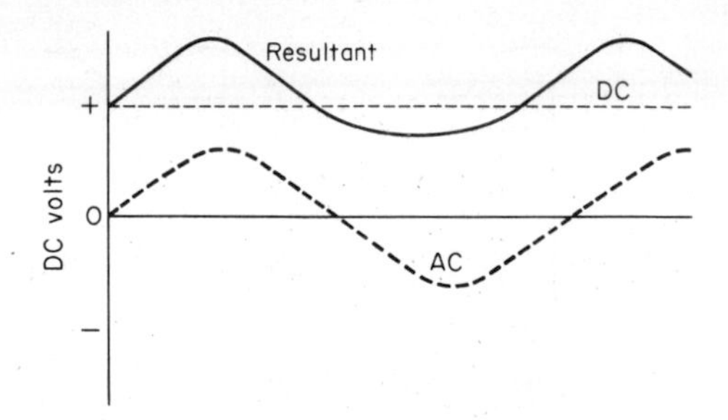

Fig. 9-5 Superimposed alternating current.

cathode efficiencies may not be the same. In any case, PR can be a useful adjunct to the technique of electroplating in some cases; it has proved especially useful in some types of copper and gold plating. The magnitude of the forward and reverse cycles in PR is usually in the order of a few seconds.

In "interrupted" direct current the dc is interrupted for short intervals so that the current passes in the same direction but not continuously. This process also has for its aim the production of smoother or brighter deposits. The interruptions—times when no current flows—are said to give time for the replenishment of the cathode film by means of diffusion and convection; thereby the cathode polarization is decreased, and the cathode efficiency may be increased. In consequence, a higher current density can be used and a given weight of metal can be deposited in a shorter total time, even though the current is not flowing continuously.

Advantages have been claimed for the application of a current form called *pulsed* (Fig. 9-7). This is somewhat similar to interrupted dc, except that the interruptions are of much shorter duration, of the order of milliseconds rather than the seconds involved in PR and interrupted dc. The beneficial effects of pulsed currents and—if they are real—the reasons for them are still under investigation.

All these forms of modified dc plating promise certain advantages. On the other hand, they require special apparatus, sometimes of considerable sophistication, to provide the forms of current required, and in each case it must be decided whether the end justifies the means. The answer, at present, must be "sometimes yes, sometimes no."

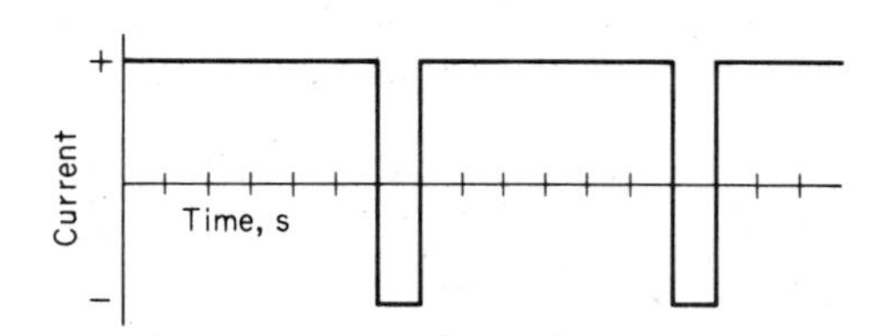

Fig. 9-6 PR plating: typical cycle.

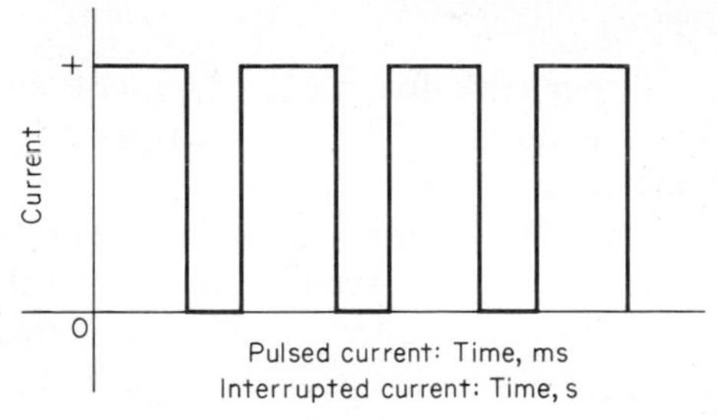

Fig. 9-7 Pulsed and interrupted current.

COSTS

The plater purchases power from the public utility in terms of the total energy supplied by the utility, measured in terms of kilowatthours (kWh); 1 kWh = 3.6 $\times 10^6$ J or 3.6 MJ. Power costs are usually a fairly small component of the total cost of plating: labor, overhead, and supplies are usually far more significant. Nevertheless, the efficient plater will wish to utilize the energy supplies to best advantage. All means of transforming high-voltage ac to low-voltage dc for plating involve some loss of power—that is, none is 100 percent efficient. The relative efficiencies of motor-generators vs. rectifiers, and of one type and make of rectifier vs. another, are to some extent controversial and cannot be discussed here. But it is relevant to point out that whatever means is used, proper maintenance is important. Rectifiers should be properly cooled, preferably should be installed in dust-free locations, and should be checked for efficient operation at regular intervals.

The several variations on ordinary dc that we have mentioned all involve some loss of total energy efficiency. They may be well worth it; it is simply a matter to be considered in each case.

Where individual tanks are supplied by individual rectifiers, the current and voltage can be regulated at the rectifier. When a rectifier serves many tanks (in electrical parallel, see Fig. 9-8, p. 164), rheostats (variable resistances) are usually required to regulate the current to each tank. Such rheostats waste energy in the form of heat. The choice, therefore, between individual rectifiers without rheostat and common rectifiers with rheostat is again an economic trade-off.

In a series circuit (Fig. 9-8) the total resistance is merely the sum of the individual resistances. $R = r_1 + r_2 + \cdots + r_n$. In a parallel circuit, the reciprocal of the total resistance is the sum of the reciprocal resistances of the components:

$$\frac{1}{R} = \frac{1}{r_1} + \frac{1}{r_2} + \cdots + \frac{1}{r_n}$$

When two resistances are in parallel, this reduces to

$$R = \frac{r_1 r_2}{r_1 + r_2}$$

INSTRUMENTATION

Two instruments are indispensable for intelligent operation of any electrolytic process: a voltmeter and an ammeter. The first measures electrical potential, the second the flow of electricity.

Both are constructed on the same principle. When a current is passed through a coil of wire surrounding the core of a fixed permanent magnet, the coil tends to rotate. If this rotation is opposed by a force such as a spring or the torsion in a wire, or by gravity, the angle of rotation is determined by the number of turns in the wire and the current passing through it. With a given coil of definite resistance, the rotation of the coil, as indicated by a needle attached to it, may serve as a measure of the current or of the voltage, since the magnitude of the former is determined by that of the latter. There are two principal types of instruments in use. In the D'Arsonval type, the magnet is fixed in position and the coil is free to move on an axis, while in the soft-iron type the coil is fixed in position and the piece of soft iron is moved by the passage of the current.

Ammeters and voltmeters are available in all degrees of accuracy and ruggedness, and although the highest degree of accuracy is usually not necessary for shop work, too much sacrifice of accuracy for the sake of low first cost is a mistake.

The essential difference between an ammeter and a voltmeter is that the ammeter has a very low resistance and the voltmeter a very high resistance. The latter high resistance is necessary so that the action of the meter itself does not draw any significant current away from the main circuit. The low resistance of the ammeter is necessary so that it does not add significant resistance to the circuit it is supposed to be measuring.

Ammeters are essentially voltmeters that measure the voltage drop over a very low constant resistance, using the familiar Ohm's law, $I = E/R$, where R is very small. Ammeters designed to measure fairly low currents may be self-contained; i.e., they have an internal *shunt*. A shunt is a low and fixed resistance, the voltage drop over which is registered by the meter. When larger amperages are concerned, many ammeters have external shunts, often several interchangeable

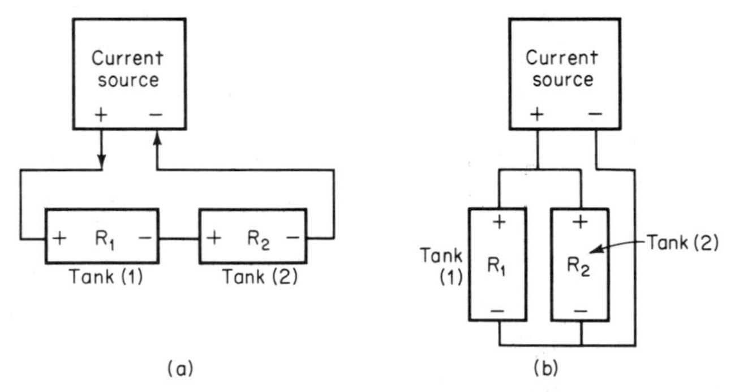

Fig. 9-8 (*a*) Series arrangement. (*b*) Parallel arrangement.

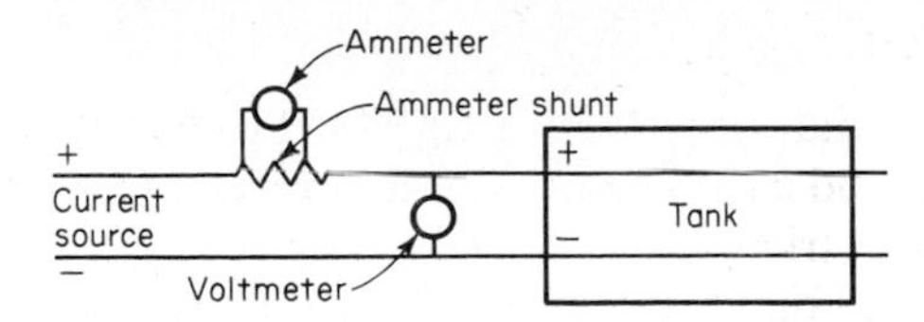

Fig. 9-9 Arrangement of ammeter and voltmeter. Ammeter may be in positive or negative line.

ones for different ranges. The principle is the same: the shunt is placed in the circuit, and the meter registers the voltage drop over it, reading directly in amperes.

Voltmeters are always placed across the circuit; ammeters or their shunts are placed in the circuit (see Fig. 9-9).

For extremely accurate work, such as in laboratory research on electrode potentials, voltmeters draw too much current (even though it is very small) and disturb the very condition being measured. Some type of potentiometer or vacuum-tube voltmeter (VTVM) must be used.

The total quantity of electricity passed during an operation should usually be known, so that the operator can estimate current efficiencies and power costs or use the figure to regulate the time the work remains in the tank. Various types of ampere-hour meters are commercially available for this purpose, and they are usually a worthwhile investment. If the amperage remains fairly constant, merely observing the reading on the ammeter and timing the operation with a clock may suffice, but this method is not especially accurate. For laboratory investigations of extreme accuracy, a coulometer should be employed; see p. 13.

Ohmmeters are available and sometimes useful. So-called clip-on ammeters are marketed; they enable one to read the current passing in a conductor without disturbing the circuit; they are convenient for determining, for example, whether a particular anode, or rack of work, is receiving its proper share of current. Their accuracy is not outstanding, but they have their special uses.

Section C

THE PLATING METALS

10

Introduction to the Plating Metals

We are ready now to consider the processes for plating the individual metals. Somewhat arbitrarily, we may group these metals into six categories, as follows:

1. Sacrificial coatings, used primarily for protection of the basis metal, usually iron and steel (sometimes called *anodic coatings,* meaning that electrochemically they are anodic to the substrate). Sacrificial denotes that the coatings "sacrifice" themselves in the act of protecting the basis metal.
2. Decorative protective coatings, used primarily for adding attractive appearance to some protective qualities.
3. Engineering coatings—a rather miscellaneous group whose members are used for specific properties imparted to the surface, such as solderability, wear resistance, reflectivity, conductivity, and many others. They are sometimes called *functional* coatings, though it would seem that protection is also a "function."
4. Minor metals—a small group of metals that are easily plated but have rather limited application.
5. "Unusual" metals—rarely electroplated, and when they are, they require special conditions, such as nonaqueous solutions.
6. Alloys—an almost unlimited number of alloys has been plated experimentally, since the possible combinations of the plateable metals, in various proportions, are innumerable. Only a few have attained commercial importance, and only these will be discussed.

The distinctions enumerated above are not by any means iron-clad nor mutually exclusive; there are many overlaps. As will be seen, the copper/nickel/

chromium group is here considered as a decorative/protective system, but all three metals have applications not involving this function. Some of the engineering metals are used for decorative applications, e.g., gold and silver. The classification, then, is really for convenience and has no fundamental significance.

Each metal will be described in general terms as to its provenance, uses, and general chemical and physical properties. Finishers can do a more intelligent job if they know something about the properties and economics of the metal they are depositing, even though such knowledge is not actually required to plate that metal. Therefore a short introduction to these matters precedes the discussion of plating methods.

The principal commercial methods of plating the metal will then be enumerated and discussed in limited detail. Only the major commercial methods will be considered; for many metals, several experimental or semicommercial methods are available. They may perhaps have advantages, but for one reason or another have not enjoyed wide market acceptance.

Two points must be understood in studying this section. (1) For almost all the metals considered, "proprietary" processes are available. The word *proprietary* connotes that a supply house offers a trade-named or trademarked process or supplies, the nature of which may or may not be known or disclosed (except perhaps in patents). This is almost universally true of additives such as brighteners. Metal finishers thus have two choices: they may make up solutions from published formulas such as are offered in this book and many others, or they may put themselves in the hands of a plating supply company which will supply all the necessary ingredients and accompanying technical literature and (presumably) will be ready to offer expert service and advice if they get into trouble. Rather obviously, metal finishers will pay for this service, in the price of "proprietary" materials or processes. Also obviously, such processes cannot be considered in this book. No inference, favorable or not, must be drawn from this omission.

Most platers, in practice, choose a compromise, purchasing well-known ingredients under their true chemical names, plus some proprietary additives to yield brightness, smoothness, or some other benefit not available from published formulas.

(2) The discussions that follow are not "cookbook" directions containing all the details necessary for successful operation of all the processes included. Emphasis is on the *principles* underlying the plating of the various coatings from the various solutions, with just enough detail to illustrate these principles. Details of what to do, when, and how are best learned by practical experience in either a laboratory or a shop, plus reliance on the type of literature which concentrates on this type of information.*

*The "Illustrated Lecture" series, available from the American Electroplaters' Society, is an excellent source of detailed directions for individual plating and finishing processes.

11

Sacrificial Coatings: Zinc and Cadmium

Zinc and cadmium coatings are used when the principal aim is protection of the substrate, usually iron or steel. Although both metals can be plated bright, or can be brightened after plating by conversion coatings or bright dipping, such bright appearance usually does not last very long in service.

Cadmium is far more expensive than zinc, and has been in short supply from time to time.* The price ratio Cd:Zn, in cost per unit weight, has varied between 10:1 and 20:1. Cadmium, therefore, is used only for those applications in which it is obviously superior to zinc for the purpose intended: it is relatively easily soldered compared to zinc; it is somewhat superior to zinc in resistance to salt atmospheres; and its corrosion products are not bulky, hence do not interfere with functional moving parts as do those of zinc. In addition, cadmium plating is somewhat less critical in control than zinc plating.

Zinc is the cheapest metal that can be used for protecting steel from corrosion. It is usually applied by hot-dipping (called galvanizing); galvanized steel is produced in very large tonnages. Steel strip may also be electroplated with zinc (electrogalvanizing), but to date the electrolytic process has been comparatively unimportant in relation to hot dipping for continuous steel strip. Nevertheless most experts consider that electrogalvanizing possesses good possibilities for future expansion.

Both cadmium and zinc protect steel galvanically or ''sacrificially''; that is, in the corrosion couple set up between the substrate steel and the coating metal, the latter is normally anodic to the substrate and corrodes preferentially. The substrate steel, being the cathode in the couple, is protected as long as there is any coating left in the immediate vicinity. The meaning of *immediate* is somewhat indefinite; under usual conditions the protection afforded by a zinc or cadmium coating may extend to a distance of 1 mm or more. For this reason the presence of pores or discontinuities in the coating is not especially important.

*Because of its extreme toxicity, cadmium plating is now actually forbidden by law in some countries.

For this reason also, the length of time during which corrosion protection persists is a function almost entirely of the thickness of the coating, and is affected little if at all by the method of application: zinc may be electroplated, hot-dipped, sprayed, or "Peen"-plated with essentially similar results as long as the thickness of the coating is the same. This is not to say that these methods of application are equivalent in all respects, only that so far as the sacrificial protection provided by the zinc coating is concerned, the thickness of zinc is the controlling factor.*

ZINC

Zinc, like cadmium, is a member of Group IIB of the periodic system; the third member is mercury. It is relatively rare, being twenty-fifth in order of abundance in the lithosphere; it is thus less abundant than such less familiar elements as zirconium, vanadium, and strontium, but somewhat more so than copper.

Zinc was known to the ancients in the form of the copper-zinc alloy brass; its recognition as a distinct metal dates from about A.D. 1000. The origin of the name is obscure, but may be related to German Zinke, a spike or tooth, derived from the manner in which it crystallizes from the melt.

Zinc is usually found in nature as the sulfide; its most important ore is zinc sulfide, ZnS, known as sphalerite or zinc blende (or simply blende) depending on its crystal form. Oxidized ores such as zinc carbonate and franklinite (a mixed zinc-iron-manganese oxide) are also of some importance.

Zinc is an essential trace constituent of plant and animal life, and zinc deficiency is the cause of various plant and animal diseases. It is an essential constituent of several enzymes necessary for metabolism.

Zinc ores are mined all over the world. Total world production (zinc content of ores) is about 5.5 million t (metric tons or tonnes). Some countries smelt their own ores, while others export the ore to smelters abroad.

After beneficiation, ores are roasted to the oxide, producing in the process large amounts of sulfur dioxide which poses a severe pollution problem, often solved by producing by-product sulfur or sulfuric acid. The resulting oxide is then either reduced by carbon (actually carbon monoxide) to metallic zinc or dissolved in sulfuric acid to form a solution of zinc sulfate, which is electrolyzed with inert anodes of lead or lead–1% silver alloy to produce zinc at the cathode and regenerate sulfuric acid which is used to dissolve more roasted ore.

Uses

The largest use of zinc is in the production of zinc-alloy die castings; this application accounts for about 41 percent of zinc consumption in the United

*There is some controversy, which cannot be entered into here, about whether this is true of the more recent proprietary acid baths for zinc plating.

States. Zinc die castings are important to electroplaters because a large proportion of zinc die castings are plated, and special preparatory methods are required: see Chap. 5. In second place is the use of zinc for galvanizing steel; this application takes about 38 percent of zinc production. By far the largest amount of galvanizing is accomplished by hot dipping, but a small proportion is electrogalvanized, and the importance of the electrolytic method is expected to increase. Brass (a copper-zinc alloy) uses about 12 percent of the zinc produced. Thus these three applications among them account for more than 90 percent of all uses of zinc. The remainder is accounted for by zinc oxide, light-metal alloys, zinc dust, batteries, chemicals, and other miscellaneous uses.

Properties

Selected properties of zinc are listed in Table 11-1. Since zinc is a soft and fairly ductile metal, seldom used for its physical properties in electroplating, the mechanical properties of electroplated zinc have not been extensively investigated and are not of primary importance to the electroplater.

Zinc is bluish white when freshly fractured. It is about as hard as copper when pure; small amounts of impurities raise the hardness. Commercial zinc is too brittle to be cold-rolled at room temperature, but at 100–150°C it can be rolled into sheet or drawn into wire. High-purity zinc is ductile at room temperature.

At room temperature, dry air does not attack zinc significantly, but at 225°C

Table 11-1 Properties of Zinc

Atomic number	30
Atomic weight	65.37
Electrochemical equivalent, mg/C	0.339
Electrochemical equivalent, g/A·h	1.22
Electronic configuration	$1s^2 2s^2 2p^6 3s^2 3p^6 3d^{10} 4s^2$
Melting point, °C	419.5
Boiling point, °C	907.0
Density, g/cm³, 25°C	7.133
Crystal structure	hcp
Electrical resistivity, μΩ-cm, 20°C	5.92
Hardness, Knoop, 400-g load, rolled	55
Hardness, Knoop, 400-g load, annealed	47
Tensile strength, rolled sheet, MPa	4.43
Tensile strength, annealed sheet, MPa	1.86
Tensile strength, electrodeposited, MPa	1–2.3
Standard potential, E^0, 25°C, V $Zn^{++} + 2e^- \rightarrow Zn$	−0.7628

and above the rate of attack increases rapidly. Moisture in the air causes attack at room temperature; presence of carbon dioxide or sulfur dioxide accelerates attack. When heated in air, zinc burns with a luminous blue-green flame, forming zinc oxide. In moist air, zinc dust can inflame spontaneously; mixtures of zinc dust with oxidizing agents are fire hazards. At red heat, steam and carbon dioxide react with zinc to form zinc oxide.

Normal atmospheric corrosion produces a hydrated basic carbonate in which the $ZnO/CO_2/H_2O$ ratio can vary. The corrosion rate is affected by the purity of the zinc: the purest metal corrodes much more slowly than commercial grades.

Zinc reacts with the halogens, particularly in the presence of moisture. Gaseous hydrohalides attack zinc rapidly, as do their aqueous solutions such as hydrochloric acid. Zinc reacts with most mineral acids; the purer the metal, the slower the reaction.

Hydrogen sulfide attacks zinc at ordinary temperatures; a coating of zinc sulfide is formed which inhibits further attack.

Zinc is an active reducing agent, as would be expected from its electrode potential; this accounts for the use of zinc dust in purifying electroplating baths of extraneous metals such as iron, as well as for its use as one of the electrodes in the LeClanché (dry) cell and similar devices. It is also used to remove and recover cadmium from impure zinc sulfate solutions:

$$Zn + CdSO_4 \rightarrow Cd + ZnSO_4$$

For all practical purposes, zinc is divalent in its compounds. Such species as Zn^+ and Zn_2^{++} (analogous to the mercurous ion) have been observed but are of little or no importance. Therefore the ion Zn^{++} (hydrated) and its complexes are the only species requiring attention for our purposes. This ion is colorless.

Formally zinc is a member of Group IIB, along with cadmium and mercury; but its chemistry in many ways more closely resembles that of magnesium than of the other members of its group. It forms many complexes: most important, with ammonia, amines, and halide, cyanide, and hydroxyl ions. The last-named complex accounts for its amphoteric properties, in which it differs from magnesium; it forms both zinc salts and zincates.

Solubilities of zinc salts are generally similar to those of magnesium. Water-soluble salts include the chloride, bromide, chlorate, sulfate, nitrate, formate, acetate, thiocyanate, perchlorate, fluosilicate, and fluoborate; and the alkali metal zincates, cyano-zinc and zinc-ammonia complexes. Insoluble salts include the oxide, hydroxide, sulfide, carbonate, fluoride, iodate, periodate, oxalate, and several salts of organic acids.

Zinc forms many covalent and coordination complexes, in which the usual coordination number is four. Oxygen, sulfur, and nitrogen are the common donor atoms.

In electroplating, the salts of principal importance are the zinc cyanide complex $Zn(CN)_4^{--}$, the hydroxo-complex or zincate ion $Zn(OH)_4^{--}$, the chloro-complexes, and simple salts such as the chloride, sulfate, and fluoborate.

Hazards

Zinc and its compounds are relatively nontoxic unless the associated ions are toxic, such as the cyanide. Lower grades of zinc metal may contain toxic quantities of lead and cadmium. Zinc compounds are permitted in contact with foods, and some are used in cosmetics and ointments. Nevertheless, hazards do exist and government regulations limit the amount of zinc permitted in plant effluents. Acidic beverages should not be stored in zinc or zinc-coated containers.

Grades

Zinc appears on the market in five recognized grades, from Special High Grade, containing at least 99.99 percent Zn, to Prime Western, having a minimum specification of 98.0 percent. Anodes for use in plating should normally be Special High Grade. Anodes alloyed with magnesium or aluminum are sometimes used to decrease chemical attack during idle periods.

ZINC PLATING

From its electronegative potential of −0.76 V, it might be expected that zinc could not be electrodeposited from aqueous solution, but in fact the high overpotential of hydrogen on most substrates does permit such deposition. On some substrates, such as graphite, which exhibit low hydrogen overvoltage it is difficult or impossible to deposit zinc from water solutions. This explains the difficulty in plating zinc on materials such as cast iron which contain discrete particles of graphite in their surfaces.

Zinc-plating baths are of three main types: alkaline cyanide, alkaline noncyanide, and acid. Each of these may be further subdivided: the cyanide into conventional and so-called low-cyanide, and the acid into sulfate, chloride, and fluoborate.

The conventional cyanide bath has been preferred for general zinc plating; characteristics that recommend it are good throwing power, ease of control, relatively wide range of satisfactory operating conditions, and many years of experience in its use. With the advent of stricter limits on effluents, however, its high cyanide content presents a problem; although means for destroying cyanide are well known, they are expensive. Several answers to the problem are proposed: use of a more dilute solution, which although it does not solve the problem, mitigates it; alkaline noncyanide baths; and the several acid baths, including proprietary "neutral chloride" solutions.

Zinc Cyanide Baths

Zinc cyanide baths contain the cyano-zincate complex ion, the zincate ion, free cyanide and hydroxyl ions, and, to balance these electrically, the necessary sodium ions. Since zinc is complexed by both cyanide and hydroxyl, the equilibria are complicated. The following equations represent the more important ones:

$$ZnO + 4\,CN^- + H_2O \rightleftharpoons Zn(CN)_4^{--} + 2OH^-$$
$$Zn(CN)_2 + 2CN^- \rightleftharpoons Zn(CN)_4^{--}$$
$$Zn(CN)_4^{--} \rightleftharpoons Zn(CN)_2 + 2CN^- \rightleftharpoons Zn^{++} + 4CN^-$$
$$Zn(OH)_4^{--} \rightleftharpoons Zn^{++} + 4OH^-$$
$$4OH^- + Zn(CN)_2 \rightleftharpoons Zn(OH)_4^{--} + 2CN^-$$
$$2OH^- + ZnO + H_2O \rightleftharpoons Zn(OH)_4^{--}$$

The positive ion is almost universally sodium, Na^+. Although potassium ion has been shown to be superior to sodium in many alkaline cyanide baths and hydroxo-complex baths (see, e.g., copper, silver, gold, and tin plating), this superiority has not extended to zinc cyanide plating, probably because of unfavorable solubility relationships.

The composition of the zinc cyanide solution is shown in Table 11-2. More critical than absolute concentrations is the ratio between the complexing agents—hydroxide and cyanide—and the zinc concentration. Several systems have been suggested for controlling this factor, among them:

1. $\dfrac{(NaCN) + (NaOH)}{(Zn(CN)_2)}$ where ( ) = normality

2. NaCN/Zn or NaOH/Zn, concentrations in grams per liter

Since analytical procedures yield grams per liter directly, it is most common to use the second criterion for control purposes. The NaCN/Zn ratio is maintained in the range 2.5 to 3.1, with an "optimum" at about 2.7, for decorative plating, and somewhat lower, 2.0 to 2.5, for applications where the main objective is protection and appearance is of less importance, such as steel strip and conduit.

Table 11-2 Cyanide Zinc Baths

	g/L		*Molarity*	
	Decorative	*Protective*	*Decorative*	*Protective*
Zinc	20–45	45–60	0.3–0.7	0.7–0.9
Total NaCN	50–140	90–150	0.7–2.0	1.3–2.2
Total NaOH	60–120	90–140	1.5–3	2.3–3.5
Na_2CO_3	20–120	30–75	0.2–1.1	0.3–0.7
Ratio NaCN/Zn	2.5–3.1	2.0–2.5	2.4–2.9	1.8–2.4
Cathode current density, A/m²		100–900		
Anode current density, A/m²		30–450		
Temperature, °C		20–50		

Because of the complicated series of interrelationships indicated by the equilibria already mentioned, it is difficult if not meaningless to determine "free" cyanide and "free" alkali or hydroxyl ion. Analytical procedures, however, easily yield total cyanide and total hydroxide, and it is these numbers that are used to determine the "ratio" used for control of the solution.

The zinc complexes provide a reservoir of metal to be plated. As long as the other bath constituents are varied proportionately, the zinc content can be varied within a wide range without detriment to the character of the deposit; however, the lower the zinc content, the more critical is the control of the other variables. The zinc content, therefore, must be set as a compromise between favorable operating characteristics and the increased chemical and waste-disposal costs entailed in higher bath concentrations.

Anode efficiency in cyanide baths is usually somewhat higher than cathode efficiency; in addition, during idle periods zinc tends to dissolve in the bath by chemical action. Therefore there is usually little occasion to add zinc salts for maintenance. As the equations on p. 176 show, it makes little difference whether zinc oxide or zinc cyanide is used for bath makeup and maintenance.

Cyanide ion complexes some of the zinc and, in ways that are not entirely understood, enables deposition of satisfactory zinc plate from alkaline baths. Although zinc can be deposited from zincate solutions without cyanide (see below), addition agents for brightening and otherwise improving the deposit appear to be more effective in the presence of cyanide. This is true even though the zinc deposited on the cathode arises mainly from the zincate (hydroxo-) complex. Much research aimed at developing cyanide-free alkaline baths has ended up in a compromise: addition of small amounts of cyanide markedly improved the operating characteristics of the baths.

The best cyanide concentration depends on the operating conditions and the concentrations of the other constituents. This is shown by the emphasis placed on the cyanide-to-zinc ratio, expressed as grams per liter total cyanide to grams per liter zinc, as a means of control. This varies somewhat with temperature, which can be in the range 20 to 40°C; the preferred ratio varies from about 2.7 at the lower temperature to about 3.0 at the higher.

When the NaOH/Zn ratio is too low, the cathode efficiency is improved but the throwing power suffers. With increase in this ratio, the cathode efficiency decreases; the decrease is greater at high current densities, resulting in improved throwing power. When the ratio is at its best level, recessed or low-current-density areas are well covered, the average cathode efficiency at moderate current-density levels is not much affected, and high-current-density areas where the loss of efficiency is most pronounced receive little excess plate. When the ratio is too high, total plating speed suffers and gas evolution is excessive.

Cyanide is consumed in several ways: by drag-out, by oxidation, and by reaction with the carbon dioxide in the air. Thus cyanide additions are required from time to time as indicated by analysis.

Hydroxide has several functions in the cyanide zinc bath: it is the main current-carrying species, it aids in anode corrosion, and it is important in regulating throwing power and efficiency, as already discussed. As stated, most of the zinc in the solution is believed to exist in the form of the hydroxo-complex (zincate).

Usually carbonate is not deliberately added to cyanide zinc baths, but all operating alkaline baths contain carbonate as a result of both carbon dioxide absorption from the air:

$$CO_2 + OH^- \rightarrow HCO_3^- \qquad HCO_3^- + OH^- \rightarrow CO_3^{--} + H_2O$$

and oxidation of cyanide:

$$2CN^- + O_2 + 4H_2O \rightarrow 2NH_4^+ + 2CO_3^{--}$$

Carbonate in the amounts usually encountered is not harmful and may even be somewhat beneficial; it is usually allowed to build up to such concentrations that drag-out balances formation, which is usually in the range of 50 to 100 g/L. If necessary, it can be frozen out at 5–10°C. Less frequently it is precipitated by addition of lime (calcium hydroxide).

Most cyanide zinc–plating baths are operated with proprietary additives for brightening the deposit; usually they are organic compounds, commonly aldehyde-bisulfite addition compounds.

Because zinc is so electronegative and thus barely plateable at all from water solution, it is not surprising that zinc-plating baths are more susceptible than most to the presence of metallic impurities. To prevent the buildup of metallic impurities in the bath, it is customary to maintain a small excess of sulfide or polysulfide ions; thus lead, cadmium, and other metals that form insoluble sulfides are precipitated continuously. Although zinc sulfide is also insoluble in water, it is soluble in excess cyanide and hydroxyl ions. The precipitated sulfides ordinarily cause no trouble and are allowed to settle with the sludge; where brightness of the deposit is of special importance, periodic or continuous filtration may be required.

The operating temperature range of zinc cyanide baths is fairly wide: 20–50°C, or even in special cases up to 65°C. Nevertheless, for each temperature of operation there is a bath concentration—zinc, cyanide, and hydroxide content—most suitable for that temperature, so that once the proper temperature has been set for a particular set of conditions, it should be maintained fairly closely. Temperatures near the low end of the range have the advantages of (usually) better stability of the addition agents used and lower rates of cyanide decomposition. Some newer addition agents are stable at higher temperatures, and increases in plating speed may be available if the temperature is raised. Where plating speed is of more importance than appearance, as in strip and conduit plating, higher temperatures are used. They also allow the use of more dilute baths with no loss in plating speed; this is of importance for waste-treatment economy.

Zinc anodes dissolve in the bath as a result of the applied current, and even when current is not flowing, they dissolve as a result of chemical action (which is also, as discussed in Chap. 2, electrochemical in nature). This yields anode efficiencies somewhat above 100 percent. To compensate for this, it is common to replace some zinc-anode area with inert steel anodes. This expedient, however, is ineffective unless the anode current density is high enough, since at low cd all the current will be carried by the zinc. The best anode current density depends on bath formulation and temperature: the higher the cyanide and hydroxide and the higher the temperature, the higher the anode cd that can be applied.

Agitation is not normally required, although it does permit some increase in current densities and can prevent thermal stratification of the bath.

Other Alkaline Baths

In an effort to reduce the costs of waste-control measures while retaining the advantages of the alkaline baths, several expedients have been proposed. A more dilute version of the standard cyanide bath has the composition shown in Table 11-3. Except that control is somewhat more critical, these "dilute" baths have operating characteristics similar to those of the conventional bath.

The use of such "dilute" baths, while it decreases heavy-metal and cyanide content of wastewaters, does not eliminate it, and as more emphasis was placed upon effluent controls in the late 1960s and through the 1970s, further steps were taken to develop a truly noncyanide process. Workers developed noncyanide alkaline (zincate) baths only to find that for best results a small amount of cyanide had to be added. Thus resulted the low-cyanide zinc, or "LCZ" baths, a typical formulation of which is shown in Table 11-4. These baths have been called *negative free cyanide* solutions, but this nomenclature is not recommended. The cyanide in these solutions is not sufficient to complex all the zinc; it is said to act as an additive rather than as an essential ligand. The composition of LCZ baths overlaps that of the "dilute" bath shown in Table 11-3.

With the development of improved brightener systems, it finally became possible to formulate alkaline zinc baths containing no cyanide (no-cyanide zinc, or NCZ) which still yielded satisfactory deposits. Typical concentration ranges are shown in Table 11-4.

The operating characteristics of these LCZ and NCZ baths are essentially similar to those of the conventional cyanide bath, with the added limitations that

Table 11-3 "Dilute" Cyanide Zinc Bath

	g/L	*Molarity*
Zinc	7–20	0.1–0.3
Total NaCN	7–50	0.1–0.7
Total NaOH	75–100	1.9–2.8
Na_2CO_3	20–100	0.2–1
Ratio NaCN/Zn	1.0–2.5	1–2.3

Table 11-4 LCZ and NCZ Alkaline Zinc Baths

	LCZ		NCZ	
	g/L	*Molarity*	*g/L*	*Molarity*
Zinc	5–15	0.08–0.2	5–15	0.08–0.2
Total NaOH	70–100	1.8–2.5	70–100	1.8–2.5
Total NaCN	5–15	0.07–0.2	0	0
Brighteners	qs*		qs*	

*A sufficient quantity.

satisfactory deposits depend to a great extent on the additives used, and that they are more critical in control.

Acid Zinc Plating

Zinc may be plated from several types of acid solutions, based primarily on either zinc sulfate or zinc chloride. Until the late 1960s most uses of acid zinc baths were confined to high-production and purely protective applications exemplified by conduit and continuous steel wire and strip. In these applications the high plating speed and low operating costs are advantageous, and the poor throwing power and lack of fully bright deposits are not serious drawbacks.

The search for cyanide-free zinc baths stimulated by the concern for waste-disposal improvements resulted in the mid-1960s in a so-called neutral chloride bath, which, with the appropriate additives, is capable of producing bright deposits and is competitive with the alkaline baths for general plating.

Typical acid zinc baths are formulated as shown in Table 11-5. The most widely used bath is based on zinc sulfate. Other baths are formulated with chloride and fluoborate; perchlorate and sulfamate baths have been recommended but are little used. Although not acid, the pyrophosphate zinc plating process may be mentioned for completeness.

Zinc sulfate or chloride may be purchased as such, or zinc metal may be dissolved in sulfuric or hydrochloric acid to make up the bath. Chlorides and sulfates such as those of sodium, ammonium, and aluminum may be added to increase conductivity; free acid also serves this purpose. Addition agents are required for the production of smooth deposits; most common are dextrin, licorice, glucose, and gelatin; β-naphthol, goulac, and many others also have been recommended.

The formulas given in Table 11-5 are approximate, and rather wide variations can be made without major effects on the operating characteristics. Use of periodic reverse (PR) current has been claimed to produce superior deposits. One advantage of these acid baths is their ability to plate directly on cast iron and carbonitrided steel parts without a prior strike.

Cathode current densities are in the range of 50 to 300 A/m². With agitation the upper limit may be raised to 1000 A/m². Anode current density is not critical. Temperature is 25–30°C. The pH range should be between 3.5 and 4.5; it should be controlled within 0.3 unit for rack-plating. For barrel plating it may be raised to about 5, but this may precipitate zinc salts unless chelating agents are used. Anode and cathode efficiencies approach 100 percent; in addition, anodes are attacked chemically. The throwing power is relatively poor, as previously stated. The baths are fairly sensitive to metallic impurities, but for the most part these tend to plate out by immersion on the zinc anodes during idle periods. The bath is best purified by use of zinc dust, but note the fire hazard in handling this material.

Since anodes dissolve at greater than 100 percent efficiency, there is seldom occasion to add zinc salts; acid must be added routinely to compensate for that consumed in the reaction with the anodes. Special anodes ("ZAM") containing aluminum and mercury as alloying constituents have less tendency to dissolve chemically than does pure zinc.

The newer neutral chloride baths, mentioned as an alternative to the conventional cyanide bath for general plating, rely on ammonium ion for some complexing of the zinc, and proprietary chelating agents are added for stronger complexing action. The formulations of these baths are as shown in Table 11-6. A typical composition may be taken as being in the middle of the range shown. Like the LCZ and NCZ baths already considered, these baths depend for their satisfactory operation in general plating on the development of appropriate brighteners. Since the neutral chloride baths are almost entirely proprietary in their formulation, no discussion of their advantages and disadvantages is possible here. Unlike the cyanide baths, they do appear to plate satisfactorily on cast iron and carbonitrided steel. Although they do not contain cyanide, there is some question

Table 11-5 Acid Zinc Baths

	Concentrations, g/L					
	I	*II*	*III*	*IV*	*V*	*VI*
$ZnSO_4 \cdot 7H_2O$	240	360	410	240	160	480
NH_4Cl	15	30				
Na_2SO_4			75		90	90
$NaC_2H_3O_2 \cdot 3H_2O$		15		15		
$NaCl$					30	30
$Al_2(SO_4)_3 \cdot 18H_2O$	30			30		
$AlCl_3 \cdot 6H_2O$			20			
H_3BO_3					20	200
Licorice	1			1		
Glucose		120				

Table 11-6 Neutral Chloride Zinc Bath

		g/L
Zinc		25–50
Chloride as Cl^-		100–165
Chelating agent*		45–90
pH	6.8–7.5	

*Complexing agent for zinc: proprietary.

whether in substituting a strong chelating agent for cyanide very much is gained, since it is necessary to remove zinc from effluent as well, and the chelating agents render this more difficult. The question is controversial, and much has been said on both sides.

Thickness and Specifications

ASTM specifications for zinc coatings on steel call for three minimum thicknesses, depending on the service life required. These coatings are classified as types GS, LS, and RS, the designations being quite arbitrary and no guidance being offered concerning appropriate environments of use. They call for 25, 13, and 3.8 μm of zinc, respectively.*

Zinc plating of high-strength steels may cause serious hydrogen embrittlement of the substrate, which must be removed by baking in critical applications.

CADMIUM

Cadmium lies below zinc in Group IIB of the periodic system. It is quite rare, its occurrence in the lithosphere being about 0.1 to 0.5 ppm, or about the same as silver and mercury. It was discovered in 1817 as a minor constituent of a zinc ore; its name reflects its provenance, since *cadmia* is an old name for calamine, a zinc carbonate. The smelting of zinc ores is the principal source of the metal; it is never mined for itself, but is always a by-product of the smelting of zinc-containing ores. For this reason the supply of cadmium is somewhat insensitive to demand, but depends on the supply of and demand for zinc. This is one reason why the price of cadmium is subject to wide swings and why it is occasionally in short supply.

Cadmium is recovered during the smelting of zinc ores by leaching the roasted

*At this writing the ASTM specification is in the process of revision; in particular, suggested service conditions for the various thicknesses will be included. As with all ASTM documents, the user should be sure to consult the latest revision.

ore with sulfuric acid and adding zinc dust to the resulting solution. The precipitated cadmium is redissolved and the solution electrolyzed using lead anodes and aluminum cathodes, to deposit relatively pure cadmium metal. It may also be recovered from complex ores by fuming it off from a mixture of the smelting residues with carbonaceous fuel and limestone, to recover a cadmium dust which is refined electrolytically.

World production of cadmium is about 16 600 t (metric tons or tonnes) annually; the United States produces about 3800 t/yr, but consumes about 50 percent more than it produces, or about 5700 t/yr. About half of this is used in electroplating; other important uses include nickel-cadmium rechargeable batteries and low-melting alloys. Because it is a good absorber of thermal (slow) neutrons, it is used in control rods in nuclear reactors. Its compounds are used in pigments and stabilizers for poly(vinyl chloride) resins and in photovoltaic cells. If solar cells become practical, the demand for cadmium compounds could escalate markedly.

Cadmium is available in a variety of shapes, including oval and ball anodes, up to 99.99+ percent pure.

Selected properties of cadmium are summarized in Table 11-7. As with zinc, its physical properties such as ductility and hardness are not of prime interest to the electroplater.

Like zinc, cadmium is divalent in all its compounds; the evidence for the existence of a few monovalent compounds is not convincing. It has more tendency than zinc to form coordinate linkages, and many of its compounds tend

Table 11-7 Properties of Cadmium

Atomic number	48
Atomic weight	112.40
Electrochemical equivalent, mg/C	0.5824
Electrochemical equivalent, g/A·h	2.097
Electronic configuration	$1s^2 2s^2 2p^6 3s^2 3p^6 3d^{10} 4s^2 4p^6 4d^{10} 5s^2$
Melting point, °C	321
Boiling point, °C	767
Density, g/cm³, 25°C	8.65
Crystal structure	hcp
Electrical resistivity, μΩ-cm, 0°C	6.83
Hardness, MPa*	300–500
Stress, MPa†	−3.51 to 21
Tensile strength, MPa*	70
Standard electrode potential, E^0, 25°C, V $Cd^{++} + 2e^- \rightarrow Cd$	−0.4029

*Electrodeposited.

†Electrodeposited; minus sign means compressive.

to be covalent rather than ionic. Unlike zinc, it is not amphoteric, i.e., does not form anionic hydroxo-complexes; but it does complex with many ligands. In electroplating, the complex of greatest importance is the cyanide, $Cd(CN)_4^{--}$. This is formed when cadmium oxide is dissolved in a sodium cyanide solution, as well as upon the addition of cyanide ion to simple cadmium salts:

$$CdO + 4NaCN + H_2O \rightarrow Na_2Cd(CN)_4 + 2NaOH$$

Cadmium is not nearly as electronegative as zinc, as shown by the precipitation of cadmium metal from solution by the addition of zinc dust. Nevertheless it is sufficiently so to be able to protect iron and steel galvanically as does zinc.

Cadmium metal dissolves in most mineral and some organic acids; salts of these acids are soluble in water. Principal insoluble salts include the ferro- and ferricyanides, arsenate, carbonate, phosphate, and oxalate.

Hazards

Unlike zinc, cadmium and its compounds are highly toxic; their effects are similar to those of mercury and arsenic. Cadmium plate must never be used on parts that will come into contact with foods or beverages. The vapor pressure of cadmium is fairly high at moderate temperatures, even below its boiling point; melting and casting of cadmium must be carried out with good ventilation, and the same is true of welding to cadmium-coated parts. Ordinary soft soldering involves temperatures too low to pose a serious hazard, but even in this case care should be taken not to breathe any vapors.

CADMIUM PLATING

Because of its high price and toxicity, cadmium is used as a protective electroplate only in those circumstances in which zinc is not equally satisfactory or where its special properties are required. Cadmium is more easily soldered than zinc; it does not form bulky corrosion products that interfere with the functioning of moving parts; it is easily deposited on cast and malleable iron and therefore is often used as a strike before zinc plating on these materials; unlike zinc it is resistant to alkalis; and it is somewhat superior to zinc in corrosion protection in strictly marine (salt) atmospheres, hence it is often called out in military and naval specifications. The high price and occasional shortages of cadmium have tended to discourage its use, and the popularity of cadmium plating is decreasing.

Cadmium is almost universally plated from the cyanide bath. As with all cyanide processes, waste-control problems have accounted for considerable research on alternative processes, leading to the development of fluoborate, sulfate, and chloride solutions not containing cyanide. But these baths have not had a success comparable to that of the noncyanide zinc baths.

Table 11-8 shows the composition and operating characteristics of typical cyanide-cadmium solutions. For bath makeup, cadmium oxide is usually used; this causes the formation of an equivalent amount of sodium hydroxide in the

Table 11-8 Cyanide Cadmium Baths

	Still	*Barrel*
Cadmium oxide, g/L	25–35	17–23
molarity	0.19–0.27	0.13–0.18
cadmium metal, g/L	22–31	15–20
Sodium hydroxide, g/L*	15–22	10–14.5
Sodium cyanide, g/L	75–90	90–100
molarity	1.1–1.3	1.3–1.45
Addition agents, g/L	0.1–15	0.1–15
Cathode current density, A/m²	150–400	——
Anode current density, A/m²	200	
Temperature, °C	20–30	

*Not added as such.

bath, as shown above. The composition of the cadmicyanide complex is not unequivocally known, though it is usually written $Cd(CN)_4^{--}$. For control purposes it is usual to determine the total rather than the "free" cyanide; metal and total cyanide are the factors requiring some degree of analytical control. Sodium hydroxide content is not considered critical; the higher the metal content, the higher the hydroxide that can be tolerated. It is usually maintained more or less automatically by the difference between the anode and cathode efficiencies. The amount resulting originally from makeup of the bath should not normally be exceeded.

Cadmium plate was probably the earliest to be found susceptible to brightening by the use of addition agents. Many types of additives function in this way; they include colloid formers such as glue, gelatin, and other proteinaceous materials, heterocyclic high-molecular-weight compounds, aromatic aldehydes, and many others. Inorganic additives such as nickel and cobalt compounds have been used.

Cathode efficiency is usually in the range of 90 to 95 percent; it is increased by increased metal content, temperature, and agitation, and decreased by excess cyanide as well as by too much brightener. Carbonate has little effect. Anode current efficiency is about 100 percent unless anodes become polarized owing to insufficient cyanide.

The throwing and covering powers of the cyanide cadmium baths are high.

Cyanide cadmium baths are among the easiest to control of all those in the electroplater's repertory. Hull cell tests and simple analytical procedures usually are sufficient. These baths are, however, quite susceptible to contamination, both metallic and nonmetallic: thallium, lead, antimony, arsenic, tin, and silver among the former, and cathodic depolarizers like nitrate and chromate among the latter. Metals can be removed by treatment with cadmium sponge or zinc dust or by low-current-density dummying.

Sodium carbonate, formed by cyanide decomposition and absorption of carbon dioxide from the air, is harmless unless it is allowed to build up beyond about 50 g/L when the Cd content is 40 g/L, or 20 g/L when the Cd content is 20 g/L. Sodium carbonate can be removed by cooling, as in the case of zinc cyanide baths.

Acid Cadmium Baths

Cadmium, like zinc, when plated from cyanide baths, can cause hydrogen embrittlement of the substrate; relief by baking is similar to zinc. When embrittlement becomes a serious problem, acid baths, especially the fluoborate, have some advantages. More recently the waste-control problems associated with cyanide solutions have renewed interest in the acid solutions. None of the acid baths, however—at least those whose composition is nonproprietary—possess the ease of control and generally satisfactory operating conditions and deposit quality available from the cyanide bath, which continues to be most used for cadmium plating.

A fluoborate bath has the composition shown in Table 11-9. A proprietary "neutral chloride" bath has been introduced which claims results comparable to those from the cyanide.

Specifications

ASTM specifies three types of cadmium deposits, designated types NS, OS, and TS. These designations (compare zinc) are arbitrary, and no guide is offered concerning appropriate service conditions. Minimum thicknesses of cadmium are 13, 8, and 4 μm respectively. It will be noted that these thicknesses are

Table 11-9 Cadmium Fluoborate Bath

	Range	*Average*
Cadmium fluoborate, $Cd(BF_4)_2$		
g/L	150–300	250
Molarity	0.5–1	0.87
Ammonium fluoborate, NH_4BF_4		
g/L	60–120	90
Molarity	0.6–1.1	0.86
Boric acid, H_3BO_3		
g/L	20–30	25
Molarity	0.3–0.5	0.4
Fluoboric acid, HBF_4, to pH	1–4	2.5–3
Licorice, g/L	1–2	1
Temperature, °C	10–40	25
Current density, A/m²	100–600	300

roughly half of those specified for corresponding zinc deposits. This reflects the experience that, *under conditions where cadmium is definitely superior to zinc,* about half the thickness will serve.

Posttreatments for Zinc and Cadmium Deposits

Bright Dipping

Cadmium plate (usually not zinc) is sometimes "bright-dipped" in solutions containing 0.5 to 1% nitric acid or, less commonly, bromic or chromic acid. Dipping time is under 1 min; a brighter finish is gained at the expense of a slight loss in thickness.

Chromate Conversion Coatings

The tendency of zinc deposits to develop "white rust," the bulky white corrosion product caused by corrosion in salt atmospheres and accentuated by salt-spray testing, can be decreased by chromate conversion coatings. Time to "white-rust" in salt-spray testing can be extended significantly by such treatments, and they are often specified, especially by the military and similar agencies. First developed for zinc deposits, they are equally useful for cadmium. These coatings are discussed further in Chap. 21.

12

Decorative/Protective Coatings: Copper, Nickel, Chromium

The only connection among the three metals discussed in this chapter is that they are often used together in combination coatings to form what is perhaps the most common electroplated finish, or at least the application of electroplating most familiar to the general public, in the form of what is loosely called "chrome plate." It would be useless to attempt to list the articles that are finished with this coating: they include furniture, electrical and plumbing fixtures, automobile, motorcycle, and bicycle brightwork, appliances, sporting equipment, toys, and notions of every kind. The coatings may be applied merely for appearance, so thin that their only value is to enhance the appearance of the article long enough for it to be salable; or they may be produced to rigid specifications so that with proper care they will ensure that the article so plated will last for a reasonable time in normal service and provide complete consumer satisfaction.

It is difficult to arrive at reliable estimates for the total amounts of the various metals used in electroplating and related branches of metal finishing, principally because electroplating is generally included under "miscellaneous" in statistical reports of metal usage. Some estimates have been published, however, which indicate that at least two metals—zinc and tin—enjoy more use as coating metals than do the three under consideration here. Both zinc and tin, however, find their main uses as coatings on continuous-strip, wire, and other large-scale applications, many of which are more appropriately considered as branches of the steel industry than of metal finishing. Leaving these two aside, copper, nickel, and chromium are the three most commonly plated metals and are generally thought of as the "backbone" of the plating trade. About 10 000 t (metric tons or tonnes) of copper are used annually in electroplating, as well as about 23 000 t of nickel. The annual consumption of chromic acid in electroplating and related operations is estimated at about 25 000 t, but it is difficult to translate this into tons of actual chromium metal plated, since the efficiency of chromic acid usage is low and variable.

The uses of copper, nickel, and chromium plating are by no means confined to the decorative/protective copper-nickel-chromium finish. Each has its independent applications. Copper, as well as being a widely used undercoat for nickel and chromium, is used in printed circuits, in electrotyping and electroforming, in wire plating, and in many other ways. Nickel, with or without a copper undercoat, has many engineering uses where its physical properties are of advantage—in electroforming and where resistance to wear and to certain corrosives is important. Chromium, in the form known as "hard chromium," is widely used for engineering applications where its extreme hardness is advantageous.

Nevertheless, to the average metal finisher, "copper-nickel-chromium" is, more or less, the standard electroplated finish, and perhaps more shops are equipped to offer this finish than any other (although again statistics are largely lacking or fragmentary). That is, to repeat, the reason for grouping these three metals here, even though they differ widely in their physical and chemical properties, and the plating processes used to produce the coating have little in common.

COPPER

Copper (Cu) was known in prehistory and has been mined for at least 6000 years. Its name and symbol are derived from Latin *cuprum,* the name of the island of Cyprus, where its early ores were found. It is widely distributed in nature and is not especially rare. Being a relatively noble element, it is found native, but its principal sources are sulfide, oxide, and carbonate ores. It forms about 0.007 percent of the earth's crust, or 70 g/t of the lithosphere; it is also present in seawater to the extent of about 0.001 to 0.01 ppm.

Although copper is mined in as many as fifty countries, the leading producers are the United States, the Soviet Union, Zambia, Chile, Canada, and Zaire. United States production is about 1,535,000 t/yr; world production is about 6,625,000 t. Native copper is insignificant as a commercial source; sulfide or oxide ores are processed by hydrometallurgical or pyrometallurgical processes, depending on the ore. Regardless of the nature of the extraction process, the result is generally a relatively impure "blister copper." This is finally refined electrolytically to produce a high-grade metal, since for most of its uses copper must be of high purity, usually 99.98 percent minimum; impurities seriously degrade one of its most valuable properties, its electrical conductivity. Most high-grade ores, at least in the United States, have been exhausted, and methods have had to be developed to extract the metal from ores containing as little as 0.25 percent Cu. Copper ores are important sources of other metals: about 25 percent of the molybdenum, 95 percent of the selenium, 75 percent of the tellurium, 30 percent of the gold, and 28 percent of the silver (in the United States) are produced in the course of refining copper ores. During the electrorefining process, some precious metals such as gold and silver, as well as selenium and tellurium, do not dissolve from the crude anodes and are collected as slimes for recovery of these elements.

Secondary or scrap copper is an important segment of the market. Since copper, unlike many other metals, finds its principal applications as the pure metal, such secondary recovery is, in principle at least, relatively simple.

Copper is the first member of the Group IB of the periodic system, the other two being silver and gold; the three are often grouped as the "coinage metals." It is one of the only two metals with a distinctive color, gold being the other.

Uses

Copper is unusual in that in most of its uses it appears as the pure metal rather than as an alloy or chemical compound, although it forms many useful alloys. Most of its uses depend in one way or another on its high conductivity for electricity and heat: the highest of any metal except silver. Its electrical conductivity, in fact, is the basis of one of the methods of expressing this property: "% IACS" stands for "percentage of the International Annealed Copper Standard," which was originally 100 percent for pure copper. Since the formulation of that scale purity has improved, and its conductivity is now given as 101.8 percent. Its principal uses include electrical equipment, light and power, building, industrial equipment, motor vehicles, and communication equipment; these add to 78 percent of total use. Its two principal alloys—brass (copper-zinc) and bronze (copper-tin)—also have many applications. Its alloys are used in coinages throughout the world; other important alloys containing more than 50 percent copper include aluminum bronze, manganese bronze, nickel silver, admiralty metal, German silver, and beryllium copper, which contains 2.8 percent or less beryllium, the most efficient hardener for copper.

Properties

Table 12-1 lists some of the principal physical properties of copper. Copper is tough, soft, and ductile. It is only superficially oxidized in air, yielding a green coating (patina) consisting of a hydroxocarbonate and a hydroxosulfate. Its reaction with sulfide-containing gases is also only superficial; but these surface films (tarnish) are sufficient to render the copper surface difficult to wet by solder, and they explain why copper used in communications equipment is often tin-plated or tin-lead–plated.

Although at room temperature the reaction with oxygen is only superficial, at red heat copper(II) oxide CuO is formed, and at higher temperatures Cu_2O. Sulfur yields Cu_2S, or a similar but nonstoichiometric compound. Copper is attacked by the halogens; in the absence of air it is not attacked by nonoxidizing or noncomplexing dilute acids. It readily dissolves in nitric acid and sulfuric acid in the presence of oxygen. Also, in the presence of oxygen, it is soluble in ammonia and alkali cyanide solutions.

Copper is classified as a member of the first transition series of elements, even

Table 12-1 Properties of Copper

Atomic number	29
Atomic weight	63.55
Electrochemical equivalent, Cu(I), mg/C	0.659
Electrochemical equivalent, Cu(I), g/A·h	2.37
Electrochemical equivalent, Cu(II), mg/C	0.329
Electrochemical equivalent, Cu(II), g/A·h	1.19
Electronic configuration	$1s^2 2s^2 2p^6 3s^2 3p^6 3d^{10} 4s^1$
Melting point, °C	1083
Boiling point, °C	2582
Density, g/cm^3, 20°C (g/m^2 for 1-μm deposit)	8.94
Crystal structure	fcc
Electrical resistivity, $\mu\Omega$-cm, 20°C	1.673
Electrical conductivity, % IACS	101.8
Brinell hardness, annealed*	43
Brinell hardness, hard rolled*	103
Tensile strength, annealed, MPa*	220–255
Tensile strength, drawn, MPa*	380–420
Standard potential, E^0, 25°C, V	
$Cu^+ + e^- \rightarrow Cu$	+0.52
$Cu^{++} + 2e^- \rightarrow Cu$	+0.34

*For massive metal; values for electrodeposited copper vary widely depending on bath and plating conditions. See Safranek in Bibliography.

though it has a filled *d* shell—(Ar shell) $3s^2 3p^6 3d^{10} 4s^1$. This is because the configuration. . . $3d^9$ is somewhat more stable than . . . $3d^{10}$, and the loss of the *d* electron leads to multiple valences, colored ions, easy complex formation, and other typical characteristics of transition metals.

Copper forms two series of compounds: monovalent and divalent, cuprous and cupric—or, in more modern terms, copper(I) and copper(II). Both series form many complexes. In aqueous solution the divalent state is much more stable, and copper(I) tends to disproportionate to copper metal and copper(II) ion:

$$2Cu^+ \rightarrow Cu + Cu^{++}$$

Thus, in aqueous environment the unipositive state exists only if the copper(I) compound is insoluble or is stabilized by complex formation. Insoluble copper(I) compounds include the chloride, bromide, iodide, oxide, sulfide, and cyanide; but water-soluble materials exist only if complexed by such ligands as cyanide, chloride, ammonia, or acetonitrile; in the presence of these groups the copper(I) form is the more stable.

Because of its highly electropositive (noble) character, copper is easily precipitated by metals higher in the emf series, such as iron or zinc. In fact, the use of scrap iron to "cement" out copper from dilute sulfuric acid is a standard method of recovering copper from sulfuric acid leachings of ores or tailings. The reduction by these metals normally proceeds all the way to metal; as stated, copper(I) ion is unstable with respect to Cu^0 and Cu^{++}.

Copper(I) compounds are colorless (unless the anion lends color) and unstable with respect to copper(II) unless they are insoluble or complexed. The relative stabilities are indicated by the potentials:

$$Cu^+ + e^- \rightarrow Cu \qquad E^0 = 0.52 \text{ V}$$
$$Cu^{++} + e^- \rightarrow Cu^+ \qquad E^0 = -0.163 \text{ V}$$

thus

$$Cu + Cu^{++} \rightarrow 2Cu^+ \qquad E^0 = -0.37 \text{ V}$$

and

$$K^* = (Cu^{++})/(Cu^+)^2 \approx 10^{16}$$

The relative stabilities of Cu(I) and Cu(II) in solution depend strongly on the nature of the anions or other ligands present, as well as on the solvent. In aqueous solution only a very low equilibrium concentration of Cu(I) (less than 0.01 *M*) can exist, and the only Cu(I) compounds stable to water are the insoluble ones such as CuCl or CuCN.

The equilibrium 2Cu(I) $\rightleftharpoons$ Cu + Cu(II) can be easily displaced in either direction. Thus with CN^- and I^-, Cu(II) reacts to give the Cu(I) compound; with anions that cannot coordinate or give bridging groups, such as perchlorate, fluoborate, or sulfate, or with complexing agents that have greater affinity for Cu(II), the latter state is favored. Thus ethylenediamine reacts with Cu(I) chloride in aqueous KCl solution to give a Cu(II) complex:

$$2CuCl + 2en \rightarrow (Cuen)_2^{++} + 2Cl^- + Cu^0 \qquad \text{(en = ethylenediamine)}$$

Binary Cu(I) Compounds

The oxide and sulfide are more stable than the corresponding Cu(II) compounds at high temperatures. The halides are insoluble: CuCl and CuBr can be made by boiling an acidic solution of the Cu(II) salt with excess copper; on dilution, CuCl or CuBr is precipitated. Addition of iodide ion to a solution of Cu^{++} forms a precipitate that rapidly decomposes to CuI and iodine. CuF is not known. The halides (except CuF) are highly water-insoluble, which may account for their stability. Solubility is enhanced by presence of excess halide ion, owing to complex formation ($CuCl_3^{--}$, etc.) and by other complexing species, e.g., cyanide, ammonia, and thiosulfate.

CuCN is also a fairly common compound, formed by adding alkali cyanide to an aqueous solution of a copper(II) salt:

$$2Cu^{++} + 4CN^- \rightarrow 2CuCN + C_2N_2$$

*K = equilibrium constant.

It is soluble in excess cyanide to form various cyanide complexes, mainly $Cu(CN)_4^{3-}$.

Copper(I) forms complexes with many organic ligands; some of these are useful in industrial chemistry, such as the use of cuprous formate to remove carbon monoxide from gas streams.

Copper(II) Compounds

In general these compounds are more stable than Cu(I); in these, one of the *d* electrons as well as the single *s* electron takes part in compound formation. [Although trivalent copper compounds, Cu(III), are known, they are few and of no practical significance.] Copper(II) ion forms a large number of soluble salts, of which the sulfate $CuSO_4 \cdot 5H_2O$ ("bluestone") is the best known and most readily available.

Most copper(II) salts dissolve readily in water to give the aquo ion, which is usually written $Cu(H_2O)_6^{++}$, but two of the water molecules are less closely associated with the central copper ion than the other four. Addition of ligands to these aqueous solutions leads to formation of complexes by successive displacement of water molecules: for example, with ammonia, the whole series $Cu(H_2O)_6^{++}$, $Cu(H_2O)_5(NH_3)^{++}$, $Cu(H_2O)_4(NH_3)_2^{++}$, . . ., $Cu(NH_3)_6^{++}$ is known, but the addition of the fifth and sixth ammonia molecules is difficult. Similar complexes are formed with amines, amine-like compounds such as EDTA, halide ions, and many others. As previously stated, when it is attempted to form the cyanide complex, the copper reverts to the Cu(I) state.

Copper forms no hydroxo-complexes and is thus not amphoteric.

From the electroplater's standpoint, the most important copper compounds are the cyanide complexes of copper(I) and such soluble simple or aquo salts of copper(II) as the sulfate, fluoborate, and chloride, and the amine, ammonia, and pyrophosphate complexes.

Hazards

Copper(I) salts, as stated, are generally insoluble or unstable. Copper(II) salts in general are not especially hazardous, although EPA regulations severely limit their discharge. Copper vessels have been used in cooking without apparent ill effects. Copper(II) sulfate has several medical and veterinary uses, internally as an emetic and topically as a fungicide. Copper sulfate is also a well-known antidote for phosphorus poisoning: the reaction $Cu^{++} + P$ yields a coating of copper metal on the phosphorus particles, rendering them inactive. In very large doses, copper(II) sulfate is definitely poisonous and can be lethal.

Ordinary good housekeeping is normally sufficient to render the handling of copper and its compounds safe.

The anion associated with the copper can, of course, modify these statements. One of the most important copper compounds for the electroplater is copper(I) cyanide; its cyanide content renders it highly toxic, like all cyanides.

Copper Plating

As would be expected from its position in the emf series, there is no difficulty in electroplating copper from aqueous solution. In fact, one of the problems is to prevent the formation of immersion deposits on less noble metals: $Cu^{++} + M \rightarrow Cu^0 + M^{++}$; as we saw in Chap. 8, p. 126, such immersion deposits are usually nonadherent and powdery. As we have also seen, such deposits can be avoided by reducing the activity of the copper ion by complexing it. The complexing agent universally used for this purpose is cyanide ion. After the initial deposit of copper from a cyanide bath has been applied, the parts can be transferred to a bath with better plating characteristics such as higher speed, better brightness or leveling, or the ability to build up thicker deposits.

In terms of tonnage, probably more copper is plated in general plating (as opposed to continuous or specialty applications) than any other metal except nickel. There are several reasons for this:

The metal is easily plated, from several types of solutions which are relatively easy to control.

Copper is the only metal that can be practically deposited directly on zinc die castings in preparation for further processing.

Copper plating baths have good coverage and in general good throwing power.

Copper is among the less expensive metals available to the metal finisher; only zinc is cheaper among the commonly plated metals. Iron and lead are also cheaper but are less generally useful as electrodeposits.

The supply of copper is generally reliable; shortages have been few and of short duration.

Copper is an excellent undercoat for further deposits; it is easily buffed to a high luster, providing a good surface for subsequent coatings. For example, it is more economical to plate steel with copper and buff the latter than to buff the original steel; solutions are available that yield bright and leveling deposits.

The physical and chemical properties of copper, already discussed, make it useful in many applications; as one example, its high electrical conductivity is exceeded only by that of silver, which is much more expensive. Steel wire is often heavily plated with copper to combine the strength of steel with the conductivity of copper.

Copper is relatively inert to most plating solutions; thus basis metals that would be attacked by many plating solutions if used directly are protected from such action by a preliminary copper plate.

Copper has a relatively favorable electrochemical equivalent (weight plated per quantity of electricity). In g/A·h the electrochemical equivalent of Cu(I) is 2.37, of CU(II) 1.18; compare with nickel, 1.09, and chromium, 0.32.

For these and other reasons, copper is one of the most useful metals in the plater's repertory. Because of the relative simplicity of obtaining highly purified copper metal and copper salts, copper has also been a favorite of academic research on electrode kinetics, since studies can be made on highly pure materials without the complications introduced by adventitious impurities or the necessity of using additives to obtain satisfactory deposits.

Three types of copper baths are in general use: cyanide (with several modifications), acid (sulfate or fluoborate), and the pyrophosphate complex bath. Each has its uses, and will be discussed separately.

COPPER CYANIDE BATHS

The only practical way to deposit adherent copper plates on active metals such as zinc and steel is to use a cyanide bath, containing the copper(I) cyanide complex $Cu(CN)_4^{3-}$. This complex is formulated also as $Cu(CN)_3^{--}$ (in fact, this assumption is used in calculating "free" cyanide—see below), but the four-coordinate formula seems the more likely. See also the discussion of "free" cyanide, Chap. 8, p. 127. As we have seen, the stability of the cuprocyanide complex is so great that the electrode potential of copper is rendered about as negative as that of zinc, so that immersion deposition is avoided. In spite of many efforts to dispense with cyanide-containing plating baths because of environmental restrictions, no practical substitute for the cyanide copper bath has been developed.

The composition of copper cyanide baths is not especially critical, but three types have emerged, each with its sphere of utility: the copper cyanide "strike" bath, the "Rochelle" bath, and the so-called high-efficiency or high-speed bath. Cyanide baths have in common the advantage of plating from the monovalent state; i.e., only one electron is involved and the electrochemical equivalent is therefore twice that of baths formulated with copper(II) salts. Their disadvantages include: allowable current densities are somewhat lower, hence plating is slower; they are somewhat more critical in control than the simple sulfate bath; their ingredients are more expensive—copper sulfate and sulfuric acid are both very cheap chemicals—and, of course, the cyanide content involves pollution-control problems.

The average compositions of the three types of copper cyanide baths are listed in Table 12-2. Their operating characteristics are shown in Table 12-3. Copper may be added as CuCN, which though insoluble in water is soluble in sodium or potassium cyanide solutions:

$$CuCN + 3NaCN \rightarrow 3Na^+ + Cu(CN)_4^{3-}$$

"Double" salts, $Na_2Cu(CN)_3$ or $K_2Cu(CN)_3$ are also available, and they have the advantage of being water-soluble.

Table 12-2 Cyanide Copper-Plating Baths

	Strike		Rochelle		High-efficiency	
Component	*g/L*	*Molarity*	*g/L*	*Molarity*	*g/L*	*Molarity*
CuCN	15	0.17	45	0.50	75	0.84
NaCN or	28	0.57	56	1.14	100* or	2.04
KCN	—	—	—	—	133*	2.05
NaOH or	—	—	—	—	30* or	0.75
KOH	—	—	—	—	42*	0.75
Na_2CO_3	15	0.14	30	0.28	—	—
$KNaC_4H_4O_6{\cdot}4H_2O$	—	—	45	0.16	optional	
By analysis:						
Cu as metal	10.5	0.17	31.5	0.50	53	0.83
"Free" NaCN† or	11	0.22	6	0.12	19	0.39
KCN†	—	—	—	—	25	0.39

*Na and K salt concentrations shown for all-Na or all-K baths; usually both are used in about equimolar proportions or all-K baths may be preferred.

†Arbitrarily taken as the NaCN in addition to that required to form $Na_2Cu(CN)_3$.

The strike bath, as is typical of baths used for the purpose of depositing a thin but adherent initial deposit to be followed by further plates, is characterized by low metal and high complexing agent concentrations, which contribute to low efficiency but ensure that immersion deposition does not occur. Strike baths are used to start plating on zinc die castings, and often on steel as well, but time of plating is usually limited to that necessary for complete coverage.

The Rochelle bath is often used for purposes similar to those for the strike. It is somewhat more concentrated, and the addition of Rochelle salts (potassium sodium tartrate) aids in anode corrosion. Heavier deposits are possible, and for some applications the Rochelle bath may be used for the complete copper deposition cycle.

Strike and Rochelle baths are normally formulated with sodium cyanide; the high-efficiency bath uses both sodium and potassium cyanide, in about equimolar amounts, or it can be formulated as an all-potassium bath. Since these baths also contain alkali hydroxide, there is considerable leeway in how they are made up: sodium cyanide can be used for cyanide content and potassium hydroxide for the potassium.

Hydroxide is not normally added to strike and Rochelle baths, although it is sometimes used in strike baths to plate steel parts, for improving conductivity. In plating zinc die castings it must be used sparingly since hydroxides attack zinc. Potassium hydroxide is added to high-efficiency baths, to improve both conductivity and throwing power.

For the Rochelle bath, proprietary additives may be used in addition to, or in place of, Rochelle salts. Citrates have been recommended instead of tartrates but are not widely used.

Excess of free cyanide is essential in all three types of bath. Free cyanide is generally assumed to be the amount of cyanide ion above that needed to form the complex $Cu(CN)_3^{--}$; this translates to 1.23 g of CN^- per gram of Cu or 2.32 g NaCN or 3.1 g KCN. Not only is excess cyanide required to prevent the immersion deposition of copper, as already discussed; it also prevents the formation of insoluble CuCN films on the copper anodes. By some mechanism not understood, cyanide ion also tends to act as an addition agent, promoting brighter deposits, especially from the high-efficiency bath.

Carbonate is added to strike and Rochelle baths, primarily to facilitate control of pH; it also reduces anode polarization. Although not intentionally added to high-efficiency baths, it is always present, both from decomposition of cyanide and by absorption of carbon dioxide from the air.

Carbonate can be controlled by precipitation as insoluble calcium or barium carbonate, by the addition of lime or barium hydroxide:

$$CO_3^{--} + Ca(OH)_2 \rightarrow CaCO_3 \downarrow + 2OH^-$$

This treatment simultaneously increases the hydroxide content of the bath. If this is objectionable, refrigeration may be used if the bath is primarily a sodium bath; sodium carbonate decahydrate is much less soluble cold than at higher temperatures. The temperature should not be allowed to fall below −3°C since otherwise copper salts would be precipitated. Barium cyanide may also be used to precipitate carbonate without increasing hydroxide content.

Current manipulation, especially PR and interrupted current, is beneficial in plating from the high-efficiency baths. Typical PR cycles are 10- to 60-s direct, 2- to 20-s reverse. Current interruption cycles typically are 10 s on, 1 s off. The major advantage of these techniques is that they yield brighter and smoother plates.

High-efficiency copper cyanide baths are sensitive to contaminants (as are the other cyanide baths, to a somewhat lesser degree); the principal ones having deleterious effects are sulfur compounds,* chromates, zinc, and residues from buffing compounds and cleaners.

*Except certain thiocyanates sometimes deliberately added.

Table 12-3 Operating Characteristics of Copper Cyanide Baths

	Strike	*Rochelle*	*High-efficiency*
Cathode cd, A/m²	100–300	150–600	100–1000
Anode cd, limiting, A/m²	100	300	500
Temperature, °C	50–63	55–70	63–82
Limiting thickness, μm	2.5	13	> 25

Copper oxide particles may cause roughness in the deposit: they are light and do not readily settle to the bottom. Upon reaching the cathode, copper oxide is reduced to the metal, forming high spots which grow rapidly to form nodules. The best remedy for this type of roughness is to bag the anodes; circulation of the electrolyte through a filter is also helpful.

Anodes

High-purity, oxide-free anodes are commercially available in several shapes and sizes. Proprietary anodes containing phosphorus are claimed to have some advantages, chiefly, however, for acid copper plating.*

Control

The composition of the electrolyte must be maintained within specified limits, especially with respect to free cyanide. If this is too low, anodes may polarize and insoluble particles may be dislodged and cause particle roughness of the deposit. Too high free cyanide restricts the bright-plating range and lowers cathode efficiency. Addition agents, if used, are usually controlled by Hull cell tests, discussed in Chap. 24.

Purification of the electrolyte by filtration on a continuous or regular basis is advisable.

ACID COPPER PLATING

Although acid copper baths, based on copper(II) salts, yield only half as much copper per unit quantity of electricity as the cyanide baths, they exhibit sufficient advantages over the cyanide to make their use not only practical but preferred for many applications. They cannot be used for direct application to steel and zinc die castings, as already explained; a cyanide strike is essential on zinc die castings, and either a cyanide or a nickel strike is necessary on steel, before entering the acid copper bath. (It goes without saying that efficient rinsing is also required between the two plating baths.)

Acid copper baths are used for all the occasions requiring a copper plate, with the exceptions already mentioned. They find use not only in plating, but also in electroforming and electrorefining. Their advantages over cyanide baths are many: the chemical cost is low, effluent control is simplified, the baths are simple and easy to control, their compositions are not critical, anode and cathode efficiencies are high, and they can tolerate much higher current densities than the alkaline cyanide baths. Because of their high conductivity, bath voltage is low; anode and cathode polarizations are low as well.

Steel wire is given a cyanide strike, then plated in acid solutions to yield a high-

*Note that OFHC (oxygen-free, high conductivity) is a trade name, but electrolytic copper of equivalent purity is widely available.

strength electrical cable. Thick (200-μm) deposits of copper are applied to steel rolls and then engraved for use in printing. Stainless steel cooking vessels are copper-plated to improve heat distribution over the outer surfaces and prevent hot spots. Copper plating in acid solutions, following a cyanide strike, is used for stopping off steel surfaces in selective carburizing.

The acid copper solutions are very simple in composition, consisting only of a copper(II) salt and the corresponding acid, i.e., copper sulfate plus sulfuric acid, or copper fluoborate plus fluoboric acid. Such solutions produce a dense mat deposit; for some of the newer "bright acid copper" baths, mentioned later, proprietary addition agents and a small amount of chloride ion are added to the sulfate bath.

With these additions, the copper sulfate bath can be made to produce bright and leveling deposits that can often be directly plated with nickel or other metals, dispensing entirely with buffing.

Copper(II) sulfate is available as the pentahydrate, $Cu(SO)_4 \cdot 5H_2O$. Copper(II) fluoborate, $Cu(BF_4)_2$, is available only in the form of a solution concentrate, containing about 45 percent $Cu(BF_4)_2$ and some excess boric acid to prevent the formation of fluorides by hydrolysis:

$$4HF + H_3BO_3 \rightleftharpoons HBF_4 + 3H_2O$$

Although copper(II) ions form weak complexes with sulfate, essentially the Cu^{++} ion in these baths is the simple aquo ion and may be regarded as uncomplexed. Fluoborate solutions have not been studied extensively from the academic standpoint, but the fluoborate ion is known to be a very poor complexing agent, and the Cu^{++} ion in this solution is most likely also the simple aquo ion.

Excess acid is needed in both baths for producing satisfactory deposits. These acids account for the high conductivity of the baths; anode and cathode polarizations are nearly negligible at low current densities, and even at cathode current densities as high as 2150 A/m^2 a 6-V current source is sufficient when the solution is agitated. Anode polarization in the sulfate bath may become troublesome above about 500 A/m^2; in the fluoborate solution, anode current densities as high as 4000 A/m^2 can be used. Agitation has a very great influence on the limiting current densities in both baths—the more violent, the higher the cd that can be used.

Anode and cathode efficiencies are nearly 100 percent at practical current densities. Throwing power is not as good as with the cyanide baths; it approximates current distribution (as would be expected from the low polarizations and flat efficiency-cd curves), but complicated shapes can be plated with the use of appropriate shields to improve current distribution.

Agitation is important if high current densities are to be used; under proper conditions very high plating rates can be attained, up to about 3.6 mm/h. With efficient agitation and good solution maintenance, there seems to be no practical limit to the thickness obtainable with these baths.

The copper sulfate bath is more widely used than the fluoborate in practice: cost per unit volume of the bath is much lower, since both copper sulfate and sulfuric acid are considerably cheaper than the corresponding fluoborates. The principal advantage of the fluoborate bath is that it can tolerate much higher current densities than the sulfate; its principal disadvantage (except for cost) is that it is not adaptable to the brighteners and levelers which have been developed for the sulfate bath to produce the modern proprietary bright acid copper baths.

Compositions and operating conditions for the two acid baths are listed in Tables 12-4 and 12-5. These compositions are not particularly critical. Both baths require excess acid to improve conductivity and to prevent hydrolysis of the copper salts, precipitating copper hydroxide (or hydrated copper oxide). In the sulfate bath, increased copper concentration leads to increased resistivity of the bath, and cathode polarization increases somewhat when the copper concentration is above 1 *M* (63 g/L Cu or 250 g/L $CuSO_4 \cdot 5H_2O$). If the copper sulfate concentration is less than about 60 g/L, the deposit quality may deteriorate slightly. Sulfuric acid decreases the solubility of copper sulfate by the "common ion" effect.

Copper fluoborate is much more soluble than copper sulfate, and the metal-ion concentration can be more than double that in the sulfate baths if high-speed plating is desired. If the acid concentration in the fluoborate bath is too low (pH above 1.7), deposits may be dull, dark, and brittle. Boric acid is added to this bath to stabilize it and prevent decomposition of fluoborate to fluoride, as previously mentioned. Commercial concentrates contain boric acid, usually in sufficient amounts, but more can be added if necessary.

Many addition agents are available for the copper sulfate bath for special purposes: increasing the limiting current density, reducing treeing, grain refining, or smoothing. Many brightening agents have been developed, including thiourea, some other sulfur-containing compounds, and others. Most of these have been superseded by newer proprietary agents, mentioned under bright acid copper, below.

Table 12-4 Composition and Properties of Copper Sulfate Baths

	Average		*Limits*	
	g/L	*Molarity*	*g/L*	*Molarity*
Copper sulfate, $CuSO_4 \cdot 5H_2O$	188	0.75	150–250	0.6–1
(Cu as metal)	48	0.75	38–63.5	0.6–1
Sulfuric acid, H_2SO_4	75	0.76	45–110	0.45–1.12
Temperature, °C	32–43		18–60	
Cathode cd, A/m^2	300–5000 depending on conditions			
Anode cd, A/m^2			to 1700	
Specific gravity, 25°C	1.165		1.115–1.21	
Resistivity, $\mu\Omega$–cm	4.2–4.3			

Table 12-5 Composition and Properties of Copper Fluoborate Baths

	Low concentration		*High concentration*	
	g/L	*Molarity*	*g/L*	*Molarity*
Copper fluoborate, $Cu(BF_4)_2$	225	0.95	450	1.90
(Cu as metal)	60	0.95	120	1.90
Fluoboric acid, HBF_4	15	0.17	30	0.34
Boric acid, H_3BO_3	15	0.24	30	0.49
pH, colorimetric	1.2–1.7		0.2–0.6	
Specific gravity, 27°C	1.17–1.18		1.135–1.36	
Cathode cd, A/m²	up to 21 000 under special conditions			
Anode cd, A/m²	to 5500 with agitation			
Temperature, °C	18–50			
Resistivity, $\mu\Omega$ –cm, 27°C	7.3			

Current manipulation has been recommended for the sulfate bath; a PR cycle of 2 s direct and 0.4 s reverse is said to be beneficial.

Anodes

Copper anodes in both sulfate and fluoborate solutions may become coated with films containing finely divided copper and copper oxide particles. Disproportionation of Cu(I) to metal and Cu(II) ions may also lead to fine particles in the anode, causing intergranular corrosion and cathode roughness; any impurities such as arsenic, tellurium, selenium, lead, and silver are insoluble and may lead to similar difficulties.

Bagging of anodes is one method of avoiding the transfer of such particles to the cathode. Addition of phosphorus, in amounts from 0.02 to 0.04 percent, is claimed to increase the tenacity of the anode film and prevent the dislodgement of particles into the solution.

The physical properties and structure of copper deposits can vary over an extremely wide range, depending on conditions of deposition. Some typical examples are shown in Table 12-6, but for more complete information the Bibliography should be consulted.

BRIGHT ACID COPPER PLATING

The copper sulfate plating bath is very old in the art, and various improvements in the shape of better addition agents and greater knowledge of the effects of plating conditions on the structure and properties of the deposits have accumulated over the years. A more recent development, which can be fairly characterized as a new process, has greatly enlarged the usefulness of the sulfate bath and

Table 12-6 Selected Properties of Copper Deposits

Desired property	*Tensile strength, MPa*	*Elongation percent, 5 cm*	*Internal stress, MPa**	*Hardness VHN_{200}, MPa*	*Electrical resistivity, μΩ −cm*
High strength	450–620	4–18	−40 to 54	1280–1560	1.75–2.02
Hardness	34–540	0–10	−40 to 30	1890–3420	1.96–4.60
Low resistivity	180–265	15–41	−0.5 to −15	470–625	1.70–1.73
Low stress	135–155	8–24	−0.7 to −0.6	550–560	1.71–1.72
Leveling	350–355	14–19	20	1250–1340	1.82
Thermal stability†	220–295	26–39	5–28	540–1040	1.73–1.76

*Negative sign indicates compressive stress.

†Change less than 0.02 percent in length after heating to 400°C.

permitted the use of acid copper plating to produce smooth, bright, and leveling deposits which not only possess new features in themselves but have been shown to change the performance of copper-nickel-chromium composite plates. They have also proved most useful in plating on plastics.

These developments are for the most part proprietary, using newly developed addition agents, which for reasons already stated cannot be discussed here. For proper action of these additives, the baths must contain minor amounts of chloride ion. A typical composition of a bright copper acid bath is shown in Table 12-7. Phosphorized anodes, as mentioned above, containing 0.02 to 0.03 percent phosphorus, are recommended. As with all acid copper baths, a cyanide strike is required for plating on zinc die castings and on steel. One supplier recommends an all-potassium cyanide strike, formulated as in Table 12-8.

Because of the proprietary nature of these baths, detailed directions must be obtained from the suppliers.

Table 12-7 Typical Bright Acid Copper Bath

	g/L	*Molarity*
$CuSO_4 \cdot 5H_2O$	195–240	0.78–0.96
(Cu metal)	50–61	0.78–0.96
H_2SO_4	45–60	0.46–0.61
Chloride, Cl^-	0.02–0.08 (20–80 ppm)	0.0005–0.002
Addition agent	as recommended	
Temperature, °C	24–32	
Anode current density, A/m²	150–300	
Cathode current density, A/m²	300–600	

Table 12-8 Cyanide Strike for Bright Acid Copper Plating

	g/L	*Molarity*
CuCN	52.5	0.59
(Cu metal)	37.5	0.59
KCN	103	1.58
Free KCN	26	0.4
Temperature, °C	50–60	
Cathode current density, A/m²	50–250	
Plating time, min	4–5	
Anodes	OFHC*	

*TM—equivalent purity acceptable.

PYROPHOSPHATE COPPER BATHS

The pyrophosphate ion, $P_2O_7^{4-}$, forms many complexes with cations; in the early history of plating, several pyrophosphate baths were published. None has survived to any extent except the copper and (to some degree) the zinc pyrophosphate baths. Although the copper pyrophosphate solution has been available for many years, it has come into general use only fairly recently, especially in the printed-circuit (PC) industry, where it has many advantages in the plating of through-holes in PC boards.

The bath has good throwing power, is noncorrosive, and is essentially nontoxic. Although the use of pyrophosphate in place of cyanide to complex the copper avoids the cyanide-disposal problem, phosphates themselves are becoming suspect as pollutants.

In addition to its use in the electronics industry, the copper pyrophosphate bath may be considered for almost any application calling for a copper deposit. In its general plating characteristics it is somewhat more comparable to the high-efficiency cyanide baths than to the acid solutions.

Copper pyrophosphate, $Cu_2P_2O_7 \cdot 3H_2O$, dissolves in potassium pyrophosphate ($K_4P_2O_7$) solutions, forming the complex ion $Cu(P_2O_7)_2^{6-}$. (Less often sodium pyrophosphate is used.) Some $CuP_2O_7^{--}$ may also be present. Control of pH is most important: at a pH above 11, copper hydroxide $Cu(OH)_2$ precipitates, and below pH 7 a copper pyrophosphate $Cu_2P_2O_7$, or $CuH_2P_2O_7$, is precipitated. Acidification below pH 7 converts the pyrophosphate ion $P_2O_7^{4-}$ to $H_2P_2O_7^{--}$ or the orthophosphate ion HPO_4^{--}, thus destroying the complex ion. The pH is lowered with pyrophosphoric acid (or proprietary materials) and raised with potassium hydroxide.

Formulation of a typical pyrophosphate copper bath is shown in Table 12-9, and operating conditions are listed in Table 12-10. The ratio of pyrophosphate to copper is an important variable and should be maintained as shown.

Table 12-9 Copper Pyrophosphate Bath: Composition

Component	*g/L*	*Molarity*
Copper, Cu	22–38	0.35–0.6
Pyrophosphate, $P_2O_7^{4-}$	150–250	0.86–1.44
Nitrate, NO_3^-	5–10	0.08–0.16
Ammonia, NH_3	1–3	0.06–0.18
Orthophosphate, HPO_4^{--}	<115	<1.20
Additives	as recommended	

Nitrate ion is present to permit a higher limiting current density since it serves to reduce cathode polarization by acting as a hydrogen acceptor:

$$NO_3^- + 10H^+ + 8e^- \rightarrow NH_4^+ + 3H_2O$$

It is usually added as ammonium nitrate, NH_4NO_3.

A small amount of ammonia is added to produce more uniform and lustrous deposits and to improve anode corrosion. Excess ammonia must be avoided since it can cause the formation of copper(I) oxide particles. Ammonia evaporates from the bath during operation, and is added on a regular basis.

Pyrophosphates and pyrophosphoric acid are inherently somewhat unstable and gradually hydrolyze to orthophosphate, PO_4^{3-}:

$$P_2O_7^{4-} + H_2O \rightarrow 2PO_4^{3-} + 2H^+ \text{ (or } 2HPO_4^{--}\text{)}$$

Some orthophosphate content is not deleterious; in fact, within limits it serves as a buffer and promotes anode corrosion. This limit is about 100 g/L. Above this range, the conductivity and bright-plating range suffer. Low pH, high P_2O_7/Cu ratio, and temperatures above 60°C contribute to the buildup of orthophosphate. There is no means of removing orthophosphate from the bath; it can be reduced only by discarding some of the solution and rebuilding it.

Vigorous agitation is required for satisfactory operation. Usual methods, including air and mechanical movement of the cathode, are satisfactory; it has

Table 12-10 Copper Pyrophosphate Bath: Operating Conditions

pH	8.0–8.8
P_2O_7/Cu ratio	7:1 to 8:1
Temperature, °C	50–60
Cathode current density, A/m^2	100–800
Anode/cathode area ratio	1:1 to 2:1
Anode and cathode efficiency, %	≈100
Agitation (air) m^3/min/m^2 surface	1–1.5

Table 12-11 Copper Pyrophosphate Bath for Printed Circuits

	Makeup		*Control limits*	
	g/L	*Molarity*	*g/L*	*Molarity*
$Cu_2P_2O_7 \cdot 3H_2O$	82.5	0.23		
$K_4P_2O_7$	338	1.0	375–450	1.13–1.37
NH_3	0.75	0.04	0.75–3	0.04–0.18
Cu			27–33	0.42–0.52
Temperature, °C			52–58	
Cathode current density, A/m²			100–800	
pH			8.2–8.8	
Brightener			qs*	
Agitation			Air	

*Sufficient quantity

been claimed that ultrasonic agitation yields uniform and adherent deposits from the pyrophosphate bath on zinc die castings.

Many organic additives have been developed and patented. Although some baths are operated without them, most operators employ proprietary materials.

Although the pyrophosphate bath contains copper in a complex anion, the complex is not stable enough to permit direct plating on zinc or steel; and, as with acid baths, a cyanide strike is required before entering the pyrophosphate solution. A pyrophosphate strike has been used with some success.

The principal present use of copper pyrophosphate baths is in the electronics industry, for plating through holes in printed-circuit boards. Both the structure of the deposits and the improved throwing power, leading to more uniform plating in the hole or a better ratio of thickness in the hole to that on the flat, account for the preference for pyrophosphate solutions in this application. See Fig. 12-1. A bath that has been recommended for this application is shown in Table 12-11. In the printed-circuit field, the pyrophosphate bath competes directly with the ductile bright acid sulfate, already discussed. The trend appears to be in favor of the latter.

NICKEL

Nickel (Ni) has been known from antiquity in the form of various alloys, but it was first isolated in 1751. The name derives from German *kupfernickel,* "copper-demon," because one of its ores appeared to contain copper, which was the desired constituent, but did not. Along with iron and cobalt it forms the first triad of Group VIII of the periodic table.

Nickel is twenty-fourth in order of abundance in the lithosphere, which contains about 0.008 percent Ni, slightly more than its copper content; the

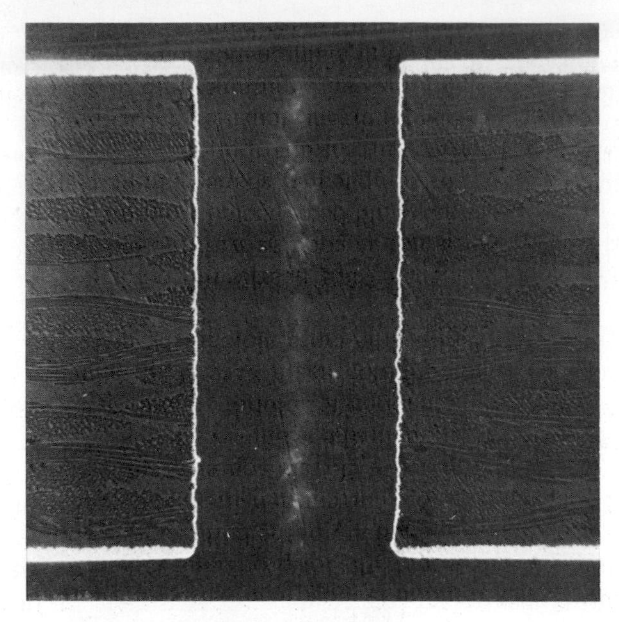

Fig. 12-1 Through-hole plating.

earth's core probably contains considerably higher amounts. Traces occur in seawater.

Nickel is widely distributed in plants and animals, but its function is not known and its compounds have no medical uses.

The cobalt-nickel pair represents the first observed case in which the periodic order of the elements was not the same as that of their atomic weights (Ni = 58.7, Co = 58.9); there can be no doubt from chemical evidence that the correct order is Fe-Co-Ni. The inversion was cleared up only with Moseley's determinations of atomic numbers by x-rays. Clearly the inversion is due to the predominance in the earth of the lighter of the two principal isotopes of Ni, 58, while cobalt is mononuclidic, pure 59.

Nickel occurs in metallic form, along with iron, in meteorites and in two very rare minerals. Although widely distributed, there are relatively few commercially important ores; the two principal ores are sulfides, and oxides or silicates, the latter known as lateritic ores. Principal countries having significant reserves of nickel ores include Canada, South Africa, the Soviet Union, and Finland. Lateritic ores, presently somewhat more difficult to process, are found also in Cuba and New Caledonia, among others.

The world's richest deposits of nickel were discovered accidentally during the surveying for the Canadian Pacific Railway, near Sudbury, Ontario. The ore was valued originally for its copper content, and the nickel values became of primary interest only later. Canada now produces about half the world's nickel; United States production is minor, although the United States consumes about 10 percent of the total world production.

The Sudbury ores are mainly of the sulfide type, principally pentlandite, a nickel-iron sulfide $(Ni,Fe)_9S_8$. The ore also contains various amounts of cobalt, gold, silver, and the platinum metals. The recovery of these precious metals is an important part of the extraction and refining process that helps to pay for the operation. Cobalt is very similar to nickel chemically, and complete separation is not usually attempted: most commercial nickel contains some cobalt.

The lateritic ores (oxides) are harder to work, but those in New Caledonia and Cuba especially now constitute the main reserves of nickel, once the Canadian ores have been exhausted.

Production

In the sulfide ores the nickel, copper, and iron sulfides occur as distinct minerals and can be treated by conventional ore-dressing methods. Lateritic ores must be treated by pyro- or hydrometallurgical methods.

The sulfide ores are roasted and smelted, producing a matte containing nickel, cobalt, and iron sulfides. This matte can be cooled slowly under such conditions that the three metals separate by gravity and can be treated separately. The nickel concentrates may be converted to an oxidized sinter for direct use by the steel industry; most is roasted, smelted, and cast into anodes for electrolytic refining. It has been found possible to use matte anodes directly in the electrorefining process; anode slimes contain economically recoverable amounts of precious metals.

Nickel is also produced by the carbonyl process, which in fact antedated the electrolytic. Impure nickel reacts with carbon monoxide, and the resulting nickel carbonyl is decomposed, forming a very pure grade of nickel from which cobalt is essentially absent.

Typical commercial nickel has the analysis listed in Table 12-12. Also included are some compositions with deliberate functional additions for use as anodes in nickel plating. For some electroplating uses, a nickel anode has been recommended which contains deliberately added sulfur.*

Uses

The major use of nickel is as an alloying element in a host of ferrous and nonferrous alloys. The pure metal also has some uses, either itself or as a cladding in reaction vessels in the chemical industry and some food-handling equipment. Except for tin and cadmium, nickel is the only metal whose electroplating applications constitute a major proportion of its consumption, usually running about 15 to 17 percent of the total.†

Total nickel consumption in the United States averages about 160,000 t annually, about 20,000 t in electroplating applications.

*S electrolytic nickel, TM International Nickel Co.

†For this reason, the major supplier of nickel (formerly International Nickel Co., now INCO) takes a great interest in nickel plating and has published valuable brochures and technical papers in this field.

Table 12-12 Typical Analyses of Some Grades of Nickel, Wt %

Element	Commercial metal	*Nickel anodes, with functional additions* Rolled depolarized	Rolled carbon	Cast carbon	"S" Nickel™
Ni	99.25–99.95	balance	balance	balance	99.95
Co	0.0005–0.45				
Cu	0.001–0.07	——	——	——	< 0.01
Fe	0.0009–0.1	——	——	——	< 0.01
S	0.0007–0.004	0.005	0.01	0.006	0.02
As	0.0001–0.0005	——	——	——	——
Pb	0.0001–0.0002	——	——	——	——
C	0.003–0.01	——	0.20	0.24	< 0.01
Si	——	0.01	0.25	0.26	——
O	——	0.15	——	——	——
Mg	——	——	0.08	——	——

—— indicates not reported.

The number of nickel alloys, in which the metal is either a major or a minor constituent, is too large for listing here. Among the more important are the stainless steels, various copper-nickel alloys, and the so-called superalloys for high-temperature uses. Its main contributions to these formulations are its corrosion resistance, magnetic properties, and resistance to high temperatures.

An interesting nickel alloy, considered separately in Chap. 16, is tin-nickel, a phase which does not appear on the equilibrium diagram of the tin-nickel system and has been prepared only by electrodeposition.

Properties

The properties of nickel are summarized in Table 12-13. The metal is ferromagnetic at ordinary temperatures, but when heated above 353°C, the Curie point, it becomes paramagnetic. The magnetic properties of nickel and some of its alloys must be taken into account when testing for coating thickness by a magnetic method (Chap. 23).

Nickel has moderate hardness and strength, good ductility and toughness, and good thermal and fair electrical conductivity. It is readily fabricated and will take and retain a high polish.

Nickel is not attacked at ordinary temperatures by dry or moist air. At red heat it slowly reacts with steam according to

$$Ni + H_2O \rightarrow NiO + H_2$$

It tarnishes in contaminated urban atmospheres; this is one of the principal reasons why most decorative nickel plate is overplated with chromium, which does not tarnish. Nonoxidizing acids slowly attack it in the cold. Dilute nitric acid dissolves it rapidly; concentrated nitric acid passivates it. Alkaline and neutral salt solutions, and most organic compounds, have little effect. Nickel resists caustic alkalis, either in solution or in the molten state; this allows the use of nickel or nickel-lined vessels for the production of sodium and potassium hydroxides in industry, and nickel crucibles for alkaline fusions in the laboratory.

The excellent corrosion resistance of nickel to so many reagents extends to many of its alloys, and the literature on this subject is very extensive.

These statements apply to massive metal. Finely divided nickel powder is pyrophoric and is a powerful catalyst, especially for hydrogenation and dehydrogenation of organic compounds; so-called Raney nickel is extensively used for such purposes.

The nickel atom has the electronic configuration $3d^84s^2$; with an unfilled d shell, it is classified as a transition metal, and in fact exhibits all valences from 1− to 4+ in at least some compounds, including zero (in the carbonyl).

For most practical purposes the only valence that needs to be considered is the divalent Ni(II). The complex chemistry of nickel is not nearly so extensive as that of its homolog cobalt. In the latter the trivalent state is stabilized by complexation; nickel is much more difficult to oxidize above the 2+ state, and its complexes are not notably stable; exceptions are the cyanides, ammonia and amines, and the chelates formed with dimethylglyoxime and its derivatives.

One practical application of a higher valence is the Edison storage battery, whose reactions are

$$Fe + 2NiO(OH) + 2H_2O \underset{\text{charge}}{\overset{\text{discharge}}{\rightleftharpoons}} Fe(OH)_2 + 2Ni(OH)_2$$

although the exact nature of the oxidized nickel species has not been elucidated.

Electrochemically nickel is slightly negative: $E^0 = -0.25$ V. However, its deposition, unlike that of most metals, involves considerable overvoltage.

Table 12-13 Properties of Nickel

Atomic number	28
Atomic weight	58.7
Electronic configuration	$1s^22s^22p^63s^23p^63d^84s^2$
Crystal structure	fcc
Melting point, °C	1453
Boiling point, °C	2730
Density, g/cm³, 20°C	8.908
Electrical resistivity, $\mu\Omega$ −cm, 20°C	6.84
Tensile strength, annealed, MPa	317
Standard potential, E^0, 25°C, V	
$Ni^{++} + 2e^- \rightarrow Ni$	−0.25

The green color of nickel salts in solution, and of the solid hydrates, is due to the Ni^{++} aquo ion; anhydrous salts are mostly yellow. Alkaline hydroxides precipitate pale green Ni(II) hydroxide from solutions at a pH about 6.7; the precipitate is insoluble in excess alkali (i.e., nickel is not amphoteric). It is readily soluble in acids and ammonia or ammonium salts, and in alkali cyanide solutions. Hydrogen sulfide precipitates a black nickel sulfide from solutions made alkaline with ammonium hydroxide; at low pH no precipitate is formed, although once formed the precipitate becomes insoluble in fairly concentrated hydrochloric acid.

Nickel compounds have two principal uses: as catalysts and in electroplating. In the plating process, although most of the nickel comes from the anodes, salts are needed for maintenance and makeup. The principal compounds, from the electroplater's standpoint, are listed below. All are simple salts of the divalent nickel aquo ion.

Nickel carbonate. The commercial material does not conform to the simple formula $NiCO_3$, but is a basic salt of somewhat indefinite composition: $x Ni(CO_3)_2 \cdot y Ni(OH)_2 \cdot z H_2O$, where $x = 2$, $y = 3$, and $z = 4$, approximately. It is used mainly for pH control.

Nickel chloride, $NiCl_2 \cdot 6H_2O$. It is a constituent of many nickel-plating baths.

Nickel fluoborate, $Ni(BF_4)_2$, is sold only as a 41% concentrate in water, containing a little free boric acid.

Nickel sulfamate, $Ni(SO_3NH_2)_2 \cdot 4H_2O$, is used in sulfamate-plating baths, primarily for electroforming.

Nickel sulfate, $NiSO_4 \cdot 6H_2O$. There exists considerable confusion in the literature as to whether the commercial material is a hexa- or heptahydrate; both exist. Both forms have been available, and the heptahydrate is the one normally formed when a water solution crystallizes at ordinary temperatures. The material of commerce, however, is the hexahydrate. It is the principal ingredient of the most widely used nickel-plating bath.

Hazards

Nickel and its salts are not poisonous, unless the associated anion is and unless very large amounts are ingested. Skin contact can cause dermatitis ("nickel itch") in susceptible individuals. One notable exception is nickel carbonyl, $Ni(CO)_4$, whose vapors are extremely toxic—much more so than carbon monoxide, so that the effect is not due entirely to the CO content. Long exposure to nickel-containing dusts has been shown to be carcinogenic. To the plater, therefore, the handling of nickel salts is hazardous unless proper precautions are taken.

Nickel Plating

Nickel is perhaps the most versatile metal in the electroplater's repertory. It is the essential element of the copper-nickel-chromium decorative/protective system; in many cases copper can be omitted without detracting from the protective value of the coating (except where it is required, as on zinc die castings), but nickel is essential.* Nickel plating has also probably been the most thoroughly investigated, from both the practical and the theoretical viewpoints, of any of the plating metals. There are at least three reasons for this interest: (1) Nickel plating is an important part of the total market for the metal (unlike most metals whose plating uses are insignificant in relation to total consumption), so that the leading producer has devoted considerable research effort to the subject; (2) nickel is, of all the plating metals, perhaps the most sensitive to the effects of additives in the electrolyte, thus stimulating academic research into electrode mechanisms: (3) as a corollary to (2), nickel-plating additives are a mainstay of the large plating supply firms, and the competition for additive business has prompted them to devote a large proportion of their research effort to this field. As a result, there exists a tremendous literature, both academic and practical, concerning nickel plating. The number of nickel-plating bath formulations is large, and the deposit's properties can be regulated within a wide range depending on the electrolyte employed and the conditions of operation. Nickel can be plated mat, semibright, or fully bright; hard or (comparatively) soft; with or without internal stress; ductile or brittle; quite corrosion-resistant or having only decorative appeal.

The principal application for nickel plating is as a bright coating under a much thinner chromium plate to provide a lustrous and protective finish for articles of steel, brass, zinc die castings, plastics (after chemical metalization), and to some extent on aluminum and magnesium alloys. The nickel plate—its thickness and type—is the main factor determining the corrosion protection afforded to the substrate. In cases where copper undercoats are not required—as they are on zinc die castings—it is still somewhat controversial whether copper adds appreciably to the performance of the composite, and whether copper can take the place of some of the nickel. At one time the chromium overplate was thought to have little function except to prevent tarnishing; it is now evident that its character and thickness also have an effect. Nevertheless it remains true that the nickel is the principal factor in protecting the substrate.

The bath chosen and the conditions of operation will depend on the use to be made of the deposit: for engineering applications, where physical properties and thickness are paramount; for purely decorative applications on inexpensive articles, where the principal aim is to achieve a bright deposit at the minimum thickness (therefore minimum cost); for such uses as automotive brightwork, where both appearance and protective value must be balanced against cost; and

*The results of the attempts to dispense with nickel, or to greatly reduce its thickness, during the severe nickel shortage of the mid-1950s were catastrophic.

so on. To serve these many markets, many different electrolytes are available, and within any given bath formulation minor variations in additives and operating conditions offer further possibilities for choice.

For purely decorative purposes, thin bright nickel may be overplated with very thin deposits of gold or brass. Nickel coatings alone are used industrially both for corrosion protection and to prevent contamination of the contents. Nickel is widely used in electroforming: of printing plates, phonograph record stampers, foil, screen, and many other items.

Nickel plating evolved slowly to its present state. The first sound nickel deposits were reported by Böttger in 1842, Adams, often called the father of nickel plating, developed the first commercially successful process in about 1870. The use of boric acid was introduced a little later, and that of chlorides to prevent anode passivity in about 1906. The present Watts bath, named for its inventor, was announced in 1916; with some improvements, principally in additives, somewhat higher concentrations, and anode compositions, it has remained the basis of most nickel-plating operations since then.

The Watts Bath

The typical formulation of the Watts bath is shown in Table 12-14. The functions of the constituents may be summarized as follows.

Nickel sulfate provides most of the nickel-ion content. It is the cheapest available nickel salt, with a stable anion that is not reduced at the cathode nor oxidized at the anode and that is nonvolatile. One of the few changes made in the original Watts formula was to increase the nickel sulfate concentration, allowing the use of higher current densities and better plate distribution.

Nickel chloride is used as the source of chloride ion, required to prevent anode passivity. Although other chlorides would serve this function, they would introduce extraneous cations; unlike those of many metals, the properties of nickel deposits are sensitive to these otherwise ''inert'' cations such as sodium, potassium, and ammonium. Chloride ion also increases the conductivity of the bath and improves throwing power.

Boric acid appears to serve as a weak buffer, controlling the pH of the cathode film. Although good deposits can be produced without it, it tends to produce whiter deposits and is compatible with the many additives that have been developed for the Watts bath, discussed below. It is obtainable in very pure form and is relatively inexpensive, stable, and nonvolatile.

Anti-pitting agents are necessary because the cathode efficiency of the Watts bath is not 100 percent; it averages about 97 percent. Thus sufficient hydrogen is discharged to produce pits caused by slowly forming hydrogen gas bubbles clinging to the cathode surface. To promote release of this hydrogen, hydrogen peroxide was used as an anti-pitter before the development of the organic bright nickels, and it is still used occasionally. When carefully controlled, mat nickel plate of good ductility can be obtained. It functions by depolarization of the cathode for hydrogen evolution and by oxidation of traces of organic contami-

Table 12-14 The Watts Nickel Bath

	Range		*Usual*	
	g/L	*Molarity*	*g/L*	*Molarity*
Nickel sulfate, $NiSO_4 \cdot 6H_2O$	225–375	0.86–1.43	330	1.25
Nickel chloride, $NiCl_2 \cdot 6H_2O$	30–60	0.13–0.25	45	0.19
(Total Ni as metal)	58–100	1–1.68	85	1.44
Boric acid, H_3BO_3	30–40	0.5–0.65	37	0.60
Anti-pit	see text			
Temperature, °C	45–65		60	
pH	1.5–4.5		3–4	
Current density, A/m^2	250–1000		500	

nants in the bath. In using the basic Watts bath without organic additives, about 0.5 mL of 30% hydrogen peroxide per liter of solution per day is recommended.

Hydrogen peroxide, however, is not compatible with the organic brighteners and levelers used in most commercial decorative nickel baths. In its stead anionic wetting agents are used to reduce interfacial tension and so to promote the release of the hydrogen bubbles before they grow too large and promote pitting. Sodium lauryl sulfate is a common surfactant in this application, but most proprietary formulas do not divulge the nature of the surfactant used.

Oil, grease, and other unwanted organic contaminants must be kept out of the bath for proper functioning of the anti-pit, whether it be hydrogen peroxide or a surfactant.

Most nickel anodes and nickel salts contain some cobalt, since manufacturers seldom try to get rid of the last traces of cobalt in commercial nickel materials. These small amounts of cobalt have little or no effect on the bath or the deposit. (Many specifications call out "nickel + cobalt" rather than nickel alone.) The deliberate inclusion of larger amounts of cobalt to deposit a true nickel-cobalt alloy has been fairly extensively investigated, as mentioned later.

Effect of Variables

As shown in Table 12-14, the composition of the Watts bath can be varied over fairly wide ranges. The average formula is suitable for average current densities of 500 A/m^2 at about 50°C; current densities from one-half to twice this value will still yield good deposits. If lower current densities are to be used, say about 200 A/m^2, the nickel sulfate and nickel chloride concentrations may be halved. For higher current densities, their concentrations may be increased somewhat, but limits are set by excessive drag-out and possible crystallization of salts. Current densities above 1000 A/m^2 can be achieved more satisfactorily by increasing the agitation, temperature, and ratio of chloride to sulfate.

In other words, all operating conditions are interrelated: current density, pH,

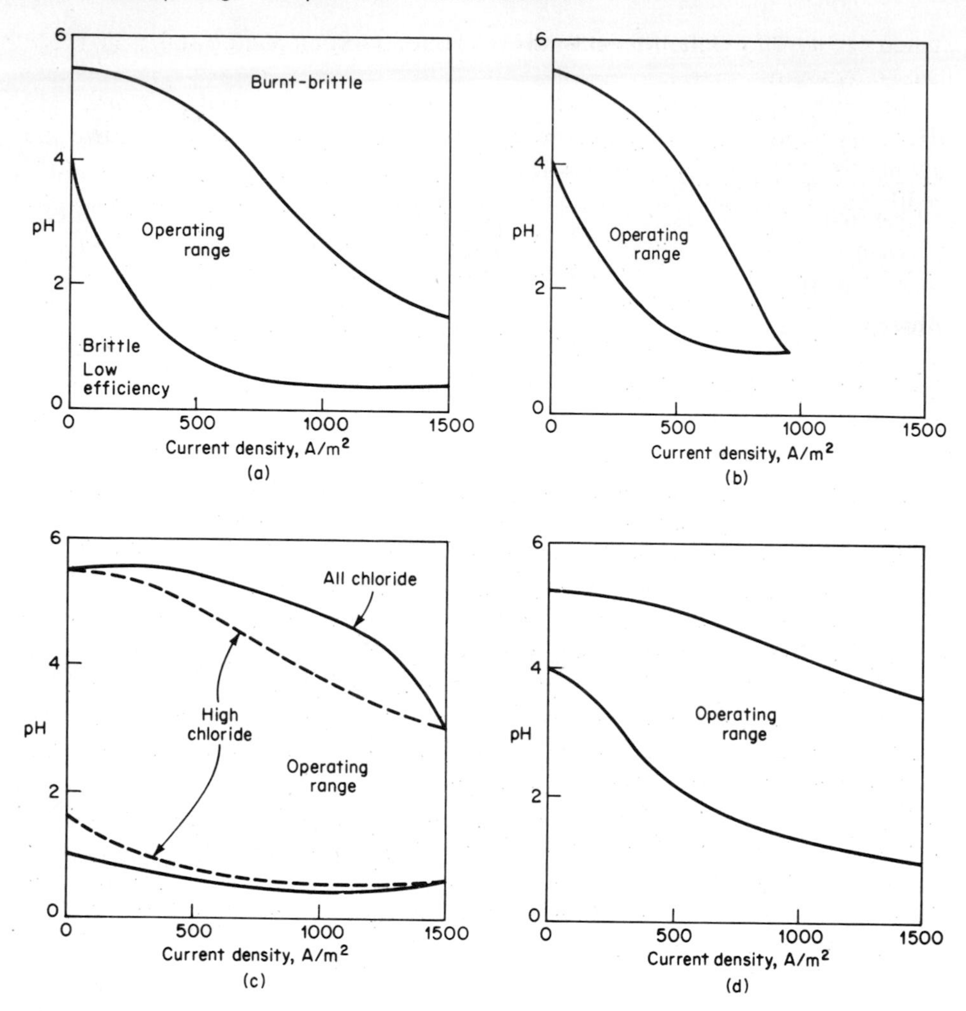

Fig. 12-2 Effects of operating variables on plating range, Watts nickel bath: (*a*) normal Watts bath; (*b*) effect of reducing temperature, from 60°C to 40°C; (*c*) effect of increasing chloride content; (*d*) effect of increased agitation.

temperature, agitation, and bath composition. The effects of some of these variables are shown graphically in Fig. 12-2. In this figure (*a*) shows the usual operating range (interrelation of pH and current density) of a normal or "average" Watts bath used under the conditions set forth in Table 12-14. If the temperature is lowered from 60 to 40°C, the operating range is much narrowed, as shown in (*b*); increasing chloride content widens the range, as (*c*) shows, and the range is also widened by more vigorous agitation, as shown in (*d*).

Low pH allows a greater range of current densities, but current efficiency and throwing power suffer. The usual pH range is 3.5 to 4.5; pH as low as 1.5 to 2 is sometimes used for a flash coat on steel to be followed by plating under the more usual conditions.

The throwing power of the Watts bath is somewhat improved by increase in pH, temperature, and nickel and chloride content. At high current densities, the low-pH Watts bath is only slightly inferior to other formulations; and by sufficiently increasing the chloride content, the throwing power can be made superior to them.

Internal Stress

The structure and mechanical properties of nickel deposits from the Watts bath, as well as from many other nickel-plating baths, are very sensitive to operating conditions, especially the pH, temperature, and current density. One of the important variables in nickel deposits is the internal stress; this property has been extensively studied, and a summary of the principal findings is given here. It will be understood that brighteners also have a significant effect on this property.

Tensile stress increases as the chloride content of the bath increases.

Temperature of deposition has variable effects depending principally on the chloride content and current density.

The effect of pH varies with bath composition; it is advisable to maintain pH below 5.0 in the Watts bath.

Current density in the range of 100 to 500 A/m^2 has only a slight effect, usually in the direction of increasing tensile stress as current density increases.

Superimposed alternating current can reduce stress.

Agitation has little effect if the bath is pure; in bright-plating baths decreased agitation will decrease stress.

Many impurities and additives act to increase stress markedly; these include hydrogen peroxide; inorganics such as lead, zinc, iron, chromium, aluminum, and phosphate ion; organic impurities including sizing from anode bags, amines from insufficiently cured rubber linings, and excessive concentrations of Class II brighteners (see p. 219).

Fluorides and fluoborates reduce tensile stress.

Many other variables have been studied, sometimes with inconsistent results; for details the literature must be consulted. Hardness, stress, and ductility are interrelated, but not necessarily always in the same direction: i.e., a hard deposit is not necessarily strong or brittle, etc.

Variations of the Watts Bath

Many variations on the basic Watts formulation are used in special applications. An all-chloride bath, included in Table 12-15, has better conductivity and throwing power than the Watts bath, and yields harder and finer-grained deposits; but it is not as amenable to modification by brighteners and other addition agents. A high-chloride, or chloride-sulfate bath, is intermediate in properties between the Watts and the all-chloride, as might be expected. An all-sulfate bath, with no nickel chloride, is used in special situations where insoluble lead anodes must be used, such as in plating the inside of a small-diameter tube. Platinum or platinized anodes may be used with conventional baths for similar purposes.

All these solutions can produce nickel deposits, with no limit on the thickness obtainable. Under average conditions, 500 A/m^2, pH of 3 to 4, and 60°C, the Watts bath produces satiny, mat deposits. These deposits are satisfactory as plated for many engineering applications; and for many years, before the development of the modern bright nickel baths, they were buffed to a high luster before application of the thin chromium deposit. The buffing operation not only brightened the plate, making it suitable for decorative applications, but undoubtedly improved its corrosion resistance by filling in or bridging over pores or discontinuities. Many years elapsed before bright nickel systems were developed that equaled in corrosion performance the buffed Watts plate.

BRIGHT NICKEL PLATING

The development of nickel-plating baths that can produce fully bright plate directly, without the necessity of buffing, has completely revolutionized the practice of decorative nickel plating (including decorative/protective nickel-chromium systems). The objections to buffing are principally economic: it is labor-intensive and requires skills that are becoming rare; it inevitably causes some decrease in metal thickness; the buffing compounds must be cleaned off before proceeding with chromium plating; and the materials used in buffing are themselves costly. For all these reasons the advent of "bright nickel" was eagerly accepted by the trade, and "buffed Watts" is now essentially obsolete in decorative nickel plating. Early bright nickel deposits sacrificed a considerable degree of corrosion performance, but later developments in "duplex" nickel plate and microdiscontinuous chromium have rendered modern bright nickel systems equal in performance to buffed Watts, at least according to most workers.

The earliest brighteners were inorganic cations, especially cadmium. The first practical bright nickel system to be commercialized in this country was the Weisberg-Stoddard bath, which contains cobalt, formaldehyde, formates, and other additives; the plate also contains 1 to 5 percent cobalt. This bath had considerable commercial success, and is still not entirely obsolete, but almost all present-day bright and semibright nickel baths rely on organic additives to the Watts bath, or to modifications of it.

Depending on the additives used, the chloride and boric acid concentrations of the Watts bath can vary. Bright, and semibright, nickel plating is a highly proprietary field, and even when the patents on which additives are based expire, metal finishers seldom attempt to formulate their own brighteners, but rely entirely on supply houses for instruction, trouble-shooting, and materials. There are many additives on the market, often tailored to the specific needs of the user: some "build brightness"* quickly for articles that are to be plated for appearance only, with no regard for corrosion performance; some are adapted to produce more corrosion-resistant deposits, at some sacrifice in the ability to build brightness; others are formulated to be relatively insensitive to contamination by zinc, which is almost inevitably present in plating on zinc die castings. Competing claims are made for economy, brightness, corrosion resistance, ease of control, receptivity to chromium, and other factors. Since no one proprietary process dominates the market, it is obvious that in all cases some compromises are made; finishers must decide for themselves among the claims.

The modern proprietary bright nickel–plating processes generally use combinations of organic addition agents for best results.

Brighteners

Nickel brighteners fall into two classes. Those designated as Class I brighteners include aromatic sulfonic acids, sulfonamides, and sulfinic acids (see Fig. 12-3). The aromatic rings (R) attached to these groups are usually benzene or naphthalene rings, but may also be various unsaturated aliphatic groups such as vinyl or allyl. These brighteners produce almost bright (hazy or cloudy) plate but cannot "build" brightness as plating proceeds. The concentrations of these additives are not critical; they are used in relatively high amounts, 1 to 10 g/L, without much effect on adhesion and limiting current density. They tend to decrease the tensile stress of the deposit and at higher concentrations can even turn the stress compressive. They introduce sulfur into the deposit, in amounts up to about 0.03 percent at a pH of 3 to 5, when used in warm baths without Class II brighteners.

Class II brighteners are used in combination with those of Class I to produce fully bright, brilliant, and leveling deposits, the luster of which increases with continued plating up to the maximum obtainable (they build brightness). They are usually unsaturated organic compounds, of many types. They introduce carbon or carbon-containing material into the plate. Most Class II brighteners must be used in conjunction with a Class I additive; if used alone, they produce brittleness and tensile stress, as well as inferior adhesion to the substrate. (Coumarin, see below, is an exception.)

The most effective Class II brighteners are organic compounds containing such unsaturated groups as aldehyde $\mathrm{H{-}\overset{|}{C}{=}O}$, as in formaldehyde $\mathrm{H{-}\overset{\overset{H}{|}}{C}{=}O}$; the

*The plate becomes fully bright in a short time at low thicknesses.

Table 12-15 Baths for Heavy Nickel Plating

Type	Ingredients*	Concentration g/L	Concentration Molarity	pH electrometric
Watts	Nickel sulfate, $NiSO_4 \cdot 6H_2O$	330	1.25	1.5–4.5
	Nickel chloride, $NiCl_2 \cdot 6H_2O$	45	0.19	
	Boric acid, H_3BO_3	38	0.61	
Hard	Nickel sulfate, $NiSO_4 \cdot 6H_2O$	180	0.68	5.6–5.9
	Ammonium chloride, NH_4Cl	25	0.47	
	Boric acid, H_3BO_3	30	0.49	
Chloride	Nickel chloride, $NiCl_2 \cdot 6H_2O$	300	1.26	2.0
	Boric acid, H_3BO_3	38	0.61	
Chloride-sulfate	Nickel sulfate, $NiSO_4 \cdot 6H_2O$	200	0.76	1.5–2.0
	Nickel chloride, $NiCl_2 \cdot 6H_2O$	175	0.74	
	Boric acid, H_3BO_3	40	0.65	
Chloride-acetate	Nickel chloride, $NiCl_2 \cdot 6H_2O$	135	1.26	4.5–4.9
	Nickel acetate, $Ni(C_2H_3O_2)_2 \cdot 4H_2O$	105	0.42	
Fluoborate	Nickel (as fluoborate)	75	1.28	2.0–3.5
	Free fluoboric acid, HBF_4	to pH →		2–3.5†
	Boric acid, H_3BO_3	30	0.49	
Sulfamate	Nickel sulfamate, $Ni(NH_2SO_3)_2$	450	1.80	3.0–5.0
	Boric acid, H_3BO_3	30	0.49	
Sulfamate-chloride	Nickel sulfamate, $Ni(NH_2SO_3)_2$	300	1.20	3.5–4.2
	Nickel chloride, $NiCl_2 \cdot 6H_2O$	6	0.03	
	Boric acid, H_3BO_3	30	0.49	
Nickel-cobalt	Nickel sulfate, $NiSO_4 \cdot 6H_2O$	240	0.91	4.7
	Nickel chloride, $NiCl_2 \cdot 6H_2O$	23	0.10	
	Boric acid, H_3BO_3	30	0.49	
	Ammonium sulfate, $(NH_4)_2SO_4$	1.5	0.01	
	Nickel formate, $Ni(CHO_2)_2 \cdot 2H_2O$	15	0.08	
	Cobalt sulfate, $CoSO_4 \cdot 7H_2O$	2.6	0.01	

*Usually + anti-pit.

†Colorimetric.

Temperature, °C	Cathode cd, A/m^2	Properties: Hardness, Vickers	Tensile strength, MPa	Elongation, %	Residual, stress, MPa
45–65	250–1000	140–160	380	30	125
43–60	200–1000	350–500	1030	5–8	300
50–70	250–1000	230–260	690	20	275–345
45	250–1000				
30–50	200–1000	350	1380	10	
40–80	400–1000	183	515	15–30	110
40–60	200–3000	250–350	620	20–30	3.5
28–60	200–2500	190	745	15–20	10
40	500	450–500			100–120

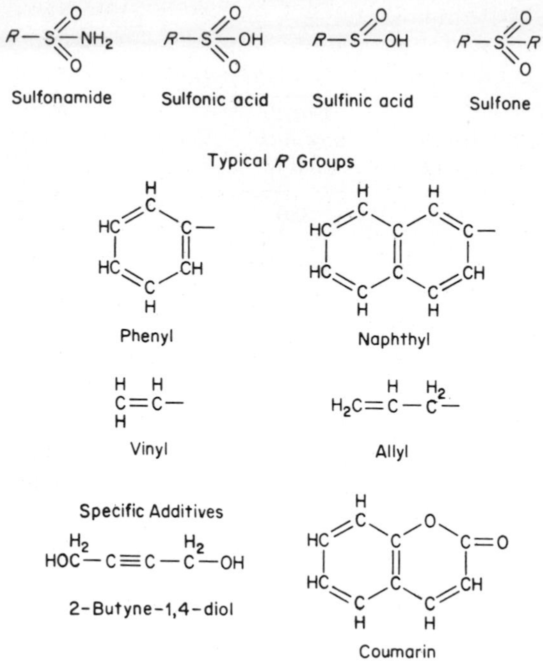

$R{-}S(=O)_2{-}NH_2$ Sulfonamide

$R{-}S(=O)_2{-}OH$ Sulfonic acid

$R{-}S(=O){-}OH$ Sulfinic acid

$R{-}S(=O)_2{-}R'$ Sulfone

Typical *R* Groups

Phenyl

Naphthyl

$HC(H)=C(H)-$ Vinyl

$H_2C=C(H)-C(H_2)-$ Allyl

Specific Additives

$HOC(H_2)-C{\equiv}C-C(H_2)-OH$

2-Butyne-1,4-diol

Coumarin

Fig. 12-3 Structural formulas of some organic nickel brighteners.

olefinic —C=C— group, as in coumarin (see Fig. 12-3): the C≡C linkage, as in butyne diol (Fig. 12-3). These are only a few of the typical groups found in Class II brighteners, almost all of which are the subject of patents and are sold under proprietary names.

Semibright and Duplex Nickel

The introduction of coumarin to yield ductile, highly leveling, and semibright deposits which are sulfur-free represented a major advance in the performance of the nickel-chromium composite. Although the deposits are not fully bright, they can be buffed easily; but the present trend is to deposit a layer of semibright sulfur-free nickel, to provide leveling, smoothing, and corrosion resistance, following this with a fully bright formulation to provide the necessary decorative appearance. The combination of semibright, sulfur-free nickel followed by fully bright sulfur-containing nickel is known as *duplex* nickel. The system is preferred at present for all applications requiring a combination of bright appearance with good corrosion resistance, the principal one being automotive brightwork. For further corrosion resistance, the chromium applied to this duplex system can be of the "microdiscontinuous" type, discussed later in this section as well as in Chap. 3.

Since the fully bright, sulfur-containing nickel is more "active," i.e., anodic, than the underlying semibright sulfur-free nickel, corrosion tends to proceed laterally in the bright layer and not to penetrate to the basis metal; see Fig. 12-4. If microdiscontinuous chromium is also used, a similar mechanism holds, spreading the corrosion laterally and further delaying penetration to the substrate.

The microdiscontinuous chromium deposits may be achieved by modifications in the chromium-plating solution, as discussed later in this chapter; they may also be induced by modifying the nickel-plating electrolyte. In one proprietary process, solid particles such as barium sulfate can be included in the nickel deposit; after practically all the required nickel is applied, the special solid-containing bath is used to produce a surface consisting of many small areas of nonconducting material. When the chromium topcoat is applied, the chromium does not plate over these particles, and the result is a "microporous" chromium (see also below). In another proprietary process, a highly stressed nickel deposit is applied as a thin layer over the main nickel coating; chromium plating over this brittle deposit cracks it, again resulting in a chromium deposit having many fine cracks.

Special-purpose Baths

There are numerous special purpose nickel-plating baths: so-called black nickel, various alloy electrolytes, and baths yielding phosphorus-containing deposits (which are more than usually amenable to heat treatment). Some are listed in Table 12-16. During a severe nickel shortage in the late 1960s, cobalt-nickel alloy deposits were widely used; cobalt is more expensive than nickel, and as far as it has been investigated, it appears to have few advantages as an electrodeposit. But the cobalt was available and nickel was not. The plating properties of the two

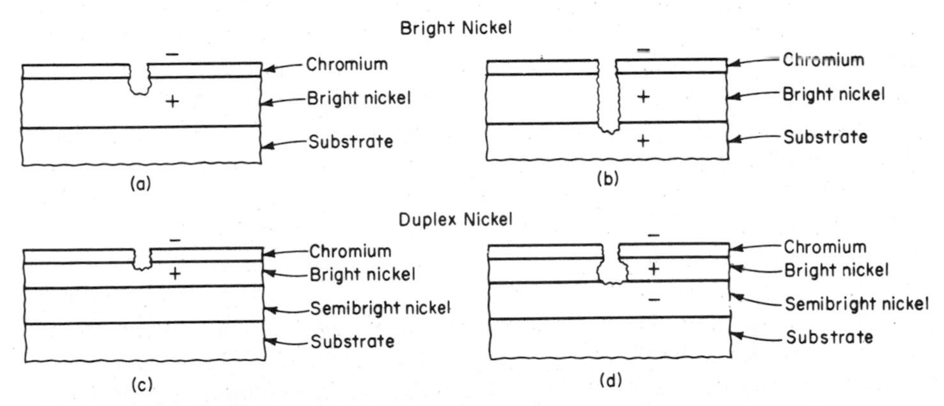

Fig. 12-4 Corrosion behavior of nickel/chromium systems. Bright nickel (top): (*a*) pit initiation; (*b*) penetration of pit. Duplex nickel (bottom): (*c*) pit initiation; (*d*) lateral spreading of pit, no penetration.

Table 12-16 Special-purpose Nickel Baths

Purpose	*Ingredients*	*g/L*	*Molarity*	*pH (electrometric)*	*Temperature, °C*	*Current density, A/m²*
Electrotyping	Nickel sulfate, $NiSO_4 \cdot 6H_2O$	70	0.27	5.6–6.0	32	100–200
	Ammonium chloride, NH_4Cl	6	0.11			
Barrel plating	Nickel sulfate, $NiSO_4 \cdot 6H_2O$	150	0.57	5.0–5.5	24–32	
	Ammonium chloride, NH_4Cl	30	0.56			
	Boric acid, H_3BO_3	30	0.49			
Black nickel	Nickel chloride, $NiCl_2 \cdot 6H_2O$	75	0.32	5.0	24–32	15
	Ammonium chloride, NH_4Cl	30	0.56			
	Sodium thiocyanate, NaCNS	15	0.19			
	Zinc chloride, $ZnCl_2$	30	0.22			
Nickel-Phosphorus alloy	Nickel sulfate, $NiSO_4 \cdot 6H_2O$	150–175	0.57–0.67	0.5–1.0	75–95	450–3500
	Nickel chloride, $NiCl_2 \cdot 6H_2O$	45–50	0.19–0.21			
	Phosphoric acid, H_3PO_4	50	0.51			
	Phosphorous acid, H_3PO_3	13–40	0.16–0.49			

metals are fairly similar (see Chap. 14), and so-called sliding conversions could be accomplished. A sliding conversion is one in which a solution of one composition is gradually converted to one of a different composition, with no interruption in production, simply by adding the substitute ingredient instead of the usual maintenance material. For example, by using cobalt anodes in a nickel-plating electrolyte, the cobalt content of the bath is built up as the nickel content decreases, and the properties of the bath and of the deposit change gradually with the change in composition. It is not necessary to dump the old bath and make up a new one.

More recently, alloys of nickel with iron have been promoted by suppliers, in this case not because of any shortage but because of the sharp rise in the price of nickel. Alloys containing up to about 30 to 35 percent iron can be produced by sliding-conversion techniques. With the appropriate proprietary addition agents these deposits are claimed to be the equivalent of nickel for many purposes, although at present their properties have not been thoroughly explored.

For the tin-nickel alloy, see Chap. 16.

Codeposition of Particles

Several types of nonconducting particles, insoluble in the nickel baths, can be dispersed in the bath and codeposited uniformly with the plate. For uniform dispersion in the bath and best results, the particles should be of a size less than about 5 μm. Various materials have been used, including alumina, titania, zirconia, silicon carbide, and barium sulfate.

Codeposition of nonconducting particles with nickel has been used to improve its wear and abrasion resistance; diamond powders have been codeposited for use in abrasive tools. Nickel has been dispersion-hardened by the codeposition of 2 to 6 percent (by volume) of very fine particles of alumina.

A more recent application of the codeposition of fine nonconducting particles has been for decorative purposes; the list of materials used is about the same as mentioned above, and their size ranges from 0.2 to 5 μm. When used as the principal deposit, this technique affords a pleasing satin-like appearance. It can also be used to cause the chromium plate deposited on it to be microdiscontinuous, as discussed above.

HEAVY NICKEL PLATING

So far we have discussed principally decorative nickel plating, which today is done almost entirely with the Watts bath, or variations of it, with proprietary additives. Another important aspect of nickel plating is its use for heavy deposits, both for electroforming applications and for corrosion and wear resistance, as well as for salvaging worn or mismachined parts. Thickness of nickel for such applications has no real upper limit; it ranges from about 75 μm upward.

Although the Watts bath, with or without additives, is used for such applications, many other formulations are available as well, and the properties of the nickel deposit can be varied almost at will, within the possible characteristics of nickel itself. The baths used for such engineering purposes are listed in Table 12-15, along with some of the principal physical properties of typical deposits. Even this list does not exhaust the possibilities of this aspect of nickel plating, since much work has been reported on the effects of various additives on these basic formulations. Many of these "engineering" deposits can be annealed or heat-treated to improve their physical properties still further.

Other special-purpose baths are listed in Table 12-16.

Finally, the so-called Woods nickel strike bath should be mentioned. This bath is not used primarily for nickel plating as such, but as preparation for plating on "difficult-to-plate" substrates, for reactivating surfaces that have become passive through long standing, and for similar uses, e.g., as a preparatory technique for applying further deposits. This bath has the following composition:

Nickel chloride $NiCl_2 \cdot 6H_2O$	240 g/L
Hydrochloric acid, concentrated, HCl	125 mL/L
Temperature	20–30°C
Current density	500–1500 A/m^2

This strike is valuable for plating on stainless steel and other metals (including nickel itself) which ordinarily will not accept an adherent plate.

ANODES

In the early days of nickel plating, anodes containing as much as 10 percent iron were used; before the development of chloride-containing baths to improve anode corrosion, the iron content appeared to prevent anode passivity. With the development of the Watts bath, the iron content was no longer required, and anodes containing at least 99 percent nickel have become the standard. These may be rolled, wrought, or cast, and some have deliberate additions of minor elements to ensure uniform dissolution of the anode in the bath; some of these formulas are included in Table 12-12.

The convenience of using anode baskets containing electrolytic chips or other types of pellets is obvious: large and heavy anodes need not be handled, no "swords" or scrap are produced, and thus anode maintenance is much simplified. The practicality of anode baskets for nickel plating had to await the availability on a commercial scale of titanium metal in workable form; other metals that might be used as anode containers either would dissolve in the bath, like iron, or are much too expensive, like platinum. Titanium has the advantage of being able to carry the current to the nickel chips, while at the same time remaining passive in the nickel bath and retaining its form and mass. The use of titanium anode baskets, filled either with electrolytic nickel (pieces of electrorefined nickel produced directly from the refinery cathodes) or with sulfur-bearing electrolytic nickel, is the preferred method of handling anodes. The sulfur-containing round "buttons" were introduced in 1972; and, according to the producer, they have many advantages for use in most nickel-plating electrolytes.

Titanium anode baskets are not universally applicable. In using baths containing ions that can activate titanium, such as fluoride, caution is called for.

Nickel deposits are sensitive not only to dissolved contaminants, but also to particulate impurities; consequently anodes usually should be bagged.

CHROMIUM

Chromium (Cr) is the first member of Group VIB of the periodic system; the other two are molybdenum and tungsten.

A mineral was found in Siberia in 1762 which was recognized as containing a new element. In 1797 this element was discovered and christened chromium, from Greek *chromos,* color, because of the many and intense colors it imparts to its compounds and minerals. The green color of emerald and the red of ruby are due to chromium in minute amounts.

Chromium ranks about twenty-first in order of abundance in the lithosphere, averaging about 0.035 percent. It is widely dispersed but never found in the uncombined state.

The only commercial ore is the mineral chromite, of ideal formula $FeO:Cr_2O_3$ or $FeCr_2O_4$, found mainly in the Soviet Union, the Union of South Africa, the Philippines, Turkey, Southern Rhodesia, and New Caledonia. Low-grade ores are mined in the United States. Chromium and its compounds have three principal uses: in metallurgy, as refractories, and in the chemical industry. Ores suitable for these uses differ and are classified accordingly as metallurgical, refractory, or chemical grade.

Although chromium was discovered in 1797, it was half a century before fairly pure samples of the metal were produced by Bunsen in 1854 by carbon reduction of chromic oxide:

$$2Cr_2O_3 + 3C \rightarrow 4Cr + 3CO_2$$

Carbon-free chromium was produced by the aluminothermic reduction (thermite process):

$$Cr_2O_3 + 2Al \rightarrow Al_2O_3 + 2Cr$$

These metals are contaminated by carbon and aluminum, respectively, and a third similar process, silicothermic reduction, yields a metal contaminated with silicon:

$$2Cr_2O_3 + 3Si \rightarrow 4Cr + 3SiO_2$$

Modern methods of producing chromium rely on electrolysis either of a chromium ammonium sulfate (chrome alum) solution in a divided cell or of chromic acid solutions, comparable to chromium-plating processes. Although electrolytic chromium is much purer than the other forms, it still contains too much oxygen for workability; this can be minimized by reduction with hydrogen or calcium vapor or by the iodide process.*

Uses

Although there are at present no commercial alloys containing chromium as a major (>50 percent) constituent, it is an important alloy addition to a host of ferrous and nonferrous metals. Up to about 3 percent chromium is added to low-alloy steels to improve mechanical properties and hardenability. Steels containing 5 to 6 percent chromium have increased resistance to corrosion and oxidation. Steels containing over 10 percent chromium are known as *stainless;* when nickel is also added, properties are improved still further.

In high-temperature cobalt- and nickel-based alloys, up to 25 percent chromium improves strength and oxidation resistance.

For ferrous alloys, pure chromium is not required; the chromium is added as ferrochromium, containing 50 to 75 percent chromium, which can be produced readily by direct smelting of chromite ore without prior chemical treatment. For the production of chromium metal, on the other hand, it is necessary first to produce a relatively pure chromium compound, as indicated above.

In refractories, chromite is useful for its high melting point, thermal expansion characteristics, and stability at elevated temperatures. Chromium chemicals are useful as pigments, both for their color and for their corrosion-inhibiting qualities. In the leather industry, chromic sulfate or chrome alum is widely used in tanning.

In metal treating, chromium chemicals are used not only for chromium plating, but also in chromizing, anodizing, chromate conversion coatings, passivation treatments, surface cleaning and etching (including pretreatment of plastics to be plated), corrosion inhibition, and stripping of deposits. Continuous strip steel is plated with very thin deposits of chromium plus chromic oxide to produce "tin-free steel" (TFS-CT) which can replace tinplate in some applications.

In the organic chemical industry, chromates are used as oxidants in a wide variety of chemical reactions.

Properties

Some properties of chromium are summarized in Table 12-17. Its mechanical properties are very sensitive to impurities, prior mechanical history, grain size,

*Impure metal is sealed in a chamber with iodine vapor and an electrically heated tungsten wire. The iodine combines with the chromium, leaving the impurities behind, in the form of an iodide, which migrates to the hot filament and is decomposed to its elements: the iodine thus is recycled, serving only as a carrier. This is known as the *Van Arkel-deBoer process,* and it is used for purifying many of the "refractory" metals. It is expensive but very effective.

Table 12-17 Properties of Chromium

Atomic number	24
Atomic weight	51.996
Electronic configuration	$1s^22s^22p^63s^23p^63d^54s^1$
Crystal structure	bcc
Melting point, °C	1875
Boiling point, °C	2199
Density, g/cm³, 20°C	7.19
Electrical resistivity, $\mu\Omega$ −cm, 20°C	12.9
conductivity, % IACS	14
Tensile strength, MPa, 500°C*	275
Standard potential, E^0, 25°C, V	
$Cr^{3+} + 3e^- \rightarrow Cr$	−0.71
$Cr^{3+} + e^- \rightarrow Cr^{++}$	−0.41

*Depends on many factors; see text. Values to 550 MPa have been reported.

and surface condition, among others. Carbon, sulfur, and oxygen in very small amounts can destroy its ductility. The metal is not ductile enough to be worked at room temperature, but can be fabricated at elevated temperatures. The ultimate tensile strength shown in the table should be viewed with these factors in mind.

Chromium is a white, hard, lustrous, and brittle metal. It resists most ordinary corrosive agents; although active in the emf series, it is passive in many environments, which accounts for its inertness.

Chromium reacts with the anhydrous halogens, hydrogen chloride, and hydrogen fluoride. The aqueous halogen acids and dilute sulfuric acid dissolve chromium. With dilute sulfuric acid, hydrogen is evolved, and with hot concentrated acid, sulfur dioxide is formed. At room temperature concentrated nitric and aqua regia do not affect chromium.

Acids such as concentrated nitric, phosphoric, chloric, and perchloric form a thin layer on chromium which confers passivity. In this condition chromium exhibits outstanding corrosion resistance. In neutral solutions, dissolved oxygen is sufficiently oxidizing to maintain passivity; but in solutions of low pH, stronger oxidizing conditions must be present, and the halogen acids must be absent to maintain this condition.

Exposed to high temperatures, 600–700°C, chromium is attacked by alkali hydroxides but not by fused alkali carbonates. Sulfides are formed when chromium is heated at 600–700°C with sulfur vapor or hydrogen sulfide; at this temperature it also reacts with sulfur dioxide. In carbon monoxide, chromium oxidizes at about 1000°C; phosphorus attacks it at about 800°C. Ammonia reacts at 850°C to form a nitride.

When heated in air, chromium forms an oxide layer of Cr_2O_3, but at room temperature it is hardly attacked by either moist or dry air. Above its melting point, chromium burns in oxygen.

Chromium Compounds

Although chromium is a member of Group VIB, the members of this group in reality have rather little in common with one another; although their chemistry is formally similar, there are more differences than similarities in their reactions. Of interest is the fact that chromium is the only one of the three that can be electrodeposited from aqueous solution in the pure state. Although both molybdenum and tungsten alloys are plateable, no convincing evidence has been brought forward of their deposition without alloying elements, principally metals of the iron-cobalt-nickel triad.

Chromium is a transition element with the electronic configuration $3d^5 4s^1$; all six electrons can take part in chemical combinations, so that chromium exhibits all oxidation states from 0 to 6+. The most important, and the only ones likely to be encountered in practice, are the 2+, 3+, and 6+ states, and especially the last two.

Chromium(II) or chromous ion is a powerful reducing agent. The hexaquochromium(II) ion is blue; it is used as an analytical reagent and is unstable in water since it can reduce hydrogen ion. It forms many complexes.

The most stable state of chromium is Cr(III). In this state chromium has a very extensive chemistry; chromic oxide Cr_2O_3 is the form in which chromium exists in its principal ores, and in refined form it is used widely as a pigment and as an abrasive in cutting and polishing compounds. The chemistry of the Cr(III) oxidation state is basically coordination chemistry. Aside from its compounds with nitrate, perchlorate, fluoborate, and the alums,* the simple hexaquo ion $Cr(H_2O)_6^{3+}$ does not exist, but there are literally thousands of complexes with all types of inorganic and organic ligands.

Cr(III) is difficult to oxidize to the VI state under acidic conditions (although such oxidation can take place, as shown later in the discussion of chromium plating); but in alkaline solution oxidation to the 6+ state as chromate ion CrO_4^{--} can be accomplished with hypochlorite, peroxides, or oxygen under pressure. In practice chromite ore is oxidized to chromate by using air in a high-temperature fusion.

The 4+ and 5+ oxidation states are known but of no present practical importance. The most important chromium compounds are the hexavalent, including chromic anhydride CrO_3 (commonly called chromic acid), the alkali chromates and dichromates, and many metal chromates such as that of zinc, used as pigments both for their color and for their corrosion-inhibiting qualities.

In some ways Cr(VI) resembles hexavalent sulfur in its chemical reactions. Like sulfuric acid, chromic acid H_2CrO_4 is a dibasic acid, strong in its first ionization and fairly weak in its second, as shown below.

High concentrations and low pH favor the formation of polyanions such as

*Alums are compounds of the type $M_2^{I}(SO_4) \cdot M_2^{III}(SO_4)_3 \cdot 24H_2O$ where M^I may be typically potassium, sodium, or ammonium, and M^{III} is typically aluminum but may be also Cr(III) and other trivalent metals.

dichromate $Cr_2O_7^{--}$, and even more condensed species such as $Cr_3O_{10}^{--}$ and $Cr_4O_{13}^{--}$ are known to exist. The coordination number of Cr(VI) is 4 in these compounds.

Chromates are used widely as oxidants in organic chemistry, as passivators and corrosion inhibitors in recirculating-water systems, and as pigments.

The anhydride of chromic acid or chromic anhydride, CrO_3, is commonly called chromic acid, although this nomenclature is, strictly speaking, incorrect. (Since in industry CrO_3 is universally called chromic acid, we shall accept this "mistake" in this book.) Chromic acid is manufactured by treating sodium chromate or dichromate with sulfuric acid:

$$Na_2CrO_4 + H_2SO_4 \rightarrow CrO_3 + Na_2SO_4 + H_2O$$

The by-product sodium sulfate has a limited market as "chrome cake."

Most chromic acid is used in metal treatment, either in chromium plating or as an ingredient of various conversion coating materials or in anodizing. It is a deep red to reddish brown crystal, which volatilizes at about 110°C. Conveniently, its formula weight is almost exactly 100. It is very soluble in water—165 g/100 g at 0°C and 206 g/100 g at 100°C—as well as in sulfuric acid and other solvents. At 250°C it decomposes into chromic oxide, Cr_2O_3, and oxygen. It is the only chromium compound used in chromium plating, at least at present.

In basic solutions above pH 6, CrO_3 forms the yellow chromate ion CrO_4^{--}; between pH 2 and pH 6, $HCrO_4^-$ and the orange-red dichromate ion $Cr_2O_7^{--}$ are in equilibrium; and at pH below 1 the main species is H_2CrO_4. The equilibria are:

$$HCrO_4^- \rightleftharpoons CrO_4^{--} + H^+ \qquad K = 10^{-6}$$
$$H_2CrO_4 \rightleftharpoons HCrO_4^- + H^+ \qquad K = 4.1$$
$$Cr_2O_7^{--} + H_2O \rightleftharpoons 2HCrO_4^- \qquad K = 10^{-2.2}$$

There are also the base-hydrolysis equilibria:

$$Cr_2O_7^{--} + OH^- \rightleftharpoons HCrO_4^- + CrO_4^{--}$$
$$HCrO_4^- + OH^- \rightleftharpoons CrO_4^{--} + H_2O$$

These equilibria, which depend on pH, are labile; on addition of cations that form insoluble chromates (e.g., Ba^{++} or Pb^{++}), it is the chromate and not the dichromate that precipitates. These equilibria hold for nitric and perchloric acid solutions; with hydrochloric acid there is conversion to the chlorochromate ion, and with sulfuric acid a sulfato complex is formed:

$$HCrO_4^- + H^+ + Cl^- \rightarrow CrO_3Cl^- + H_2O$$
$$HCrO_4^- + HSO_4^- \rightarrow CrO_3(SO_4)^{--} + H_2O$$

Acid solutions of dichromates are strong oxidizing agents:

$$Cr_2O_7^{--} + 14H^+ + 6e^- \rightarrow 2Cr^{3+} + 7H_2O \qquad E^0 = 1.33 \text{ V}$$

In basic solutions the oxidizing power is not so strong:

$$CrO_4^{--} + 4H_2O + 3e^- \rightarrow Cr(OH)_3 + 5OH^- \quad E^0 = -0.13 \text{ V}$$

Although some polyanions are known, chromium(VI) does not give rise to the extensive series of polyacids characteristic of molybdenum and tungsten.

Hexavalent chromium in the chromates can be reduced to the trivalent form by several reducing agents; the most important practically is sulfur dioxide or sulfite ion, used in the finishing industry as a preliminary to precipitating chromium hydroxide in wastewater purification:

$$2Cro_4^{--} + 3SO_2 + 4H^+ \rightarrow 2Cr^{3+} + 3SO_4^{--} + 2H_2O$$

$$\text{or} \quad 2CrO_4^{--} + 3HSO_3^- + 7H^+ \rightarrow 2Cr^{3+} + 3SO_4^{--} + 5H_2O$$

On raising the pH of the solution, chromic hydroxide is precipitated:

$$Cr^{3+} + 3OH^- \rightarrow Cr(OH)_3 \downarrow$$

Hazards

Pure chromium, chromite, and trivalent chromium compounds are relatively nontoxic. Hexavalent compounds, on the other hand, can be seriously hazardous. They irritate, corrode, and possibly poison the tissues of the body. They can cause denaturation of tissue proteins. Contact affects principally the skin and the respiratory tract. Skin contact may result in ulcers and dermatitis; inhalation of chromate dust or chromic acid spray and mist may cause ulceration or perforation of the nasal septum, as well as chronic irritation of the nasal passages. Carcinogenic factors are also suspected. Some individuals are more susceptible than others to the dermatitic effects, but even nonsensitive persons may be sensitized by continued contact.

The maximum allowable exposure to dusts and mists of hexavalent chromium measured as CrO_3 is 0.1 mg/m^3 for 8-h exposure. (Standards are subject to change with increasing governmental emphasis on environmental factors.) Hexavalent chromium is one of the more strictly controlled contaminants in plant effluents.

Chromic acid also poses a fire hazard when in contact with organic chemicals; ethyl alcohol, for example, dropped onto chromic acid crystals will catch fire.

Chromium Plating

Chromium is the most recent important addition to the list of metals that can be practically electroplated*; commercially successful chromium-plating processes were introduced about 1925, and chromium now ranks with nickel as one of the two most important plating metals. The invention of a reliable chromium-plating

*Rhenium and some of the less common platinum group metals were introduced later, but they can hardly be compared in importance to chromium.

process is generally credited to Fink and his coworkers.* Improvements and refinements have been made in the intervening half-century, but the present chromium-plating bath does not differ in essentials from the one then disclosed.

There are two main types of chromium plating: decorative, in which thin coatings (usually less than 0.75 μm) serve as a nontarnishing and durable surface finish; and industrial or "hard" chromium, in which much heavier coatings are used to take advantage of the unusual combination of properties possessed by chromium, including resistance to heat, wear, corrosion, and erosion, and low coefficient of friction. In the latter type, which is finding ever-increasing uses, the chromium is usually deposited directly on the basis metal without intermediate coatings, while decorative chromium usually serves as a final or topcoat on other electroplated coatings, typically copper-nickel. Sometimes, especially on cutting tools, hard chromium coatings are little, if any, thicker than decorative coatings.

All practical chromium plating is done from a solution consisting mainly of chromic acid, CrO_3, plus small but critical quantities of an anion, usually sulfate or a complex fluoride, which for want of a better name is called a *catalyst*. The only variations are in the concentration of the bath, the amount and nature of the catalyst, and the operating conditions.

Plating from such a bath has many disadvantages: reduction takes place from the highest valence of the metal, leading to a very low electrochemical equivalent: 52/6 = 8.67 g/$\mathfrak{F}$, only 0.09 mg/C or 0.32 g/A·h. Cathode current efficiency is low—only about 12 percent for conventional baths and perhaps twice that for some mixed-catalyst solutions—reducing still further the actual electrochemical equivalent and at the same time giving rise to voluminous formation of hydrogen gas, which in turn causes a chromic acid mist or spray that must be vented for safety of workers. Higher than usual voltages are required to drive the high current densities that must be used; the bath is somewhat viscous and difficult to rinse off. Soluble anodes cannot be used, so that continuous chemical additions must be made for maintenance. (This entails no additional expense, however, for chromium in the form of CrO_3 is much cheaper than pure chromium metal—an exception to the general rule.)

All this is not said to disparage chromium plating; on the contrary, it is evidence of the tremendous importance of chromium as a deposited metal that platers are willing to tolerate such disadvantages for the sake of the very significant advantages offered by chromium deposits. If this were not so, the trade would have shrugged off chromium plating as impractical.

Other than the detailed studies that improved our understanding of, and ability to control, the process introduced in 1925, no major advances were achieved until the introduction in about 1950 of the so-called SRHS† systems, which employed a different catalyst system and exhibited significant improvements in current efficiency and some other operating characteristics.

Until fairly recently, the thin chromium topcoat in the composite copper-

*Although a court decided that Udy shared the credit.
†Trademark ("Self-regulating high-speed").

nickel-chromium was thought of as only protecting the nickel from tarnishing and scratching; it was believed that corrosion protection was mainly a function of the type and thickness of the nickel. Since the 1950s attention has been directed to the contribution of the chromium itself to the performance of the composite, and it is now obvious that when properly applied, chromium can add significantly to this performance. Chromium can be deposited in a form termed *microdiscontinuous,* mentioned later; in this form its contribution to the total system is an important one.

It would seem logical to plate chromium from its 3+ state, using one of the innumerable simple or complex compounds formed by this ion. Chromium is indeed recovered by the electrolysis of chromic sulfate baths, as mentioned above; but no practical process has yet appeared for plating bright chromium from trivalent baths, nor is this from lack of trying. Such baths would have obvious advantages: at least twice the electrochemical equivalent, perhaps higher efficiency, use of the much less toxic Cr^{3+} in place of Cr(VI), and thus fewer waste-disposal problems. Such processes are indeed announced from time to time,* and perhaps one day a truly practical substitute for the chromic acid bath will find commercial acceptance. For the present, we confine our attention to the chromic acid baths, while advising the reader to be alert to current literature for new developments.

CHROMIC ACID BATHS—GENERAL CONSIDERATIONS

The only practical chromium-plating bath consists essentially of chromic acid (CrO_3) in a concentrated solution in water, so that the bath itself consists of a solution of H_2CrO_4 and perhaps polyacids, as already discussed. But chromium cannot be deposited from a solution containing only CrO_3 and water: there must be present a small and relatively critical amount of an acid radical which acts as a catalyst to enable cathodic deposition of chromium metal. The most commonly used are sulfate and fluoride, the latter usually in the form of fluosilicate (or silicofluoride), SiF_6^{--}. Simple fluorides are operative but so critical in concentration that they are difficult to control. For practical operation, the ratio (by weight) of CrO_3 to total catalyst radicals must be maintained within well-defined limits, usually about 100:1 in the case of sulfate.

It is usually unimportant in what form the acid or catalyst radical is added, as long as it is soluble. Sulfuric acid is the common form for the addition of sulfate, though sodium sulfate may be used. Sulfuric acid is convenient because its formula weight (98) and that of the sulfate anion (96) are so nearly identical that no calculations are required to determine the correct amount of sulfuric acid to add. For the fluoride-catalyzed baths, fluosilicic acid and alkali fluosilicates are the most common sources of the anion.

Although the current efficiency of chromium plating is low (generally in the

*Several "commercial" processes were announced in the early 1970s, and at least one of them appears to show some promise in early commercial trials. Awareness of current literature is required.

range of 10 to 25 percent for bright plate, and often nearer the lower than the upper range), a fairly high rate of deposition is achieved owing to the high current densities used. The voltages required to obtain these current densities are higher than in most plating operations—from 4 to 12 V, depending on conditions. Thus the generator or rectifier capacity required for chromium plating is higher than for most metal plating.

The throwing power of chromium-plating baths is poor; measured by the Haring-Blum technique (Chap. 8), it is negative. Good coverage can be obtained if the optimum ratio of CrO_3 to catalyst is maintained and anodes are carefully designed. This includes the use of conforming or bipolar anodes, and often shielding of high-current-density areas as well.

The conductivity and specific gravity of chromic acid solutions are shown in Fig. 12-5. Small amounts of Cr(III) and some other cations decrease conductivity; but the specific gravity is little affected, and measurements of this property can be used as a control measure to monitor the CrO_3 concentration of the bath if supplemented by occasional analysis and if metallic impurities are not too high. The maximum conductivity of the bath is not reached until a concentration of 400 to 500 g/L is reached. Commercial chromium-plating baths usually contain from 200 to 400 g/L of chromic acid, to obtain as good conductivity as is compatible with good current efficiency, satisfactory deposits, and stability of solution composition.

ELECTRODE REACTIONS

Although, as we have seen, chromic acid exists in several forms in these highly concentrated and very acidic solutions (even the "dilute" bath is concentrated by usual standards, being about 2.75 *M*), it is convenient to assume that most of the chromium is in the dichromate ion $Cr_2O_7^{--}$ under these conditions. At least three reactions at the cathode are possible, and all occur:

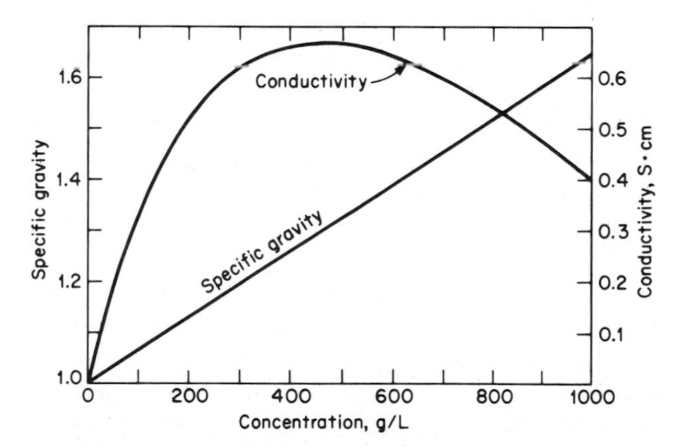

Fig. 12-5 Conductivity and specific gravity of chromic acid solutions.

Deposition of chromium: $Cr_2O_7^{--} + 14H^+ + 12e^- \rightarrow 2Cr + 7H_2O$
Evolution of hydrogen: $2H^+ + 2e^- \rightarrow H_2$
Formation of Cr(III): $Cr_2O_7^{--} + 14H^+ + 6e^- \rightarrow 2Cr^{3+} + 7H_2O$

It must be emphasized that these equations portray only the overall stoichiometry of the reactions; the kinetics or mechanisms by which they take place are, in spite of considerable research, still in dispute. It is difficult to visualize, for instance, that a dichromate ion could react simultaneously with as many as twelve electrons and fourteen hydrogen ions: there must be intermediate steps.

Unfortunately, it is the second reaction, hydrogen evolution, that is predominant and consumes from 80 to 90 percent of the power supplied to the system. Actual deposition of chromium metal uses only 10 percent to somewhat less than 20 percent of the power; and the formation of Cr^{3+} consumes a small but not negligible proportion. The bath can tolerate a certain amount of Cr^{3+}, and under proper conditions of operation this is held to manageable proportions by reoxidation at the anodes.

Chromium anodes dissolve inefficiently under the conditions of electrolysis; even if they were practical in this sense, chromium metal is far more expensive than chromium in the form of CrO_3. For this reason insoluble anodes are used; the only satisfactory anode metal is lead or lead containing up to 10 percent of alloying elements, usually tin, antimony, or both. These anodes, besides completing the circuit, perform other equally important functions.

As at the cathode, three simultaneous reactions take place at lead anodes:

Evolution of oxygen: $2H_2O \rightarrow O_2 + 4H^+ + 4e^-$
Oxidation of chromic ion: $2Cr^{3+} + 6H_2O \rightarrow 2CrO_3 + 12H^+ + 6e^-$
Production of lead dioxide on the anode: $Pb + 2H_2O \rightarrow PbO_2 + 4H^+ + 4e^-$

Most of the power is consumed in oxygen evolution. But the other two reactions are very important: reoxidation of Cr(III) at the anode helps to balance its production at the cathode and maintain the Cr^{3+} level at acceptable values; and for proper operation of chromium-plating baths, the lead anode must be covered with a layer of lead dioxide, as shown. If this film is lost or does not form, lead chromate will form instead, and the anode will not fulfill its function of regulating the Cr^{3+} concentration of the bath.

Anode Maintenance

Idle anodes, in a nonworking bath, soon lose their oxide coating and react with the plating solution to form a film of yellow lead chromate. As soon as current is turned on, the lead dioxide film begins to re-form—or should do so. If it does not, something is wrong and should be corrected; in a working bath the anodes should show the dark, almost black, coating of lead dioxide. Lack of this film in a working bath indicates poor contact of the anode with the bus bar or anode hook, indicating that the anode is not receiving its share of the current; or the anode may have been short-circuited by touching a submerged conductor. The lead chromate film may be removed by removing the anode, cleaning the bus bars and hooks, and scratch-brushing the anode.

CHROMIUM-PLATING THEORY

Any satisfactory theory of chromium plating from chromic acid baths must account for the role of the so-called catalyst, as well as the kinetics involved in the total stoichiometric equations already cited. In spite of much effort over the years, no such theory has been advanced. Many of the investigations are vitiated by failure to take into account the sulfate already present in almost all commercial chromic acid, so that claims for chromium deposition from "pure" chromic acid must be discounted: the chromic acid used already contained sufficient sulfate radical to permit deposition. When little or no catalyst is present, either no deposit is formed or it is iridescent and nonmetallic.

Other questions remain similarly unanswered: what is the role of the catalyst and why does current efficiency vary so widely with temperature, current density, concentration of chromic acid, and the CrO_3:catalyst ratio? Why do different catalyst anions give such widely disparate results? Why do fluoride anions give only dark, dull deposits when no sulfate is present? What distinguishes the "bright range" from the conditions outside this range that yield deposits, but not bright ones?

All these questions must somehow be related to conditions in the cathode film, but relationships are far from clear. It was early concluded, on the basis of radioactive tracer studies, that the chromium deposits directly from the hexavalent state, without intermediate Cr(III) formation. Thermodynamically it is slightly easier to reduce CrO_3 than $\frac{1}{2}Cr_2O_3$ to metal: the free energy of formation of the latter is −530 kJ vs. −507 kJ for CrO_3, indicating slightly easier decomposability for chromic acid than for the trivalent salts. One theory proposes that what is deposited is not chromium metal at all, but a hydride that fairly rapidly decomposes to the metal.

Much work on the composition of the cathode film is open to one objection or another; surely research on such highly oxidizing and acidic solutions is difficult. It can only be restated here that no satisfactory explanation of the mechanism of chromium plating is available. This is surely unfortunate from the academic standpoint, but fortunately sufficient practical knowledge has been accumulated so that chromium plating can be carried out with few difficulties. Theoretical considerations are discussed and well referenced by Dubpernell ("Modern Electroplating," cited in the Bibliography).

DECORATIVE CHROMIUM

Chromium is the almost universal final finish for nickel and copper-nickel decorative-protective systems as described in the previous sections. It has a pleasing blue-white color* and high reflectivity, which it maintains under most conditions of exposure; it resists tarnish very well—this was the original reason for its application over nickel plate; it has good corrosion resistance and under proper conditions can enhance the protection offered by the composite plate; it resists wear and scratching.

Chromium is seldom applied to the substrate directly for decorative purposes, but almost always as the final coat following nickel or copper-nickel. The principal exception is the chromium plating of stainless steel; this is done only for color matching, because the color of chromium and that of stainless steel differ, and where parts are to be used on the same assembly, it is preferred that the colors match.

The thickness of chromium plate applied over nickel depends to some extent on the exposure conditions to be met and the general quality level of the product; the more severe the exposure, the heavier the chromium should generally be. Household appliances for use under indoor and dry conditions may carry thicknesses up to about 0.1 μm,† interior automobile trim somewhat more, and exterior trim up to 1.25 μm or more. Thickness depends to some extent on the type of chromium; e.g., microporous chromium must not be too thick or it will bridge over the induced porosity (see later). ASTM specifications are cited in the following section; see p. 247.

Marine fixtures call for the same thicknesses as automotive exterior trim. In truth, marine fixtures for salt-water service probably should not be plated at all, but should be of solid bronze or other corrosion-resistant metal. Few economically practical plating systems will resist seawater exposure with any degree of satisfaction.

There are two principal types of chromic acid–plating baths: ''conventional,'' in which the catalyst ion is sulfate; and mixed-catalyst baths, in which in addition to sulfate the catalyst contains fluoride, usually in the form of a fluosilicate.

CONVENTIONAL BATHS

These sulfate-catalyzed baths contain varying amounts of chromic acid, from 250 to 400 g/L, plus an amount of sulfuric acid such that the CrO_3/SO_4 weight ratio is about 100:1. This ratio is more important than the actual concentrations; it can be varied between about 75:1 and about 150:1, but more commonly is maintained at 100:1, as stated. Further, although the chromic acid concentration can be varied within the limits cited, it is usually held at one extreme or the other. The 250-g/L bath is known as the *dilute* bath and the 400-g/L bath as the *concentrated* bath.

*Admittedly, *pleasing* is a subjective term; some would disagree.

†Plumbing fixtures call for about the same thickness as household appliances. Unfortunately the plumbing industry appears not to have set rigid specifications, as all who have used the fixtures in public washrooms can attest.

Table 12-18 Basic Chromium-Plating Baths

	Dilute bath		*Concentrated bath*	
	g/L	*Molarity*	*g/L*	*Molarity*
Chromic acid, CrO_3	250	2.5	400	4.0
Sulfate, SO_4^{--}	2.5	0.026	4	0.042
Ratio CrO_3/SO_4	100		100	

Each has its advantages and disadvantages. Table 12-18 lists the usual formulations for decorative plating.*

The dilute bath features lower makeup costs, somewhat better cathode efficiency (faster plating speed), and lower drag-out, resulting in lower costs for effluent treatment; the concentrated bath requires a lower operating voltage, is less sensitive to contamination, and gives better coverage. Thus the choice between them depends on the type of work being plated, the plating time available, and economic factors.

Operating conditions for both baths vary within fairly narrow limits. The usual operating temperature is 43°C, with a range of 32 to 50°C. Average cathode current density is about 1450 A/m², with a range of 700 to 3600 A/m². Although the conductivity of the solutions is high, to maintain these high current densities requires a source of dc somewhat more powerful than usual in plating; this cannot be specified exactly but ranges from 4 to 12 V or even higher.

At least two factors must be considered in determining the best operating conditions for a chromium-plating bath for decorative purposes: the plate must be bright, and its coverage should be as complete as possible. At best the throwing power and the covering power of the chromium-plating bath are mediocre, but under some conditions they are better than under others.

Bright Range

Under any given set of conditions, chromic acid baths have a range of operating conditions under which the plate is bright, and beyond which it is hazy, cloudy, or simply dull. For decorative purposes the plate must be bright: buffing is out of the question except perhaps to salvage a few misplated parts out of a production lot. Plating speed is also a factor, since time in the plating bath is usually set as low as possible, and in effect this means that current efficiency should be as high as feasible.

In Fig. 12-6 is plotted the effect of temperature and current density on the

*In an effort to reduce waste-treatment costs, there has been some trend toward the use of more dilute baths than the 'dilute'' bath in the table. Such baths are somewhat more critical in control but otherwise may be quite satisfactory.

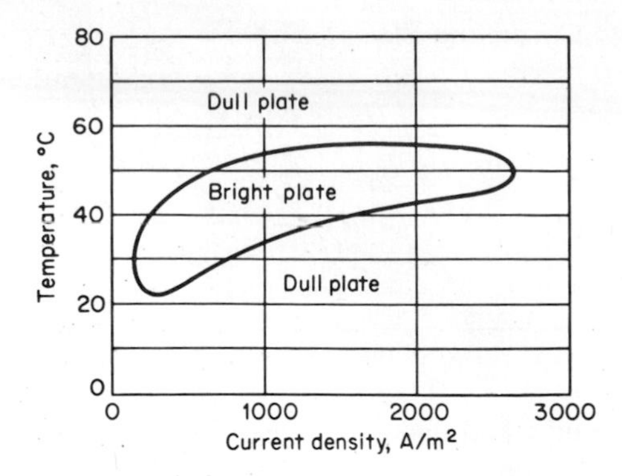

Fig. 12-6 Bright range of chromium-plating solutions.

bright range of chromic acid–plating baths. The area within the curve delineates the conditions under which the plate will be bright and outside of which it will be unsatisfactory from a decorative standpoint. This area is affected somewhat by whether the dilute or the concentrated bath is used, but the curves are very close together. The effects of the variables can be summarized for either case: at temperatures between about 40 and 55°C the bright range extends from 500 to almost 2500 A/m², while at the lower temperature of 30°C the range is limited to 200 to 800 A/m². This latter range is too narrow for plating of shaped parts, indicating the necessity of elevated temperatures for chromium plating.

The cathode efficiency of chromium plating, and thus the plating speed at any given current density, depends on current density, temperature, bath concentration, and both the catalyst (sulfate) concentration and the CrO_3/SO_4 ratio.

The dependence of plating speed on current density for the dilute and concentrated baths is illustrated in Figs. 12-7 and 12-8, at the optimum 100:1 ratio.

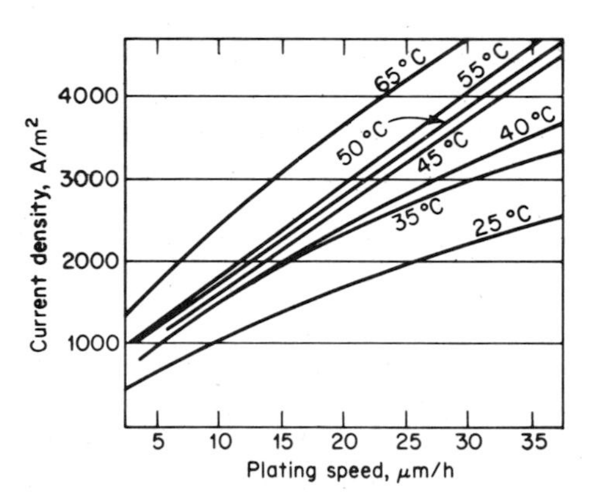

Fig. 12-7 Plating speed, dilute solution; ratio 100:1.

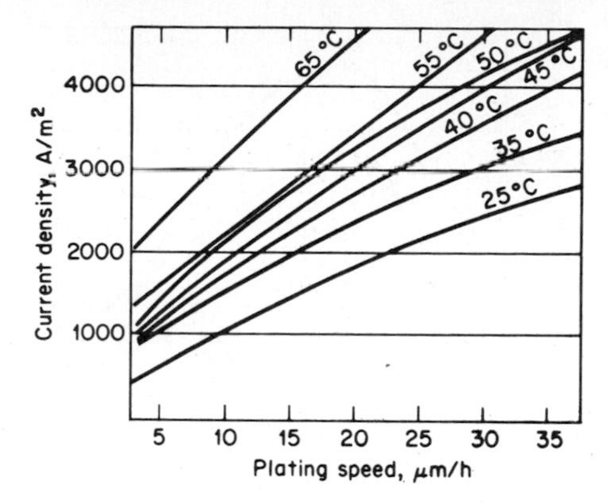

Fig. 12-8 Plating speed, concentrated solution; ratio 100:1.

IN HARD Cr PLATING, CURRENT DENSITY IS A FUNCTION OF HOW FAST YOU WANT TO PLATE A WORKPIECE. IN SHOPS WHERE TIME IS NOT OF THE ESSENCE, LOW CURRENT DENSITIES CAN BE USED WITHOUT SLOWING PRODUCTION THROUGHPUT.

Figure 12-9 illustrates the effect of sulfate concentration on cathode efficiency for both baths at a particular condition of temperature and current density, and Fig. 12-10 shows the effect of the CrO_3/SO_4 ratio.

The acidity of chromium-plating baths is very high and is not usually measured or controlled. Such measurements as have been reported indicate pH values off the usual scale, at small negative values.

Throwing Power

The throwing power and the covering power of chromic acid–plating baths are poor, perhaps the poorest of any commonly used plating solution. The conductivity is high but is reduced by many impurities such as iron and copper. The cathode polarization during deposition is relatively constant. The major variables governing throwing power in bright chromium plating are the current efficiency

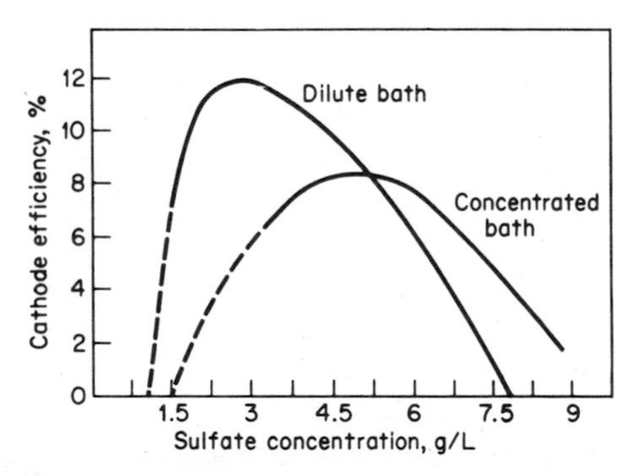

Fig. 12-9 Cathode efficiency vs. sulfate concentration; 45°C, 1000 A/m².

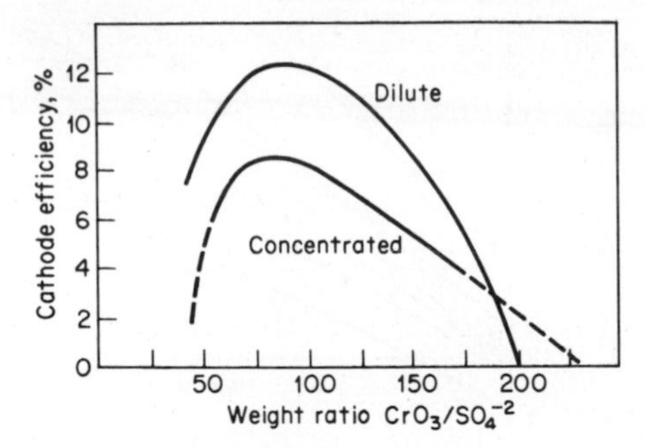

Fig. 12-10 Cathode efficiency vs. catalyst ratio, 45°C, 1000 A/m².

and the bright-plating range. Throwing power is maximized by employing the highest practical current density for a given set of conditions.

In the throwing-power box (Haring-Blum cell), cyanide copper or other baths considered to have a good throwing-power rate about 20 to 40 percent, and most nickel and acid copper baths yield figures near zero. Using the same test, chromium-plating baths give results varying from −13 to as low as −100 percent. This disadvantage can be overcome only by careful design of fixtures such as racks with auxiliary anodes; by appropriate use of current shields and sometimes bipolar electrodes; and by avoidance of designs having deep recesses or other shapes requiring good throwing power.

Impurities

Cations often found in chromium-plating baths include Cr(III), iron, copper, nickel, zinc, and sodium. As we have seen, some Cr(III) results naturally from the cathodic reduction reaction, and it can build up to unacceptable values if the cathode area is too large with respect to the anode area, since, as we have also seen, one of the anodic reactions is reoxidation of Cr(III) to Cr(VI). Introduction of oxidizable organic matter can also cause formation of Cr(III) in the bath. The means for controlling Cr(III) content are obvious from these facts: use of a large anode area (or, when this is not practical, dummying* for a while with a small cathode area) and keeping organic matter out of the bath. Although it has been claimed that a certain amount of Cr(III) is beneficial, such claims have not been substantiated.

Iron, copper, nickel, and zinc and other metallic impurities may enter the bath with the work or in other ways. None has any noticeable beneficial effect on operating characteristics, and many have serious detrimental effects.

*Dummying is operating a plating solution for reasons other than production, e.g., "breaking in," purification, or other adjustment.

Maintenance and Control

Chromium-plating baths are stable in use, and their composition can be maintained constant readily by physical or chemical analysis. If anodes are in proper condition, as explained, the Cr(III) formed at the cathode is reoxidized to Cr(VI) at the anode and is maintained at a low and acceptable figure; if the area of lead anodes is insufficient, this may not be true and Cr(III) can rise to the point where it is deleterious. Very occasionally anodes other than lead are used for special purposes: these do not perform this oxidizing function, and unless absolutely necessary should be avoided.

The most useful control test for chromium plating is perhaps the Hull cell test, which gives in one measurement the coverage and the bright range; if these are not normal for the nominal conditions of operation, experience will tell the operator in which direction to make adjustments.

Specific gravity is a fair indication of total chromic acid concentration, provided metallic impurities are absent or low. In old baths, where such impurities may have built up to fairly high levels, specific-gravity measurements based on charts such as Fig. 12-5 may yield erroneous results. The sulfate content may be determined in several ways. If it is in excess, it can be precipitated by addition of barium carbonate; or additional chromic acid may be added to return the ratio to the desired value, if the bath can tolerate such addition.

Allied to throwing power is covering power, or coverage. This can be evaluated by a Hull cell or bent-cathode test (see Chap. 24). It is a frequent problem in bright chromium plating. Operating at maximum allowable current densities, entering the bath with a "hot lead,"* and use of an initial "surge" of current are some of the ways coverage can be improved. Detailed trouble-shooting manuals should be consulted for further discussion.

A widely used test for both coverage and crack patterns in chromium plate is known as the Dubpernell test, discussed here because it is peculiar to chromium plate and not a general plating test. Copper from a dilute acid copper sulfate bath, operated at low voltage, will not plate on chromium but will do so on exposed substrates such as nickel. The part is thus plated in the copper sulfate bath; any areas seen to be plated with copper are not covered by chromium.

Mist Suppression

The large amount of gases evolved at both anode and cathode during chromium plating causes the formation of much chromic acid mist and spray. As already explained, such mist is irritating and potentially toxic, and government regulations strictly limit the concentration permitted in the work place. This spray is usually prevented from entering the working area by means of forced ventilation;

*Hot lead: the current is applied before the part enters the bath, so that current begins to flow as soon as the part touches the solution; it is known also as hot contact.

suction vents must be positioned over the tank in such a way as to sweep away essentially all this mist. (How the mist then may be separated from the air stream, and the chromic acid possibly reclaimed for reuse, is under study.)

An obvious way of suppressing mist and spray is the use of wetting agents to form a foam blanket, which permits quiet release of gases while keeping the liquid in the tank. However, wetting agents are organic compounds, and until fairly recently no such compounds were known that would be stable under the severely oxidizing conditions of a chromium-plating bath. The development of stable fluorocarbon surfactants has permitted the use of some organic fume suppressants; their efficacy and stability must be evaluated by users under their particular conditions. Holding a piece of white filter paper a short distance above the solution and observing the extent to which it is discolored by chromic acid is a good method of evaluation, although surface tension measurements may be used for control purposes.

Floating of hollow polyethylene spheres over the solution to break up the mist is somewhat effective. In spite of these measures, ventilation may be mandated by government regulations.

Preparation for Chromium Plating

Most decorative chromium is plated over bright nickel, and if the work proceeds direct from the nickel-plating bath to the chromium-plating bath with only the necessary rinses and without undue delay, this usually presents no problem. If the nickel surface is allowed to become passive, such as by long standing, by use of contaminated nickel baths, or by some other cause, activation before chromium plating may be required to prevent blotchy, hazy, or nonadherent chromium plate. Such activation usually consists of a simple dip in dilute sulfuric acid, or for more stubborn cases this may be preceded by cathodic cleaning of the nickel surface (never anodic cleaning). Chromium plating direct on stainless steel also requires that the substrate be activated by cleaning followed by an activating acid dip.

Current for Chromium Plating

In Chap. 9 we noted that direct current from various sources is subject to a certain degree of "ripple"; i.e., it is not absolutely steady but deviates from the average by amounts varying according to the dc source and its wiring. For most plating metals, this ripple is of little or no significance; but for plating bright chromium, a ripple factor of 6 percent or less is usually specified. A high ripple factor may make it impossible to achieve a bright chromium deposit. For perhaps similar reasons, current interruptions are undesirable. This fact makes the design of chromium-plating barrels much more difficult than barrels for other metals, because by the very nature of barrel-plating the current to any given item is intermittent. The problem has been solved by special designs, but barrel chromium plating is far less common than barrel plating with most other metals.

SELF-REGULATING AND MIXED-CATALYST BATHS

Although the superior activity of silicofluoride and other fluorides as catalysts was known, the difficulty in controlling these baths because of the extreme sensitivity of fluoride and analytical problems militated against their commercial use. In the early 1950s a series of patents revealed how to use the advantages of such catalysts while to a large extent avoiding their disadvantages; these systems were introduced as SRHS (self-regulating high-speed) systems.

Their advantages included higher efficiency (hence shorter plating times); a "self-regulating" feature which, it was claimed, made regular analysis unnecessary; and better activating properties so that partial passivity of the preceding nickel plate was not so serious as with conventional baths.

Silicofluoride (or fluosilicate) gives somewhat higher current efficiencies than sulfate when used as the catalyst. The new baths used both sulfate and silicofluoride, hence the term *mixed catalyst*. The self-regulating feature was obtained by proper use of compounds having such solubilities that when the catalyst ratio became too low (too much catalyst), some of the anion was precipitated automatically, and when it became too high, some of the catalyst ion dissolved in the bath—both from a reservoir of solid compounds maintained in the bath at all times. Typical mixtures consisted of chromic acid (about 97 to 98 percent of the total) with minor additions of strontium sulfate, $SrSO_4$, and potassium fluosilicate, K_2SiF_6; or chromic acid, potassium bichromate, strontium chromate, potassium fluosilicate, and strontium sulfate. Several variations on these basic formulas were disclosed. The relative solubilities of strontium chromate, strontium sulfate, potassium fluosilicate, and the other possible compounds in the system were said to regulate the ratio of chromic acid to catalyst ions more or less automatically, even as temperature and total concentration were altered.

Since the introduction of these baths, other complex fluorides have been proposed, including those of rare-earth metals, fluoaluminates, fluozirconates, and others.

These baths are operated in essentially the same way as conventional chromium-plating solutions. In addition to higher efficiency, as shown in Fig. 12-11, these systems are claimed to have several other advantages over the sulfate-catalyzed baths:

> They have a faster plating speed, a direct result of the higher efficiency. It should be remembered that although the efficiency of the mixed catalyst bath approaches only about 25 percent, this is still a significant improvement over the 10 to 15 percent of sulfate baths.
>
> Better coverage is claimed.
>
> The bath is superior in plating over somewhat passive surfaces. While it remains true that passive nickel will give poor results in any chromium-plating bath, the requirement is not quite so critical in the mixed-catalyst bath, which has some activating properties.

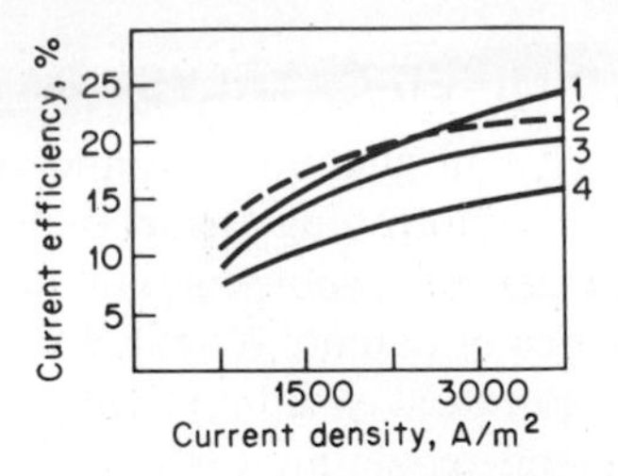

Fig. 12-11 Current efficiency—mixed catalyst vs. "conventional" baths at 45°C: curve 1—mixed, 400 g/L bath; curve 2—mixed, 250 g/L bath; curve 3—conventional, 250 g/L bath; curve 4—conventional, 400 g/L bath.

It is said to be less sensitive to current interruption; this is to some extent a corollary of the above.

The bright range is wider, and the bath is somewhat less sensitive to many impurities.

Microcracked deposits (see below) are more readily produced from these baths.

The principal disadvantage of the mixed-catalyst system is that in the plating of hard chromium (see below) some difficulty may be encountered with etching the substrate in low-current-density areas.

TANKS

For many years lead-lined steel or acid-brick tanks were used as the only materials that would stand up against the strongly oxidizing chromic acid solutions; tin-lead alloys were also used, with some gain in corrosion resistance. Now that superior organic coatings are available, some synthetic resins can be used as linings; many are limited to temperatures below 60°C, and the suitability of any particular resin should be assured before it is installed. Heating and cooling may be by lead-alloy coils; tantalum coils and heat exchangers are a more recent development. Quartz-sheathed electric heaters are satisfactory in sulfate-catalyzed baths.

CRACK PATTERNS

Chromium as normally plated tends to be cracked, in an irregular pattern probably caused by severe internal stresses as it deposits. The nature of the cracks has been studied extensively, and there is a copious literature. For many years this cracking was merely accepted as a necessary concomitant of chromium plating, and it was thought to limit the practical thickness of decorative chromium, since it was believed to decrease protective qualities.

Chromium itself is an extremely corrosion-resistant metal, and if it could be plated in such form that it would be free of cracks, it seemed reasonable that the resulting deposit would be essentially noncorroding in most environments. Processes for so regulating the bath and conditions of operation for producing crack-free deposits were introduced in about 1953; the deposits were dull but could be buffed to a high luster. Somewhat later it was learned how to produce crack-free deposits that were bright as plated. Unfortunately, although the deposits were indeed crack-free, this property did not stand up against mechanical abuse such as impingement of road debris on an automobile bumper. Crack-free chromium, however, has remained in use on such parts as washing machine shafts, high temperature–resistant coatings, gun barrels, and similar items, where appearance is less important than performance and impingement attack is unlikely.

The search for more corrosion-resistant nickel-chromium systems then took the opposite tack: provision of many cracks, of regulated size and number, so as to distribute the corrosion attack among many small sites instead of among a few large ones. This principle was illustrated in Chap. 3. Such chromium may be "microcracked" or "microporous," and the term *microdiscontinuous* is applied to both forms, which are about equivalent in performance.

Microporous chromium is achieved by modifications in the nickel undercoat, as we have seen in the preceding section. Microcracked chromium, which was developed somewhat earlier, is achieved by modifications in the operating conditions and composition of the chromium-plating bath itself. Crack patterns are somewhat critical. ASTM specifications call for at least 30 cracks per millimeter measured in any direction. For microporous chromium, there must be at least 100 pores per mm^2, invisible to the unaided eye. The crack or pore pattern can be evaluated readily by the Dubpernell test described above or by microscopic observation.

Duplex chromium coatings, consisting of a layer of bright crack-free chromium followed by one of microcracked chromium, have also been recommended.

HARD CHROMIUM PLATING

Hard chromium plating is the name adopted by industry for what might be better termed chromium plating for engineering rather than decorative purposes. There is no evidence that "hard" chromium is necessarily any harder than decorative chromium.

The solutions used are the same: chromic acid plus sulfuric acid, in a ratio of about 100:1, and the concentrations also about the same: 250 or 400 g/L CrO_3, plus 2.5 or 4 g/L sulfuric acid. A third, intermediate, solution is also often used: 300 g/L CrO_3, 3 g/L sulfuric acid.

Since deposits are usually much heavier than decorative chromium deposits, plating times are much longer and a knowledge of plating speed is necessary.

Such graphs as those shown in Fig. 12-7 and 12-8 are helpful in determining the plating time required for any given thickness of chromium from the dilute and concentrated baths at various temperatures. The usual temperature range chosen is 49 to 55°C.

The mixed-catalyst baths are somewhat faster and exhibit better coverage than the sulfate-catalyzed solutions. They can introduce an etching problem, in low-current-density areas; this may or may not be serious, and by proper stopping off of areas not to be plated it can be overcome. It is largely a matter of the finisher's preference whether the faster plating and better coverage are sufficient to make solving the etching problems worthwhile.

Chromium as a surface finish offers a rare combination of physical and mechanical properties: low coefficient of friction, high hardness, good corrosion resistance, high heat resistance, and anti-galling properties. It can be used over many different types of substrate, whose properties, however, must be taken into account in determining the thickness of the chromium deposit. For example. there is little point in applying a hard chromium deposit over a soft substrate unless the deposit is thick enough to mask the properties of the basis metal.

The anodes are also the same as in normal decorative chromium plating: lead or lead with up to 10 percent tin or antimony. For plating complex shapes, anodes can be used in wire form for conforming to the shape of the cathode.

The principal way in which hard chromium plating differs from decorative is that the deposits are usually much thicker and their physical properties much more important. As a rule, the product is destined for rough usage: slight errors that might be tolerated in decorative plating cannot be permitted. The parts being plated are often large and valuable: rejects can be catastrophic rather than simply annoying. Since plating times are much longer, proper conditions must be maintained over longer periods, which requires better control and more careful monitoring.

The principles are the same; thus success in hard-chromium plating depends more on extreme care, good background of experience, and perhaps apprenticeship in a shop already performing the work. Hence a full discussion of the details of hard chromium plating belongs in shop manuals rather than in a text such as this.

OTHER CONSIDERATIONS

The structure of chromium plate has a direct bearing on its suitability for various applications. Many variations in structure and physical properties can be obtained by appropriate adjustment of plating conditions. Several monographs are concerned with this aspect of the subject and should be consulted for details.

Hardness and wear resistance also are subject to wide variation. Hardness is usually given as 8800 to 9800 MPa. This number depends, like all hardness measurements, on the method used to determine it.

One of the most useful properties of chromium plate is its low coefficient of friction against other materials; this accounts for its utility in shafting, piston rings, internal-combustion-engine cylinders, and similar uses.

Chromium plate generally reduces the fatigue strength of steels upon which it is deposited; shot peening and other methods of preparing the steel for plating can reduce this effect. Heat treatment after plating can also be useful.

MODIFICATIONS OF CHROMIUM PLATE

Porous Chromium

Chromium plate can be modified, by means of special mechanical or chemical treatment, to a form containing macroscopic pits or channels which retain oil films well. This form is known as *porous chromium* (not to be confused with microporous chromium).

Colored Chromium

Black and colored chromium processes have been introduced from time to time, with varying degrees of success. They are not pure chromium metal but contain oxides and undoubtedly other ingredients. Most of these processes are proprietary, and their virtues cannot be evaluated here.

Specifications for Decorative Nickel-Chromium Coatings

Most large companies and many standardizing bodies have established minimum specifications for nickel-chromium or copper-nickel-chromium coatings on various substrates and for various types of articles, depending on the severity of the exposure to which they will be subjected. As typical of such specifications, ASTM specification B 456 is summarized here.

The type and thickness of both the nickel and the chromium are specified according to the service condition that the article is expected to withstand. These are defined as follows (SC stands for service condition):

> SC4 (Very severe). Likely damage from denting, scratching, and abrasive wear in addition to exposure to corrosive environments; e.g., automobile exteriors, boat fittings.
>
> SC3 (Severe). Includes frequent wetting by rain or dew or possibly strong cleaners or saline solutions; e.g., porch and lawn furniture, bicycles, hospital furniture and fittings.
>
> SC2 (Moderate). Indoors where condensation may occur; e.g., kitchens and bathrooms.
>
> SC1 (Mild). Indoors in normally dry atmospheres, with coating subject to little wear or abrasion.

Types of nickel. b = fully bright nickel; p = dull or semibright nickel containing less than 0.005 percent sulfur and passing a ductility test—may be buffed if desired; d = double-layer (duplex) nickel (or triple-layer) of which the bottom layer contains less than 0.005 percent sulfur and the top layer contains more than 0.04 percent sulfur. The thickness of the bottom layer must be at least 60 percent of the total thickness on nonferrous metals and at least 75 percent of the total on steel.

Types of chromium. r = "regular" (conventional), at least 0.25 μm thick, except for SCl when it may be only 0.13 μm thick; mc = microcracked, having more than 30 cracks per millimeter in any direction over the whole surface, and having a minimum thickness of 0.8 μm; mp = microporous chromium, having a minimum thickness of 0.25 μm and containing a minimum of 100 pores per mm^2 invisible to the unaided eye.

In designating coating systems, ASTM has adopted a classification system as follows:

1. The chemical symbol for the basis metal (Fe for steel, etc.)
2. The chemical symbol for nickel (Ni)
3. A number indicating the minimum thickness of the nickel, in micrometers
4. A letter designating the type of nickel, as outlined above
5. The chemical symbol for chromium (Cr)
6. A letter or letters designating the type of chromium and its minimum thickness in micrometers, the latter already being included in the definition of chromium type, as outlined above.

For example: FeNi40dCrr indicates a ferrous substrate plated with duplex nickel 40 μm thick, followed by conventional chromium at least 0.25 μm thick; ZnNi25dCrmc indicates a zinc alloy substrate, plated with at least 5 μm of copper or brass (understood), 25 μm of duplex nickel, and microcracked chromium at least 0.8 μm thick.

Table 12-19 summarizes ASTM B 456 specifications for nickel-chromium deposits on steel, zinc alloy, and copper or its alloys for the various service conditions outlined. On zinc alloy it is understood that an undercoating of copper or yellow brass at least 5 μm thick precedes the specified coatings.

With respect to steel substrates, it is now stated that copper may be used between the steel and the coating, but cannot take the place of any nickel. This prohibition is under some criticism within the standardizing body; some workers feel that the advent of the newer bright acid copper deposits should modify the

Table 12-19 ASTM B 456: Nickel-Chromium Coatings*

Service condition number	*Classification numbers*		
	On Steel	*On zinc alloy*	*On copper and its alloys*
SC4	FeNi40dCrr	ZnNi40dCrr	CuNi30Crr
	FeNi30dCrmc	ZnNi30dCrmc	CuNi25Crmc
	FeNi30dCrmp	ZnNi30dCrmp	CuNi25Crmp
SC3	FeNi30dCrr	ZnNi30dCrr	CuNi25dCrr
	FeNi25dCrmc	ZnNi25dCrmc	CuNi20dCrmc
	FeNi25dCrmp	ZnNi25dCrmp	CuNi20dCrmp
	FeNi40pCrr	ZnNi40pCrr	CuNi25pCrr
	FeNi30pCrmc	ZnNi30pCrmc	CuNi20pCrmc
	FeNi30pCrmp	ZnNi30pCrmp	CuNi20pCrmp
			CuNi30bCrr
			CuNi25bCrmc
			CuNi25bCrmp
SC2	FeNi20bCrr	ZnNi20bCrr	CuNi15bCrr
	FeNi15bCrmc	ZnNi15bCrmc	CuNi10bCrmc
	FeNi15bCrmp	ZnNi15bCrmp	CuNi10bCrmp
SC1	FeNi10bCrr	ZnNi10bCrr	CuNi5bCrr

*See text for meanings of symbols

specification to the extent of allowing some substitution of copper for a proportion of the nickel. Since ASTM specifications are published annually, the reader should be sure to have access to the latest revision.

Other ASTM specifications relevant to this chapter, but not discussed in detail here, include:

B 177. Chromium plating on steel for engineering use.

B 343. Preparation of nickel for electroplating with nickel.

B 503. Use of copper and nickel electroplating solutions for electroforming. This document includes a useful summary of the physical and mechanical properties of nickel plated from the Watts and the sulfamate bath and of copper plated from the sulfate and the fluoborate baths.

B 558. Preparation of nickel alloys for plating.

13
The Engineering Metals

INTRODUCTION

We classify under this heading the precious metals gold, silver, and the platinum group, as well as tin and lead. Like our other classifications, this is largely arbitrary. Gold and silver are used decoratively as well as for their physical and chemical properties; but while their decorative applications tend to decrease or remain static, their importance as industrial coatings is increasing, especially in electronics and allied industries. Of the six platinum metals, only rhodium has much use as a decorative coating; the others, insofar as they are plated at all, are used for engineering applications. Tin and lead are seldom used for decorative appeal, and their classification as industrial coatings is logical.

Except for tin and lead, the group considered here does share some common attributes. All are "precious," i.e., expensive and rare; all are "noble," i.e., have highly positive electrode potentials and are relatively inert in most environments. Again excepting tin and lead, they occupy contiguous places in the periodic table: gold and silver in Group IB and the platinum group in Group VIII, adjacent to Group IB horizontally.

Tin and lead occupy vertically adjacent places in Group IVB of the periodic system; their chemistries have much in common. Lead is not a common plating metal, but tin-lead alloys, of varying compositions, are increasing in importance, again owing to the demands of the electronics industry.

THE PRECIOUS METALS

Of the engineering metals being considered in this chapter, the so-called precious metals form a well-defined subgroup. They have certain aspects in common which differentiate them from all the other plating metals; these aspects are primarily economic, but there are also chemical similarities. And the economic factors have some corollaries which influence the practical aspects of the plating of these metals.

The chemical similarities result from the position of the precious metals in the periodic system. All are members of either Group IB or Group VIII, which are neighbors in the table, and thus they exhibit both horizontal and vertical relationships. These will be discussed in their proper order when we consider the individual metals.

On the whole, purely economic considerations are out of place in a text on the fundamentals of metal finishing; there is no fundamental difference between plating a cheap metal like zinc and an expensive one like silver. But economics cannot be entirely ignored, and it has some effect on the practice of plating the precious metals. These metals differ not only in degree but in kind from the other plating metals because their prices differ from those of the others by orders of magnitude.

Table 13-1 lists the prices of the more common plating metals during a particular week in 1975. Metal prices are subject to wide and frequent variation, and the figures in the table are only representative; they may have changed drastically by the time this book is published, or when you happen to be reading this page. But the relationships among them will probably not have changed very much, and the table only illustrates a point.

If nickel be taken as a typical plating metal, the table shows that the other metals above the line differ from it in price by less than an order of magnitude, while the precious metals are in an entirely different category. This causes the plater applying these metals to take a somewhat different approach to the processes.

Table 13-1 Metal Prices During September 1975

	Dollar basis		*Volume basis**	
Metal	*Price, \$/kg*	*Price relative to Ni = 100*	*Specific gravity*	*Price, volume basis, Ni = 100*
Cadmium	7.50	145	8.65	141
Chromium (in CrO_3)	2.45	47	7.14	38
Copper	1.43	28	8.93	28
Lead	0.44	9	11.34	11
Nickel	5.16	100	8.90	100
Tin	7.23	140	7.31	115
Zinc	0.86	17	7.14	14
Gold	4 920	95 350	19.3	207 000
Palladium	2 572	50 000	12.0	67 500
Platinum	5 144	100 000	21.45	240 000
Rhodium	11 250	218 000	12.5	306 000
Silver	146	2 900	10.5	3 400

*An indication of the weight required to cover unit area to unit thickness.

In most plating sequences, platers are under some governmental pressure—which differs from one metal to another and by geographic region—to limit the discharge of certain metals in the waste stream from their plants. If they are plating platinum or gold, however, loss of any significant amount of these metals in the effluent will put them out of business without government intervention. Metal *must* be recovered from rinses; rejects must be stripped and the stripping solution preserved; and the metal must be either recovered or sold to a refiner to recover at least a part of its value.

Tanks are usually smaller than those for the more common metals. Platers tend to specialize, because the problems are qualitatively different from those of the copper-nickel-chromium plater, for example. Although there are many "job shops" offering precious-metals plating, more of it tends to be "captive," i.e., carried out in the finishing department of the firm that manufactures the items being plated.

In most plating, metal cost tends to be a relatively minor factor compared to items such as labor and overhead. Though it is more expensive to plate nickel than to plate zinc, the factor is not simply the ratio between the metal costs, but much less, because labor and overhead tend to be about the same.

For the precious metals, metal cost is perhaps the major factor; thus the emphasis on control and operation of the shop is altered.

Precious metals, like others, often (or usually) must be plated to a minimum-thickness specification. One way to meet such a specification with, say, zinc, is to overplate slightly to be sure that all parts meet the requirement. This expedient is not available to the precious-metal plater, who cannot afford to "give away" metal that is not being paid for. Therefore control over thickness and thickness distribution must be far more stringent than is the case with the others, and the methods of measuring thickness must be as accurate as the state of the art permits.

If buffing is called for, buffing wheels and used compounds must be saved for reclamation of their metal content. Floor sweepings must be similarly conserved if there is any chance that they contain bits of metal.

One example of the difference in emphasis is in rinse reclamation. One of the methods of recovering metals from rinses and effluents is ion exchange, the principles of which have been considered in Chap. 7. After the metal is concentrated on the ion-exchange resin, it is washed out in a "regeneration" cycle, in more concentrated form, and the resin is prepared for re-use. This method can also be used for, say, gold. But it does not pay to go to the trouble of regenerating the resin; the gold-containing resin is simply ignited to recover the gold: the cost of the resin is insignificant compared to the value of the gold.

Security measures become of great importance. Although thefts of nickel and tin anodes have not been unknown, especially in times of "gray markets" and shortages, such thievery involves the transport of relatively heavy and bulky material and is usually the work of professional criminals. A worthwhile amount of gold, on the other hand, can easily fit in one's pocket, meaning that security must be more vigilant and extreme; bookkeeping and inventory control must be highly accurate.

In these, and perhaps other, ways plating of the precious metals poses special problems; while not part of the fundamentals of electroplating, they deserve passing mention.

Finally, the increasing importance of gold, especially in the electronics and communications industries, along with its escalating price, has led to many investigations in a search for acceptable substitutes and for ways of minimizing its use. Thus the "preciousness" of the precious metals has had a positive effect on research in the metal-finishing field.

SILVER

Silver (Ag), with atomic number 47, is the middle member of Group IB of the periodic system, which comprises the so-called coinage metals copper, silver, and gold. The members of main Group I (IA), the alkali metals, are much alike and for many purposes can be discussed as a group. But the coinage metals exhibit as many differences as similarities and require separate consideration; about the only thing they share with the alkalis is the group valence of 1^+.

Copper has already been described (Chap. 12). All three coinage metals have been known since ancient times. The origin of the word *silver* is not known; the symbol Ag derives from Latin *argentum*.

Silver is rare; it occurs in the lithosphere to the extent of about 5×10^{-6} percent. Its chief mineral is the sulfide argentite, Ag_2S, which is usually associated with other sulfides such as those of lead and copper. Several other minerals of lesser importance are known, as is native silver. Silver is usually a by-product of the mining of other metals, principally lead, copper, and gold; a few mines are operated for their silver values.

Silver is mined in many countries; the main ones are Canada, The United States, Mexico, Peru, the Soviet Union, both Germanys, and Australia.

World production is lower than world consumption; this has led to rising prices, hoarding of coinage, discontinuance of the use of silver in the coinage of most countries, and increasing importance of recovery of silver from secondary sources.

More than half the silver produced is a by-product of the extraction of lead and copper; thus its availability is influenced by the supply-demand situation of those metals. Once fixed by the United States Treasury Department, the price of silver is now free to find its own level, which is far above the previous U.S. Treasury ceiling. Silver is no longer used in U.S. coinage. The drop in its use in coinage is shown dramatically by the consumption figures for 1969 and 1970: 605 t (metric tons) in 1969 and 22 t in 1970.

Statistics of price, consumption, etc., for silver, as well as for the other precious metals, are usually quoted in troy ounces: 1 troy ounce = 31.1 g, and 1 t = 32 150 troy oz.

Production

The extractive metallurgy of silver is tied to that of copper, lead, and zinc; silver follows these metals through concentrating and smelting. Where silver occurs in sufficient concentration to be treated by itself, cyanidation or amalgamation may be used to extract it.

In copper refining, silver separates from the main stream in the anode slimes of the electrolytic refinery. The slimes are leached with hot sulfuric acid to remove some extraneous metals; the treated slimes are smelted in a "doré" furnace with fluxes such as calcium carbonate, silica, borax, and calcium fluoride. Most of the base metals separate in the slags; the "doré metal" is cast into anodes and electrolyzed in a nitrate solution to produce silver. In lead refining, the details differ but the final steps of casting into doré anodes and electrorefining are the same.

The recovery of silver from secondary sources is significant: scrap, waste photographic fixing solutions and film, alloys, electroplating rejects, and the like. Large users such as film manufacturers refine their own waste materials; smaller users deal through refiners, who use methods similar to those described.

Uses

For most of its uses silver is too soft and is hardened by addition of various percentages of copper. Sterling silver contains 92.5 percent silver, 7.5 percent copper; coinage silver contained 90 percent silver and 10 percent copper. Silver brazing alloys contain amounts of silver ranging from 1 to almost 100 percent, depending on use; they are used in many industries for metal joining.

Silver is an effective element in both primary and secondary batteries. It is too expensive for civilian applications, but a silver chloride–magnesium cell is an effective primary battery which is activated by seawater and finds uses in sonar and related devices. Several secondary batteries are also of interest, including a silver-cadmium and a silver-zinc cell.

Silver and its compounds have several catalytic uses in the chemical industry. The manufacture of mirrors was at one time carried out by the deposition of silver, either chemically or by vapor deposition. This use is being replaced by aluminizing.

The largest single use of silver is in photography, which accounts for about 30 percent of total consumption. Electroplating is an important application, accounting for almost 10 percent of the total consumption. Other significant uses include sterling ware, jewelry, and dental and medical supplies.

Properties

Physical properties of silver are listed in Table 13-2. Silver has a brilliant white metallic luster, much admired for tableware and ornamental articles. Pure silver has the highest electrical and thermal conductivity of all metals, as well as the lowest contact resistance; the last is often neutralized by its tendency to tarnish in the air.

Silver almost always shows the group valence of 1+, utilizing for chemical combination only the 5*s* electron. Unlike copper and gold, there is little tendency for one or more of the 4*d* electrons to take part. A few divalent compounds are known, notably AgF_2; and although a higher oxide, AgO, is sometimes cited as an example of divalency, it is now thought to be a rare example of trivalency: $Ag(I)Ag(III)O_2$. For practical purposes, only the 1+ state need be considered.

The standard electrode potential of silver stands between those of copper and gold: Cu, +0.52 V; Ag, +0.799 V; and Au, +1.68 V. Copper and silver, but not gold, are readily tarnished by atmospheres containing sulfur compounds; but unlike copper, silver is not oxidized.

The silver–silver chloride half-cell, with a standard reduction potential of +0.222 V, is a valuable reference electrode, often preferred to the more familiar calomel half-cell because it presents no potential problem of mercury contamination and is very rugged.

Table 13-2 Properties of Silver

Atomic number	47
Atomic weight	107.868
Electronic configuration	$1s^2 2s^2 2p^6 3s^2 3p^6 3d^{10} 4s^2 4p^6 4d^{10} 5s^1$
Density, g/cm³, 20°C	10.491
Melting point, °C	960.8
Boiling point, °C	2212
Electrical resistivity, μΩ cm, 20°C	1.59
Conductivity, % IACS, pure	108.4
Ordinary 999 fine	104
Crystal structure	fcc
Standard potential, E^0, 25°C, V	
$Ag^+ + e^- \rightarrow Ag$	+0.7991
Reduction potentials, V, vs. NHE*	
$AgCl + e^- \rightarrow Ag + Cl^-$	+0.222
$AgCN + e^- \rightarrow Ag + CN^-$	−0.017
$AgI + e^- \rightarrow Ag + I^-$	−0.151
$Ag(CN)_2^- + e^- \rightarrow Ag + 2CN^-$	−0.31

*Normal hydrogen electrode.

Silver is unique in its ability, when molten, to absorb large quantities of oxygen and to liberate most of it on resolidification. At temperatures just above its melting point in air, silver dissolves 20 times its own volume of oxygen.

Silver, as might be expected from its standard potential, is inert to many substances. Its principal reaction is with nitric acid, yielding silver nitrate and a mixture of NO and NO_2. The reaction takes place at all concentrations of nitric acid; that most used is about 1:1 acid, hot, for etching silver and for production of silver nitrate, which is the main industrial silver compound and the starting point for producing others.

At elevated temperatures, all the halogens combine with silver, quantitatively and exothermically. Silver is dissolved by alkali metal cyanide solutions: either anodically, as in electroplating, or in the presence of oxygen, to form the complex cyanoargentate (or argentocyanide):

Anodic: $$Ag + 2CN^- \rightarrow Ag(CN)_2^- + e^-$$

With oxygen: $$2Ag + 4CN^- + H_2O + \tfrac{1}{2}O_2 \rightarrow 2Ag(CN)_2^- + 2OH^-$$

The reaction with cyanide and air is important in the extraction of silver from its ores.

Most of the other reactions of silver should be classed as corrosion rather than rapid chemical reactions. A list of reagents that react with silver at various, but usually relatively slow, rates, includes alkali metal cyanides, peroxides, sulfides; bromine; chlorine oxo-acids; chromic acid; ferric sulfate; molten glass, including borax; hydrogen sulfide; mercury salts and metal; peroxysulfates and permanganates; sodium thiosulfate; sulfur; and concentrated sulfuric acid.

Silver forms many insoluble compounds and many stable complexes with inorganic and organic ligands. It forms a few simple aquo ions, with nitrate, perchlorate, and fluoborate. Its complexes with the halides, cyanide, ammonia, and thiosulfate are of principal importance in electroplating. Most of the insoluble silver compounds are soluble in solutions containing complexing agents; for example, the very insoluble silver bromide is soluble in thiosulfate and cyanide solutions. Silver chloride is somewhat soluble in hydrochloric acid; the chloride ion first diminishes the solubility owing to the common ion effect and then increases it as the complexes $AgCl_2^-$ and $AgCl_3^{--}$ are formed. The principal exception is silver sulfide; this is so insoluble ($K_{sp} \approx 10^{-51}$) that even cyanide and thiosulfate solutions do not react with it. But it is readily soluble in strongly oxidizing acids like nitric and hot sulfuric.

Grades

The purity of silver is quoted in parts per thousand rather than percent: commercial fine silver is ''999'' fine, or 99.9 percent pure. Three recognized grades are 999, 999.5, and 999.9 (ASTM Specification B 413).

Hazards

The handling of silver metal presents few if any hazards. Soluble silver salts are poisonous by ingestion or injection; and of course the cyanide is poisonous, as are all cyanides. Silver nitrate (once called "lunar caustic") is used medicinally, but only externally.

Silver Plating

Silver was probably the first metal to be plated commercially; an Elkington (British) patent dated 1840 described a cyanide bath essentially not very different from that in use today. Most historians date the beginning of electroplating as an industry from this date.* The bath consisted of the double cyanide of silver with sodium cyanide, $NaAg(CN)_2$, with excess free cyanide. Although many other baths have been proposed, none has been commercially successful,† and the cyanide bath is the only one considered here. Other electrolytes proposed include an iodide complex bath and baths containing the simple aquo silver ion such as sulfamate and fluoborate. These baths have had little or no practical success; the silver, being uncomplexed, is affected by light, and plating tends to give large-grained, coarse deposits with relatively poor distribution. In the high-cyanide electrolytes, silver is more noble than gold and tends to deposit by immersion on most basis metals. Therefore a cyanide "strike" is almost always necessary before plating. Bright silver plating is done entirely from cyanide electrolytes, containing for the most part proprietary brighteners.

The largest use of silver deposits, in spite of the trend toward stainless steel flatware, is in the flatware–hollowware trade, especially the latter; use in flatware is declining. The second largest user is the electronics industry, where large amounts are plated onto conductors, waveguides, and similar items to take advantage of the unsurpassed electrical conductivity of silver. In most of these applications, silver is plated over copper and copper alloys, such as brass and nickel silver.‡ The principal application for plating silver directly on steel is in bearings; this use is declining as the reciprocating aircraft engine gives way to jet engines. Some chemical reactors are of steel plated with silver.

Improvements in silver plating since its inception have not involved any basic changes; they have come about through development of improved brighteners

*In fact, a dictionary published around the turn of the century defined electroplating as "application of a deposit of silver upon another metal."

†A proprietary noncyanide bath has been announced.

‡"Nickel silver" contains no silver; it is an alloy of copper (65 percent), nickel (18 percent), and zinc (17 percent).

and improvements in plating speed made possible by minor changes in formulation of the bath.

Metal cyanides became commercially available toward the beginning of the twentieth century; previously platers had to prepare their own salts by a roundabout method including dissolving silver, precipitating the chloride, dissolving this in sodium cyanide solution (which frequently they were forced to prepare for themselves also). Later, the availability of silver cyanide made it possible to prepare solutions by dissolving silver cyanide in sodium or potassium cyanide solutions. Still more recently, "double salts"—potassium silver cyanide—became available, and they are more convenient in use.

Plating for decorative and for industrial purposes is done from the same types of bath; the latter are usually formulated for higher speed and thus have thicker deposits. Typical bath compositions are listed in Table 13-3.

Alkali cyanide serves to complex the silver, thereby minimizing the tendency for immersion deposition. Free cyanide is necessary for good anode corrosion, to dissolve the AgCN film on the anodes, and improves the conductivity of the bath. It increases cathode polarization and thereby increases throwing power.

The preference for potassium cyanide over the sodium salt is due to many factors. The commercial salt is of higher purity. Potassium ion is more conductive than sodium; potassium carbonate is much more soluble than sodium carbonate and allows higher concentrations of carbonates without deleterious effects. Most bright baths operate better with potassium salts, and higher plating speeds are possible . Since the metal is contained in the anion $Ag(CN)_2^-$, it is not entirely clear why potassium baths yield better results on the whole, but it is a fact, not only in silver plating but in other anionic complex baths such as copper, gold, and tin.

Carbonate and hydroxyl ions help to improve conductivity; they are required for proper functioning of some proprietary brighteners, and the hydroxyl ion aids in preventing breakdown of cyanide by maintaining an alkaline pH (alternatively, by acting as an acceptor for atmospheric carbon dioxide). Without it, carbon dioxide from the air can react with cyanide as follows:

$$CO_2 + 2CN^- + H_2O \rightarrow CO_3^{--} + 2HCN$$

since, weak as it is, carbonic acid is stronger than hydrocyanic.

Potassium nitrate, used in some baths, is helpful in extending the bright range. In others it can be harmful, depending on the nature of the brighteners. It improves anode corrosion.

Low silver content and high cyanide favor improved throwing power, and higher concentrations permit the use of higher current densities, as would be expected. Since silver is expensive, bath concentrations are held as low as practical considering the requirements of the job. When thin deposits are wanted and speed is not essential, rather dilute baths are used; when thicker deposits and higher speeds are desired, more concentrated solutions are justified in spite of their higher cost.

Silver-plating baths may be operated at room temperature (20–27°C) at current

Table 13-3 Silver-plating Baths

	Decorative types				*Industrial types*			
	Conventional		*High speed*		*Bath I*		*Bath II*	
Ingredients	*g/L*	*Molarity*	*g/L*	*Molarity*	*g/L*	*Molarity*	*g/L*	*Molarity*
Silver cyanide, AgCN	30–55	0.22–0.41	45–150	0.34–1.12	45–50	0.34–0.37	75–110	0.56–0.82
(Ag as metal)	24–44	0.22–0.41	36–121	0.34–1.12	36–40	0.34–0.37	60–89	0.56–0.82
Total potassium cyanide, KCN	50–78	0.77–1.2	70–235	1.1–3.6	65–72	1.0–1.1	85–143	1.3–2.2
Free KCN	35–50	0.54–0.77	45–160	0.69–2.5	45–50	0.69–0.77	50–90	0.77–1.4
Potassium carbonate, K_2CO_3	15–90	0.1–0.65	15–90	0.1–0.65	45–80	0.33–0.58	15 min	0.1 min
Potassium nitrate, KNO_3	—	—	40–60	0.4–0.6	40–60	0.4–0.6	—	—
Potassium hydroxide, KOH	—	—	4–30	0.07–0.5	10–14	0.18–0.25	0–30	0–0.53
Brighteners	q.s.		q.s.		q.s.		q.s.	
Temperature, °C	20–28		38–50		42–45		38–50	
Current density,* A/m^2	50–150		50–1000		500–1000		500–1000	

*With agitation.

densities from 50 to 150 A/m^2. Agitation, recommended, is by cathode rod movement or by mechanically moving the solution. For higher-speed plating, higher temperatures and vigorous agitation are required; at a temperature of 50°C, with high silver content of the bath, and with vigorous agitation, current densities to 1000 A/m^2 can be realized.

Intermittent or continuous filtration is recommended, with provision of treatment by activated carbon, for removal of organic impurities.

Anode and cathode efficiencies are close to 100 percent; therefore the solution tends to remain in balance over long periods, and only cyanide and additives usually require close control. Many metallic impurities have little effect, but iron contamination, introduced from steel anodes or drag-in from strike solutions, can cause off-color deposits and interfere with the action of brighteners. Iron can be removed by cooling to about 3°C and filtering off the precipitated ferrocyanide.

Anodes

High-purity, at least 999 fine, anodes should be used. "Black anodes" are experienced by most silver platers at times; they can be due to low free cyanide, low pH, or too high a current density. Small amounts of iron, bismuth, lead, antimony, sulfur, selenium, and tellurium also can be responsible. The film may be quite tenacious, but during idle periods will slough off, causing roughness; hence the requirement for intermittent or continuous filtration or the use of anode bags.

Steel and stainless steel anodes are used in strike solutions, and carbon or platinum anodes are used occasionally for special situations.

Silver Strike Solutions

Substrate preparation is conventional (see Chap. 5) except that a silver strike is necessary before entering the plating bath. Most metals are less noble than silver (including even gold in the cyanide system); thus they will precipitate silver by immersion from the plating baths listed in Table 13-3, and deposits will be poorly adherent. Strike baths contain low silver and high free cyanide to avoid this immersion deposition as far as possible, and they provide a surface that will accept an adherent silver deposit. The strike sequence depends on the nature of the substrate. For steel, a double strike is used, first in a solution containing some copper cyanide and then in a conventional ("second") strike solution. The second strike is used as the first and the only strike for substrates such as copper and its alloys, nickel, or nickel-silver. Typical compositions for silver strike baths are shown in Table 13-4.

The strike serves another purpose: to cover work consisting of more than one metal, such as soldered parts and assemblies, Time is 8 to 25 s for bright decorative silver and 15 to 35 s for thicker deposits. Where possible, the work should enter the strike solution with a "hot lead." A rinse between the strike and plating solutions may or may not be necessary, depending on the degree of contamination of the strike bath.

Table 13-4 Silver Strike Baths

Ingredient	*Bath I**		*Bath II*†	
	g/L	*Molarity*	*g/L*	*Molarity*
Silver cyanide, AgCN	1.5–2.5	0.01–0.02	1.5–5	0.01–0.4
Copper cyanide, CuCN	10–15	0.11–0.17	—	—
Potassium cyanide, KCN	75–90	1.15–1.4	75–90	1.15–1.4
Temperature, °C	22–30		22–30	
Current density, A/m^2	150–300		150–300	
Potential, V	4–6			

*First strike for steel.

†Second strike for steel and strike for nonferrous metals.

Decorative Silver Plating

Most decorative silver plating concerns small items such as hollowware, flatware, jewelry, candlesticks, and the like. Basis metals differ widely and are often combinations of two or more metals joined with solder or brazed. Much decorative silver plating involves replating of used items; usually the silver must be stripped, the item often is repaired with solder and then replated. The skills of the practical plater are very much involved, but fundamentally the operation does not differ from what has already been described. Silver strikes are usually required, as mentioned above; their composition varies according to the nature of the work, but they are always high in free cyanide and low in silver content: free cyanide up to 90 g/L, silver 1.25 to 3.5 g/L.

As would be expected, decorative silver plating usually involves brighteners. Carbon disulfide, once standard, is still used to some extent, as is thiosulfate; neither gives fully bright deposits, but ones that are easily buffed. Modern brighteners are invariably proprietary, and information must be obtained from vendors.

Hardeners such as antimony have been recommended, for greater wear resistance of the deposits; it does harden them, but there is considerable disagreement about whether hardness and wear resistance necessarily go together.

Flatware is subject to federal specifications governing the thickness of silver (actually the weight of silver per unit area) required in order that the plate may be labeled ''quadruple plate,'' ''triple plate,'' etc.

Engineering Applications

The excellent mechanical properties of silver suggest its use as a bearing material; this led to the development of high-speed plating baths for silver on steel-backed sleeve bearings. Silver has been used widely as an intermediate material for heavy-duty bearings and for prevention of galling or seizing of metal surfaces under light loads; anti-galling applications include silver plating on stainless steel bolts and titanium compressor blades and as a sealing medium for hot gas seals. Aerospace specifications govern this type of application.

The principal requirements for industrial silver plating are adhesion, ductility, and deposit soundness for thick deposits, up to 1.5 mm. Adhesion is, of course, of prime importance, and preplate procedures are accordingly critical. One recommended sequence for steel is: degrease, anodic alkaline clean, anodic sulfuric acid etch, and activate in 1.2 *N* hydrochloric acid; follow with a nickel strike, then the conventional silver strike, and finally the silver plate.

Because of the heavy deposits demanded, high-speed plating is of more importance than in decorative work, and the baths are accordingly more concentrated. Typical formulations are included in Table 13-3. Agitation is mandatory.

The brightener usually used in these baths is ammonium thiosulfate, at a rate of 0.02 to 0.05 g/L every 24 h. Potassium thiosulfate has been preferred by some.

The thickness required for the various applications varies from about 7.5 μm or less for anti-galling applications—no thicker than some decorative deposits—which can be plated from conventional decorative baths, to 25 to 38 μm for pressure-activated hot gas seals in rocket engine systems, and upward for special situations.

Silver plate is used in the electrical and electronic industries because of its unsurpassed electrical conductivity. While this does not reach the theoretical for massive silver, it is still higher than that of other metals.

The principal drawback of silver for contact use is its tendency to tarnish and form sulfide films, which increase contact resistance. Thus silver is sometimes overplated with overlays of gold or rhodium for such service.

Silver plate is used in the production of radar waveguides, because radio-frequency conductivity is directly related to electrical conductivity. The high throwing power of silver baths also recommends them for these very complicated shapes.

The phenomenon known as *silver migration* has militated against the use of silver in printed circuits. Under a positive dc potential within a damp resin component, silver will "migrate" across the insulation; on drying, silver metal will be found in the body of the insulation, creating a low-resistance leakage path. Silver plating cannot be used when circuit boards must meet certain military standards.

Postplate Treatments

As is well known, silver tarnishes readily in most atmospheres. The dark brown to black sulfide stain not only causes deterioration in appearance, but diminishes solderability and increases contact resistance.

Many methods have been proposed for preventing or delaying the formation of tarnish films; these have included alloying of the silver with other metals and several types of postplate treatments. Neither method has had any conspicuous success; although alloying may serve its stated purpose, it usually degrades other useful properties. Such postplate treatments as chromating may delay, but not prevent, tarnishing, while also interposing an electrical resistance almost as bad as the tarnish itself.

Overlays of gold or rhodium are efficacious, but of course they change the appearance of the deposit. Thick overlays of gold can be successful in an economic sense: that is, it is less expensive to plate 15 μm of silver followed by 10 μm of gold than to plate 25 μm of gold.

Passivation treatments, including chromating and coating with a colloidal beryllium oxide, have been promoted. Their success in delaying tarnish is controversial. For decorative applications, as in flatware and hollowware, most users have simply tolerated the tarnish and resorted to well-known cleaning methods to remove it when necessary. Contact with aluminum in a hot alkaline solution such as trisodium phosphate is probably the easiest method, although many silver cleaners are marketed.

GOLD

Gold (Au), atomic number 79, is the third and last member of Group IB of the periodic system, below copper and silver. See also the preceding section, Silver. In its chemical reactions it resembles the platinum group about as much as it does silver.

Gold was probably the first metal known to and used by humans, since it occurs in nearly pure form, and its color, weight, and brightness make it easily distinguishable from the sand and gravel with which it is often associated.

Lexicographers trace the word *gold* to an old Gothic term meaning yellow; the symbol Au is from Latin *aurum*.

The indestructibility, appearance, and rarity of gold have been important in world history. The thirst for gold, which was synonymous with wealth, prompted many of the early voyages of discovery and conquest, and "gold rushes" opened up our own West and helped to settle Alaska. The possession of gold has been a symbol of wealth from the earliest times.

Alchemy, the search for the "philosopher's stone" which would transmute base metals into gold, while futile, contributed, if not theoretical, at least factual knowledge to the advancement of chemistry, metallurgy, and medicine.

Gold is distributed widely throughout the world, normally in very low concentrations and generally in the native form. It is usually alloyed with silver, and often with some copper. The only compounds of gold found in nature are the tellurides, containing varying amounts of gold, silver, and tellurium. The total occurrence in the lithosphere is very low, estimated at about 5×10^{-7} percent. Seawater contains gold, the amounts depending on depth and location. The average is estimated to be about 0.02 μg/L, but the enormous amount of seawater in the world results in a total of perhaps 70 million tons of gold in the oceans. Furthermore, off-shore exploration of the continental shelf indicates that gold is present on the ocean floor. The total is conservatively estimated to be perhaps 10 billion tons of gold. There is no practical means at present of exploiting these resources, but the possibilities for the future should not be entirely discounted.

Native gold is found in both lode and alluvial deposits. The largest gold deposits are in the Witwatersrand and the Orange Free State of the Union of South Africa; here the gold is present in deep mines, as veins and stringers in a matrix of pyrite and quartz sand. Many pyrite and pyrrhotite minerals which are worked for their copper, lead, zinc, or nickel content also contain some gold, which is recovered as a valuable by-product. In some areas, erosion of primary deposits over geologic time has freed native gold, resulting in native nuggets and flakes found in the sand and gravel of river beds. Discovery of such deposits stimulated the famous gold rushes of California and the Yukon.

Gold is found in nearly every country of the world. The leading producer by far has been the Republic of South Africa. Canada, the United States, the Soviet Union, and Australia are also important producers, and many other countries produce smaller but significant amounts. In the United States, about 5 percent of total production is from placer mining (the panning for gold so well known to devotees of "Westerns"), about 55 percent from gold and silver ores, and 40 percent as a by-product of mining for base metals such as copper and lead.

Production

Gold usually occurs in veins or placer deposits, which, if not already of fine size, must be crushed; it can then be separated by gravity concentration, by amalgamation, or by cyanidation. In some ores, where the gold is present as the telluride or intimately mixed with sulfides, preliminary roasting is required.

In the amalgamation process, the pulp of crushed ore and water is passed over amalgamated copper plates, to which the particles of gold adhere. The gold amalgam is scraped off at intervals, the excess mercury squeezed out, and the remainder distilled off.

In cyanidation, the gold is dissolved in sodium cyanide solution by bubbling air through the solution; the reaction is similar to that described for silver:

$$2Au + 4NaCN + \tfrac{1}{2}O_2 + H_2O \rightarrow 2NaAu(CN)_2 + 2NaOH$$

The cyanide also dissolves silver and some base metals. The gold is precipitated from solution by zinc dust or occasionally aluminum.

Gold recovered by any of these methods must be refined. It is melted under oxidizing conditions, to remove most of the copper and other base metals, leaving gold plus silver. If the silver content is low, the gold can be recovered by refining in a chloride solution; this is known as the *Wohlwill process*. The gold dissolves anodically, and the silver chloride, which would otherwise coat the anode, is made to slough off by superimposing an alternating current on the electrolytic current.

Another refining process involves chlorination of the precious-metal mixture; base metals are converted to chlorides which volatilize; silver forms the chloride which is molten at the temperature of the reaction and can be poured off; gold remains. The process (Miller process) is not suitable for materials containing platinum metals.

Much gold is recovered from scrap, sweepings, and other secondary sources; methods vary according to the nature of the scrap. The gold may be converted to the chloride by treatment with aqua regia:

$$Au + 3HNO_3 + 4HCl \rightarrow HAuCl_4 + 3NO_2 + 3H_2O$$

Nitrogen oxides are removed by heating, and gold is precipitated by reduction with sulfur dioxide or ferrous sulfate.

Doré metal (see Silver) containing more silver than gold is treated by electrolysis in a nitrate solution. Silver is recovered at the cathode; gold does not dissolve and is retained in anode bags.

Several other refining methods are used; the choice depends on the nature of the starting material and its gold content. Recovery of gold from floor sweepings and other scrap may be extremely complicated, since almost anything may be present; the skill of the refiner is a major factor.

Gold, like silver and the other precious metals, is quoted and sold in troy ounces: 1 troy oz = 31.1 g. Its purity may be expressed, like that of silver, in "fineness," that is, in parts per thousand. But a special system is also used, in which the unit is the karat (k or kt). A karat is a unit of purity equal to 1/24 of 100 percent; i.e., pure gold is 24 karat (k); 18-k gold is 18/24, or 75 percent, pure gold, etc.

Uses

The principal use of gold is as a monetary standard, even though many countries are not officially on the "gold standard." Gold is universally prized, the amount available is relatively constant, and its value is high enough that large quantities are not needed for most transactions. It is estimated that about half the world's gold is in the hands of governments or is held by private persons as a hedge against inflation. The United States, reversing a long-standing policy, now allows private ownership of gold bullion.

Thus much of the gold supply is unavailable for use by industry; this poses an economic problem, because industrial uses, especially in the electronics and allied industries, are increasing, and the price of gold has risen to unprecedented heights. Once controlled by the U.S. Treasury at $35/troy oz ($1.125/g), the price of gold has increased from that level by 4 or 5 times. In spite of this, and in spite of many efforts to find cheaper substitutes, gold remains indispensable in many electronic and electrical components.

In 1972, nonmonetary uses of gold totaled about 227 t (metric tons), of which about 60 percent was in jewelry and the arts, 10 percent in dentistry, and 30 percent in industrial applications. Both jewelry and industrial applications include gold used in electroplating.

Jewelry is made of gold ranging from 21-k (87.5 percent Au) down to 10-k or even lower. Gold can also be bonded to base-metal cores, producing "gold-filled'' or rolled gold plate. The process is called cladding. If the material is to be used functionally, it is known as clad stock. If it is used in jewelry, it may be referred to as gold-filled and is usually designated on a ratio basis such as 1/10 14 k, meaning a 14-k alloy bonded to a suitable base metal in the ratio of at least 10 percent of the weight of the article as 10-k alloy. No article having a karat surface of less than 5 percent of the total weight can be marked gold-filled. Electroplating is used extensively for similar purposes; strict federal regulations govern the use of terms, indicating to the customer the amount of gold that can be expected in the product.

Electrical and electronic devices use gold-plated electrical contacts to ensure reliable performance. The telephone system is a large user of relays requiring extreme reliability, long life, and low electrical noise. Gold plating is the surest known means of securing these properties.

The dental profession is a major user of gold; its noncorrodibility in oral media and its good casting qualities recommend it for this use.

The chemical industry uses some gold for corrosion-resistant equipment. Gold-palladium alloy thermocouples are used in jet aircraft engines and in industrial equipment.

Gold leaf, produced mechanically, by vacuum evaporation, or by electroplating, is widely used for decorative purposes.

Much of the gold used in these applications is recoverable when the item has reached the end of its useful life; owing to the high price of gold such recovery has become increasingly important.

Gold has been used in medicine for the treatment of arthritis, and radioactive colloidal gold has had some use in cancer therapy.

Properties

Some properties of gold are summarized in Table 13-5. Gold is an extremely noble metal, as shown by its standard electrode potential. It does not react with oxygen, sulfur, selenium, nitrogen, or carbon at any temperature, and hydrogen is essentially insoluble in it. Tellurium forms a number of intermetallic compounds, which appear in some ores.

Halogens, if completely dry, show little or no reaction with gold at room temperature, except for bromine. In the presence of moisture, bromine reacts vigorously, chlorine and iodine somewhat less so. Fluorine reacts strongly at red heat but not below 110°C.

Gold resists attack by most acids: pure sulfuric below 250°C, hydrofluoric if it is free of oxidizing agents, and hydrochloric below its boiling point. But if oxidizing agents are present or gold is made anodic, reaction with hydrochloric acid is rapid. Nitric acid, with specific gravity less than 1.46 (80 percent HNO_3), if free of halogens, does not attack gold. The most active solvent for gold is aqua regia, a mixture of 1 part nitric and 3 parts hydrochloric acid.

Table 13-5 Properties of Gold

Atomic number	79
Atomic weight	196.9665
Electronic configuration	$1s^22s^22p^63s^23p^63d^{10}4s^24p^64d^{10}4f^{14}5s^25p^65d^{10}6s^1$
Crystal structure	fcc
Melting point, °C	1063*
Boiling point, °C	2809
Density, g/cm³, 25°C	19.302
Electrical resistivity, $\mu\Omega$-cm, 0°C	2.06
Tensile strength, annealed, MPa	124
Hardness, Vickers	25
Standard potential, E^0, 25°C, V	
$Au^+ + e^- \rightarrow Au$	+1.68

*Calibration point on the thermometric scale.

Hydrocyanic acid reacts very slowly in the absence of oxygen, but if oxygen is present, the reaction is rapid. Phosphoric acid solutions, even hot, have negligible reaction. The only single acid which dissolves gold is selenic at about 225°C.

Gold resists attack by alkali hydroxides and carbonates at all temperatures; but alkali cyanide solutions in the presence of oxygen dissolve it, as has been shown above. Fused nonoxidizing salts other than cyanides have little or no reaction. Fused nitrates do not react, but sodium peroxide corrodes gold.

Mercury readily amalgamates with gold; contact with mercury and mercury salts (in either the 1+ or 2+ state) should be avoided, since they may cause cracking of stressed gold alloys; ferric chloride and nitric acid solutions may have similar effects.

All three coinage metals—copper, silver, and gold—have a single *s* electron outside a complete *d* shell. In spite of this similarity in electronic structure, there are few similarities among the three metals, and it is not clear why.

The only stable cationic species in the group valence state of 1+ (apart from complex ions) is Ag^+; by contrast the Au^+ ion is unstable with respect to the disproportionation

$$3Au^+(aq) \rightarrow Au^{3+}(aq) + 2Au(s) \qquad K \approx 10^{10}$$

The chemistry of gold in both its oxidation states, 1+ and 3+, is essentially the chemistry of complexes; few simple compounds exist. There is no good evidence for gold(I) oxide, Au_2O, but gold(III) oxide, Au_2O_3, can be made though it is very unstable. Gold(III) fluoride is known; both the trivalent chloride and bromide are dimers, Au_2X_6, where X = Cl or Br.

On the other hand, there are many gold(I) complexes stable in aqueous solution, the most important of which is the cyanide $Au(CN)_2^-$; others include the chloride and thiosulfate complexes. The cyanide complex is very stable, $K = 4 \times 10^{28}$; crystalline compounds such as $KAu(CN)_2$ are commercially important, and the free acid $HAu(CN)_2$ can be isolated.

Many other gold(I) complexes are known, with substituted phosphine, arsine, and sulfide ligands as well as carbon monoxide.

Gold(III) is always complexed in solution, usually as anionic species such as $AuCl_3OH^-$. Gold(II) exists in one or two organogold compounds but otherwise is known only as a transient intermediate.

Gold(III) has the same electronic structure as platinum(II), and its complexes show many similarities with those of divalent platinum.

The tetrahaloaurate(III) ions are easily prepared, and their alkali salts can be isolated. When gold is dissolved in aqua regia, or gold(III) chloride is dissolved in hydrochloric acid, and the solution is evaporated, chloroauric acid (III) is obtained as yellow crystals of $(H_3O)^+(AuCl_4^-)\cdot 3H_2O$. Other water-soluble salts such as $KAuCl_4$ and the sodium analog also are easily prepared. Other anions including cyanide, nitrate, and sulfate form complexes.

Gold(III) also forms a number of four-coordinate cationic complexes with pyridine, phenanthroline, diethylenetriamine, etc. Several organogold compounds have been prepared, of both gold(I) and gold(III).

Gold compounds are readily reduced to metal; if the reduction takes place in aqueous medium, colloidal gold may result. The reduction of gold chloride with stannous chloride yields a gold sol known as "Purple of Cassius." In this reaction, gold is reduced to metal and tin is oxidized to $H_2Sn(OH)_6$ which acts as a protective colloid. The reaction is very sensitive and can be used analytically.

Few gold compounds are of commercial importance. Gold chloride is really $HAuCl_4 \cdot xH_2O$ or tetrachloroauric(III) acid; similarly gold bromide is tetrabromoauric(III) acid. The most important gold compounds are the cyanides: aurous or gold(I) cyanide, AuCN; potassium gold cyanide or potassium dicyanoaurate(I), $KAu(CN)_2$; and the analogous sodium salt. The last two are by far the most important gold compounds in terms of the amount sold.

A number of organogold compounds form the basis of various compositions used in the decorative arts, to be sprayed or silk-screened onto ceramics, glass, etc., and thermally decomposed to form a decorative gold film. Organogold compounds have been developed that can be fired at temperatures as low as 250°C, thus making possible the application of gold films to plastics for printed circuits and other advanced technology components.

Hazards

As is evident from the use of gold in medicine and dentistry, gold and its compounds present no hazard to health unless the radical associated with the gold is poisonous. The principal gold compound of interest to electroplaters, potassium gold cyanide, is of course poisonous because of its cyanide content.

Gold Plating

Gold plating is used both decoratively and for industrial purposes. The latter include primarily the electronics, communications, and aerospace applications.

Until fairly recent years, although many gold-plating baths were used, the alkaline cyanide systems predominated, and the most important use for gold plating was decorative: for jewelry, flatware, hollowware, and similar items. Since the early 1950s, the requirements of the electronics and communications industries called for thicker coatings, with greater resistance to wear and corrosion, lower porosity, and greater hardness. These requirements spurred much research and resulted in the development of new processes, capable of meeting stringent specifications. In the decorative field also, improvements have resulted in more ductile systems and ones capable of yielding deposits of many colors and various karat contents.

Printed circuits, contacts, and connectors are representative of the items that are gold-plated for the electrical industry. Gold is used primarily because of its good electrical contact properties and corrosion resistance. Although not so conductive as silver or copper (resistivity, in $\mu\Omega$-cm; Ag, 1.5; Cu, 1.9; Au, 2.2), gold is preferred because it does not oxidize or tarnish to form interfering films which affect contact resistance and solderability.

Transistor bases and integrated-circuit components are gold-plated for use in the electronic industry. Gold is used in electronic components because it withstands corrosion in semiconductor production, is easily soldered and welded, is resistant to oxidation, is ductile, and has good electrical properties. Gold is one of the few metals that will not "poison" semiconductor materials.

For aerospace applications, gold has the highest values of infrared emissivity and reflectivity. This property protects both people and instruments against radiation in space. Corrosion resistance, again, is another factor.

In the chemical industry—where the products are valuable enough to justify its use—the chemical corrosion resistance of gold is used in construction of reactors, which may be electroformed. It is used in heat exchangers in condenser tubes for water desalination.

Decorative applications are, of course, older and were the original reason for gold plating of jewelry, watch cases, writing instruments, and the like. With the advent of more modern gold-plating systems, it is possible to apply the gold, once confined to mere "flashes," to almost any thickness desired, so that "gold-plated" need no longer imply a product that "was the first thing to turn green in the spring."* Intricate shapes that could not be handled by mechanical means can be electroplated. Appearance and corrosion protection are the main considerations in decorative gold plating; appearance includes color, since by minor modifications it is possible to produce a large number of colors to suit the taste of the customer.

*"Abe Martin," a popular humorist of the early twentieth century.

GOLD-PLATING SYSTEMS

Practically all the systems for plating gold, for either decorative or industrial purposes, are based on cyanide baths; the gold(III) chloride solution, used in refining, has had occasional application for decorative purposes but is little used and according to most authorities has little to recommend it. Only the cyanide systems will be considered here. Some exceptions, such as sulfite complex baths, have been introduced, but it is too early to predict their commercial acceptance.

Supplying materials and systems for gold plating is a highly proprietary and fiercely competitive field. One vendor offers no less than 27 different solutions, yielding gold deposits of varying purity and physical and chemical properties, appropriate for various applications; and other suppliers offer similar assortments. Since it is not possible to comment here on the comparative virtues of proprietaries, only some general principles can be discussed. It is possible for finishers to formulate their own solutions from published sources, but few do.

Gold(I) cyanide systems all contain the cyanide complex $Au(CN)_2^-$; the systems fall into three categories: alkaline, neutral, and acid. Unlike other cyanide-complex solutions, the gold cyanide complex is stable under acid conditions, down to a pH of about 3; but this unusual property was not made use of until about the mid-1950s. All other cyanide-plating baths must be maintained at an alkaline pH, since if the solutions become acid, the cyanide complexes are decomposed and poisonous hydrocyanic acid is liberated. As noted previously, however, the compound cyanoauric(I) acid $HAu(CN)_2$ can be isolated as a stable and identifiable material.

Gold-plating baths have been classified in various ways. Four distinct types are recognized, but according to some authorities two or more of these may be merely subclasses of the same type. Three are built around baths containing sodium, or more frequently potassium gold(I) cyanide. They are the unbuffered alkaline cyanides operated in a pH range of 8.5 to 13, a neutral buffered system operated at pH 6 to 8.5, and an acid buffered system operated at pH 3 to 6. The fourth system is a noncyanide type, where the gold is complexed by agents other than cyanide. According to some, however, the distinction between neutral and acid buffered systems is based on history rather than on any real difference in basic chemistry, and the neutral and acid cyanide types are lumped together as "buffered" systems. The alkaline cyanide bath is clearly different, in that it contains a quantity of free uncomplexed cyanide ion.

The source of gold in all systems is gold(I) cyanide, AuCN, which is commercially available in three forms: gold cyanide itself, sodium gold cyanide $NaAu(CN)_2$, and potassium gold cyanide $KAu(CN)_2$. The last-named is most commonly used: unlike AuCN, it is water-soluble, and it is much more so than the sodium salt. The fact that potassium salts are somewhat more expensive than their sodium counterparts is insignificant in the case of an expensive metal like gold.

Potassium gold cyanide [or potassium cyanoaurate(I)] provides the ion $Au(CN)_2^-$, a very stable complex. Few free Au^+ ions are present in the solution, and as with other metal cyanide complexes, it is believed that deposition takes place directly from the complex ion, although studies on mechanism have not provided any firm answers to the kinetics involved.

Other compounds present in gold-plating baths fall into three principal categories: current-carrying species, buffers, and brighteners. The last category includes both metallic ions and organic compounds.

Basis-metal preparation for gold plating is conventional in the sense that the substrate must be clean and free of oxides and other materials that might hinder good adhesion of the deposit. For gold plating, all substrates require a gold strike before the main deposit (see later). Because of the value of gold, a stagnant rinse is essential to recover the gold dragged out of the plating bath.

Each type of bath has its advantages and disadvantages. The alkaline bath, pH 9 to 13, possesses good throwing power and is less liable to codeposition of base-metal impurities in the bath.

The neutral bath, pH 6 to 8, permits alloy formation via codeposition of base metals. Its neutral pH causes little or no attack on delicate substrates.

The acid baths, pH 3 to 6, permit alloy formation via base-metal codeposition, and like the neutral bath do not attack delicate substrates. They can produce the purest gold deposits available.

All the baths can be used for both rack-plating and barrel plating. The choice depends on the functional requirements of the gold coating and the substrate upon which it is deposited.

In preparing all gold-plating solutions, it is good practice to dissolve all the components except the gold salts and brighteners and to filter the solution through activated carbon to purify it. Gold salts and brighteners are then added.

Alkaline Cyanide

This system is historically the oldest. The gold is in solution as $Au(CN)_2^-$ ion, at a concentration of about 8.3 g/L of gold. The solution contains excess cyanide and operates at a pH above 8.5 and at a cathode current efficiency approaching 100 percent. These baths can deposit either pure gold or alloys containing silver, nickel, copper, and cobalt, which alter the color and physical properties of the deposit. The structure of the gold is columnar; the deposits contain appreciable amounts of carbon, the source of which has been the subject of considerable research and is not entirely clear as yet. At lower pH the cyanide baths give deposits containing a dark polymeric derivative of cyanogen, C_2N_2, or a partially hydrolyzed polyhydrocyanic acid, originating at the anode. In some formulations phosphate is added to ensure against a drop in pH at the anode. Thickness for thickness, alkaline cyanide systems yield deposits much more porous than those from the newer buffered acid or neutral systems. Deposits from alkaline cyanide systems are useful for most applications, except for sliding electrical contacts, where they tend to gall and fail quickly. The alkaline system is very tolerant of impurities and changes of composition before the quality of deposits begins to suffer.

The alkaline cyanide baths have excellent throwing power, but the alkalinity and high cyanide content may limit the types of substrates that can be handled. Ceramic and plastic substrates may be attacked.

Typical formulations of alkaline baths are listed in Table 13-6. Plating conditions can vary widely; the lower the temperature, the lower the current density permissible.

Potassium gold cyanide is the source of the gold metal content. Additional potassium cyanide stabilizes the complex, retards codeposition of base metals, and tends to minimize the tendency for immersion deposition. It promotes anode corrosion when gold anodes are used and increases conductivity and throwing power.

Potassium carbonate increases conductivity and improves throwing power; but if it is allowed to increase too much, the effect on throwing power is reversed. Potassium phosphate has similar functions and also acts as a pH buffer. Potassium hydroxide increases conductivity and retards the anodic decomposition of cyanide ion. It is necessary to maintain the solubility of some types of organic and inorganic brighteners.

Without brighteners, this bath produces sound but mat deposits. Brighteners, almost always proprietary, promote grain refinement; most codeposit with the gold. Metallic brighteners will codeposit and affect not only the appearance but also the hardness of the deposit. Organic brighteners may or may not codeposit, and may or may not affect hardness. All brighteners tend to reduce the density of the deposits.

Table 13-6 Alkaline Gold Baths

	Mat finish		*Bright finish*	
Ingredient	*g/L*	*Molarity*	*g/L*	*Molarity*
Potassium gold cyanide, $KAu(CN)_2$*	3–17.5	0.01–0.06	6–23.5	0.02–0.08
(Gold as metal)	2–12	0.01–0.06	4–6	0.02–0.08
Potassium cyanide, KCN	15–45	0.23–0.69	15–90	0.23–1.4
Potassium carbonate, K_2CO_3	0–45	0–0.33	0–30	0–0.22
Potassium phosphate, K_2HPO_4	0–45	0–0.26	0–45	0–0.26
Potassium hydroxide, KOH	10–30	0.18–0.53	10–30	0.18–0.53
Brightener	0	—	0.1–10	—
Temperature, °C	50–70		15–25	
Current density, A/m^2	10–50		30–150	
Anodes	Pt, SS,† Au		Pt, SS	
pH	11–13		9–13	
Agitation	moderate		rapid	
Filtration	intermittent		continuous	

*68 percent Au (in theory 68.37 percent).

†Stainless steel.

Table 13-7 Neutral Gold Baths

	Mat finish		*Bright finish*	
Ingredient	*g/L*	*Molarity*	*g/L*	*Molarity*
Potassium gold cyanide, $KAu(CN)_2$	6–23.5	0.02–0.08	3–15	0.01–0.05
(Gold as metal)	4–16	0.02–0.08	2–10	0.01–0.05
Potassium phosphate, KH_2PO_4	0–90	0–0.66	0–90	0–0.66
Chelates	15–90	—	50–150	—
Brighteners	0		0.1–30	—
Temperature, °C	25–70		35–60	
Current density, A/m^2	20–100		50–200	
Anodes	Pt, SS, Pt/Ti		Pt, SS, Pt/Ti	
pH	6–8		6–8	
Agitation	moderate		rapid	
Filtration	continuous		continuous	

Gold anodes can be used, but recent practice favors platinized titanium or stainless steel anodes.

BUFFERED SYSTEMS

These may be further divided into neutral and acid electrolytes. The deposits from these baths, which are much more recent developments than the alkaline cyanide, exhibit a wider range of properties than is obtainable from the alkaline system. Just as in the alkaline baths, the gold is present as $Au(CN)_2^-$ ion, which is stable at a pH as low as 3.0 [note that the free acid $HAu(CN)_2$ can be isolated, as mentioned previously]. Deposits from these baths may be pure or may contain codeposited metal. The alloy baths usually operate around pH 4, while pure gold deposits are obtained from baths in the ranges around 4 or 7. The flexibility in operating pH allows the use of photoresist materials superior to those possible with the alkaline system, which attacks many such materials.

Neutral Baths

Typical formulations for neutral baths are listed in Table 13-7. No free cyanide is added, because in the pH range of 6 to 8 where these baths operate free cyanide is unstable. Neutral gold baths readily permit alloy deposits because of the absence of free cyanide, which tends to discourage alloy deposition. They show little attack on plastic or ceramic substrates.

Neutral gold baths, without brighteners, can produce pure mat gold deposits or alloyed deposits which may be semibright to bright. For the unbrightened baths, cathode efficiencies are in the range of 95 to 100 percent; for bright baths much lower, about 20 to 50 percent.

As before, the gold content of the bath is provided by potassium gold cyanide. Potassium phosphate acts as a buffer, as well as increasing conductivity and improving throwing power. Chelating agents (proprietary or undisclosed) retard the deposition of metallic impurities that may be present, and are necessary for proper action of the brighteners. Metallic brighteners result in alloy deposits of increased hardness. All brighteners reduce the density of the deposit.

Gold anodes are not used in neutral baths because they are inert. The gold content is usually replenished by the addition of potassium gold cyanide. With use, the pH tends to rise and is maintained by the addition of dilute phosphoric acid. If necessary to raise the pH, 20% potassium hydroxide solution may be used.

Anodes are typically platinized titanium or gold-plated platinized titanium. The gold plating is inert, but it does tend to keep the voltage down. Stainless steel anodes introduce contamination, as do carbon anodes, and their use is not recommended. Baths for pure gold deposits often show cathode efficiencies of about 100 percent. The efficiency falls significantly if the bath becomes contaminated with iron or other transition metals. As little as 0.1 percent iron in the deposit may cause the cathode efficiency to drop to 80 percent, and the deposit will have a poor appearance and high porosity and stress.

Acid Baths

Typical formulations are listed in Table 13-8. The acid gold systems are the latest addition to the repertory of the gold plater; it required some years for the realization that "cyanide baths must be maintained on the alkaline side" was not true of gold solutions. In some ways, acid gold baths are the most versatile. They may be formulated to produce either mat or bright deposits. The latter facilitate codeposition of many base metals with the gold to any practical thickness, with a wide variety of colors, ranging from pale yellow to deep orange to light violet.

The acid baths do not attack delicate substrates like ceramics and plastics. The bath formulated for a mat finish can produce the purest gold deposits commercially obtainable; purity of 99.999+ percent has been claimed, with a density practically theoretical.

Plating conditions vary widely, depending on the result desired. For the mat-finish systems, cathode efficiency approaches 100 percent; for bright systems it is only about 25 to 35 percent.

In use, the pH of the bath rises and is maintained by the addition of dilute phosphoric or sulfuric acid. The lower practical limit of pH is 3.0; lower pH for any appreciable time will precipitate AuCN. Potassium hydroxide in 20% aqueous solution is used to raise pH.

Gold anodes are not used because they are inert in this system; bath replenishment is as usual, by addition of potassium gold cyanide. The appropriate phosphate is used as a buffer; it also increases conductivity and improves

Table 13-8 Acid Gold Baths

Ingredient	*Mat finish*		*Bright finish*	
	g/L	*Molarity*	*g/L*	*Molarity*
Potassium gold cyanide, $KAu(CN)_2$	3–23.5	0.01–0.08	6–17.5	0.02–0.06
(Gold as metal)	2–16	0.01–0.08	4–12	0.02–0.06
Potassium phosphate, KH_2PO_4	0–100	0–0.73	0–100	0–0.73
Chelates	10–200	—	10–150	—
Primary brighteners	0		0.1–20	—
Secondary brighteners	0–10	—	0.1–10	—
Temperature, °C	40–70		25–50	
Current density, A/m^2	10–50		80–200	
Anodes	C, Pt, Pt/Ti		C, Pt, Pt/Ti	
pH	3–6		3–5	
Agitation	moderate		rapid	
Filtration	continuous		continuous	

throwing power. Chelating agents have many functions: they improve throwing power, increase conductivity, act as inhibitors for the deposition of any base metals present, and may serve as buffers. In some cases they are necessary for proper action of the brighteners.

Phosphate buffers are, in fact, suitable in the acid, neutral, and alkaline pH ranges; deposits can be pure and soft, but can also be hard. With continued use under alkaline conditions, the phosphate baths will accumulate cyanide and finally will resemble the alkaline cyanide systems.

The structures of gold from buffered systems are fine-grained and not resolvable with a conventional metallograph at 600X. The hard golds with alloying additions are laminar. It has been reported that deposits from alloy baths, containing nominally 99.8 percent Au and 0.2 percent Co or Ni, actually contain up to 0.2 percent carbon. Nitrogen, oxygen, and hydrogen were also found. The source of this carbon is still under study; it may be the cyanogen polymers previously mentioned.

A large number of inorganic and organic additives, for brightening, hardening, and alloying of the deposit from these buffered baths, have been patented. Almost universally these systems are proprietary, and information must be obtained from suppliers of the systems.

Anodes

The type of anode used in gold plating depends on both the bath type and economic considerations. Gold anodes are usable in alkaline cyanide baths and serve to replenish the gold content; they are not used for bright alkaline cyanide baths because of their effects on the brightener systems. Gold anodes are expensive and add to inventory costs, as well as being a temptation to theft.

Platinum anodes can be used for all the baths mentioned. They are also very expensive; they are not quite as great a temptation to theft because platinum is not as easily disposed of in illegal channels as gold. However, they are used only when performance is critical.

Platinized titanium anodes can be used for the neutral and acid baths. Their cost is much less than that of pure platinum, and they are almost as satisfactory. They are recommended as being preferable in most cases to stainless steel or carbon.

Stainless steel anodes can be used in neutral and bright alkaline systems. Carbon anodes are used only in the bright acid plating baths. They are not recommended for neutral or alkaline baths because they deteriorate in use and can cause deposit roughness.

NONCYANIDE SYSTEMS

Several noncyanide systems are described, patented, or both. They include a complex thiomalate system and baths based on gold(III) chloride complexes.

Baths based on sulfite complexes appear to be the most recent contributions to the large number of systems available to the plater. One authority, at least, believes that such systems may soon rival the cyanide in commercial acceptance. For the present, however, the cyanide systems—alkaline, neutral, or acid—dominate the field of gold plating.

GOLD STRIKE BATHS

As has been mentioned, a gold strike must almost always precede the main gold deposit. Two systems are used: the alkaline cyanide and the acid. Their formulations are listed in Table 13-9. Gold strikes serve two functions: they promote adhesion of the subsequent gold deposit by eliminating the possibility of an immersion deposit forming in the gold-plating bath, and they reduce contamination of the bath by either the basis material or solution drag-in from preceding steps. Thickness of gold strikes varies between 0.025 and 0.25 μm.

The selection of a gold strike is determined by the type of substrate to be plated. In general, the acid strike baths are used for plastic-bearing basis material such as printed circuits or potted connectors. Low metal content, about 0.5 to 4 g/L, generally is required in a strike bath so that low efficiencies will result; this entails copious hydrogen evolution which is said to aid in cleaning the surface.

Alkaline strike baths are generally operated at a pH between 8 and 13, acid strikes at pH 3 to 7. Temperature of the alkaline strike is between 40 and 60°C, and of the acid strike 30 to 60°C. These strikes are generally controlled by voltage: 3 to 8 V for alkaline and 4 to 12 V for acid strikes.

Anodes for the alkaline strike are stainless steel; for the acid, platinum,

Table 13-9 Gold Strike Baths

Ingredient	*Alkaline*		*Acid*	
	g/L	*Molarity*	*g/L*	*Molarity*
Potassium gold cyanide, $KAu(CN)_2$	0.75–3	0.0025–0.01	0.75–6	0.0025–0.02
(Gold as metal)	0.5–2	0.0025–0.01	0.5–4	0.0025–0.02
Potassium cyanide, KCN	15–90	0.23–1.4	0	0
Potassium phosphate, K_2HPO_4	15–45	0.09–0.25	0	0
Potassium phosphate, KH_2PO_4	0	0	0–45	0–0.33
Chelates	0		15–90	—
Temperatures, °C	40–60		30–60	
pH	8–13		3–7	
Voltage, V	3–8		4–12	
Anodes	SS		Pt/Ti, Pt, C	
Agitation	moderate		moderate	

platinized titanium or carbon. Agitation is moderate. Time in the strike varies from 5 s to 5 min, depending on the bath and whether the operation is rack-plating or barrel plating.

NOTES ON CONTROL AND OPERATION

Contaminants

Metallic contamination of gold-plating baths can arise in two ways: drag-in of plating solution from preceding steps and attack on the basis metals being plated. Most metals are complexed or chelated by the components of the gold baths and do not affect the deposit; but this is not true of all metals in all systems. Silver, cadmium, zinc, lead, and copper will codeposit with gold from the alkaline cyanide and neutral baths; lead, zinc, and silver from acid baths.

The presence of these impurities (as distinguished from deliberately added alloying metals) in the deposit cannot be tolerated for most uses. There is no good way of removing them either chemically or electrolytically (by ''dummy-ing''); it is necessary to reclaim the gold from the bath, eliminate the source of contamination, and start afresh.

Organic contamination has several possible sources. It may arise from drag-in from prior steps, dissolution of rack coatings and stopoffs or plastic substrates, and by electrolytic decomposition of cyanides or organic additives present in the bath.

Use of proper rinsing procedures can reduce or eliminate drag-in, and proper selection of rack coatings, stopoffs, and plastic substrates can minimize contamination from these sources. Electrolytic decomposition will always occur and cannot be entirely prevented. Carbon treatment removes most, but not all, organics. The carbon will absorb some gold (10 to 15 percent of its weight) and therefore should be preserved and refined for its gold content.

For control purposes, Hull cell tests are very useful if the gold content of the bath is high enough; for baths very low in gold, the test itself changes the composition of the bath so much that the test is not helpful. In such cases the bent-cathode test will tell the experienced operator much about the condition of the bath.

Decorative Applications

Color gold "flash" coatings are defined as coatings less than 0.2 μm thick, and are usually used as a finish over base metals or gold-filled items to impart a specific color to them. Almost all decorative flash coating is carried out in cyanide baths with very low metal content and various combinations of metallic additives, depending on the color desired. Operating characteristics are similar: temperature about 60°C, current density about 300 A/m^2, 7.5 to 15 g/L free cyanide, and stainless steel anodes. Metallic additives yield various colors; for example, a 24-k color is obtained from a pure gold cyanide bath of conventional composition; if a pale-yellow gold (so-called Hamilton) is wanted, nickel and copper cyanides are added; silver gives a green gold; copper and nickel, in different proportions, yield a "green" gold, etc. Cadmium may be used to increase the green shade. Antique-gold effects are attained by adding lead, cadmium, or both to the bath. Other alloying elements sometimes used include antimony and indium.

Decorative gold plating is also performed to greater thicknesses, up to 40 μm, for such items as watch cases that are subject to considerable abrasive wear. The deposit should not only possess the desired appearance, but also be ductile, hard enough to resist scratching, and pure enough to resist tarnishing. Deposit purity is usually in the range of 16 to 18 k. Processes for this type of plating are almost entirely proprietary.

Most decorative gold plating remains in the category of art rather than science, and depends to a large extent on the experience of the operator.

Thicknesses

For many applications in industry, physical properties such as hardness, ductility, and solderability are important. These properties will be specified by the purchaser, and vendors offer solutions emphasizing one or another. In addition, various thicknesses of gold are considered appropriate for varying uses. They may be summarized as follows:

For flash decorative coatings: 0.02 to 0.2 μm

On contacts and connectors for solderability: 0.2 to 0.75 μm

On contacts and connectors for solderability and weldability: 0.75 to 1.25 μm

Contacts and printed circuits for etch resistance and solderability: 1.25 to 2.5 μm

Standard coating for engineering and decorative use where increased abrasion resistance is required: 2.5 to 5 μm

Exceptional resistance to corrosion and wear by handling: 5 to 7.5 μm

Electron emission characteristics: 12.5 to 38 μm

Electroforming: 38 μm and up

"Spot" plating

Because of the high price of gold, the industry is finding ways of "selective plating" or "spot plating," i.e., putting the gold only where actually needed instead of over the whole part. For example, instead of plating a complete contact pin, only the end where the gold is functional is plated. These efforts constitute a continuing research program with large consumers.

SPECIFICATIONS

There are many military and other governmental specifications covering gold deposits, as well as intracompany documents. ASTM designation B 488, covering gold coatings for engineering uses, may be summarized as representative.

Gold coatings covered contain at least 95.0 percent gold; they are usually used for their corrosion resistance, nontarnishing characteristics, low and stable contact resistance, solderability, or thermal reflectivity. Coating characteristics considered are (1) purity, (2) hardness, and (3) thickness. Depending on projected use, these three variables are important for proper performance of the coating.

Hardness and purity are interrelated. The hardness of gold coatings used for engineering purposes falls into four ranges, which for convenience are identified by letters, as follows:

Letter code	*Knoop hardness range*
A	90 or less
B	91–129
C	130–200
D	over 200

The relationship between purity and hardness is shown as follows:

Purity (% gold)	*Hardness code*
99.9 min.	A only
99.5–99.8	A, B, or C
95.0–99.4	B, C, or D

Suggested purity and hardness for specific applications are as follows:

Semiconductor components, nuclear engineering: coatings of 99.9 percent purity, code A hardness; or 99.8 percent purity, code A, B, or C hardness, to any thickness.

Wire-wrap connections: 99.5 percent purity or higher, code A hardness, any thickness.

Solderability: coatings of any listed purity, although higher-purity deposits are more desirable since oxidation of some codeposited metals could interfere with solderability. Thickness should generally not exceed 1.3 μm, since there is some evidence that thicker coatings may lead to formation of brittle joints with tin-lead solders. (Gold is readily soluble in molten tin, and some of the alloys are brittle.)

Printed-circuit boards: Boards that are to be sheared or cropped should not be plated with heavy thicknesses of hard gold because of the possibility of delamination. Coatings with purities of 99.0 to 99.7 percent are used at hardnesses B or C, and at thicknesses usually not more than 2.5 μm.

Static separable connectors: Any combination of purity and hardness is used, depending on specific requirements. Thicknesses up to about 5 μm are general.

For these applications, which are usually over copper or copper alloys, the usual recommended procedures for preparation of the substrate should be applied; some copper alloys, such as those containing more than 10 percent zinc, 1 percent beryllium, 2 percent tellurium, or 1 percent lead present special problems, and undercoatings such as nickel may be used to prevent diffusion into the final gold coating. Silver is not recommended as an undercoating; copper may be used in some cases. In any case, a gold strike, as already described, should be used preceding the final gold coating.

THE PLATINUM METALS

The six platinum metals constitute the second and third triads of Group VIII of the periodic system; the first triad is the iron-cobalt-nickel group.

The platinum metals, their symbols, atomic numbers, and atomic weights are shown in Fig. 13-1; included for orientation are the metals of the iron triad and the neighboring elements in Groups VIIB and IB.

The six platinum metals resemble each other in many ways; separations within the group are tedious and complex, and historically there was considerable confusion in identifying them as separate entities. But the resemblance to the Fe-Co-Ni triad is purely formal, limited to some slight similarities between nickel and palladium, their unusual catalytic activity, and their tendency to form many and stable complexes. The chemistry of the platinum complexes was one of the cornerstones of the Werner coordination theory, which forms the basis of much of modern inorganic chemistry.

Vertical similarities within the group are somewhat more notable: the pairs ruthenium-osmium, rhodium-iridium, and palladium-platinum show many similarities.

All six are rare. There is no exact agreement among geochemists as to their abundance in the lithosphere; ranges quoted are, in ppm: ruthenium, 0.001 to 0.005; rhodium, 0.001; palladium, 0.005 to 0.01; osmium, 0.001 to 0.005; iridium, 0.001; and platinum, 0.005 to 0.2. Prices are accordingly high and tend to fluctuate widely. As a very rough guide, price per unit weight, taking that of platinum as 1.0, tends to be: Os 2; Rh, 2; Ir, 1.1; Pt, 1.0; Ru 0.45; and Pd, 0.28. A typical price for platinum is about $5/g. As with the other precious metals, prices are normally quoted in troy ounces; 1 troy oz = 31.1 g.

All the metals are "noble," i.e., have positive electrode potentials and are relatively inert. Accordingly they are found free in nature. Native platinum was known to the pre-Columbian Indians of South America, but its identity as an

	Group VII B	Group VIII			Group IB	Period
Element	Manganese	Iron	Cobalt	Nickel	Copper	4
Symbol	Mn	Fe	Co	Ni	Cu	
At. No.	25	26	27	28	29	
At. Wt.	54.9380	55.847	58.9332	58.71	63.546	
Element	Technetium	Ruthenium	Rhodium	Palladium	Silver	5
Symbol	Tc	Ru	Rh	Pd	Ag	
At. No.	43	44	45	46	47	
At. Wt.	98.9062	101.07	102.9055	106.4	107.868	
Element	Rhenium	Osmium	Iridium	Platinum	Gold	6
Symbol	Re	Os	Ir	Pt	Au	
At. No.	75	76	77	78	79	
At. Wt.	186.2	190.2	192.22	195.09	196.9665	

Fig. 13-1 The platinum metals in the periodic system.

element was not established until the early nineteenth century. It then turned out that native platinum was a mixture or alloy; palladium was isolated from platinum in 1803 by Wollaston, who also isolated rhodium. A year later, in studying the insoluble residue remaining after treating crude platinum with aqua regia, Tennant obtained two new metals, iridium and osmium. And finally in 1844, Claus reported the discovery of ruthenium.*

There are few minerals of the platinum metals; the more important are: sperrylite, $PtAs_2$; cooperite, PtS; braggite, (Pt, Pd, Ni) S; and michinerite or froodite, $PdBi_2$. Most sources are various alloys of the metals themselves, which occur as either primary or placer deposits.

Native platinum may consist of 70 to 90 percent Pt, containing smaller amounts of the other platinum metals and iron. Osmiridium or iridosmine—the name depends on which is the major constituent—is an important source of these metals.

Major changes have occurred over the years in the commercial sources of the platinum metals. The first source, Colombia, is still important; but at present the major supplier is the Soviet Union, which probably produces over half the world's supply. The only domestic source is in Alaska, which produces a crude platinum and iridosmine.

Platinum arsenide, sperrylite, was discovered in the copper-nickel ores of the Sudbury district of Ontario by Sperry in 1875. The platinum metal content of these ores is minute—hardly more than the radium content of pitchblende. But the large quantities of ore processed for copper and nickel recovery, coupled with the tremendous concentration factor inherent in the electrolytic refining process, have made this a major source of supply, third in importance after the Soviet Union and the Union of South Africa. The noble metals, including silver and gold as well as the platinum group, end up in the anode slimes of the refinery, whence they are recovered. These ores could not be profitably worked for their platinum metal content; it is a welcome by-product of nickel and copper recovery.

The most recent entry in platinum metal production is the Republic of South Africa, which now ranks second to the Soviet Union. Unlike the Canadian deposits, these ores are rich enough to be worked for the sake of their platinum content.

*Derivation of the names of the platinum group:

Ruthenium: Latin *ruthenia,* Russia; first discovered in an ore from the Ural mountains

Rhodium: Greek *rhodon,* rose; from the color of its compounds

Palladium: the asteroid Pallas, discovered about the same time

Osmium: Greek *osmē,* stink; from the odor of the tetroxide

Iridium: Latin *iris,* rainbow; from the variety of colors of its compounds in solution

Platinum: Spanish *platina,* "little silver"

Recovery of secondary platinum metals is a very important sector of the total market. In many of their applications these metals are not "used up" to any major extent, but are merely damaged mechanically: spinnerets, catalysts, anodes, laboratory ware, and the like are scrapped because of mechanical damage rather than chemical attack, and can easily be re-refined so as to be fully equivalent to "virgin" material. Secondary platinum accounts for 25 to 30 percent of the total United States market.

Total world production of the platinum metal group amounts to about 110,000 t. That consumed by industry in the United States totals about 45,000 t. The relative importance of the six metals may be estimated by the following rough figures: of the total, ruthenium accounts for 1.3 percent; rhodium, 3.7 percent; palladium, 56 percent; osmium, 0.1 percent; iridium, 1 percent; and platinum, 38 percent.

Production

The platinum metals occur in many forms; hence the methods of production and refining differ considerably. Native grains may be panned by hand, in a manner somewhat similar to panning for gold; this method is still used by the natives of South America. Rich ores, such as occur in South Africa, require only flotation and gravity separation before being sent to the refinery. In the copper-nickel ores of Canada, the copper and nickel refining process concentrates the extremely small percentages in the ores many-fold in the anode slimes, or the residues from the carbonyl purification process. These residues become the starting point for the refining and separation of the precious metals.

The refinery process itself is a long, multistep operation, necessitated by the chemical similarity of the six platinum metals. Flow sheets are very complicated and differ according to the starting material.

Uses

Almost all the applications of the platinum metals depend on their nobility—i.e., corrosion resistance—and their unusual catalytic activity. Physical properties such as hardness and wear resistance account for some applications.

Platinum and palladium are the most abundant; together they account for about 95 percent of the total United States consumption. The other four have more limited and specialized applications. Intragroup alloys also have many uses.

Platinum is used for its corrosion resistance in applications where no other metal is suitable. The ability to clad cheaper metals with a coating of platinum renders it economically feasible to use in cases where the pure metal would be prohibitively expensive—in chemical reactors, for example. Its use in the laboratory for crucibles, dishes, electrolytic anodes, and the like is familiar to all chemists.

The catalytic activity of the metals was recognized early. Platinum and palladium are the most often used; but others have special applications, and often an alloy of two or more is better than a single one. A familiar example is the platinum-rhodium alloy gauze used in the oxidation of ammonia to produce nitric acid. Platinum catalysts are used in many other chemical industries, including petroleum refining. Palladium is preferred for the hydrogenation and dehydrogenation of many natural products. The platinum group metals are prominent in the development of automotive exhaust catalysts for pollution abatement.

Platinum and palladium are used in jewelry, both for their intrinsic noncorroding properties and because their rarity and high price render them valuable.

The platinum metals have many applications in the field of high-temperature technology, both singly and as alloys. Thermocouples, such as the platinum–platinum-rhodium, and the platinum resistance thermometer are only two of many similar examples. Platinum is used for lining furnaces for high-quality glass and for crucibles for growing crystals for lasers and other optical devices. Various platinum metal alloys are used as spinnerets in the production of rayon by the viscose process. Almost all the metal used for this purpose can be reclaimed.

Ruthenium catalysts have been used in organic synthesis; electrodeposits of ruthenium are very hard and are useful for sliding contacts. However, its main use is as an alloying constituent to impart hardness to other platinum metal alloys.

Rhodium is used principally as an alloy with platinum, in the 10 percent Rh alloys already mentioned. Of the six, it has had the most application in metal finishing: as a noncorrosive decorative finish and for its reflectance in mirrors and searchlights.

Osmium has some catalytic chemical uses; it is also used in the hard alloys used for fountain-pen nibs and phonograph styli.

Iridium is used principally as a hardening addition to other platinum metals. Osmiridium (30 percent Os, balance Ir) has been used for fountain-pen nibs and phonograph styli and pivots.

Although all the metals can be electrodeposited, the only ones having had appreciable use in metal finishing are platinum, palladium and rhodium.

Properties

Some properties of the platinum metals are listed in Table 13-10. Note the high densities of the members of the second triad, osmium, iridium, and platinum. Iridium is the heaviest known metal, and platinum and osmium are only a little lighter.

All six are transition metals and exhibit their typical properties: variable valence, colored ions, and great tendency to complex formation; in fact, simple aquo cations of the metals are very rare.

The electronic configurations of the metals are shown separately in Table 13-11. Energy differences among these outer levels are small, leading to the mentioned multiple valences. The most common oxidation numbers of the metals are

Table 13-10 Properties of the Platinum Metals

	Ru	*Rh*	*Pd*	*Os*	*Ir*	*Pt*
Atomic number	44	45	46	76	77	78
Atomic weight	101.07	102.9055	106.4	190.2	192.22	195.09
Melting point, °C	2310	1960	1552	3045	2443	1769
Boiling point, °C	4100	3700	2900	5000	4500	3800
Density, g/cm³, 20°C	12.45	12.41	12.02	22.5	22.65	21.45
Electrical resistivity, μΩ-cm, 0°C	6.6	4.33	9.93	8.8	4.71	9.85
20°C	7.2	4.5	≈10.4	9.5	5.11	10.6
Tensile strength, annealed, MPa	540	755	165	—	1100	137
Elongation, %	—	30	24–30	—	—	25–40
Vickers hardness, annealed	220	120	38	400	500	39
Crystal structure	hcp	fcc	fcc	hcp	fcc	fcc

included in the table; most blank spaces could be filled in with a few compounds, but the table includes those most often encountered.

Some standard electrode potentials are listed in Table 13-12.

The chemical properties of the metals depend largely on their state of subdivision. In compact form they are extremely inert and react with few common reagents. They are also known in the form of sponges or "blacks"; in this form they may be highly reactive. Table 13-13 summarizes the reactions of the compact metals with many common acids; note the outstanding resistance of iridium and ruthenium to all those listed. Of the six, palladium is the least resistant to attack by acids.

Many molten oxides, hydroxides, and salts and salt mixtures react with the platinum group. Fused peroxides, hydroxides, and nitrates of Group IA (the alkali metals), oxides of Group IIA (the alkaline earth metals), mixtures of these, and fused alkali cyanides attack the metals.

Phosphorus, arsenic, silicon, sulfur, selenium, tellurium, and carbon attack the metals at red heat. Fluorine and chlorine react at high temperatures; moist chlorine and bromine attack palladium even at room temperature. Hot gases containing sulfides tarnish platinum and palladium.

The reactions taking place when the metals are heated in air or oxygen are complicated and not entirely understood. Platinum loses weight in air or oxygen at 1000°C, but not in nitrogen or argon. The obvious explanation—the formation of a volatile oxide—is probably not correct because no such oxide has been

Table 13-11 Electronic Configuration and Oxidation Numbers of the Platinum Metals

Element	*Electronic configuration*	*Most common oxidation numbers*							
		+1	*+2*	*+3*	*+4*	*+5*	*+6*	*+7*	*+8*
	$1s^2 2s^2 2p^6 3s^2 3p^6 3d^{10} 4s^2 4p^6$-								
Ru	$-4d^7 5s^1$		X	X	X		X		X
Rh	$-4d^8 5s^1$	X		X					
Pd	$-4d^{10}$		X		X				
	$1s^2 2s^2 2p^6 3s^2 3p^6 3d^{10} 4s^2 4p^6 4d^{10} 4f^{14} 5s^2 5p^6$-								
Os	$-5d^6 6s^2$			X	X		X		X
Ir	$-5d^7 6s^2$	X		X	X				
Pt	$-5d^9 6s^1$		X		X				

Table 13-12 Some Electrode Potentials of the Platinum Metals

Half-reaction	*Potential, V*	*Half-reaction*	*Potential, V*
$RuCl_3 + 3e^- \rightarrow Ru + 3Cl^-$	+0.68	$OsCl_6^{3-} + 3e^- \rightarrow Os + 6\ Cl^-$	+0.71
$RuCl_5^{--} + 3e^- \rightarrow Ru + 5Cl^-$	+0.4	$HOsO_5^- + 4H_2O + 8e^- \rightarrow Os + 90H^-$	+0.02
$RuO_4^{--} + 8H^+ + 6e^- \rightarrow Ru + 4H_2O$	+1.93	$OsO_4^{--} + 8H^+ + 6e^- \rightarrow Os + 4H_2O$	+0.994
$RhCl_6^{3-} + 3e^- \rightarrow Rh + 6Cl^-$	+0.431	$IrCl_6^{3-} + 3e^- \rightarrow Ir + 6Cl^-$	+0.77
$Rh^{3+} + 3e^- \rightarrow Rh$	+2.261	$Ir^{3+} + 3e^- \rightarrow Ir$	+1.448
$Pd^{++} + 2e^- \rightarrow Pd$	+0.987	$Pt^{++} + 2e^- \rightarrow Pt$	+1.2
$PdCl_4^{--} + 2e^- \rightarrow Pd + 4Cl^-$	+0.621	$PtCl_4^{--} + 2e^- \rightarrow Pt + 4Cl^-$	+0.73
$Pd(OH)_2 + 2e^- \rightarrow Pd + 2OH^-$	+0.07	$Pt(OH)_2 + 2e^- \rightarrow Pt + 20H^-$	+0.15

Table 13-13 Resistance of Platinum Metals to Some Common Acids

Acid	*Temperature, °C*	*Ru*	*Rh*	*Pd*	*Os*	*Ir*	*Pt*
Aqua regia	20	A	A	D	D	A	D
	100	A	A	D	D	A	D
Hydriodic, 60%	20	A	A	D	B	A	A
	100	A	A	——	C	A	D
Hydrobromic, 60%	20	A	B	D	A	A	B
	100	A	C	D	C	A	D
Hydrochloric, 36%	20	A	A	A	A	A	A
	100	A	A	B	C	A	B
Hydrofluoric, 40%	20	A	A	A	A	A	A
Nitric, 70%	20	A	A	D	C	A	A
Nitric, 95%	20	A	A	D	D	A	A
	100	A	A	D	D	A	A
Perchloric, 65%	20	——	——	A	——	——	A
	100	A	A	C	——	——	A
Phosphoric, 10%	20	A	A	A	——	A	A
	100	A	A	B	D	A	A
Sulfuric, 98%	20	A	A	A	A	A	A
	100	A	B	C	A	A	A

A = no attack; B = slight attack; C = significant attack; D = rapid attack; —— means no data.

identified. Palladium is superficially oxidized at 700°C. Rhodium is superficially oxidized at red heat in air, as is iridium. Ruthenium forms a stable oxide, RuO_2; and osmium loses weight rapidly owing to the formation of the volatile OsO_4.

Any complete description of the extremely complex chemistry of the platinum metals would in itself require a textbook. With their multiple valences, their ready formation of complexes with ligands of all types, their overall similarity and minor differences, which render separations long and tedious, this group probably represents the most complicated in all inorganic chemistry with the possible exception of the lanthanides (rare earths). Since the platinum metals are relatively unimportant to the metal finisher, such a discussion is not justified here; a few of the more common characteristics of the metal compounds are outlined below.

Ruthenium

Ruthenium exists in at least eight oxidation states. It is not significantly soluble in any single acid, and even aqua regia has little effect. The only octavalent compound is RuO_4, which is volatile and poisonous. It can be formed by distillation from either alkaline or acid solution under strongly oxidizing conditions. Dissolved in alkali, it forms a green solution of a perruthenate(VII), $MRuO_4$, which is easily reduced to the orange ruthenate(VI), M_2RuO_4. Direct chlorination of the metal yields $RuCl_3$. Ruthenium forms many complex ions; its affinity for the nitrosyl group NO is especially notable. Alkali chlorides form complex salts of the type $M_2Ru(NO)Cl_5$. Ruthenium also forms carbonyls and a large number of amine complexes. Simple cations are rare or nonexistent.

Rhodium

Compact rhodium metal is practically insoluble in all acids at 100°C, including aqua regia. Its most stable valence is 3+. Heated in air, it forms Rh_2O_3. When freshly precipitated $Rh(OH)_3$ is dissolved in hydrochloric acid at controlled pH the solution first formed contains a cationic aquo chloro–complex. When this solution is boiled with excess hydrochloric acid, the hexachlororhodate(III) anion is formed.

When a solution of $RhCl_3$ is treated with sodium nitrite, the very soluble sodium hexanitritorhodate(III), $Na_3Rh(NO_2)_6$, is formed. The salt is soluble in alkaline solution; the corresponding potassium and ammonium salts are relatively insoluble.

Rhodium forms many complexes with ammonia, amines, cyanide, chloride, bromide, and polynitrogen and polyoxygen chelating agents.

Palladium

In some ways palladium is similar to nickel and silver, and more so to platinum. It is the least refractory of the group; it dissolves quickly in aqua regia, and slowly even in hydrochloric acid. Finely divided, it is soluble in all strong acids. In air at red heat the monoxide PdO is formed; chlorine and fluorine form the dihalides.

Palladium(II) chloride is formed when a solution of the metal in hydrochloric acid is evaporated. Divalent palladium forms many complexes in which it has a coordination number of 4; it can be oxidized to Pd(IV) with chlorite or chlorate. The addition of ammonium chloride to such a complex precipitates ammonium hexachloropalladate(IV).

Many palladium complexes are important in metal finishing. Pd(II) has a great affinity for nitrogen-containing ligands, and palladium plating makes use of several complex ions. The use of palladium(II) chloride solutions (more probably an anionic chloro-complex) is important in plating on plastics (Chap. 19).

Osmium

The most important osmium compound is the tetroxide OsO_4, which is volatile and very poisonous. Like all the platinum metals, it forms many complexes, notably with nitrite, oxalate, carbon monoxide, and thiourea.

Iridium

The most common valences are 3+ and 4+. Trivalent iridium forms many cationic and anionic complexes; the ammines are extremely stable.

Platinum

Platinum is the most thoroughly studied of the group. Its principal valences are 2+ and 4+, the former being somewhat more stable. In compact form platinum is inert to all mineral acids except aqua regia; under oxidizing conditions fused alkalis attack it slightly. In hot aqua regia or hydrochloric acid containing chlorates or hydrogen peroxide, platinum dissolves to yield a solution of hexachloroplatinic(IV) acid H_2PtCl_6, which is often called platinic chloride.

When a solution of this acid is boiled for a long time with sodium hydroxide, the chloride ions are replaced by hydroxyl ions, yielding sodium hexahydroxyplatinate(IV), $Na_2Pt(OH)_6$.*

The above-mentioned hexachloroplatinic(IV) acid is the most common platinum compound; its sodium salt is soluble but the ammonium and potassium salts are insoluble, and this property is used analytically for the determination of potassium, as well as of platinum.

Divalent and tetravalent platinum form as many complexes as any metal. The tetranitritoplatinum(II) complexes are soluble in alkaline solution; tetranitritoplatinum(II) ion is formed when a solution of platinum(II) chloride is boiled at

*Interestingly, this compound is important in tin chemistry. It is isomorphous with sodium stannate, and this is one piece of evidence that the formula of the latter is $Na_2Sn(OH)_6$ rather than Na_2SnO_3.

neutral pH with an excess of sodium nitrite. The ammonium salt may be explosive.

Most platinum(II) complexes have a coordination number of 4. Complexes with olefins, cyanides, nitriles, halides, isonitriles, amines, phosphines, arsines, and nitro compounds are among the many which have been prepared.

Platinum(IV) has a coordination number of 6; it also forms complexes, with halides, nitrogen and sulfur compounds, and many other donors, but to a somewhat less extent than platinum(II).

Hazards

The metals in their compact form are nontoxic. In the finely divided state (sponges, powders, blacks) their great catalytic activity suggests caution in handling: explosive mixtures of gases could be ignited by contact with them. Osmium tetroxide is both volatile and highly toxic; ruthenium tetroxide is also poisonous, although not so volatile.

The other compounds of the platinum metals, so far as they have been investigated, do not appear to pose any hazards. Some persons appear to be allergically sensitive to them; the only cure for this is to remove such persons from sources of contact.

Plating The Platinum Metals

The baths for plating the platinum metals differ in several ways from usual plating baths. Some of the factors responsible for these differences are the extremely noble potentials of the metals, their marked tendency to form complex ions, and their high costs.

The high cost of the platinum metals limits the metal concentration of the plating bath in order to decrease the capital investment required. For instance, the typical rhodium-plating bath contains only about 2 g/L of metal. Higher metal content permits faster plating, and may be used where the purpose is technical rather than decorative.

Since the platinum metals readily form very stable complexes, there is little tendency for them to form immersion deposits even on base metals (with some exceptions). Baths containing these complexes can frequently be used for plating directly on such metals as copper and nickel. However, often a strike from a noncorrosive bath such as gold cyanide is desirable or even essential. The most reactive base metals will nearly always require a preliminary strike of nickel in addition to a gold strike.

In most platinum metal baths, the metal does not dissolve anodically and insoluble anodes are used. The metal plated out must be replaced chemically. Palladium and platinum anodes do dissolve in some acid chloride baths, and there soluble anodes are preferred.

Cost is also a factor in limiting the thickness of deposits to that absolutely necessary for the purpose. The range of thicknesses usually encountered is from about 0.1 to no more than 5 μm. In spite of these low thicknesses, the platinum metals can be very effective: as little as 0.1 to 0.2 μm of rhodium can prevent sticking of silver-plated seals in Apollo rocket engines or prevent tarnishing of costume jewelry, and 1.2 μm of platinum plated on titanium is sufficient to yield an effective insoluble anode in gold plating (see the section Gold Plating).

In at least one case, a platinum metal offers economy in spite of its high price: palladium has been suggested, and used, as a substitute for gold, being not only cheaper than gold on a weight basis, but also only about half as dense, so that a given weight goes about twice as far.

Deposits from molten cyanide electrolytes of the platinum metals are superior to those from aqueous baths with respect to soundness, thickness plateable without cracking, and freedom from stress. Bath temperatures of 600°C, the need for an argon atmosphere, and the precautions required have severely limited the interest in and use of such baths, and they will not be considered here. Their possibilities should, however, be kept in mind in special circumstances which might make their use justifiable.

Electroless plating baths have been developed for several of the platinum metals.

Plating of the platinum metals is not widely practiced commercially, except for the decorative applications of rhodium and its use as a tarnish-preventive coating on mirrors and searchlights. Palladium plating has been of interest as a partial substitute for gold in some electronic applications, where it is about equally satisfactory and considerably less expensive on both a volume and a weight basis. On the other hand, there is much less background of experience in platinum-metal plating than in gold plating, and hence some reluctance to embark on the necessary research and development programs required to render the processes as inherently trouble-free as the more conventional gold-plating processes.

One large computer manufacturer has reported the substitution of palladium for gold in some components, but the process used was not disclosed.

To the extent that it is used at all, platinum-metal plating is best carried out from proprietary solutions, the nature of which is usually not disclosed in any detail. Supplies generally are purchased in the form of concentrates, containing a guaranteed percentage of precious metal; the plater dilutes these concentrates as prescribed by the vendor and follows the directions.

As has been mentioned, all the platinum metals form very stable complexes with a large number of ligands; thus the free platinum metal-ion concentration in most solutions is extremely small. This tends to overcome the nobility of these metals and their tendency to deposit by immersion on most substrates. Some of the complexes are so stable that it is difficult to plate metal at all from their solutions, and cathode efficiencies may tend to be low. From many baths, only thin deposits can be obtained before cracking occurs, and deposits are often highly stressed.

One of the factors involved in choosing a precious metal for an electroplating application is price vs. density. The densities of the precious metals fall into two groups, centering around 12 g/cm^3 for the light group and 21 g/cm^3 for the heavy group. Density is important because metals are purchased on a weight basis and used on a volume basis. Thus, other things being equal, the metals of the light group—silver, ruthenium, rhodium, and palladium—are more economical than those of the heavy—gold, osmium, iridium, and platinum.

Melting point is a factor in some applications; high melting point is favorable in some electric contacts, preventing sticking of contacts by cold welding. A high-melting-point metal is often superior as a barrier deposit to prevent diffusion of the base metal through the precious-metal deposit to the surface.

Although gold and silver have much lower electrical resistivities than the platinum metals, the latter are nevertheless usable in some applications where their other superior properties overcome this disadvantage.

In some applications—mirrors, searchlights—reflectivity is important. Silver is outstanding in this respect, until it tarnishes; rhodium is also quite white and much more stable in most atmospheres.

Hardness is sometimes a factor; here ruthenium, osmium, and iridium excel. In general, strength and wear resistance parallel hardness, while ductility varies inversely. For example, the harder metals, osmium and ruthenium, are relatively brittle. The intermediate metals, rhodium and iridium, are strong but sufficiently ductile to permit the production of thin strip and small-diameter wire. The softer metals, palladium and platinum, are not particularly strong but are quite ductile.

Because of their price and scarcity, the platinum metals are used only when no other metal will do, or as a substitute for the even more expensive gold. Some typical applications may be cited.

For electric contact purposes, the deposit should be highly resistant to corrosion and wear, have good electrical conductivity and high melting point. Gold comes nearest to meeting these requirements, but in specific cases palladium provides better hardness and wear resistance, and is cheaper than gold on both a volume and a weight basis.

Insoluble anodes provide another use for the platinum metals. Every chemist is familiar with the use of platinum anodes in the laboratory. On the larger scale of the shop, such anodes would be extremely expensive, but platinum can be plated on titanium, and platinized titanium anodes are useful both for gold-plating baths (see Gold) and for anodes in cathodic protection systems. Even through any slight pores in the platinum deposit, the titanium itself is anodized, becoming passive and not passing current; the combination is almost as serviceable as pure platinum, with a few exceptions in solutions that dissolve the anodic film on the titanium.

High-melting-point platinum metals can serve as a barrier to diffusion when gold, for example, is plated over copper. Palladium is most often used in such applications.

Wear resistance of switches can be enhanced by palladium plating, and even more so by rhodium and ruthenium.

Rhodium has been used as a reflecting surface on mirrors of electroformed nickel.

For jewelry, where color and appearance as well as good corrosion resistance are desirable, silver and gold are of course the standard deposits; but rhodium plating has the advantage over silver of being nontarnishing, and it is still pleasing in appearance.

Palladium and platinum are seldom chosen for decorative applications; the deposits tend to be rather dull and uninspiring.

Patented or proprietary plating processes are available for almost all the platinum metals, and some are available in several different formulations to tailor the type of deposit—hardness, stress, ductility—to the desired result. This is especially true of palladium, platinum, and rhodium; osmium and iridium are seldom used as electrodeposits, and ruthenium, though available, remains a seldom used specialty.

Because most formulations are proprietary, and because relatively little research has been reported concerning the deposition of the platinum metals, it is difficult to discuss fundamentals. Hence this discussion is confined to mentioning a few of the uses of the deposits of these metals, and offering some formulations that have been reported. In many cases, these formulations are based on "beaker plating" and have not been put to the test of large-scale commercial application.

Most of the progress in recent years in plating the platinum metals has been the result of work by vendors of plating supplies, and their results have appeared in the form of proprietary formulations, usually patented but seldom discussed in the technical literature. Although formulas are published, and a few are outlined in what follows, the finisher wishing to make use of a platinum metal deposit is probably best advised to consider such a proprietary process, from one of the several suppliers offering them.

INDIVIDUAL PLATINUM METALS

Ruthenium

The main reason for interest in ruthenium plating is that it is substantially cheaper than rhodium, which it resembles in many properties. It is not widely used, however, and though several baths have been published (see Table 13-14), the extent of commercial experience and background with these baths is limited.

Rhodium

Rhodium plating was introduced in the 1930s, when a bath capable of depositing brilliant white, tarnish-resistant, wear-resistant, and hard coatings was developed. The bath was used to produce thin (0.1 to 0.2 μm) finish coatings on such items as white gold, platinum, silver, or nickel-plated costume jewelry and similar decorative pieces. There followed engineering applications that made use of the good contact characteristics and high and stable reflectivity of rhodium.

Table 13-14 Baths for Ruthenium Plating

	Concentrations, g/L				
Bath:	*A*	*B*	*C*	*D*	*E*
Source of Ru metal	$(NH_4)_3[Ru_2NCl_8(H_2O)_2]$	$K_3[Ru_2NCl_8(H_2O)_2]$	Ru nitrosyl sulfamate	Ru nitrosyl chloride	Ru sulfamate
Ru, as metal	12	10	5	2	5
Hydrochloric acid, HCl	to pH 1	—	—	—	—
Ammonium sulfamate, $NH_4NH_2SO_3$	10	—	—	—	—
Ammonium formate, NH_4CO_2H	—	10	—	—	—
Sulfamic acid, NH_2SO_3H	—	—	40	—	5.3
Sulfuric acid, H_2SO_4	—	—	—	21	—
Temperature, °C	70	70	70	55–75	26–50
Current density, A/m^2	100	80	400	200–300	100–250

Baths were modified to permit the application of thicker coatings. Heavy deposits for high-temperature protection of molybdenum and tungsten have been produced from molten cyanide electrolytes.

Most rhodium baths are proprietary; the finisher purchases a concentrate, with a guaranteed metal content, for preparing the bath. The essential ingredient is usually a rhodium sulfate or rhodium phosphate compound. The concentrate is added to the corresponding acid solution to make up the bath. The nature of the rhodium complex in these solutions is not known, and with most base metals there is the danger of an immersion deposit. The danger is less for nickel than for other substrates; furthermore, nickel so dissolved in the bath is not particularly deleterious to the rhodium deposit. Consequently, nickel, silver, or gold plate is generally used as an undercoat for rhodium plating. Present practice tends to favor the phosphate over the sulfate baths; both types are listed in Table 13-15. Sulfate baths are preferred for barrel plating and for thicker deposits, above 2.5 μm.

Proprietary solutions are available for the deposition of low-stress deposits. In general they are similar to the sulfate bath but contain stress reducers such as selenic acid or magnesium sulfate; the mechanism of these additives is not known.

One vendor offers the following guide to the requisite thickness of rhodium deposits in its various applications.

Application	*Thickness, μm*
Decorative	0.025–0.125
Reflectors	0.025–0.375
Tarnish prevention	0.125–0.250
Light-duty electrical contacts	0.125–0.5
Corrosion prevention	0.25 min. (not a really practical use)
Severe corrosion conditions	2.5 or more
Heavy-duty electrical contacts	2.5 and up
Severe-wear electrical contacts	5 and up
Infrequently used contacts	5 and up

Palladium

Palladium is a good contact material, and is cheaper on both a weight and volume basis than gold; there is considerably more practical background in palladium and rhodium plating than for the other platinum metals. A large number of solutions are available for plating palladium.

Many of the electrolytes are based on palladium "P" salt, which is diammine-palladium(II) dinitrite $Pd(NH_3)_2(NO_2)_2$, in an ammoniacal medium and with various additions: ammonium phosphate, ammonium sulfamate, ammonium formate, and others, as conducting salts.

Table 13-15 Baths for Rhodium Plating

	Concentrations, g/L					
Bath:	*F**	*G**	*H**	*I*†	*J*‡	*K*§
Source of Rh metal	PO_4^{3-} concentrate	SO_4^{--} concentrate	PO_4^{3-} concentrate	SO_4^{--} concentrate	SO_4^{--} concentrate	SO_4^{--} concentrate
Rh, as metal	2	1.3–2.1	2	5.25	1	2.5–5
Phosphoric acid, H_3PO_4, 85%	55–115	—	—	—	—	—
Sulfuric acid, H_2SO_4	—	50–160	50–160	50–100	160	160
Voltage: V	4–8	3–6	3–6			
Temperature, °C	38–50	38–50	38–50	44–50	44–50	44–50
Current density, A/m²	200–1500	200–1000	200–1000	1000–3000	50–200	50–200
Anodes	Pt	Pt	Pt	Pt	Pt	Pt

*Thin decorative.

†Heavy industrial.

‡Barrel, decorative.

§Barrel, industrial.

Coatings from these baths, shown in Table 13-16, are usually bright or semi-bright up to about 5 μm, with hardness about 200 to 300 DPH,* and are suitable as electric contacts and diffusion barriers.

Another type of electrolyte is based on palladium chelates with *N-N'*-cycloalkane diaminetetraacetic acid, containing about 5 to 10 g of Pd, and buffered with monopotassium phosphate. This type of solution can, it is claimed, give mirror-bright deposits over a range of conditions and pH from 4 to 12. Acid electrolytes for bright plating are based on dinitrosulfatopalladous(II) acid and its salts.

If coatings of high ductility are desired, electrolytes based on palladium(IV) ammine salts, $Pd(NH_3)_4X$, are preferred. These solutions are relatively high in metal content, and rates of deposition are correspondingly high; internal stress is low, so that this type of bath is suitable for electroforming.

An acid chloride electrolyte has also been used for the production of thick deposits and for electroforming. This solution attacks base metals; hence a barrier such as gold or platinum is required before palladium plating.

Osmium

The high melting point of osmium and its high hardness are reasons for potential interest in plating it for electric contacts. Very little, however, has been published on osmium plating†; an alkaline bath using a complex formed by the reaction between osmium tetroxide and sulfamic acid has been proposed. Commercial use is minimal or lacking.

Iridium

Really sound deposits of iridium have been obtained only from molten salt baths. Aqueous solutions so far reported are capable of only very low cathode efficiencies, and deposits thicker than about 1 μm were cracked. Little or no iridium plating is practiced commercially.

Platinum

Platinum plating is useful mainly for the production of platinized titanium anodes for gold plating and for cathodic protection and other electrochemical processes. The inertness of platinum to most electrolytes, even as anode in chloride solutions, makes it useful for these applications. The utility of platinized titanium anodes depends both on the corrosion resistance of platinum itself and the tendency of metals such as titanium and tantalum on which it is deposited to anodize and thus insulate themselves in any pores or cracks which might be

*See Chap. 23.

†See J. N. Crosby, *Trans. Inst. Metal Finishing,* **54,** 75 (1976).

Table 13-16 Baths for Palladium Plating

	Concentrations, g/L					
Bath:	*L*	*M*	*N*	*O*	*P**	*Q**
Source of Pd	$Pd(NH_3)_2(NO_2)_2$	$Pd(NH_3)_4(NO_2)_2$	"P" salt	?	$Pd(NH_3)_4Br_2$	$PdCl_2$
Pd, as metal	4	10–25	10	2–5	21–32	52
Ammonium nitrate, NH_4NO_3	90					
Sodium nitrite, $NaNO_2$	11 3					
Ammonium hydroxide, NH_4OH	to pH 9		to pH 7.5–8.5		to pH 9–10	
Potassium hydroxide, KOH				75–225		
Ammonium bromide, NH_4Br					32	
Ammonium chloride, NH_4Cl						22–38
Hydrochloric acid, HCl						to pH 0.1–0.5
Temperature, °C	43–55	38–60	26–32	26–50	43–55	38–50
Current density, A/m^2	50–300	50–200	20	50–200	200–400	50–100
Anodes		Pt, Pt/Ti, Pt/Ta		SS	Pt	Pd

*For heavy deposits.

present or develop in the platinum coating. Thick coatings of platinum are interesting for protecting refractory metals from high-temperature oxidation.

The most important platinum-plating baths are based on ammine, nitrito, or hydroxo-complexes, or on the acid chloride. Platinum "P" salt is diamminedinitroplatinum(II). Some typical formulations are shown in Table 13-17.

Electroless platinum-plating baths have been developed.

TIN

Tin (Sn, atomic number 50) falls between germanium and lead in Group IVB of the periodic table. It is not especially rare, constituting about 0.004 percent of the lithosphere, but in commercially workable concentrations it is found in relatively few places.

The origins of the name *tin,* and the Latin *stannum* from which the symbol Sn is derived, are not known. It was probably one of the earliest metals known. Tinning of iron (the predecessor of the modern tinplate) did not appear until the fourteenth century in Bohemia.

The important tin-producing areas, which account for 90 percent of production, are Malaysia, Bolivia, Thailand, Indonesia, Nigeria, Zaire, and China. Smaller amounts are found in Australia, the United Kingdom, Burma, Japan, Canada, Portugal, and Spain. Historically, the mines in Cornwall, England, were important; some of these are being reactivated, but their contribution to the total world supply will remain minor. The United States has no workable deposits of

Table 13-17 Baths for Platinum Plating

	Concentrations, g/L				
Bath:	*R*	*S*	*T*	*U*	*V*
Source of Pt	"P" salt	"P" salt	$H_2Pt(NO_2)_2SO_4$	$K_2Pt(OH)_2$	H_2PtCl_6
Pt, as metal	10	40	5	12	20
Ammonium nitrite, NH_4NO_2	100				
Sodium nitrite, $NaNO_2$	10				
Ammonia, NH_3, 28%	40				
Sulfamic acid, NH_2SO_3H		80			
Sulfuric acid, H_2SO_4			to pH 2		
Potassium hydroxide, KOH				15	
Hydrochloric acid, HCl					300
Temperature, °C	90	75	40	75	65
Current density, A/m^2	40	150	50	75	—

tin, although Alaska reports occasional production. Therefore tin is high on the list of strategic materials.

The principal tin ore is cassiterite, SnO_2. Tin sulfide minerals such as stannite, $Cu_2S \cdot FeS \cdot SnS_2$, and other similarly complex sulfide ores are found in lode deposits in Bolivia and elsewhere, associated with cassiterite and granitic rocks. Most cassiterite, however, is found in low-grade alluvial or placer deposits.

Alluvial deposits are mined by dredging, followed by beneficiation using flotation and other standard methods. The final concentrate, ready for direct smelting, contains 70 to 77 percent tin, i.e., almost pure tin oxide SnO_2 (in theory 78.9 percent).

Underground lode deposits in Bolivia are found at altitudes of 3600 to 4600 m (2.25 to almost 3 mi) above sea level; in Cornwall, ores are 400 m below sea level. The ore is mined by conventional underground mining techniques: blasting, drilling, crushing, and grinding preceding concentration by usual methods.

Some Bolivian ores contain up to 9 percent tungsten. This is a mixed blessing: the tungsten content is valuable, but it greatly increases the complications in smelting and refining the tin.

Total free-world production of tin is about 200,000 t annually; that of the Iron Curtain countries is significant but not accurately known.

Secondary tin is an important source, especially in the United States. Many tin alloys such as type metal can be remelted and reused. Scrap tinplate resulting from the operations of the can manufacturing industry is detinned, with both the steel and the tin made available for re-use. About 2000 to 3000 t of tin is recovered by detinning plants annually in the United States, much entirely equivalent to "virgin" tin and some in the form of tin compounds. In fact, detinning plants are the principal source of tin compounds in the United States. Many tin alloys are similarly recovered for re-use. In all, secondary tin accounts for 25 to 30 percent of total United States tin consumption.

Production

In principle the extractive metallurgy of tin is simple. The main source, cassiterite, is essentially SnO_2, and it suffices to reduce this to elemental tin with any suitable reducing agent, the obvious one being carbon. There are practical complications: the temperature required to reduce tin oxide with carbon is high enough to reduce the oxides of other metals present, and reduced iron forms compounds with tin called "hardhead." At smelting temperatures tin is extremely fluid; it seeks out the most minute openings for escape from the furnace and soaks into porous refractories. Tin reacts with both basic and acidic linings, forming slags which contain appreciable amounts of tin and silica which must be re-treated.

Electrolytic refining can produce a higher grade of tin than pyrometallurgical methods, but the capital cost is much higher, and there is little demand for tin of higher purity than the 99.8 percent obtainable by pyrometallurgy. Electrolytic refining, where practiced, utilizes an acid electrolyte containing stannous sulfate, cresolsulfonic or phenolsulfonic acid, free sulfuric acid, and β-naphthol or glue as an addition agent necessary to promote sound deposits. The electrolyte operates at room temperature, at a current density of 80 to 100 A/m^2, and at a cell voltage of 0.3 V.

Tinplate mills and can-manufacturing plants generate a considerable amount of waste tinplate: "skeletons," resulting from punching circular can ends from square sheets, and side trimmings. This waste tinplate is handled in detinning plants. At one time the detinning process consisted of passing dry chlorine gas through carefully dried tin scrap, taking advantage of the fact that when perfectly dry, chlorine reacts with tin but not with iron if the temperature is carefully controlled. The tin tetrachloride produced was distilled off for use as such or conversion to other tin compounds, and the steel remaining constituted a high grade of steel scrap for the open-hearth furnaces of the steel mills. Present practice in detinning uses solutions of sodium hydroxide with an oxidizing agent such as sodium nitrate or sodium nitrite; these solutions dissolve tin to form sodium stannate, while the steel remains passive. The sodium stannate solution, after purification, may be electrolyzed with steel anodes to produce on the cathodes an exceptionally pure grade of tin, or it may be processed into other tin compounds. In an electrolytic detinning process, the tin scrap itself is made anode in a sodium hydroxide solution, and tin sponge is deposited on the cathodes; this is periodically removed and worked up into ingots.

Uses

Tin is used relatively rarely in its unalloyed form, except as a coating on other metals. Tinfoil is used for electrical condensers, bottle-cap liners, gun charges, and food wrappings ("tinfoil" is often not really tin; more likely it is aluminum). Tin wire is used for fuses and safety plugs. Extruded tin pipe and tin-lined brass pipe are the preferred means for conveying distilled water and carbonated beverages, especially beer, and sheet tin is used to line storage tanks for distilled water. Collapsible tubes are used for packaging a few specialty pharmaceutical preparations. Tin powder is used in powder metallurgy, for coating paper, and for solder pastes. A pool of molten tin is used in the float-glass process for making plate glass.

The largest single use of tin is for coating steel to make tinplate, the starting material for the "tin can." This use accounts for over half the primary tin consumed in the United States. Tinplate was formerly manufactured by the hot-dip process, in which sheets of steel are dipped into molten tin; this process is all but obsolete. It has been superseded by the electrolytic process, in which continuous coils of steel strip are passed through cleaning, pickling, plating, rinsing, and posttreatment tanks at high speed (up to 650 m/min); as one coil is exhausted, the next is welded on, and the process is completely continuous and nearly automatic. The electrolytic process enables the use of much thinner deposits than can be obtained by hot dipping; now that aluminum and "tin-free steel" are threatening some markets of tinplate, economy in the use of tin is necessary for economic survival. These competitive materials are replacing tinplate in some applications, but there is no present sign that tinplate is obsolescent for its major application, the packaging of food in the "sanitary can."

Tin coatings are useful in many other applications, as will appear.

Many tin alloys are widely useful in industry; solder, a tin-lead alloy containing from 2 to 70 percent tin; bronze, a copper-tin alloy containing 5 to 15 percent tin; bearing metals or babbitts, of many compositions including tin, antimony, copper, and lead; pewter, a tin-antimony-copper alloy that is enjoying renewed popularity for tea and coffee services, trays, and the like; and type metals which contain varying proportions of tin, antimony, and lead. With the decline in letterpress printing in favor of offset and other methods, this use of tin is declining.

Tin-alloyed cast iron is increasing in importance, and a tin-aluminum alloy is used in some bearing applications.

The consumption of tin in the form of its compounds is relatively minor, but tin compounds are important in many industries, usually as minor additives to impart specific properties, such as the use of organotin compounds to stabilize polyvinyl chloride plastics against heat and light degradation.

Properties

Selected properties of tin are listed in Table 13-18. Tin is soft, low-melting, ductile, and easily cold-worked by rolling, extrusion, and spinning. Its color is silvery white, but as cast it may have a yellowish tinge caused by a thin film of oxide. In fact, this yellowish hue is insisted upon by some users as a sign of good-quality metal. This opinion appears to have no basis in fact: the color depends on the temperature of casting and the cooling rate. Highly polished tin has good reflectivity for light, and it retains its brightness well in both indoor and outdoor exposure.

With its low melting point and high boiling point, tin has a liquid range exceeded by few metals. It readily alloys with many metals and forms several intermetallic compounds of commercial importance. Copper, nickel, silver, and gold are soluble in liquid tin. Molten tin wets and adheres readily to clean iron, steel, copper, and copper-base alloys.

There are two allotropic forms of tin: white (β) and gray (α). White, or ordinary, tin is the familiar form, and the only one having the properties that make tin useful; it crystallizes in the body-centered tetragonal system. Gray tin has a diamond lattice; it is considerably less dense than β-tin and is nonmetallic in appearance and properties; it is a semiconductor. The allotropic transformation is of considerable practical importance, because the transition temperature is 13°C (or, according to some authorities, 18°C); this means that metallic tin is actually unstable on a cool day. Fortunately, the transition is extremely slow under most conditions; in fact, the existence of gray tin was not even discovered until the mid-nineteenth century, when some tin organ pipes in Moscow were found to have disintegrated during an exceptionally cold winter. The allotropic change is known as "tin pest" or "tin disease," probably because it appears to

Table 13-18 Properties of Tin

Atomic number	50
Atomic weight	118.69
Electronic configuration	$1s^2 2s^2 2p^6 3s^2 3p^6 3d^{10} 4s^2 4p^6 4d^{10} 5s^2 5p^2$
Melting point, °C	231.9
Boiling point, °C	2270
Density, g/cm³, α (gray)	5.77
Density, g/cm³, β (white, ordinary)	7.29
Transition temperature, $\alpha \rightleftharpoons \beta$, °C	13.2
Resistivity, $\mu\Omega$-cm, 20°C, β form	11.5
Electrical conductivity, % IACS	15
Tensile strength, cast, 15°C, MPa	14
Brinell hardness,* 10 kg, 20°C	3.9
Crystal form, α (gray)	cubic
Crystal form, β (white)	body-centered tetragonal
Electrochemical potentials, E^0, V	
$Sn^{++} + 2e^- \rightarrow Sn$	−0.136
$Sn^{4+} + 2e^- \rightarrow Sn^{++}$	+0.15
$Sn(OH)_6^{--} + 2e^- \rightarrow Sn(OH)_4^{--} + 2OH^-$	−0.90
$Sn(OH)_4^{--} + 2e^- \rightarrow Sn + 4OH^-$	−0.79

*See Chap. 23; Brinell hardness is not readily converted to DPH.

spread from centers of "infection"; i.e., inoculation with particles of gray tin can greatly accelerate the transformation. The transformation is of concern only if (1) the tin is exceptionally pure, and (2) exposure to very low temperatures is prolonged. Since electrodeposited tin is likely to be considerably purer than hot-dipped tin coatings—electrolysis acts to some extent as a refining step—it is only fairly recently that the formation of gray tin has been noticed in commercial applications. The transformation is inhibited or prevented by the incorporation into the tin of a few tenths of a percent of antimony, lead, or bismuth, and a plating process for codepositing the required amount of bismuth into the tin has recently become important.

Tin is amphoteric: it reacts with acids and bases, and is relatively resistant to neutral solutions. The overvoltage of hydrogen on tin is quite high, about 0.75 V, so that attack by acids and bases is slow unless an oxidizing agent is present to depolarize the evolution of hydrogen. Distilled water has no effect on tin, which has been the preferred medium for preparing and storing it.

The tin-iron galvanic couple is of great commercial importance because of its action in the tin can. In this couple iron is normally anodic, and would corrode through pores; but under the anaerobic conditions inside the can, and in the presence of organic food acids, the potential of the couple reverses and tin becomes anodic, thus protecting the steel from attack. This property, in effect, makes the canning industry possible, at least in its present form.

Nitrogen, hydrogen, carbon dioxide, and gaseous ammonia do not react with tin. Tin is attacked by moist sulfur dioxide. The halogens readily react to form the tetrahalides; the halogen acids also react, especially when hot and concentrated. Hot sulfuric acid dissolves tin; cold dilute nitric acid reacts slowly, but hot concentrated nitric acid converts tin to an insoluble hydrated stannic oxide, often called metastannic acid. Other acids that rapidly attack tin include sulfurous, chlorosulfonic, and pyrosulfuric; phosphoric acid reacts somewhat more slowly. Organic acids such as lactic, citric, tartaric, and oxalic attack tin slowly if air or oxidizing agents are present.

Dilute ammonium hydroxide and sodium carbonate have little effect on tin; strong alkalis such as sodium and potassium hydroxide solutions react readily to form soluble stannates; the reaction is slow if the tin is in massive form, faster if it is powdered, and faster still if a metal such as iron is present that can form a galvanic couple with tin.

Oxidizing salt solutions such as potassium peroxysulfate, ferric salts, and stannic salts dissolve tin. Nonaqueous organic solvents, lubricating oils, and gasoline have little effect.

The outer electron shell of the tin atom is $5s^2 5p^2$; thus there are four electrons available for bonding, and accordingly tin is tetravalent in many of its compounds. As with its homologs germanium and lead in Group IVB, however, the $5s$ electrons may act as an "inert pair," and tin is also divalent. In this respect tin is intermediate between germanium and lead; with the former, bivalency is uncommon, and lead exhibits tetravalency only in its organic compounds (with a few exceptions). For tin, the two valence states are almost equally stable and readily interconvertible. Solutions of stannic tin or tin(IV) are readily reduced to stannous tin or tin(II) by many reducing agents, especially metals such as antimony and nickel. Solutions of tin(II) are just as readily oxidized to tin(IV) by all the common oxidants, including air.

In alkaline media, tin(IV) is the more stable; alkaline stannite, or stannate(II), solutions disproportionate according to

$$2Sn(OH)_4^{--} \rightarrow Sn(OH)_6^{--} + Sn + 2OH^-$$

This reaction is important in plating from alkaline stannate solutions, as will be shown.

In acidic solutions, tin(II) compounds probably exist in the form of the Sn^{++} aquo ion, but Sn^{4+} probably does not exist as such and is either hydrolyzed or complexed, as in $SnCl_6^{--}$ and $Sn(OH)_6^{--}$. All tin compounds tend to hydrolyze in aqueous solution: alkaline solutions must be stabilized by the presence of excess alkali, acid solutions by excess acid.

Solutions of Sn(II) are easily oxidized by air; they may be stabilized by rigorous exclusion of air, or by keeping a few pieces of tin in the container:

$$Sn^{4+} + Sn \rightarrow 2Sn^{++}$$

Tin, either bivalent or tetravalent, forms relatively few complexes; the most important for present purposes are the hydroxo- and halo-complexes: stannate(IV) $Sn(OH)_6^{--}$ and various halides such as SnF_4^{--} and SnF_6^{--}. It forms no cyanide complex.

Tin compounds important to the metal finisher include the following:

Stannic hydroxide, $SnO_2 \cdot xH_2O$, or α-stannic acid, when freshly precipitated, is readily soluble in dilute mineral acids or alkali hydroxides; aging or heating converts it to the so-called β-stannic acid, which is insoluble in dilute acids but soluble in concentrated hydrochloric acid. Freshly precipitated stannic hydroxides are easily peptized to colloidal sols by sodium, or preferably potassium, hydroxide solutions; such sols are used in tin plating.

Stannous chloride, $SnCl_2$, is available both as the anhydrous salt and as the dihydrate $SnCl_2 \cdot 2H_2O$, called "tin crystals." This compound is the basis of a widely used bath (the "Halogen" process) for producing electrolytic tinplate. The anhydrous compound has the advantage of greater stability in storage. It is a white solid, melting at 246.7°C, that is easily soluble in water and some organic solvents. Stannous chloride is also the source of tin in the tin-nickel–plating electrolyte. It is used in the process of plating on plastics as a sensitizer.

Stannous sulfate, $SnSO_4$, is a white crystalline powder, soluble in water with the usual hydrolysis. It is used in most acid electrolytes for tin plating, including the Ferrostan process for electrolytic tinplate, and the bright acid tin-plating baths.

Stannous fluoborate, $Sn(BF_4)_2$, is made by the reaction of stannous oxide SnO with fluoboric acid; it is available only in the form of a solution concentrate, like other metal fluoborates. It can be used for tin plating, but is more generally employed for plating the tin-lead alloys.

Sodium and potassium stannates, $M_2Sn(OH)_6$ (M = Na or K)* are the important ingredients of the alkaline tin-plating baths. Both are by-products of the detinning industry, much refined before marketing. The potassium salt is now preferred for plating. There is a British specification for both sodium and potassium stannates for plating use, but no corresponding U.S. document. The only important grade is technical, which is sufficiently pure for electroplating purposes. For this application, the impurities usually present—alkali hydroxide and carbonate, and moisture—are harmless. Absence of deleterious impurities such as lead and antimony is desirable; otherwise the only

*Often written $Na_2SnO_4 \cdot 3H_2O$. It is almost certain that the water is not water of hydration but is present in the complex $Sn(OH)_6^{--}$.

requirement is that the materials contain sufficient of the relatively expensive metal to justify their price. Commercial potassium stannate carries a maximum specification of about 1 percent sodium. Technical sodium stannate normally contains about 41 percent tin (in theory 44.5 percent) and potassium stannate about 38.5 percent (in theory 39.7 percent).

Grades and Specifications

Pig tin is usually marketed by brand name rather than by specification. For many years the market standard has been so-called Straits tin, indicating its origin in the Malay peninsula area, but many brands can meet this standard of 99.8 percent minimum purity. For plating purposes, this grade is sufficiently pure as long as specific impurities do not exceed the maximum in the specification. About 1 percent aluminum is deliberately added to one brand of tin anodes for alkaline tin plating; this should not be regarded as an impurity.

The most deleterious impurities in tin for use as anodes in plating are cadmium, nickel, and antimony. The plater should specify grade A or better as anodes, unless the so-called high-speed anodes containing 1 percent aluminum are to be used.

Hazards

Tin and its inorganic compounds are essentially nontoxic; otherwise the canned food industry could not exist in its present form. This property is unusual for a heavy metal; compare its near neighbors in the periodic system such as lead, antimony, thallium, arsenic, and cadmium. Any significant toxicity in inorganic tin compounds is a function of their individual nature: the stannates are highly alkaline, the tetrachloride fumes in air to produce hydrogen chloride gas, stannous chloride solutions are acidic, etc.

Many organotin compounds are toxic in varying degrees, but are not likely to be encountered in metal-finishing operations.

Tin Plating

Tin is seldom plated for decorative purposes; it owes its principal applications in metal finishing to its compatibility with foods, i.e., nontoxicity, and its excellent solderability. Softness and ductility account for some uses, and it possesses fair bearing properties. Under usual atmospheric exposure conditions, it is cathodic to steel, and thus does not offer the sacrificial protection provided by zinc and cadmium. In itself it is tarnish- and corrosion-resistant, but to exhibit these properties, coatings must be thick enough to be essentially nonporous.

The largest use for tin plating is in the production of electrolytic tinplate, which, as has been mentioned, has essentially superseded the hot-dip product. Applications more familiar to the average metal finisher include coatings on refrigerator evaporators, dairy and other food-handling equipment, washing machine parts, builders' hardware, radio and electronic components including printed circuits, copper wire, large pipe couplings, and pistons (sometimes also piston rings) for internal-combustion engines.

Tin coatings on copper wire protect the wire from the sulfur in the insulation as well as providing solderability. Although it is not one of the best bearing materials, tin can be used for this purpose and is easily applied: aluminum alloy automotive pistons are tinned by aqueous immersion. The tin prevents scoring and seizing of the cylinder walls during the running-in period.

Coatings of tin and tin-rich tin-lead alloys ("solder plate") are used in the electronics industry, where their solderability and resistance to various chemical etchants are desirable.

Tin may be plated either mat or bright. Until a few years ago, attempts to produce bright tin deposits met with little success, but now there have appeared several proprietary processes for producing bright tin deposits from acid baths. It is still a matter of some controversy whether there is much point in bright tin plating for most purposes, since its application is functional rather than decorative; nevertheless, bright tin plating is commercially successful.

Tin may be plated from either alkaline stannate baths or from acid baths, based usually on either stannous sulfate or stannous fluoborate. These baths produce mat deposits; with appropriate additives, the sulfate baths constitute the proprietary bright-plating processes.

Mat deposits can often be brightened by a process variously known as flow brightening, reflowing, or flow melting. Almost all electrolytic tinplate is thus treated, as are some other tin-plated items. In the electrolytic tinplate industry, many methods are used for bringing the tin coating up to a temperature slightly above its melting point of 232°C and quickly quenching it to produce the bright appearance; these include induction, resistance, and radiant heating. Only one method is applicable in general plating: immersion in hot oil or fat. The oil is generally a long-chain fatty acid ester of glycerin such as palm oil, tallow, or partially hydrogenated oils of various types. The medium should have a sufficiently high flash point for the temperature involved, as well as some free fatty acid to act as a flux. The melting should be carried out as soon as possible after plating.

Successful reflowing demands that the tin coating be neither too thick nor too thin. Too thin a deposit will not come out bright, and too thick a coating will tend to dewet—gather up into beads. Average limits for brightening by this method are 2.5 to 7.5 μm; this depends largely on the size and shape of the article and the nature of the substrate. Large, flat areas have more tendency to dewet than wires or rounded shapes. Successful reflowing also depends on good cleaning, plating, and rinsing practice; unsound deposits that might pass in the as-plated condition will be revealed by reflowing.

Deposits of tin, like those of many other metals, can grow "whiskers" which can cause malfunction of miniaturized electronic components by causing electric short circuits. This is a problem only in low-voltage applications: high voltage will quickly burn off such whiskers. Reported methods of eliminating or reducing this hazard include codeposition of about 1 percent lead and stress relief after plating by heating.

Both "tin whiskers" and "tin pest," though real enough, are often exaggerated as problems in the use of tin deposits. Many cases of supposed "tin pest" have turned out to be merely corrosion, and suspected "tin whiskers" sometimes prove to be merely lint or some other harmless material.

There are two general types of tin-plating electrolytes: acid and alkaline. Each has advantages and disadvantages, and the choice will depend on which characteristics are most important in the individual case.

Alkaline baths are based on sodium or potassium stannate, preferably the latter. Advantages of this bath include simplicity of formulation, wide range of concentrations within which satisfactory deposits are obtained, no need for addition agents, wide tolerance for impurities, ability to use insoluble anodes where desirable (such as plating the inside of pipe), ease of analytical control—only two constituents requiring determination, excellent throwing power (perhaps the best of any common plating bath), and the hot alkaline bath can tolerate some carelessness in cleaning.* Disadvantages are: tin is plated from the tetravalent state, so that the electrochemical equivalent of tin is only half that from acid electrolytes; cathode efficiencies are less than 100 percent (sometimes considerably less), still further reducing the plating speed; operating temperatures are elevated, rendering the bath unsuitable for plating on some delicate substrates and requiring expenditure for heating; higher voltages are required; and the bath cannot produce bright coatings except by reflowing. Operation with soluble anodes requires rather critical control and improper anode conditions can cause completely unsatisfactory deposits.

Acid baths are based on either stannous sulfate or stannous fluoborate. Advantages of these electrolytes include: deposition from the bivalent state, so that the electrochemical equivalent of tin is twice that from the alkaline baths; lower bath voltage, and usually better cathode efficiencies, approaching 100 percent; ability to yield bright deposits by the use of appropriate additives; and operation at or near room temperature, so that delicate substrates are not affected. Compensating disadvantages include: they are much more critical in composition and operating conditions for satisfactory deposits; they require addition agents even for mat plate (without additives the deposits are trees and nodules with no coherence); insoluble anodes cannot be used; and throwing power is inferior to that of the alkaline baths (though still quite good as acid baths go).

*This is not meant to imply that carelessness in cleaning is recommended, only that the alkaline stannate bath is forgiving in this respect.

ALKALINE TIN PLATING

Sodium stannate first became available commercially in the 1920s, and it was quickly adapted to tin plating, resulting in the early 1930s in the sodium stannate bath of Wernlund and Oplinger. These workers realized that for satisfactory operation the tin must be in the tetravalent form, and they included hydrogen peroxide in the formulation as a means of ensuring this condition. Their bath has remained virtually unchanged up to the present; by 1940 it had virtually displaced all older formulations.

Work since then has elucidated the unusual behavior of tin anodes in an alkaline medium, has shown that the originally recommended addition of sodium acetate, though harmless, was unnecessary, and has afforded factual background to the empirical findings of the inventors. More recent developments in alkaline tin plating have been the demonstration of the superiority of potassium stannate over sodium stannate; the introduction of a better anode, capable of a wider operating range than pure tin; and the development of a process using insoluble anodes, with chemical replenishment, along with the ability to codeposit a small amount of bismuth with the tin to overcome the tin pest in those relatively few cases where it is a problem.

Typical formulations of alkaline stannate baths are listed in Table 13-19. Both sodium and potassium stannate solutions are in wide-scale commercial use; the choice between them is largely a matter of balancing first cost vs. operating efficiency. Potassium-formulated solutions are more versatile, because the solubility of the potassium salt is much higher than that of sodium stannate; furthermore, sodium stannate is one of those unusual compounds having a "reverse" temperature coefficient of solubility; i.e., it is more soluble cold than hot. At any given concentration and current density, the cathode efficiency of the potassium bath is higher than that of the sodium; its tolerance for carbonate is greater, and the cathode current efficiency–current density curve falls less steeply. In short, the permissible operating cathode current density of the potassium bath is higher than that of the sodium. All stannate baths tend to produce a certain amount of sludge, consisting of a hydrated tin oxide, which represents an economic loss of tin; but this tendency is greater in the sodium bath. Potassium salts are more conductive than sodium, so that the conductivity of the potassium bath is higher; this is especially important in barrel plating. Potassium salts are more expensive than their sodium analogs, but the additional cost is usually justified.

There is no difference in quality of deposit.

The formulations of the stannate baths as shown in Table 13-19 are only typical; provided the proper balance is maintained between temperature, current density, and free-alkali content (which determines anode current density), there is nothing critical about any of these concentrations. This may become clearer as the functions of the ingredients, and the effects of operating conditions, are outlined.

Mixed baths—potassium hydroxide—sodium stannate or vice versa—are

Table 13-19 Alkaline Stannate Baths

*Bath**	*Makeup*		*Control limits*	
	g/L	*Molarity*	*g/L*	*Molarity*
1. Potassium stannate				
$K_2Sn(OH)_6$	100	0.34	95–110	0.32–0.36
(tin, Sn)	40	0.34	38–43	0.32–0.36
Free potassium hydroxide, KOH	15	0.27	13–19†	0.23–0.34
Cathode current density, A/m²	300–1000			
Temperature, °C	65–90			
2. Potassium stannate	210	0.67	195–220	0.63–0.72
(tin)	80	0.67	75–85	0.63–0.72
Free potassium hydroxide	22	0.39	15–30	0.27–0.54
Cathode current density, A/m²	to 1600			
Temperature, °C	70–90			
3. Potassium stannate	420	1.35	390–450	1.25–1.43
(tin)	160	1.35	150–170	1.25–1.43
Free potassium hydroxide	22	0.39	15–30	0.27–0.54
Cathode current density, A/m²	to 4000			
Temperature, °C	80–90			
4. Sodium stannate, $Na_2Sn(OH)_6$	100	0.35	95–110	0.33–0.38
(tin)	42	0.35	39–45	0.33–0.38
Free sodium hydroxide, NaOH	10	0.25	7.5–11.5†	0.19–0.29
Cathode current density, A/m²	50–300			
Temperature, °C	60–85			

*Baths 1 and 4 are for average conditions; baths 2 and 3 are essentially high-speed baths, and anode current densities cannot approach those at the cathode.

†For barrel plating free alkali may be higher.

operable but not recommended, since there is no background of experience in their use.

There are only two essential ingredients of the stannate baths: alkali stannate and the corresponding alkali hydroxide. Other compounds will be formed in use.

Alkali stannate—$Na_2Sn(OH)_6$ or $K_2Sn(OH)_6$—provides the tin content of the bath. Its concentration is not critical. If it is too low, the cathode efficiency, or permissible cathode current density, will suffer; if it is too high, drag-out and other losses will be unnecessarily costly. It should be kept at such a level that the desired plating speed is obtained. In the potassium bath, but not the sodium bath, the stannate concentration may be greatly increased if extraordinarily high plating speeds are required—up to 450 g/L for current densities to 4000 A/m^2—but such high tin contents are unnecessarily concentrated for average conditions.

Alkali hydroxide—NaOH or KOH—has several functions, and its control is much more important than control of tin content. The hydroxyl ion is the principal conducting medium in the bath. A reservoir of hydroxyl ion is also necessary to act as an acceptor for the absorption of carbon dioxide from the air. In its absence, this would decompose stannate, according to

$$Sn(OH)_6^{--} + CO_2 \rightarrow SnO_2 \downarrow + CO_3^{--} + 3H_2O$$

The effect of carbon dioxide may also be attributed to its lowering the pH of the solution; the result is the same. Hydroxyl ion also tends to repress the decomposition of stannate ion:

$$Sn(OH)_6^{--} \rightarrow SnO_2 \downarrow + 2OH^- + 2H_2O$$

Finally, hydroxyl ion is necessary for satisfactory corrosion of tin anodes; its concentration must be that appropriate to the anode current density and the temperature. Increase in free alkali tends to decrease cathode efficiency.

Other ingredients of the alkaline stannate bath are not essential, but usually present:

Sodium or potassium acetate. In the original formulation of the stannate bath, sodium acetate was specified at a level of about 15 g/L. We now know that it is unnecessary though harmless. It cannot, as was once claimed, act as a buffer at the pH of the solution, which is well above 13. Since acetic acid is often used to control the free-alkali content, alkali acetate is formed in the bath and is found in many operating solutions.

Alkali carbonate. Carbonates are not deliberately added to either bath, but absorption of carbon dioxide from the air produces them during operation. Carbonate appears to be quite inert; it has been claimed that it improves the throwing power of the sodium bath, but the evidence is questionable. Carbonate may be removed from the sodium bath by cooling, as with other alkaline solutions. This is not possible with the potassium bath owing to the high solubility of potassium carbonate; nor can barium or calcium salts be used, since they would precipitate stannates as well. But removal of carbonate is seldom necessary, since potassium carbonate appears to have no deleterious effect and can simply be allowed to build up.

Hydrogen peroxide. The presence of even small amounts of bivalent tin (usually due to improper anode conditions) can cause spongy and highly unsatisfactory deposits; in fact, it is the only common cause of difficulty in these baths. Hydrogen peroxide constitutes an immediate "first-aid" measure to oxidize this bivalent tin and allow plating to continue, but its routine use should not be necessary—the improper anode conditions should be corrected. Hydrogen peroxide decomposes to water, leaving no residue; but if too much is used, it constitutes a waste of money, and cathode efficiency will be low or zero until it has decomposed.

Addition agents. In alkali stannate plating the electrode efficiencies are usually below 100 percent, and the resulting gas evolution can cause a caustic mist or spray which is annoying to operators and corrosive to shop equipment. A small amount of foaming agent will correct this difficulty; deliberate addition is seldom necessary, since commercial stannates usually contain a little soapy material. If required, a little sodium oleate or other common soap or surfactant can be added. Brightening or smoothing agents for this bath are neither available nor needed for satisfactory mat deposits.

Increase in temperature improves both anode and cathode efficiencies; in the potassium stannate bath there is no real upper limit to the operating temperature except that set by steam consumption and excessive evaporation. The sodium stannate bath is seldom operated much above 80°C.

Increase in stannate content improves cathode efficiency or broadens the range of cathode current densities. It has little or no effect on anode operation.

Increase in free alkali increases anode efficiency, or anode current density range, and decreases cathode efficiency.

Other bath ingredients have little effect.

There is no substantial upper limit to the thickness of tin deposits obtainable from the stannate bath; usual thicknesses are in the range 5 to 25 μm.

Anodes in the Stannate Bath

The only critical factor in the operation of the stannate bath is the condition of the tin anode. Anodes may be of pure tin, of various shapes and sizes; tin alloyed with 1 percent aluminum, marketed as "high-speed" anodes; or inert materials such as steel or nickel alloys.

The high-speed anode has a wider useful current-density range and maintains its efficiency at higher current densities than pure tin; in either case, however, anodes must be "filmed." Lack of the proper film will cause the tin to dissolve as bivalent tin, which is the main cause of unsatisfactory deposits.

In order to film tin anodes, a higher-than-normal current density—a "surge" of current—must be impressed on them for a few seconds to a minute; then the current is reduced to its operating value. If this process is carried out properly, there will form on the anodes a yellowish film (on pure tin) or an olive-green one (on 1 percent aluminum) which, if conditions remain constant, will be maintained as long as current flows. The film dissolves fairly quickly on shutdowns and must be re-formed on start-up. If current density is too high for the operating conditions, the film will thicken and turn black; the tin is now covered with an impervious film of oxides and is essentially inert and not dissolving. If the current density is too low, the film will be lost and tin(II) will form in the bath.

Once proper anode operation is learned, practically all the sources of trouble with the stannate bath will be obviated. The action of tin anodes in a stannate bath is unusual, and has been treated in some detail in several references.*

Anode Shapes

Tin anodes are offered in various shapes in an effort to provide more surface, but they have never demonstrated any real improvement over standard round, elliptical, or slab anodes. The use of ball anodes in steel containers is not recommended because most of the current will be carried by the steel. Titanium-anode baskets, however, are satisfactory.

Inert anodes in the stannate bath have several advantages: the filming difficulties are avoided, and the evolution of oxygen at the anode ensures that stannate(II) cannot form. They do not change shape in use, and in some cases—plating the inside of pipe, for example—they are required. There is no limit—upper or lower—on the anode current density usable, and the inventory cost of tin is avoided. Their use, however, implies chemical replenishment of the bath; and the continual addition of alkali stannate, with the plating out of tin and the neutralization of the alkali thus formed by the addition of acetic acid, will eventually result in a bath so concentrated in salts that no more can be dissolved:

$$\begin{array}{lll} \text{Cathode:} & K_2Sn(OH)_6 + 4e^- & \rightarrow Sn + 2KOH + 4OH^- \\ \text{Anode:} & 4OH^- & \rightarrow O_2 + 2H_2O + 4e^- \\ \text{Total:} & K_2Sn(OH)_6 & \rightarrow Sn + 2KOH + O_2 + 2H_2O \end{array}$$

Therefore, the use of inert anodes generally has been confined to instances where the tin bath was used only occasionally or for special jobs.

A process enabling the use of inert anodes, with chemical replenishment by a tin oxide sol, has now made the use of inert anodes practical. The sols consist essentially of hydrated tin oxide, $SnO_2 \cdot xH_2O$, peptized by KOH, the total formula being approximately $SnO_2 \cdot xH_2O \cdot \frac{1}{2}KOH$; thus only about one-fourth as much potassium ion is added as when potassium stannate is used for replenishment. This amount can usually be tolerated since it is taken care of by drag-out. The tin oxide sol is converted to stannate ion by the free hydroxyl ion in the bath, and plating proceeds normally.

These tin oxide sols have been modified by the addition of a bismuth compound to yield a deposit containing about 0.1 to 0.2 percent Bi; this is sufficient to inhibit the formation of gray tin, or tin pest, where it is of concern.

These sols are usable with the potassium stannate system only; in the sodium system they do not convert readily to stannate and eventually sludge out.

*See F. A. Lowenheim, "Modern Electroplating," cited in the Bibliography.

ACID TIN PLATING

The principal use for acid tin plating is in the production of electrolytic tinplate. Baths based on stannous sulfate and stannous fluoborate are also used in general plating. Recently developed additives have made practical the production of bright deposits from the sulfate electrolytes.

All acid tin baths contain the tin in the bivalent tin(II) form, which is an advantage as far as electrochemical equivalent is concerned. Since tin(II) is easily oxidized, even by air, most acid tin baths require the use of some form of anti-oxidant or other precautions to prevent undue oxidation; further, the use of insoluble anodes is precluded. The Sn(IV) so formed is not harmful, but since it is essentially inert in the bath, it represents an economic loss of tin.

All acid tin electrolytes require addition agents for the production of sound deposits; plating from solutions containing only a stannous salt and excess acid will produce only treed, macrocrystalline, and nonadherent deposits.

Stannous Sulfate

Typical formulations are listed in Table 13-20. Common addition agents are cresol- or phenolsulfonic acid, β-naphthol, and gelatin. Proprietary additives are also offered. The cresol- and phenolsulfonic acids tend to inhibit oxidation of tin; the others yield smooth and useful deposits. The bath is worked at room temperature, and anode and cathode efficiencies approach 100 percent. Control of the

Table 13-20 Stannous Sulfate Bath

	Makeup		*Control limits*	
	g/L	*Molarity*	*g/L*	*Molarity*
Stannous sulfate,* $SnSO_4$	72	0.34	54–90	0.25–0.42
(Tin: Sn)	40	0.34	30–50	0.25–0.42
Free sulfuric acid, H_2SO_4	50	0.51	40–70	0.41–0.71
Phenolsulfonic acid,† $HOC_6H_4SO_3H$	40	0.23	30–60	0.17–0.34
Gelatin	2	—	2	—
β-naphthol, $C_{10}H_7OH$	1	0.007	1	0.007
Temperature, °C			≈20	
Cathode current density, A/m^2				
Still			100	
Agitated			to 1000	
Anode current density, A/m^2			200	

*Most commercial stannous sulfate does not contain the theoretical percentage of tin; enough should be used to provide the tin content specified.

†Cresolsulfonic acid is an alternative.

bath is somewhat more critical than control of the stannate solutions, owing to the addition agents required, which must be kept in balance.

Cast or rolled anodes of Grade A tin or better should be used. A certain amount of anode slime forms, and the use of anode bags and continuous filtration is advisable. Control of the addition agents is the most critical factor in operation of the solution. Hull cell tests, or sample cathodes plated direct in the tank, are useful to the experienced operator.

The throwing power of the bath is good, though not as high as that of the stannate solutions.

Bright tin–plating baths are based in general on the sulfate electrolyte, with proprietary addition agents. The bright deposits are about equivalent in properties to the ordinary mat deposits in solderability, corrosion resistance, and leveling characteristics. Claims that they are superior to mat tin in resistance to whisker growth and tin pest should be viewed with some skepticism.

Since deposits from the acid baths—both mat and bright—contain some codeposited organic matter from the additives, occasionally one encounters some difficulty in soldering, because of the codeposited organics bubbling out of the plate. This is not common, and under most circumstances the quality of tinplate, whether from the stannate or the various acid baths, is about equivalent.

Stannous Fluoborate

The principal advantage of the fluoborate over the sulfate bath is the higher current density that can be employed—a property shared by most fluoborate electrolytes. The high solubility of metal fluoborates in general, including tin, makes possible baths with high metal content, with its accompanying possibility of faster plating rates. In other respects the fluoborate electrolyte is in principle not different from the sulfate bath. Formulation and operating conditions are listed in Table 13-21.

Table 13-21 Stannous Fluoborate Bath

	g/L	*Molarity*
Stannous fluoborate, $Sn(BF_4)_2$*	200	0.68
(Tin)	81	0.68
Fluoboric acid, HBF_4	100–200	1.14–2.28
Gelatin	6	——
β-naphthol	1	0.007
Temperature, °C	20–40	
Cathode current density, A/m^2		
Normal, no agitation	250–1250	
Vigorous agitation	2500–4000	

*Based on 100 percent stannous fluoborate; but available only as a concentrate.

Table 13-22 The Halogen Bath*

	g/L	*Molarity*
Stannous chloride, $SnCl_2$	63	0.33
or $SnCl_2 \cdot 2H_2O$	75	0.33
(Tin)	39	0.33
Sodium fluoride, NaF	25	0.60
Potassium bifluoride, KHF_2	50	0.64
Sodium chloride, NaCl	45	0.77
Addition agents	1–2	—
Current density, A/m^2	4500	
pH	2.7	

*Not in use for general plating; high-speed strip application.

Continuous Electrotinning of Steel Strip

The plating of tin on continuous coils of strip steel for the manufacture of electrolytic tinplate is the largest single application of electroplating, at least in terms of tonnage of product. The process was developed rather hastily during World War II as a tin-conservation measure, although experimental and pilot lines had been operating considerably earlier. Three electrolytes were in use: the alkaline stannate, a modified stannous sulfate (Ferrostan), and the so-called Halogen electrolyte, developed especially for this application; a typical formula for the last-named is shown in Table 13-22. Figure 13-2 is a schematic diagram of the three types of lines used in the steel mills.

The stannate electrolyte, although it produced good-quality plate, perhaps even superior to that of the other baths, has gone out of favor on economic grounds, and most tin mills now use one of the acid electrolytes. The stannous fluoborate solution is used for electrolytic tinplate in Germany, but not in the United States, where it is, however, in use for the somewhat analogous process of continuous wire plating.

Since the average metal finisher will not encounter the necessity for producing electrolytic tinplate—it is an enterprise of the large steel mills, with capital investments in the millions of dollars—no further details will be given here. Interested readers are referred to the literature.*

However, electrolytic tinplate production affords a good example of the effect of violent agitation on limiting current densities. Electrolytic tinplate can be

*W. E. Hoare, E. S. Hedges, and B. T. K. Barry, "The Technology of Tinplate," St. Martin's Press, New York, 1965. This book is a detailed account of the whole process of tinplate manufacture, from production of the steel, including both electrolytic and hot-dip tinplate. A good, short summary is found in: J. B. Long, *Plating*, **61**, 938 (1974).

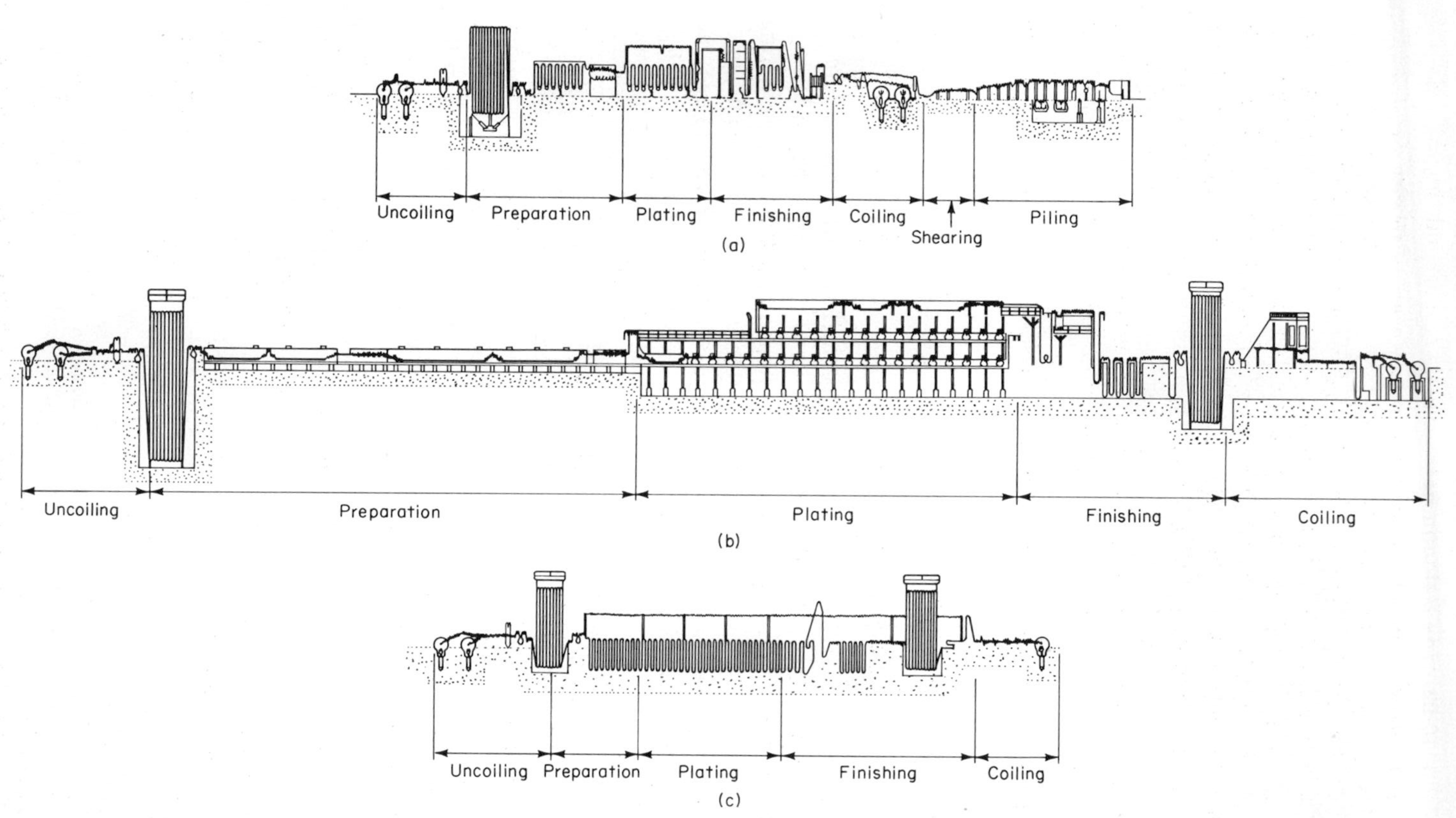

Fig. 13-2 Schematic diagram of three types of electrotinning lines: (*a*) Ferrostan line; (*b*) Halogen line; (*c*) alkaline line.

produced economically only if production speeds are high, and this entails high speeds of strip travel through the "line." To deposit the required amount of tin in minimum time, in turn, demands very high cathode current densities while retaining satisfactory deposits and cathode efficiencies. Current densities may be as high as 6500 A/m^2 or more with the Halogen electrolyte. (The alternative would be an uneconomically long tank.) Such high current densities would not be attainable with still or slow-moving strip; see the discussion of limiting current densities, Chap. 8, pp. 140 and 144. This constitutes a good illustration of the effect of the dimensions of the diffusion layer upon limiting current densities. At the very high strip speeds used, the flow of electrolyte at the cathode surface is highly turbulent, and the diffusion layer is reduced to extremely small dimensions.

Specifications

ASTM designation B 545 specifies thicknesses required for the various uses of electrodeposited tin. Service conditions (SC) are described as follows:

> SC4—very severe conditions require a complete cover of tin free of pores. Examples of SC4 are exposure to abrasion or corrosive liquids or gases; thicker coatings, up to 125 μm, may be required for water containers, threaded steel couplings in oil-drilling rigs, or exposure to seacoast atmospheres.
>
> SC3—severe, includes exposure to dampness and mild corrosion from the atmosphere in industrial locations. Examples are fittings for gas meters and automotive accessories such as air cleaners or oil filters.
>
> SC2—moderate, includes interior atmospheres; electronic applications, e.g., transformer cans and the like; chassis frames; and solderable connectors.
>
> SC1—mild, includes precoating nonsolderable basis metals to facilitate soldering of electronic components, as a surface preparation for protective painting, as a stopoff in nitriding.

For solderability, one of the principal uses of tin coatings, zinc-containing basis metals such as brass must have a nickel or copper undercoat at least 2.5 μm thick before tin plating, to prevent migration of the zinc into the tin coating.

Table 13-23 indicates the thicknesses specified by ASTM for the various service conditions described above.

Table 13-23 ASTM Specifications for Tin Deposits (B 545)

	Minimum thickness			
	On steel		*On copper and copper-base alloys**	
Service condition	*Local, μm*	*Average, μm*	*Local, μm*	*Average, μm*
SC4	30	45	30	45
SC3	20	30	15	20
SC2	10	15	8	12
SC1†	5	8	5	8

*An undercoat of at least 2.5 μm of copper, bronze, or nickel must be used on alloys containing appreciable zinc (e.g., brass).

†May be flow-brightened.

LEAD

Lead, atomic number 82, was one of the earliest metals known. The English word is probably cognate with German *Lot*, a plummet or sounding-lead, and traces back to the same Teutonic root. The symbol Pb is from Latin *plumbum*.

Lead has been mined and worked in all civilizations from earliest times. In the third century B.C. the Romans operated the lead mines of Spain in the Rio Tinto area, and these mines are still in operation.

Lead is fairly low in the scale of abundance in the lithosphere—about 0.0016 percent. Its comparatively low price is due to the relatively concentrated locations of its ores and the relative ease of recovery. Major producing countries are the United States, Mexico, Canada, Peru, Bulgaria, the Soviet Union, Yugoslavia, China, and Australia. Lead is mined in more than 50 countries, and total world production approximates 4 million t.

The isotopes of lead are of more than usual interest, since they each represent the end product of the decay of the radio-active elements: ^{206}Pb is the end of the uranium series; ^{208}Pb is the end of the thorium series; and ^{207}Pb is the end of the actinium series. The determination of the atomic weight of lead from different radioactive sources contributed to our knowledge of isotopes and atomic theory. By the same token, the usually accepted atomic weight of lead, 207.2, may not be true of all specimens, depending on their source.

Although there are many lead minerals, by far the most important is galena, PbS; others of some commercial interest are lead sulfate, or anglesite, and lead carbonate, or cerussite. Lead occurs in minor concentrations in a host of other minerals and ores.

The chief ore of lead, galena, is found in nature associated with sulfides of other metals such as those of iron, zinc, copper, bismuth, arsenic, antimony, and tin. Separation processes vary according to the nature of the ore and the other metals present; selective flotation processes can often achieve a relatively pure lead sulfide containing up to 80 percent Pb (in theory 86.6 percent). More generally, lead concentrates average about 50 percent Pb and contain as impurities iron, silica, zinc, arsenic, silver, gold, and bismuth. The ore is smelted in blast furnaces after roasting for sulfur removal. Roasting converts the lead and most other metals to oxides; the furnace processes, which may be quite complicated, result in a matte, a lead bullion, slags, and "speiss," an intermetallic compound containing iron, antimony, and other metals originally present in the ore. Slags and speiss are reworked for their metal content; the lead bullion, quite impure at this point, is refined by either pyrometallurgical or electrolytic processes, or sometimes both.

Secondary lead is an important source, accounting for more than 40 percent of total U.S. consumption. Many lead alloys such as type metal and battery plates can be recovered easily for re-use.

Uses

Lead is an important metal industrially, surpassed in tonnage only by iron, copper, zinc, and aluminum. World consumption is estimated at about 4×10^6 t annually; it is about 1.4×10^6 t annually in the United States.

By far the most important application is in the lead-acid storage battery: it accounts for about 50 percent of the total, about equally divided between the metal and the oxide. Tetraethyllead takes about 19 percent; alloys including solder, lead ammunition, sheet and pipe, type metal, terne, weights and ballast, cable sheathing, collapsible tubes, and others account for about 20 percent. The remainder is divided among many miscellaneous uses including pigments and a few other compounds. The addition of lead to brass produces a free-machining metal; it is often added to steel and aluminum for similar purposes.

Four grades of lead are recognized as standard; the purest grades are known as "corroding lead" and "chemical lead."

Lead forms alloys with most nonferrous metals, and the addition of a few tenths of 1 percent of some metals markedly improves some of its properties. For example, 0.1 to 0.2 percent arsenic is added for making shot, to improve sphericity and hardness. Antimony imparts the hardness required for some types of ammunition, and also the property of expanding on solidification required for type metal and casting alloys; it also hardens lead. Tin improves castability in lead-antimony alloys; bearing metals consist of lead and antimony, sometimes with addition of copper, tin, and zinc.

As little as 0.2 percent calcium produces an age-hardening alloy and improves its fatigue strength several-fold. Cadmium, tellurium, and other nonferrous metals are also used in small amounts to impart various desirable properties to lead-base alloys. Terne plate, used in roofing and automobile gas tanks, contains up to 20 percent tin, with the balance lead.

Solders contain lead and tin in various proportions, depending on whether a "pasty" stage is wanted or the alloy should solidify sharply. The tin-lead alloy diagram (Fig. 4-5) shows that a 62 percent tin alloy freezes sharply; this (or compositions close to it) is the solder most used in the electronics industry. Lead is important in chromium plating (Chap. 12) since lead or high-lead alloys are almost universally used in this process.

Properties

Physical properties of lead are listed in Table 13-24. The softness and low strength of lead make it unattractive as a structural material; it flows even at room temperature. On the other hand, this property makes it easy to fabricate, and very light mills are sufficient to roll lead sheet from large castings. It may be extruded under very light pressure through small openings. In construction, lead must be supported by other stiffer materials; it can be stiffened to some extent by alloying additions. It is readily welded and extruded: the latter process is widely used for extruding lead sheathing around an insulated copper cable. Its low melting point makes it useful for low-temperature fusible alloys, as in sprinkler systems and fuses.

Lead has a relatively high boiling point, making it useful for gold and silver assaying and as a liquid heating bath in drawing or tempering steel.

Lead is the densest common metal, making it useful for weighting and ballast. It is relatively impermeable to most types of radiation, and so it is used in x-ray shields, nuclear reactors, and other applications where radiation protection is required.

Lead follows tin in Group IVB of the periodic system; like tin, its outer electron shell consists of two s and two p electrons, in this case $6s^2 6p^2$. The tendency for the s electrons to act as an "inert pair," not taking part in chemical

Table 13-24 Properties of Lead

Atomic number	82
Atomic weight	207.2
Electronic configuration	$1s^2 2s^2 2p^6 3s^2 3p^6 3d^{10} 4s^2 4p^6 4d^{10} 4f^{14} 5s^2 5p^6 5d^{10} 6s^2 6p^2$
Melting point, °C	327.43
Boiling point, °C	1740
Crystal structure	fcc
Electrical resistivity, $\mu\Omega$-cm, 20°C	20.65
Brinell hardness, cast	4.2
Tensile strength, MPa	14
Standard electrode potential, E^0, V	
$Pb^{++} + 2e^- \rightarrow Pb$	−0.12

combinations, is even more marked than with tin (see Tin), and in most of its compounds lead is divalent only. Exceptions are a few, mostly relatively unstable, tetravalent halides and some anionic complexes such as the plumbates, and its organic compounds such as lead tetraethyl. Most of its inorganic compounds find it in the oxidation state of 2+. It is amphoteric, forming both lead(II) salts with acids and hydroxo-complexes such as plumbites and plumbates.

Lead tarnishes readily in moist atmospheres, but not in perfectly dry air or in water which is free of air. When melted in air, it oxidizes; at dull red heat it is converted to lead oxide, PbO (litharge or massicot). If the heating is continued at 430°C, more oxygen is taken up, and the PbO converts to Pb_3O_4, which on further heating at 550°C dissociates again into PbO and oxygen. Other oxides known are a sesquioxide Pb_2O_3 and a dioxide PbO_2, which also decompose at high temperatures to PbO and oxygen; Pb_3O_4 is called minium* or red lead, and it behaves chemically like a mixture of PbO and PbO_2, but the crystal contains $Pb^{IV}O_6$ octahedra linked in chains. Lead dioxide, PbO_2, is often called lead peroxide, but there is no evidence that it is a true peroxide—i.e., that it contains peroxide ions as does, say, sodium peroxide.

Lead dissolves readily in nitric acid, more rapidly in dilute than in concentrated acid. Dilute hydrochloric and sulfuric acids have little effect, since coatings of the insoluble $PbCl_2$ and $PbSO_4$ protect it from further attack. Its resistance to sulfuric acid at all concentrations and temperatures up to about 200°C makes lead extremely useful in chemical reactors, in tank and chamber linings for the manufacture of sulfuric acid and other reactions where such resistance is desirable. Above 200°C the action of sulfuric acid becomes stronger, and at 260°C lead is dissolved.

Acetic, citric, and other similar organic acids attack lead slowly in the presence of air. Sulfur dioxide also attacks lead at high temperature, as does hydrofluoric acid; but the action of the latter is quickly checked by a coating of insoluble PbF_2, and the acid can be stored in lead vessels.

Many lead compounds are of industrial importance. Lead oxide, PbO (litharge or massicot), is extensively used in the glazing of ceramics and in making glassware. Minium or red lead, Pb_3O_4, is a bright red pigment used in corrosion-protective paints, especially as a first coat for steel.

Lead dioxide is a strong oxidizing agent, used in the manufacture of matches and in lead-acid storage batteries. It is formed and decomposed during the charge and discharge cycles.

Lead oxides dissolve readily in acids to form lead(II) salts. Plumbates, $Pb(OH)_6^{--}$, are formed by reaction with concentrated alkali hydroxides and by fusion with alkaline earth hydroxides. Concentrated hydrochloric acid forms $PbCl_2$ even with PbO_2; at very low temperatures $PbCl_4$ is formed but is unstable with respect to $PbCl_2$.

*Hence the word *miniature* for a type of painting, from coloring with red lead; basically, the word has nothing to do with the size of the painting.

Basic lead sulfate and lead carbonate (white lead) are used as white pigments; and lead chromate, $PbCrO_4$, is an important yellow pigment known as "chrome yellow." The use of lead pigments is slowly declining as it is forbidden by governmental regulations in many applications.

Soluble lead salts include the nitrate, $Pb(NO_3)_2$, and acetate, $Pb(C_2H_3O_2)_2{\cdot}3H_2O$. This is unusual in that it is only slightly ionized in aqueous solution; as a result, several slightly soluble lead salts dissolve readily in solutions containing the acetate ion.

Insoluble lead salts include the sulfide, sulfate, chromate, and carbonate. The halides are only slightly soluble; the order of solubility is $PbCl_2 > PbBr_2 > PbI_2 = PbF_2$.

The fluosilicate, fluoborate, perchlorate, and sulfamate are formed by the action of the corresponding acid on litharge or lead carbonate. All are soluble and important in electroplating, especially the fluoborate.

Lead has an extensive organic chemistry, and at least two organolead compounds are industrially important: tetraethyllead and tetramethyllead are widely used anti-knock compounds in gasoline. Whether governmental pollution-abatement regulations will force the abandonment of these additives is at present an unanswered question.

Hazards

Lead compounds are invariably toxic, essentially in proportion to their solubility. The use of lead compounds as insecticides has been practically eliminated, not because of their toxicity to plants but because by such use some lead residues may be transferred to human and animal food. Similarly, the use of lead-containing paints has been strictly regulated.

Inhalation of lead fume or dust, or dusts of its compounds, is dangerous; thus extreme care is required in "lead burning" or joining the slabs of lead used to line tanks and other containers. The use of lead compounds as anti-knock agents in automotive fuel is under study by federal agencies and may be forbidden. There has been some concern that emission of lead compounds from automotive exhaust may be dangerous to the population at large. These lead anti-knocks also "poison" some of the emission-control catalysts used on late-model cars.

Lead is a cumulative poison, which adds to its hazards. Lead metal itself, at room temperature, poses little hazard, but lead compounds should be handled with due regard for their dangers.

Ingestion of lead compounds has various effects, depending on the degree and time of exposure. Most common are gastrointestinal disturbances, loss of appetite, nausea, and vomiting. More serious cases result in general weakness, pains in the joints, dizziness, and headache. Advanced cases exhibit the so-called lead line on the gums. Infants who eat lead-containing paints sometimes become mentally retarded.

Lead Plating

Lead is dull in appearance and is seldom used for decorative applications. Its atmospheric corrosion resistance is good, and its corrosion products tend to be insoluble and to protect the metal from further attack, even though lead does not afford "sacrificial" corrosion protection as do zinc and cadmium. Deposits of lead, or 5 percent tin-lead alloys, on steel have been shown to have excellent corrosion-protective qualities in atmospheric exposure, especially when undercoated by thin deposits of copper. Terne plate (a 5 to 20 percent tin, 95 to 80 percent lead alloy) is widely used on steel for roofing, although most of this product is made by hot dipping rather than electroplating. Its principal use is for automobile gas tanks; outdoor signs also employ terne plate.

The principal applications of lead deposits include the following: protection of metals from some corrosive liquids like dilute sulfuric acid, where the lead sulfate produced prevents further attack; lining of brine refrigeration tanks; chemical apparatus; metal gas shells; barrel plating of nuts and bolts; parts for storage batteries; and plating of bearings. In many of these applications, especially the last, lead is frequently plated as an alloy or coated with other metals such as indium that may be diffused into the lead by heat treatment.

Because so many lead compounds are insoluble, the choice of electrolytes is somewhat limited. For example, sulfate and chloride solutions are unavailable, and anion-complex solutions comparable to the stannate tin solution are unstable because tetravalent lead reverts to the divalent state. Plumbite [Pb(II)] solutions have been investigated, without much success. The number of lead-plating electrolytes investigated is large, but only a few have survived in commercial use.

The soluble salts of lead which might be considered for plating purposes include the nitrate, fluosilicate, sulfamate, fluoborate, and perchlorate. Nitrate solutions are generally inapplicable because the anion is reduced at the cathode instead of the metal. Perchlorate solutions appear to yield satisfactory deposits, but because of the (probably unjustified) prejudice against the handling of perchlorates, they have had no commercial uses. The fluosilicate bath is used for lead refining (the Betts process) but for plating purposes has been generally superseded by the fluoborate electrolyte. Lead sulfamate solutions have enjoyed a certain amount of commercial use, but the potential hydrolysis of sulfamate to sulfate has been regarded as a disadvantage since lead sulfate is insoluble:

$$NH_2SO_3^- + H_2O \rightarrow NH_4^+ + SO_4^{--}$$

Like almost all acid baths of simple cations, lead-plating solutions require the presence of organic or colloidal addition agents to yield acceptably smooth and adherent deposits. Animal glue, gelatin, or peptone are those most generally specified, though others have been recommended. For present purposes, the only electrolyte requiring discussion is the fluoborate. Lead-tin alloys, of various compositions, are presently more widely used than pure lead deposits, are also plated from fluoborate electrolytes, and are mentioned later.

Lead Fluoborate

This compound, easily prepared by the reaction between lead oxide or carbonate and fluoboric acid, is available only as a concentrate, containing about 51 percent $Pb(BF_4)_2$.

Although many metals are plated from fluoborate solutions, on occasion (such as nickel and copper for electroforming and high-speed plating applications), the principal use for fluoborates in electroplating is in the deposition of lead and tin-lead alloys. Some mention of the chemistry of fluoborate solutions is therefore appropriate.

The production of fluoboric acid by the reaction between hydrofluoric and boric acids is generally formulated as

$$4HF + H_3BO_3 \rightarrow HBF_4 + 3H_2O \qquad (13\text{-}1)$$

This reaction is reversible, and to prevent the hydrolysis of fluoborates to form fluorides, excess boric acid is required. However, the reaction in reality is not so simple; several steps are involved, since hydroxo-substituted fluoboric acids are also known. They are related to HBF_4 by substitution of hydroxyl groups for fluorine: HBF_4; $HBF_3(OH)$; $HBF_2(OH)_2$; $[HBF(OH)_3]$; and finally $[HB(OH)_4]$. The last two compounds are not known, but the mono- and dihydroxy fluoboric acids are isolable, and it is probable that in the production of the acid the first reaction is

$$H_3BO_3 + 3HF \rightarrow H(BF_3OH) + 2H_2O \quad \text{(rapid)} \qquad (13\text{-}2)$$

followed by

$$H(BF_3OH) + HF \rightarrow HBF_4 + H_2O \quad \text{(slow)} \qquad (13\text{-}3)$$

Therefore in fluoborate solutions, the known hydroxyfluoborates may be present as well as the true fluoborate, which might better be termed *tetrafluoboric acid* to distinguish it from the others.

Fluoboric acid has not been obtained in the free condition, though aqueous solutions containing 80 to 85% concentrations can be prepared. The usual commercial concentrate is considerably weaker, containing 50% HBF_4 and some excess boric acid. Its hydrolysis to produce fluoride ion [the reverse of Eq. (13-1)] also probably takes place in steps, with stepwise substitution of hydroxyl groups for fluoride, going through the hydroxyfluoboric acids.

Aqueous solutions of fluoboric acid can also be prepared by the hydrolysis of boron trifluoride, but this is not a commercial method.

The fluoborates are similar in structure and in solubilities to the perchlorates, though they have been less extensively investigated. Perchlorate is known to be a very poor complex former—most metal perchlorate solutions are regarded as uncomplexed—and the same is thought to be true of metal fluoborates. Most

Table 13-25 Lead Fluoborate–Plating Baths

	*Bath A**		*Bath B**	
	g/L	*Molarity*	*g/L*	*Molarity*
Lead fluoborate,† $Pb(BF_4)_2$	220	0.58	440	1.16
(Lead, Pb)	120	0.58	240	1.16
Free fluoboric acid, HBF_4	30	0.34	60	0.68
Free boric acid, H_3BO_3	13.5	0.22	27	0.44
Animal glue	0.2		0.2	
or peptone	0.2–1.0		0.2–1.0	
Temperature, °C	25–40		25 – 40	
Cathode current density, A/m^2	50–500; avg. 200		50–700; avg. 300	
Anode current density, A/m^2	100–300		100–300	

*Bath A: deposits up to 25 μm, fairly low cd; bath B: barrel plating, heavy deposits (to 1.25 mm).

†100 percent basis.

metal fluoborates are very soluble, which renders them useful in plating; and they do not share the supposed explosive hazard associated with perchlorates.* Insoluble fluoborates parallel those of perchlorate ion: potassium perchlorate is insoluble, as is potassium fluoborate.

Freshly prepared fluoboric acid solutions do not attack glass, but do so on standing or heating, unless excess boric acid is present, since the active agent is the HF formed by hydrolysis.

In the lead fluoborate solution, the lead compound is generally written $Pb(BF_4)_2$; however, if a 1.0 *N* HBF_4 solution is saturated with lead carbonate, more lead is dissolved than would correspond to this formula, actually about 1.2 *N;* thus some soluble hydroxyfluoborates are no doubt also formed.

Formulations of lead fluoborate solutions are listed in Table 13-25. The bath is operated with excess free HBF_4 to increase its conductivity; it also yields finer-grained deposits with less tendency to treeing. Excess boric acid has little effect on the character of the deposit, but, as mentioned, is required to suppress hydrolysis of the fluoborate; any fluoride formed would precipitate the insoluble lead fluoride.

Commercial lead fluoborate concentrates usually contain sufficient free fluoboric acid and excess boric acid, so that upon dilution to plating strength the resulting solution will contain about the right amounts of these materials; if it does not, of course, they may be added.

The water used should be low in sulfate and chloride, since these lead salts are insoluble.

*In truth the only hazard in handling perchlorate involves very concentrated solutions in contact with organic matter; but a few disasters, caused by utter disregard of elementary precautions, have given it a bad name, at least among metal finishers.

Addition agents are necessary for both lead and lead-tin–alloy plating. There is little demand for bright lead deposits, so that most agents are used only for obtaining smooth, fine-grained, adherent plates. Most commonly used are peptone, glue or gelatin, and resorcinol.

Lead anodes may be either "corroding" or "chemical" lead. Impurities in the anode tend to form sludges or slimes, which enter the bath and increase the tendency to rough deposits.

Corrosion Resistance

Freedom from porosity rather than thickness is the main criterion for the protective value of lead deposits; such porosity tests as the ferroxyl test may be used. Hydroquinone as addition agent was claimed to yield less porous deposits. For severe service, deposits should be at least 75 μm thick, and up to 125 μm is necessary for some applications. Lead-plated nuts and bolts usually are plated to a thickness of 13 to 25 μm. The deposit on storage-battery parts varies from 8 to 75 μm.

A series of corrosion tests conducted by ASTM* gave the following results for the protective value of lead coatings on steel.

1. The type of atmosphere has a major effect. A deposit 25 μm thick protected steel for more than 10 years in New York City, but for only 4 to 7 years at Kure Beach, North Carolina, a seacoast atmosphere. Lead was better than zinc in New York but not at Kure Beach.

2. The corrosion rate of lead itself (as distinguished from its protective value) varied similarly.

3. There seems to be no relation between the corrosion of bare steel and the corrosion protection afforded to steel by lead coatings.

4. The relation between protective value and thickness of deposit is direct.

5. A copper flash is beneficial for thicker lead deposits but detrimental for thin ones.

The corrosion rate of lead and its corrosion products are obviously important factors. In some atmospheres corrosion products tend to fill up any pores and inhibit further corrosion; this may account for the protection afforded. Fine-grained deposits offer greater corrosion protection than coarse ones. Deposits containing about 5 percent tin seem to be superior both to pure lead and to deposits higher in tin.

*For details see appendix to ASTM B 200.

Lead-Tin–Alloy Plating

The fluoborates of both lead and tin are highly soluble, and presumably the metals are present in the form of simple (aquo), uncomplexed ions. Their standard electrode potentials can then be used directly as a guide to the possibility of codeposition. E^0 for $Sn^{++} \rightarrow Sn$ is -0.136 V, and for $Pb^{++} \rightarrow Pb$ it is -0.126 V. Tin is therefore the less noble metal, but only by 10 mV, and the metals are easily codeposited in all proportions from fluoborate baths.

The alloys have been used extensively for bearings and for printed circuits; it is the latter use which has prompted the more recent interest in tin-lead plating. For bearings, the alloy usually contains 6 to 20 percent tin; a 5 to 15 percent tin alloy is plated on strip steel for lubrication, paint adhesion, and solderability. The possibility of electrodeposition of terne plate on continuous steel strip has been investigated.

Lead or lead–7% tin has been plated inside the bores of 105-mm gun tubes, to act as a lubricant during swaging. Round copper bars have been plated with heavy (3.2-mm) deposits of lead and lead-tin alloys for use as anodes in chromium plating; the copper core increases the current-carrying capacity as well as the stiffness of the anodes.

Lead-tin alloys containing small amounts of copper are sometimes used for bearings. The solutions are the same as for lead-tin; copper fluoborate is added slowly according to an ampere-hour schedule or another, similar, one.

For printed circuits, alloys containing 50 to 70 percent tin are preferred; thicknesses range from 7.5 to 18 μm for optimum solderability; but see also ASTM B 579, discussed later.

Preparation of the tin-lead–alloy solutions is essentially the same as that of lead fluoborate solutions. The tin fluoborate concentrate can be added to the lead solution, or the latter can be operated with tin anodes to introduce the necessary tin content. Typical formulations are listed in Table 13-26.

Tin-lead–plating baths so formulated have relatively poor throwing power, and are not suitable for through-hole plating in multilayer printed-circuit boards. A "high-throw" bath* claiming to overcome these difficulties is shown in Table 13-27.

For many uses the tin/lead ratio in the alloy is somewhat critical; the deposit, of course, can be analyzed by usual chemical methods, but a simpler determination is to set up a lead-plating cell in series with the plating tank and weigh the lead and the alloy deposits after a convenient period; then

$$\text{Tin (\%)} = 134.1 \left(\frac{\text{wt of lead deposit}}{\text{wt of alloy deposit}} - 1 \right)$$

Lead is determined by difference.

*U.S. patent 3 564 878.

Table 13-26 Tin-Lead–Plating Baths

	7 percent tin (bearings)		*60 percent tin (solder)*	
	g/L	*Molarity*	*g/L*	*Molarity*
Stannous fluoborate,* $Sn(BF_4)_2$	14.5	0.05	129	0.44
(Tin)	6.0	0.05	52	0.44
Lead fluoborate,* $Pb(BF_4)_2$	160	0.42	53	0.14
(Lead)	88	0.42	30	0.14
Free fluoboric acid, HBF_4	100–200	1.14–2.28	100–200	1.14–2.28
Peptone	0.5		5.0	
Cathode current density, A/m^2	300		300	
Temperature, °C	15–38		15–38	
Agitation	mild (cathode rod)†			
Anode/cathode ratio	2/1			

*100 percent basis.

†Air agitation cannot be used since it would oxidize Sn^{++}.

The principal factor influencing the tin/lead ratio in the deposit is the ratio in the bath. The nature of the addition agent and its concentration have some effect, however, and since tin is slightly less noble, its percentage in the alloy tends to increase with increases in current density, except for solutions containing peptone, which alters the cathode polarization.

Tin content is increased also by an increase in addition-agent concentration, and of course by an increase in tin content of the bath.

Anodes should be of the same composition as that of the alloy being deposited. Anode and cathode efficiencies both approach 100 percent.

Proprietary processes for "bright solder" plating have been promoted.

Specifications

ASTM B 200 concerns lead coatings on steel and offers detailed results on exposure tests conducted in industrial, marine, and rural atmospheres, as summarized on p. 328. Requirements for meeting salt spray resistance (where this is considered meaningful) are as follows:

Hours in salt spray	*Pb coating thickness, μm*
200	38
96	25
48	12
24	6

Table 13-27 "High-throw" Solder-Plating Bath*

	Optimum		*Range*	
	g/L	*Molarity*	*g/L*	*Molarity*
Stannous fluoborate,† $Sn(BF_4)_2$	38	0.13	29–50	0.10–0.17
(Tin)	15	0.13	12–20	0.10–0.17
Lead fluoborate,† $Pb(BF_4)_2$	19	0.05	15–27	0.04–0.07
(Lead)	10	0.05	8–14	0.04–0.07
Free fluoboric acid, HBF_4	400	4.6	300–500	4.0–5.7
Peptone	5		2.7	
Temperature, °C	24		16–38	
Current density, A/m²	150		100–250	
Agitation	mild (cathode rod)			
Filtration	optional			
Anodes	60/40 Sn/Pb, bagged with Dynel or polypropylene			

*U.S. Patent 3 554 878.

†100 percent basis.

ASTM B 579 refers to tin-lead–alloy coatings (solder plate) containing a minimum of 50 percent and a maximum of 70 percent tin. Service conditions are described as follows.

> SC4, very severe conditions, require a completely pore-free coating. If the coating is subject to abrasion or is exposed to corrosive liquids or gases, a deposit 30 to 125 μm thick may be required.

Table 13-28 ASTM Specifications for Solder Plate; B 579

	Minimum thickness, μm	
Service condition	*On steel*	*On copper, copper alloys,* and nonmetallics†*
SC4	30‡	30
SC3	20‡	15
SC2	10	8
SC1	5§	5§

*On brass containing 15 percent or more zinc, at least 2.5 μm of copper or nickel is required; the same holds for beryllium copper.

†Suitably sensitized and metalized.

‡Undercoat of 2.5-μm copper is recommended.

§May be flow-brightened.

SC3, severe conditions, include exposure to dampness and industrial atmospheres. This service condition includes preservation of solderability on fairly long storage (up to 9 months) and also includes the use of the coating as an etch resist in printed-circuit boards.

SC2, moderate conditions, include dry or interior atmospheres; this includes retention of solderability for shorter periods than SC3.

SC1, mild service, includes less severe requirements than SC2 such as preserving solderability for short periods, up to 3 months.

Requirements of this specification are listed in Table 13-28.

14 Minor Metals

The three metals considered in this chapter have little in common except that, although important in various contexts, they are not widely used as electroplated finishes. Surely that is the only reason for including iron, for of course iron and steel, far from being "minor" metals, are the indispensable structural metals of modern civilization, and the steel industry is regarded as a basic indicator of the health of the economy. Nevertheless, there is little occasion to electrodeposit iron; rather it is the substrate upon which electrodeposits are applied.

The second metal to be considered, cobalt, possesses some properties that make it useful in special applications, but on the whole and for most electroplating purposes it has few advantages over nickel, which is considerably cheaper and usually more available.

The third metal of this group, indium, is truly rare and as far as price is concerned should be reckoned among the precious metals. The use of indium plating is reported to be increasing to some extent, but for the average metal finisher it remains distinctly in the category of a minor metal.

All three metals, however, can be electrodeposited from aqueous solutions that have proceeded beyond the experimental stage and are ready for application when required. In this they differ from most of the metals in the following chapter, which are not only rarely electroplated, but are still largely experimental. In most cases the finisher cannot simply adopt a developed technique but must probably do a bit of experimentation before being able to produce satisfactory electrodeposits.

IRON

Iron (Fe, atomic number 26) has been known from antiquity and in the form of its alloys, the steels, is the most important industrial metal. The intentional smelting of iron ores probably dates back only to about 1300 B.C.; the prototype of the modern blast furnace was developed in the fourteenth century A.D.

The symbol Fe is from Latin *ferrum*.

Iron is the fourth most abundant element in the lithosphere, constituting about 5.1 percent of the earth's crust and most of its core; it ranks behind only oxygen, silicon, and aluminum in abundance. It is found, in at least trace amounts, in almost every material in the earth's crust; it is essential to human, animal, and plant life.

By far the most important use of iron is in the making of steel, an alloy of iron with small and controlled amounts of carbon, plus several other elements depending on final use. Steels higher in various alloying elements are also of great importance.

Really pure iron—of so-called "four nines" purity—is a laboratory curiosity; 99.997 percent pure iron has been prepared, but only as a research project. The properties of iron, as reported in this chapter and in various handbooks, depend to a large extent on its purity, and thus many of the figures are at best approximations, since many determinations have been made on relatively impure samples.

Native iron is not unknown; it occurs in meteorites and in two very rare minerals of no commercial importance. The list of iron-containing minerals is extremely long, but only a few are of industrial importance: magnetite, Fe_3O_4; hematite, Fe_2O_3; limonite or goethite, $Fe_2O_3 \cdot xH_2O$; siderite, $FeCO_3$; and pyrite, FeS_2. The last is used only if the sulfur content is also to be recovered as sulfuric acid.

Iron ores occur all over the world; availability of iron ore is a fundamental factor in the development and maintenance of modern industrial societies. The impurities in the ores may determine their usefulness in the process of making iron and steel; these are principally phosphorus, sulfur, manganese, silica, and alumina.

Production

The most important process for producing iron is the blast furnace, which is essentially a large chemical reactor into which are charged iron ore, coke, and limestone. The main product of the blast furnace is pig iron, most of which is processed further to make steel. Some is used to make the various grades of cast iron. This is not the place to discuss the details of the steel-making process, which of course is a basic, large-scale industry. The reaction

$$2Fe_2O_3 + 3C \rightarrow 4Fe + 3CO_2$$

expresses merely the end result of a series of complicated reactions taking place in the furnace. Texts devoted to the subject should be consulted for details.

There are countless grades and specifications for iron and steel. One ASTM specification alone (A 43) tabulates 259 different grades of pig iron as to their contents of minor elements such as silicon, manganese, phosphorus, and sulfur, with maximum limits of impurities for 16 other elements. Specifications for other grades of iron, steel, and ferrous alloys in general are too numerous to count.

Steel is an alloy containing less than 2 percent carbon; it may contain minor or major amounts of many other elements, including, but not confined to, nickel, chromium, molybdenum, tungsten, cobalt, and titanium.

Cast irons are alloys of iron with carbon and silicon, often with other elements present in lesser amounts. They are differentiated from steel by their higher carbon content, which modifies their properties largely. White cast iron may contain up to 6 percent carbon; it is hard, but not usefully malleable. Slow cooling leads to partial graphitization of the carbon; this improves malleability and ductility, resulting in malleable iron, ductile iron, and gray cast iron.

Wrought iron is produced by squeezing out the slag from a pasty mass. Some slag remains in the product, which is, however, considerably purer than cast iron. It is no longer manufactured in the United States.

Commercially pure iron (''Armco ingot iron'') is the purest form of iron sold in large quantities; it is produced by refining mild steel. Electrolytic iron powder is produced by electrolysis of ferrous sulfate solutions, under conditions such that the deposit is spongy and nonadherent—just the opposite of the deposit required by the electroplater. Iron powder can also be made by hydrogen reduction of iron oxide, a grade known as *ferrum reductum.* A highly pure grade is carbonyl iron, made by the decomposition of iron pentacarbonyl; the product is almost perfectly spherical, which lends it special magnetic properties.

Properties

As mentioned above, really pure iron is a research curiosity; the properties listed in Table 14-1 are those of ''relatively pure'' iron. Minor amounts of alloying constituents, prior mechanical and thermal history, and other factors have a major effect on many of the listed numbers. As is well known, steels can be fabricated with almost any combination of properties desired by the user. But the property of iron and steel of most interest to the metal finisher is its comparative lack of resistance to most corrosive agents; this property, in effect, makes the metal-finishing industry both necessary and possible.

Table 14-1 Properties of Iron

Atomic number	26
Atomic weight	55.847
Electronic configuration	$1s^2 2s^2 2p^6 3s^2 3p^6 3d^6 4s^2$
Melting point, °C	1532 ± 5
Boiling point, °C	3000 ± 150
Density, g/cm³	7.87
Crystal structure	bcc
Electrical resistivity, μΩ-cm, 20°C	9.71
Tensile strength, MPa	240–275
Brinell hardness	82–100

Iron is the first member of the first triad of Group VIII in the periodic table, consisting of iron, cobalt, and nickel. Like all members of this group, it exhibits the multiple valences, colored ions, and relative ease of complex formation typical of transition elements. Iron in its compounds has five oxidation states: 0, 2+, 3+, 4+, and 6+, of which only 2+ and 3+ are practically important.

In the 2+, iron(II), or ferrous state, ferrous hydroxide is slightly amphoteric, dissolving easily in acids but also in concentrated sodium hydroxide. In uncomplexed form, Fe(II) is somewhat more stable than Fe(III); complexation tends to stabilize the 3+ state. Ferrous salts are, however, easily oxidized to ferric.

In the 3+, iron(III), or ferric state, ferric hydroxide is also amphoteric; as an acid, it forms, with some difficulty, the ferrites $MFeO_2$. In complex compounds, the 3+ state is the more stable, and Fe(III) forms many complexes with inorganic and organic ligands. The complex chemistry of ferric iron is almost as extensive as that of chromium.

Valences of 4+ and 6+ are rare: the ferrates(IV) such as Ba_2FeO_4 and a similar strontium compound are almost the only representatives. Under very strongly oxidizing conditions, iron can be raised to the 6+ state: ferrates(VI), M_2FeO_4, where M is a univalent metal, are isomorphous with sulfates; they are very strong oxidizing agents but of little practical significance.

Because of the great importance of iron and steel and their tendency to corrode under most conditions, their electrochemical properties have been studied extensively. Some of the principal electrode potentials of iron and its compounds are listed in Table 14-2.

From this table and other knowledge, it is seen that oxidation of iron is easier in the presence of groups that precipitate or complex the oxidation product. Ferric iron is fairly easily reduced to ferrous, but under alkaline conditions ferrous hydroxide is readily oxidized to ferric. Since ferric hydroxide (or, more accurately, hydrated ferric oxide) is much more insoluble than ferrous, acidic pH is required in order to maintain a solution in the reduced form and to prevent the precipitation of $Fe(OH)_3$.

Table 14-2 Electrochemical Properties of Iron

Half-reaction	*Standard potential, E⁰, 25°C*
$Fe^{++} + 2e^- \rightarrow Fe$	−0.440
$Fe(OH)_2 + 2e^- \rightarrow Fe + 2OH^-$	−0.877
$Fe(CN)_6^{4-} + 2e^- \rightarrow Fe + 6CN^-$	−1.5
$Fe^{3+} + e^- \rightarrow Fe^{++}$	+0.771
$Fe(OH)_3 + e^- \rightarrow Fe(OH)_2 + OH^-$	−0.56
$Fe(CN)_6^{3-} + e^- \rightarrow Fe(CN)_6^{4-}$	+0.36

The cyanide complexes of both the ferrous and ferric states, ferro- and ferricyanide [hexacyanoferrate(II) and hexacyanoferrate(III) in strict nomenclature] are both extremely stable and have no applications in iron plating. They exhibit the reactions of neither iron nor cyanide ion.

The reactions of iron with air or oxygen are complicated; presence or absence of moisture has a major effect. Several oxides are possible, among them the minerals already mentioned and ordinary rust and scale; many are nonstoichiometric, and depending on conditions the rust formed may be protective—i.e., prevent further corrosion—or not.

Iron dissolves readily in dilute mineral acids; if the acids are nonoxidizing and air is absent, Fe^{++} is formed; with air present, or with warm dilute nitric acid, some Fe^{3+} is formed. Very strong oxidizing media such as nitric and chromic acids passivate iron.

Ferrous and ferric ions in solution are easily interconverted by redox reactions. Ferrous ions are green to colorless; freshly precipitated ferrous hydroxide is white, but in the presence of air it turns green, then black, and finally forms the reddish-brown ferric hydroxide. Ammonia precipitates iron only partially, because ammine complexes are formed.

Ferric ion in solution is usually reddish brown, perhaps because of complex formation; ferric ion is said to be colorless. Ferric ion forms complexes with cyanide; the mixed-valence compounds ferrous ferricyanide and ferric ferrocyanide are deep blue, and are known as Prussian Blue and Turnbull's Blue as well as by other names, depending on their mode of formation and their composition. (The formulations $Fe_3^{II}[Fe^{III}(CN)_6]_2$ and $Fe_4^{III}[Fe^{II}(CN)_6]_3$ are oversimplifications of what is a very complicated series of compounds.)

Iron is cheap and abundant, and most of its inorganic compounds are so also. In fact, ferrous sulfate is a waste product (known as copperas) of the pickling of iron and steel with sulfuric acid, and its disposal is a major problem to the steel industry; this is one reason for the trend to pickling with hydrochloric acid, a process which can be made "closed loop." Ferrous sulfate used in commerce is $FeSO_4 \cdot 7H_2O$.

Other iron compounds of interest in metal finishing include ferrous chloride, $FeCl_2 \cdot 4H_2O$ and the ferrous salts of fluoboric and sulfamic acids.

Ferric chloride, $FeCl_3 \cdot 6H_2O$ (also obtainable anhydrous) is not used in plating, but is applied in etching of printed circuits and similar fields, because of its oxidizing properties.

Hazards

Iron compounds are essentially free of hazard, unless the associated radicals are hazardous. Since most simple iron compounds hydrolyze and must be stabilized by excess acid, the acidic nature of the solutions involves the dangers common to all acids.

Iron Plating

Iron is the cheapest metal available, and the physical properties of electrodeposited iron are of potential interest in many applications. Furthermore, it is easily electroplated from several types of electrolytes. Despite these advantages, iron is not widely used as an electrodeposit. It has no decorative applications, and even though electrodeposited iron is somewhat more corrosion-resistant than ordinary iron, probably because of its higher purity, it remains subject to most of the corrosion problems that beset iron and steel and that require that they be protected. Uses of iron plating therefore depend entirely on applications in which its physical properties are paramount.

Iron plating continues to be of interest, as evidenced by a trickle of papers in the literature; but the interest appears to be sporadic, and few applications have found a permanent place. Some of the factors which may militate against the wider use of iron plating include these. The properties of the deposit are extremely sensitive to the purity of the electrolyte and the operating conditions, especially the pH and concentration of ferric ion; the processes appear to work really well only if in continuous production: intermittent operation creates problems in restoring the electrolyte to proper operating condition. Many baths are highly corrosive to ordinary equipment. Pitting and roughness of the deposit are major problems.

Uses of iron plating that have survived include some electroforming applications, such as production of phonograph record stampers and iron molds for rubber, glass, and plastics; iron sheet with special magnetic properties; and restoration of worn parts. Since solder wets but does not alloy with iron (at least at soldering temperatures), copper soldering tips are often iron-plated. Iron electrotypes have long been used by the U.S. Bureau of Printing and Engraving for printing currency and bonds.

There is considerable interest in nickel-iron alloys containing up to 40 percent iron, deposited from proprietary solutions, as a means of replacing relatively expensive nickel with the much cheaper iron without too much sacrifice of the properties of bright nickel; see Nickel Plating, p. 221.

Iron is plated from baths containing the ferrous or iron(II) ion. Presence of ferric or iron(III) ion in more than trace concentrations may lower cathode efficiency (since current is used in the reaction $Fe^{3+} + e^- \rightarrow Fe^{++}$) and may also lead to brittle, stressed, and pitted deposits. Precipitation of Fe(III) hydrate in the cathode deposit is probably the cause of some of these difficulties.

The most commonly used iron-plating baths are the ferrous sulfate and the ferrous chloride baths, or mixtures of the two. Iron fluoborate and sulfamate solutions also have been used. Conducting salts are normally added.

The ferrous sulfate bath (Table 14-3) produces smooth but brittle deposits; it can be operated at room temperature, which is an advantage. The sulfate normally is added as the double salt ferrous ammonium sulfate; but ferrous

Table 14-3 Iron-plating Baths

	Sulfate		Double sulfate		Chloride		Sulfate-Chloride		Fluoborate		Sulfamate	
	g/L	*Molarity*	*g/L*	*Molarity*	*g/L*	*Molarity*	*g/L*	*Molarity*	*g/L*	*Molarity*	*g/L*	*Molarity*
Iron, Fe^{++}	48	0.86	36–51	0.64–0.92	84–125	1.5–2.25	60	1.08	55	1.0	75§	1.35
Ferrous sulfate, $FeSO_4 \cdot 7H_2O$	240	0.86	—	—	—	—	250	0.90	—	—	—	—
Ferrous ammonium sulfate, $FeSO_4(NH_4)_2SO_4 \cdot 6H_2O$	—	—	250–360†	0.64–0.92	—	—	—	—	—	—	—	—
Ferrous chloride, $FeCl_2 \cdot 4H_2O$	—	—	—	—	300–450	1.5–2.25	36	0.18	—	—	—	—
Ferrous fluoborate, $Fe(BF_4)_2$	—	—	—	—	—	—	—	—	227‡	1.0	—	—
Calcium chloride, $CaCl_2$	—	—	—	—	300	2.7	—	—	—	—	—	—
Sodium chloride, NaCl	—	—	—	—	—	—	—	—	10	0.17	—	—
Ammonium chloride, NH_4Cl	—	—	—	—	—	—	20	0.37	—	—	—	—
Ammonium sulfamate, $NH_4SO_3NH_2$	—	—	—	—	—	—	—	—	—	—	30–38	0.25–0.32
pH*	2.8–3.5		low 2.8–3.5 high 4.0–5.5		1.2–1.8		3.5–5.5		3.0–3.4		2.7–3.0	
Temperature, °C	32–65		24–65		85 min.		27–70		57–63		50–60	
Current density, A/m^2	400–1000		200–1000		200–800		200–1000		200–900		500	

*Adjust with appropriate acid.

†Or equivalent $FeSO_4 \cdot 7H_2O + (NH_4)_2SO_4$.

‡As 41 percent concentrate containing about 10 percent Fe, 1 percent HBF_4, and 3 percent H_3BO_3.

§As sulfamate concentrate.

sulfate may be used alone or with additions of the sulfates of sodium, magnesium, or aluminum.

Two pH ranges are in use; pH regulation is important because iron(III) hydroxide precipitates at a pH of about 3.5, while iron(II) hydroxide does not precipitate until a pH of about 6 is reached. In the low-pH range, even a well-reduced bath contains some Fe^{3+}, and operation at a pH too close to 3.5 results in dark, stressed deposits, caused by inclusion of basic Fe(III) salts in the deposit; however, if the pH is too low, cathode efficiency suffers.

In the high-pH range, Fe(III) hydroxide is always present as a sludge, but is not included in the deposits unless they are thick. The high-pH sulfate bath has improved covering power and yields deposits that are less stressed than those from the low-pH bath. Optimum pH is 4 to 5.

The chloride bath has the advantage of producing ductile deposits when operated at temperatures of 85°C or above. The bath, often referred to as the *Fischer-Langbein solution,* is usually formulated with ferrous and calcium chlorides; various modifications have been made to this basic formulation over the years (see Table 14-3). This bath must be operated at high temperature for acceptable results; deposits at 25°C are dark, hard, and highly stressed. Many additives have been proposed to modify the properties of the deposit: manganese(II), aluminum, beryllium, and chromium(II) chlorides have been reported to produce more ductile and softer deposits.

Fluoborate and sulfamate baths have been proposed. The advantages of iron fluoborate baths are those they share with other fluoborate solutions: higher permissible current densities.

Continuous filtration is desirable, but care must be exercised to prevent pumps from aspirating air, which aggravates the problem of oxidation. Filtration of ferric hydroxide, which is gelatinous, is difficult and usually requires the use of filter aids. Alternatively, tanks can be made deep enough so that sludge of Fe(III) hydroxide can be allowed to settle to the bottom without disturbing the bulk of the electrolyte.

Anodes of high-purity (''Armco'') iron are preferred, but steel and cast iron have been used. Since no commercial iron is pure, anode bags should be used to retain the resulting slimes and sludges. Few materials will resist the extremely corrosive conditions of the chloride bath; glass fiber is usable, as are Orlon and Dynel if the temperature is not too high.

Equipment for the chloride and fluoborate baths must be of corrosion-resistant material; the high acidity and elevated temperatures of operation render these solutions extremely corrosive.

Iron is deposited only for its physical properties, which are therefore of major interest. Tensile strength and hardness of deposits tend to increase with a decrease in operating temperature and an increase in current density. Lower temperature also results in more brittle and stressed deposits. Brittleness can be reduced by heat treatment at 150–200°C for 10 to 15 min. Heat treatment at 480–540°C anneals the iron deposit to a soft condition. Electrolytic iron is fairly pure, and so it cannot be hardened by heat treatment. Tensile strength of deposits varies between 310 and 480 MPa as deposited, hardness from 150 to 400 Vickers, and elongation from 5 to 25 percent. These figures are only rough guides.*

COBALT

Cobalt (Co, atomic number 27) is the middle member of the iron-cobalt-nickel triad of Group VIII. It was discovered in 1735. The name is from German *Kobalt,* a mischievous spirit (cf. nickel) because its arsenical ores injured the hands and feet of the miners (Greek *cobalos,* mine).

Although widely diffused, cobalt is only about one-third as abundant as nickel, and few ores are mined for their cobalt content; most of the metal is recovered as a by-product of the recovery of nickel and copper from their ores. Most of its minerals are sulfides, arsenides, or oxidized forms of these.

By far the chief producing country is Zaire; other countries with fairly large production include Canada, Norway, the Soviet Union, and Zambia. These countries produce about 45 percent of the world's production of approximately 50,000 t annually; the rest is scattered among many smaller producers.

Recovery of cobalt depends on the type of ore being processed. Final recovery is either electrolytic or by carbon reduction in an electric furnace, after appropriate chemical separations from other metallic constituents.

Uses

Metallurgical uses of cobalt account for about 75 percent of its consumption; the rest is accounted for by chemical applications. The pure metal is little used; the principal use at present is as the artificially radioactive isotope ^{60}Co, used in radiation therapy and other radiative applications as a source of gamma rays.

Cobalt-base alloys are among the so-called superalloys used in gas turbines and other applications requiring resistance to extremely elevated temperatures involving also abrasion and other severe conditions. Superalloys based on cobalt contain major quantities of chromium, tungsten, and nickel; tantalum and niobium are also present in some formulas.

Hard-facing and other wear-resistant alloys also are based largely on cobalt, with major additions of chromium, nickel, and tungsten.

A wide range of commercial magnetic materials makes use of the magnetic properties of cobalt; these alloys usually have included iron and nickel also, but new alloys of cobalt with the rare earth metals have even more striking magnetic properties.

*For extended data see Safranek, cited in the Bibliography.

Table 14-4 Properties of Cobalt

Property	Value
Atomic number	27
Atomic weight	58.9332
Electronic configuration	$1s^2 2s^2 2p^6 3s^2 3p^6 3d^7 4s^2$
Melting point, °C	1495
Boiling point, °C	2802
Density, g/cm³	8.85
Electrical resistivity, $\mu\Omega$–cm, 20°C	5.8
Curie point, °C	1121
Crystal structure, α	hcp
Crystal structure, β	fcc
Transition temperature, $\alpha \rightleftharpoons \beta$, °C	417
Hardness, Vickers, 20°C	225
Tensile strength, MPa	≈240
Standard electrode potential, E^0, V	
$Co^{++} + 2e^- \rightarrow Co$	−0.277

In its other major applications, cobalt is used in the form of its compounds. These are used as ground-coats in porcelain enamels, as pigments, and as catalysts. Cobalt salts of organic acids such as the resinate, oleate, and their analogs are used extensively as drying agents for paints, inks, and varnishes, as dressing for fabrics, and as catalysts.

Cobalt is one of the micronutrients needed by most plants; as a rule it is added to some fertilizers as the sulfate.

Properties

Some physical properties of cobalt are listed in Table 14-4. The naturally occurring element is almost mononuclidic, ^{59}Co, but the artificially radioactive ^{60}Co is widely used in radiation therapy, as has been mentioned. Cobalt is a hard magnetic metal, resembling iron and nickel in appearance. It is, along with iron and nickel, one of the three ferromagnetic elements; alloys with the other two members of the group have exceptional magnetic properties.

The properties of cobalt are sensitive to its structure, since it exists in at least two allotropic modifications: α-Co is stable below 417°C and β-Co is stable above that temperature,* but the transition is very sluggish, and many samples consist of mixtures of the two allotropes. Therefore there is considerable variation in its reported physical properties, depending on the nature of the particular sample studied.

*In some publications this nomenclature is reversed.

Chemically, cobalt is intermediate in properties between iron and nickel. In general it is more corrosion-resistant than iron but less so than nickel. The finely divided metal is pyrophoric, but massive metal is not attacked by air or water below 300°C. It combines with the halogens to form the corresponding halides and with most other nonmetals to form binary compounds. It does not combine with nitrogen, but decomposes ammonia at high temperature to form a nitride.

Cobalt dissolves readily in dilute mineral acids, but only slowly in hydrofluoric acid, to form cobaltous, or cobalt(II), salts. Strong oxidizing agents such as the dichromates passivate it. Ammonium and sodium hydroxides and dilute acetic acid attack it slowly.

Cobalt is not particularly resistant to oxidation. Between 300 and 900°C the scale consists of two layers: a thin layer of Co_3O_4 on the outside and a layer of CoO next to the metal. Above 900°C the Co_3O_4 decomposes to CoO.

Iron, as we have seen, can have valences from 0 to 6+, even though valences higher than 3+ are rare; with cobalt, the valence states higher than 3+ are rarer still. (The progression of properties in the group proceeds to nickel, where even the 3+ state is relatively rare.) Cobalt(II) is the more stable state in simple salts; in fact, the Co^{3+} ion cannot exist as such in water solution since it evolves oxygen. Complexation stabilizes the 3+ state, and cobalt(III) has an extensive complex chemistry. In fact, the trivalent cobaltic or cobalt(III) ion is one of the most prolific complex formers known. The important donor atoms are, in order of decreasing donor tendency, nitrogen, carbon in the cyanides, oxygen, sulfur, and the halogens. The most numerous complexes are the ammines.

The trivalent complexes of cobalt are of little interest in electroplating; most cobalt-plating solutions are fairly strictly analogous to nickel-plating solutions, and the cobalt salts of interest to platers are the sulfate, $CoSO_4{\cdot}7H_2O$, and the chloride, $CoCl_2{\cdot}6H_2O$. The hydrated ion is light pink, whereas anhydrous salts are mostly blue; this color change has been used in many humidity and moisture-indicating devices.

Hazards

So far as is known, cobalt compounds are not toxic.

Cobalt Plating

There are few uses for electrodeposits of cobalt; it is more expensive than nickel (usually) and less resistant to most types of corrosion. Since it closely resembles nickel in many properties, it can be used as a complete or partial substitute; and during a period of severe shortage of nickel during 1969–1970, there was great interest in the substitution. Many of the brighteners used in nickel plating are adaptable to cobalt or cobalt-nickel–alloy plating, and several companies undertook a "slide conversion" by placing cobalt anodes in their nickel baths, thereby depositing cobalt-nickel alloys of various compositions, apparently with satisfaction. However, as soon as the nickel shortage ended, interest in cobalt plating among general metal finishers ebbed.

One of the earliest successful bright nickel deposits was actually a 15 to 18 percent Co alloy*; the cobalt content was later reduced to 5 percent or less. Although this deposit had excellent properties in many respects and was not to be equaled in outdoor corrosion resistance until the advent of duplex nickel, it has been superseded almost entirely by the organic bright nickels.

Cobalt and cobalt-alloy deposits are of interest in certain specialized applications, for their magnetic properties. These are valuable in electronic memory devices used in computers and similar apparatus.

Cobalt-tungsten and cobalt-molybdenum alloys are of potential interest because of their hardness and wear resistance, especially at high temperatures. Neither tungsten nor molybdenum can be deposited from aqueous solutions in the pure state, but their cobalt alloys can be electrodeposited fairly easily. These observations have not, however, resulted in any major applications so far.

Some cobalt-tin alloys have been promoted as replacements for decorative chromium. (See Chap. 16.)

Cobalt plating is analogous to nickel plating, and the formulations of cobalt-plating solutions are similar, as shown in Table 14-5. Some cobalt alloy–plating baths of interest for the magnetic or wear-resistant properties of the deposits they yield are listed in Table 14-6.

INDIUM

Indium (In, atomic number 49) is a member of the main Group III (or IIIA) of the periodic table, below boron, aluminum, and gallium and above thallium. It is widely scattered in many minerals, but there are no commercial indium ores. Its abundance in the lithosphere is estimated at about 0.1 ppm.

Indium was discovered as a new element in 1863, by means of a strong blue line in the spectrum of a zinc ore; the name *indium* derives from this line in the *indigo* part of the spectrum.

Commercial sources of indium are lead and zinc ores; it is recovered electrolytically from the flue dusts and residues from the smelting of these ores. Zinc concentrates are roasted and leached with a dilute solution of sulfuric acid; treatment with zinc dust or aluminum precipitates indium along with silver, copper, and other metals. The precipitate is treated with sulfuric acid; silver and copper are removed by treatment with hydrogen sulfide, and indium is recovered

*The Weisberg-Stoddard process.

Table 14-5 Cobalt-plating Solutions

	Chloride [a]		*Sulfamate*		*"Double" sulfate*		*Sulfate*	
	g/L	*Molarity*	*g/L*	*Molarity*	*g/L*	*Molarity*	*g/L*	*Molarity*
Cobalt as metal, Co	22–26	0.38–0.44	106	1.8	26–29	0.44–0.5	69–118 [e]	1.17–2.0 [e]
Cobalt chloride, $CoCl_2 \cdot 6H_2O$	90–105	0.38–0.44	—	—	—	—	45 [d]	0.19 [d]
Cobalt sulfamate, $Co(SO_3NH_2)_2$	—	—	450	1.8	—	—	—	—
Cobalt ammonium sulfate, $Co(NH_4)_2(SO_4)_2 \cdot 6H_2O$	—	—	—	—	175–200	0.44–0.5	—	—
Cobalt sulfate, $CoSO_4 \cdot 7H_2O$	—	—	—	—	—	—	330–565	1.17–2.0
Boric acid, H_3BO_3	60	1.0	—	—	25–30	0.4–0.5	30–45	0.5–0.75
Formamide, $HCONH_2$	—	—	30 [c]	0.9	—	—	—	—
Sodium chloride, NaCl [f]	—	—	—	—	—	—	17–25 [d]	≈0.3
Wetting agent	—	—	0.3–0.5	—	—	—	—	—
pH [b]	2.5–3.5		—		5.0–5.2		3.0–5.0	
Temperature, °C	50–55		20–50		25		35–38	
Current density, A/m^2	300–400		160–480		100–300		215–500	

[a] Normally plus stress reducer and brighteners.
[b] With appropriate acid.
[c] mL/L.
[d] Optional.
[e] Omitting $CoCl_2$.
[f] Or KCl.

Table 14-6 Cobalt Alloy–Plating Solutions

	*50 percent nickel**		*40 percent molybdenum*		*5 percent phosphorus*		*30 percent tungsten*	
	g/L	*Molarity*	*g/L*	*Molarity*	*g/L*	*Molarity*	*g/L*	*Molarity*
Cobalt sulfate, $CoSO_4 \cdot 7H_2O$	29	0.1	85	0.3	—	—	—	—
Cobalt chloride, $CoCl_2 \cdot 6H_2O$	—	—	—	—	180	0.75	100	0.42
Nickel sulfate, $NiSO_4 \cdot 6H_2O$	280	1.1	—	—	—	—	—	—
Nickel chloride, $NiCl_2 \cdot 6H_2O$	50	0.2	—	—	—	—	—	—
Sodium molybdate, $Na_2MoO_4 \cdot 2H_2O$	—	—	48	0.2	—	—	—	—
Sodium tungstate, $Na_2WO_4 \cdot 2H_2O$	—	—	—	—	—	—	45	0.14
Phosphoric acid, H_3PO_4	—	—	—	—	50	0.5	—	—
Phosphorous acid, H_3PO_3	—	—	—	—	15	0.18	—	—
Boric acid, H_3BO_3	30	0.5	—	—	—	—	—	—
Sodium citrate, $Na_3C_6H_5O_7 \cdot 5H_2O$	—	—	100	0.3	—	—	—	—
Rochelle salt, $KNaC_4H_4O_6 \cdot 4H_2O$	—	—	—	—	—	—	400	1.4
Ammonium chloride, NH_4Cl	—	—	—	—	—	—	50	0.9
To pH, with	3.7–4.0, H_2SO_4		10.5, NH_4OH		0.5–2.0, $CoCO_3$		8.5–8.7, NH_4OH	
Temperature, °C	66		25		75–95		82–90	
Cathode current density, A/m^2	380		1000		500–4000		220–500	
Anodes	Co, Ni, separate circuits		Co		Co		Co/W, 70/30	

*Plus a wetting agent.

electrolytically. Other methods are also used, including precipitation of indium as the phosphate, and chlorination of lead-zinc-indium mixtures.

Uses

Indium is expensive (somewhat more than half the price of silver) and in limited supply, so that its uses are correspondingly limited. Since there is only one domestic producer, accurate statistics on consumption are not published. It forms many low-melting alloys with bismuth, tin, lead, and cadmium. Some of these alloys are used as solders for connecting lead wires to germanium in transistors. It also finds use in III-V semiconductors such as indium arsenide. Indium halides are used in mercury-vapor lamps. Some indium alloys are used in the nuclear energy field. In the area of surface finishing, indium is plated on cadmium- and lead-bearing alloys for surface protection.

Relatively little indium is used, and its applications are subject to variation with time, as new uses are found and older uses discontinued in favor of more economical materials.

Properties

Some physical properties of indium are listed in Table 14-7. Indium is a very soft, low-melting metal, extremely ductile but having little strength.

Indium does not react with air at ordinary temperatures; at red heat it burns to form the oxide In_2O_3. It dissolves in mineral acids, but does not react with alkalis or boiling water. It combines directly with the halogens and with sulfur when heated.

The outer electronic configuration of indium, as of the other main Group III elements, consists of two *s* and one *p* electrons, all three of which are available

Table 14-7 Properties of Indium

Atomic number	49
Atomic weight	114.82
Electronic configuration	$1s^2 2s^2 2p^6 3s^2 3p^6 3d^{10} 4s^2 4p^6 4d^{10} 5s^2 5p^1$
Melting point, °C	157
Boiling point, °C	≈2000
Density, g/cm³, 20°C	7.31
Crystal structure	face-centered tetragonal
Electrical resistivity, $\mu\Omega$-cm, 20°C	8.37
Tensile strength, MPa	2.6
Brinell hardness	0.9
Standard electrode potential, E^0, 25°C, V	
$In^{3+}(aq) + 3e^- \rightarrow In$	−0.34

for chemical combination. But as the group is descended, there is a greater and greater tendency for the two *s* electrons to behave as an "inert pair"; although this designation may not explain the phenomenon, it describes it briefly. This tendency is more marked in Group IV (see Tin and Lead), but it is evident also in this group with thallium, which is both univalent and trivalent. Indium is almost always trivalent, but univalent and divalent compounds are known. The tendency to complex formation is small but significant. Simple aquo ions are present in perchlorate (and presumably fluoborate) solutions, but it is probable that solutions of the chloride, sulfate, and other salts are to some extent complexed in such ions as $InSO_4^+ (aq)$. The tendency to complex formation may explain the relatively low conductivity of solutions of indium compounds: thus rather than the simple ionization

$$InCl_3 \rightarrow In^{3+}(aq) + 3Cl^-$$

which forms four ions, only two ions are formed:

$$InCl_3 \rightarrow InCl_2^+ + Cl^-$$

Solid indium salts are not generally hydrated.

Electrochemically, indium is slightly more noble than cadmium. Reactions of indium compounds, including electrode reactions, tend to be sluggish and to depend more on kinetic than on thermodynamic factors.

Hazards

The toxicity of indium and its compounds has not been extensively investigated. Animal tests indicate some degree of hazard, but for normal electroplating applications usual good housekeeping practices should be sufficient. Indium should not be used in contact with food products, since its solubility in food acids is high.

Indium Plating

An early use of indium plating was for diffusion alloying in aircraft engine bearings; the metal was plated on the bearing surface and subsequently diffused into it by heat treatment. Corrosion by oxidation products of lubricating oils was materially reduced by the indium deposit. This application has declined with the introduction of jet aircraft to replace propeller-driven planes. Indium may be used also to improve the corrosion resistance of other substrates by diffusion; these include copper, brass, zinc, and lead.

In the electronics industry, electroplated indium is used as a doping agent for the production of *p*-type germanium transistors.

Indium has been plated from four types of solution: sulfate, fluoborate,

Table 14-8 Indium-plating Solutions

	High-pH cyanide		*Sulfate*		*Fluoborate*		*Sulfamate*	
	g/L	*Molarity*	*g/L*	*Molarity*	*g/L*	*Molarity*	*g/L*	*Molarity*
Indium as metal, In	15–30*	0.13–0.26	20	0.17	72	0.63	20‡	0.17
Indium chloride, $InCl_3$	29–57	0.13–0.26	—	—	—	—	—	—
Indium sulfate, $In_2(SO_4)_3$	—	—	88	0.17	—	—	—	—
Indium sulfamate, $In(SO_3NH_2)_3$	—	—	—	—	—	—	70‡	0.17
Indium fluoborate, $In(BF_4)_3$	—	—	—	—	235	0.63	—	—
Dextrose	30–40	—	—	—	—	—	8	—
Potassium cyanide, KCN	140–160	2.1–2.5	—	—	—	—	—	—
Potassium hydroxide, KOH	30–40	0.5–0.7	—	—	—	—	—	—
Sodium chloride, NaCl	—	—	—	—	—	—	45	0.77
Sodium sulfate, Na_2SO_4	—	—	10	0.07	—	—	—	—
Boric acid, H_3BO_3	—	—	—	—	22–30	0.35–0.5	—	—
Ammonium fluoborate, NH_4BF_4	—	—	—	—	40–50	0.4–0.5	—	—
Sulfamic acid, HSO_3NH_2	—	—	—	—	—	—	25	0.25
pH	—		2.0–2.5		1.0		0–0.2	
Temperature, °C	20–30		20–30		20–30		20–30	
Current density A/m^2	150–300		200		100–1600		200–1000	
Cathode efficiency, %	≈50		30–70		50		90	
Anodes	steel		In + Pt†		In + Pt†		In	

*Added as $InCl_3$ or $In(OH)_3$.

†Anode efficiency is greater than cathode efficiency, so some inert anode area is required.

‡Minimum.

cyanide, and sulfamate. Present preference appears to be for the cyanide or fluoborate electrolyte. There are at least three types of cyanide bath: the original formulation, stabilized with D-glucose; a "high-pH" type; and a proprietary formulation.* Only the second is detailed in Table 14-8.

Many other solutions have been proposed; but since indium plating remains so far a specialty practiced by few and installations tend to be small, it is difficult to arrive at a consensus regarding preferences and experience with the various baths.

Some typical formulations, and operating conditions, are listed in Table 14-8.

The source of indium in these baths may be a freshly precipitated indium hydroxide, prepared, for example, by precipitation from an indium chloride solution followed by filtration. In the case of the high-pH cyanide bath, the $InCl_3$ may be used directly; or the appropriate compound may be purchased if available. Concentrates are also commercially available.*

*Indium Corporation of America.

15 Uncommon Metals

So far we have discussed, in more or less detail, 18 metals which are fairly commonly electrodeposited. Of the 90 or so elements in the periodic system (not counting trans-uranium elements), all but 19 are metals or have some metallic character, so that most of the metals have not been mentioned yet.

As can be seen from Fig. 8-4, many of these metals cannot be deposited from aqueous solutions, either because their electrode potentials are too negative, even allowing for hydrogen overvoltage, or for kinetic reasons. The latter is probably the case with metals like molybdenum and tungsten, for example, since from what we know about their electrode potentials they should be capable of deposition.

When aqueous solutions are not suitable, two alternatives present themselves: organic electrolytes and fused salts. Neither is especially attractive to the average metal finisher, but if the end result is sufficiently rewarding, they may be considered.

Organic electrolytes pose many problems: probable fire and explosion hazard; requirement for inert atmospheres in some cases and at least moisture-free atmospheres in all; low conductivity compared to water solutions; higher costs; and possible environmental problems. Nonetheless, these problems have not prevented the development of a practical, if somewhat exotic, process for aluminum plating, which is discussed briefly. Although some other metals, such as beryllium and magnesium, might seem to offer promise of deposition from appropriate organic solvents, no commercial or even satisfactory beaker-scale process has been reported.

Fused-salt electrolytes are by no means unheard of in industry: magnesium, aluminum, sodium, and several other metals are recovered by fused-salt electrolysis on an extremely large scale. In these cases, however, the cathode metal is molten, the temperature of electrolysis being above the melting point of the metal. When the electrolysis of fused salts takes place at temperatures below the melting point of the metal, yielding a solid deposit, most metal deposits are treelike or dendritic, having just enough adhesion to permit their recovery as a sponge or powder, but being useless as electroplates. The electrodeposition of a few of the "refractory" metals from fused-salt baths in a useful form has been developed, and will be discussed briefly; but such baths, too, are not particularly attractive to the average metal finisher. The high temperatures of operation, large heat input, corrosive nature of the electrolytes, hazards, and costs render such processes specialties for the few.

In this chapter we consider only a small number of the "uncommon" metals not discussed so far. A literature review of all the metals that fall under this heading is available.*

These metals fall into three categories: (1) easily plateable from aqueous solution but not widely used; (2) plateable from organic electrolytes but not from water; (3) plateable only from fused salts. These categories represent current conditions; it is always possible that someday a way to plate tungsten, for example, from aqueous solution may be developed. On the other hand, there is no chance that metals as electronegative as aluminum or magnesium will ever be plated from water solutions, and nonaqueous solutions offer the only possibilities.†

Aqueous Solutions

ARSENIC, ANTIMONY, AND BISMUTH

These three elements are the last three members of main Group V (VA) of the periodic system, of which the first two members—nitrogen and phosphorus—are not metals, although the latter does have a rare metallic allotrope. Arsenic generally is not classed as a metal, but it has some metallic properties; antimony is somewhat intermediate between metals and nonmetals; bismuth is almost entirely metallic in character. Some properties of the three elements are listed in Table 15-1.

All have an outer electronic configuration consisting of two *s* and three *p* electrons. Accordingly they exhibit oxidation numbers of 3+ and 5+; and in some cases, especially the lighter elements can gain three electrons to attain the configuration of the next inert gas, as in arsine, AsH_3, and stibine, SbH_3. The elements become more and more metallic in character as we descend the group, as is typical. Thus arsenic and antimony seldom exist in solution as simple As^{3+} or Sb^{3+} ions, but almost always in coordinate complexes; but bismuth shows a fair tendency to form Bi^{3+} *(aq)*. As is also typical, raising the elements to higher

*M. L. Holt, "Modern Electroplating," cited in the Bibliography.

†For present purposes we neglect the deposition into a mercury cathode, as in the preparation of alkali metal amalgams, or into a dropping mercury electrode as in polarography. While this may be "electrodeposition," it certainly is not "electroplating."

Table 15-1 Properties of Arsenic, Antimony, and Bismuth

	As	*Sb*	*Bi*
Atomic number	33	51	83
Atomic weight	74.9216	121.75	208.9804
Electronic configuration	(Ar) $3d^{10}4s^24p^6$	(Kr) $4d^{10}5s^25p^3$	(Xe) $5d^{10}6s^26p^3$
Melting point, °C	817*	630.5	271.3
Boiling point, °C	613†	1635	1560
Density, g/cm³	5.72‡	6.697	9.8
Electrical resistivity, $\mu\Omega$-cm, 20°C	33.3‡	39.1	106.8
Crystal form	hex.	hex.	rhombohedral
Common oxidation numbers	3−, 3+, 5+	3−, 3+, 5+	3+ (5+)
Standard electrode potential, E^0, 25°C, V			
$MO^+ + 2H^+ + 3e^- \rightarrow M + H_2O$		+0.212	+0.32

*2.8 MPa pressure.

†Sublimes.

‡α, metallic form.

oxidation states becomes more difficult as the series is descended, and Bi(V) is rare although As(V) and Sb(V) are fairly common.

Compounds of all three elements are toxic to various degrees; bismuth is the least so, and some authorities rate it as nontoxic.

All three can be electrodeposited from aqueous solutions, but are little used. Arsenic is occasionally plated for decorative antique effects; antimony has attracted sporadic attention because it is relatively inexpensive and has some corrosion-resistant properties that might be useful. Bismuth is expensive considering its lack of outstanding properties as a deposit.

Arsenic may be plated from the following solution, among many others:

Arsenic trioxide, As_2O_3	120 g/L
Sodium hydroxide, NaOH	120 g/L
Sodium cyanide, NaCN	4 g/L
Temperature 20–45°C	
Current density, 30 to 200 A/m²	
Steel anodes	

Antimony can be electrodeposited from many types of solution; those most favored by investigators have been based on either antimony fluoride, SbF_3, or on hydroxycarboxylic acids (citric, tartaric, etc.) in which the antimony is present in the form of the antimonyl ion, SbO^+. One problem in antimony plating has been the deposition of so-called explosive antimony, which results when the bath contains chloride ion and which has been shown to be the result of codeposition of chlorides. A second, and more serious, problem is the brittleness of antimony deposits and their general lack of adhesion to the cathode. One bath proposed has the following composition:

Antimony trioxide, Sb_2O_3	50 g/L
Potassium citrate, $K_3C_6H_5O_7 \cdot H_2O$	150 g/L
Citric acid, $H_3C_6H_5O_7$	180 g/L
pH	3.5–3.7
Temperature	55°C
Current density	250 A/m^2
Anodes antimony or carbon	

However, no commercial experience with this bath has been reported, and there is no evidence that it has overcome the objections already mentioned. It should be stated that although antimony deposits may appear to be ductile soon after plating, they often become brittle after standing for a week or two; satisfaction should not be assumed until the deposits have been aged.

Bismuth can be plated from solutions of its complex chloride, $BiCl_4^-$, and from perchlorate solutions such as the one given here:

Bismuth oxide, Bi_2O_3	40 g/L
Perchloric acid, $HClO_4$	104 g/L
Glue	0.03%
Cresol	0.08%

It is stated that bismuth deposits have some minor uses in the electronics and nuclear energy fields, and a comprehensive review of the literature, supplemented by further investigations to determine optimum conditions, is available in the same article.*

MANGANESE AND RHENIUM

Manganese (Mn, atomic number 25) and rhenium (Re, atomic number 75) are the first and third members of Group VIIB of the periodic system. Manganese has been known for two centuries and is very abundant, being twelfth in order of abundance in the lithosphere or eighth if only metals are considered; it is present in the earth's crust to the extent of about 0.1 percent. On the other hand, rhenium

*W. Dingley, J. S. Bednar, and G. R. Hoey, *Plating and Surface Finishing*, **63**(4), 26 (1976).

Table 15-2 Properties of Manganese and Rhenium

	Mn	*Re*
Atomic number	25	75
Atomic weight	54.9380	186.2
Melting point, °C	1244	3180
Boiling point, °C	2120	5885
Density, g/cm^3	7.30	21.04
Electrical resistivity, $\mu\Omega$-cm	185	19.3
Electronic configuration	(Ar) $3d^5 4s^2$	(Xe) $4f^{14} 5d^5 6s^2$
Crystal structure	cubic	hcp
Standard potential, E^0, 25°C, V	$Mn^{++} + 2e^- \rightarrow Mn$ −1.18	$ReO_4^- + 8H^+ + 7e \rightarrow Re + 4H_2O$ +0.363

is extremely rare; although predicted (''zwi-manganese'') by Mendeleeff, it was not discovered until 1925.* The abundance of rhenium is estimated at about 10^{-9} percent; it is found not in manganese ores, as might be expected, but in some molybdenum and copper ores; its production is entirely a by-product of the mining and recovery of these metals, especially molybdenum.

Both are transition metals, exhibiting variable valences and other typical properties; but whereas the higher oxidation states of manganese are strongly oxidizing, e.g., potassium permanganate, and the simple aquo ion Mn^{++} is quite stable and forms many salts, with rhenium it is the higher oxidation states that are more stable. Perrhenate ion ReO_4^- is not a notably strong oxidizing agent, and simple Re^{++} ions are almost unknown.

Some properties of manganese and rhenium are listed in Table 15-2. Rhenium has the highest melting point of any metal except tungsten; it is therefore the highest melting metal plateable from water solution.

Both metals can be plated from aqueous solution; manganese is the most electronegative metal thus depositable. The plating baths are quite different in character: manganese is plated from divalent manganese salts, $Mn^{++}(aq)$, whereas rhenium is plated from the perrhenate complex.

Manganese is recovered by an electrolytic process on a large scale; the electrolyte is a solution of manganese(II) sulfate, ammonium sulfate, and sulfuric acid at a pH of 1 to 1.5; sulfur dioxide is added to the catholyte of the divided cells. Although electrowinning of manganese is an established process, manganese plating is not practiced commercially. The metal has no decorative appeal, and deposits so far obtained have not proved attractive for either

*The middle member of the group, element 43, was apparently discovered at the same time as rhenium, by Noddack and coworkers; but although the discovery of rhenium was amply confirmed, their ''masurium'' or eka-manganese was shown to be nonexistent, and this void in the periodic system was not filled until 1939 by the totally synthetic element technetium.

corrosion protection or their physical properties. Another drawback is that manganese, as plated, is the allotrope γ-manganese, which is ductile; but on standing this transforms to the form stable at room temperature, which is brittle.

A manganese-plating bath has the following composition:

Manganese sulfate, $MnSO_4 \cdot 2H_2O$	100 g/L
Ammonium sulfate, $(NH_4)_2SO_4$	75 g/L
Ammonium thiocyanate, NH_4CNS	60 g/L
Temperature	25°C
pH	4–5.5
Cathode current density	2400 A/m^2

Rhenium, unlike manganese, is both expensive and rare. Its deposition is warranted only for special applications, and no established uses are recorded. In spite of this, at least two companies offer proprietary formulations for rhenium plating. Typical of formulas published in the open literature is the following:

Potassium perrhenate, $KReO_4$	11 g/L
Sulfuric acid (conc.), H_2SO_4	3.3 g/L
Temperature	25–45°C
Cathode current density	1300 A/m_2

Phosphoric and citric acids may take the place of sulfuric acid, and perrhenic acid may take the place of potassium perrhenate. At least one proprietary bath is operated at the much higher pH of 5.7. A review of rhenium plating is offered by Camp.* Later workers used proprietary baths whose formulas were not published. There is some indication that rhenium deposits may be useful for their wear resistance, but "problems of high cost, long deposition time, inferior macrothrowing power, and very low cathode current efficiencies" are definite drawbacks. A silver-rhenium–alloy deposit is said to be useful as a solid-film lubricant for high-temperature applications.†

Organic Electrolytes

There are two types of metals that cannot be deposited from aqueous solution: those like aluminum that are so electronegative that their deposition potentials

*E. K. Camp, *Plating* **52,** 413 (1965).

†E. W. Turns and R. D. Krienke, *Plating,* **52**, 1149 (1965); E. W. Turns, ibid., **58,** 127 (1971).

cannot be reached in the presence of dischargeable hydrogen, and those like tungsten and molybdenum which from purely thermodynamic considerations should be depositable, but which apparently form such stable complexes or for other kinetic reasons simply do not deposit, at least from any combination of conditions yet attempted. For the former—the electronegative metals—there is some hope that organic solvents may prove satisfactory; for the latter, there is little or no reason why organic solvents should offer any improvement over water solutions, and such is indeed the case.

Organic solvents have been extensively investigated theoretically: studies of dielectric constant, acidity and basicity, reaction rates, and the like have been reported. But few reliable researches have appeared concerning their suitability as electrolytic solvents for the electrodeposition of metals; and even fewer have paid attention to such electrodeposition in the form of satisfactory electroplates. Many organic solvents tend to decompose under the influence of an electric current. It is true that water also decomposes, but in the case of water the decomposition products, hydrogen and oxygen, pass off harmlessly as gases at the electrodes, whereas many organic solvents produce reaction products such as polymers, degradation products containing carbon, or similar materials that are far from harmless to the continuation of satisfactory electrolysis.

Brenner* has pointed out some of the shortcomings of previous work in this field and has mentioned the factors to be considered in choosing an organic solvent for electrodeposition of metals that cannot be plated from aqueous solution.

The dielectric constant, for example, is not a criterion for choice. Diethyl ether, with a dielectric constant of only 4.5, is a better solvent for electrodeposition than hydrogen cyanide, with a dielectric constant of 95 (water = 80). Conductivity of the pure solvent is also irrelevant: the conductivity of water is about 10^6 that of ether, but both are good solvents for the purpose.

On the positive side, the organic media which are potentially useful for depositing the metals under consideration are those that have a "weak coordinating center," for example diethyl ether's oxygen atom or an aromatic hydrocarbon having double bonds. This type of solvent may form a loose coordination compound with the solute. Polar solvents, on the other hand, tend to form such stable complexes with the solutes that they are no more effective than water in permitting the deposition of these metals.

ALUMINUM

The desirability of depositing aluminum on ferrous substrates has been taken for granted by many workers. From its position in the emf series, aluminum should be even more protective to steel than zinc; its corrosion products, unlike those of zinc, are not bulky and unsightly. Presumably, an aluminum deposit could be anodized for further corrosion resistance. On a volume basis, aluminum is cheaper than zinc, and accordingly would be the cheapest metal that can be

*A. Brenner, *Advances in Electrochemistry and Electrochemical Engineering*, **5**, 205, (1967).

deposited for corrosion protection. Hot-dipped coatings of aluminum on steel are produced commercially. Throughout the history of electroplating, therefore, there has been great interest in the possibility of aluminum plating; nor have (unverifiable) claims for its success been lacking in the older literature.

Some properties of aluminum are summarized in Table 15-3.

There can be no doubt that the electrode potential of aluminum is too negative to be reached in aqueous solution, even allowing for hydrogen overvoltage (we except the deposition into a mercury cathode, as in polarography where the hydrogen overvoltage is extraordinarily high and the reaction is further favored by the energy released in the formation of amalgams). This fact has not deterred investigators from reporting the successful deposition of aluminum from aqueous solutions; many such reports, and even patents, appeared in the literature up to about 1930. All are unfounded.

Aluminum is produced by electrolysis of molten salts, and attempts have been made to use molten-salt electrolysis to produce useful coatings on steel and other substrates. Although some such processes may have been technically successful, none has proved commercially attractive. The other possible approach to the electrodeposition of such an electronegative metal is the use of nonaqueous electrolytes, e.g., organic solvents; and to the extent that aluminum plating is practiced at all, it is this approach that has been favored so far.

The choice of organic solvent is important. Most organic liquids are not good ionizing solvents; few common metal salts are soluble in organic solvents, and those that are often are not ionized and form nonconducting solutions. The solvent must be "aprotic," i.e., must not contain ionizable hydrogen atoms in its molecular constitution. This limitation rules out alcohols, for example. Early attempts to electrodeposit aluminum from organic electrolytes made use of ethyl

Table 15-3 Properties of Aluminum

Atomic number	13
Atomic weight	26.98154
Melting point, °C	660
Boiling point, °C	2452
Crystal structure	fcc
Density, g/cm^3, 20°C	2.6989
Electrical resistivity, $\mu\Omega$-cm, 20°C	2.65
Electrical conductivity, % IACS, 20°C	64.94
Electronic configuration	$1s^2 2s^2 2p^6 3s^2 3p^1$
Tensile strength, MPa	48
Brinell hardness	12–16
Standard electrode potential, E^0, 25°C, V	
$Al^{3+} + 3e^- \rightarrow Al$	−1.66

bromide–benzene mixtures, and a process using aluminum chloride dissolved in alkyl pyridinium chlorides or bromides has been studied. The only significant large-scale application of aluminum deposition to survive, however, is the hydride process invented at the National Bureau of Standards. Although not in wide commercial use, the process, or a modification of it, is used by at least one manufacturer to produce aluminum-plated fasteners and other small parts. It has also been used to produce fairly large electroforms. If the finisher is willing to observe the rigorous precautions required, there seems little doubt that the process can produce satisfactory results.

The solvent is diethyl ether (usually called simply ether), $(C_2H_5)_2O$. The hydrogens in this solvent are not ionizable, and therefore there is no tendency for hydrogen evolution when solutions are electrolyzed. The solute is a mixture of aluminum chloride, $AlCl_3$, (anhydrous) and lithium aluminum hydride, $LiAlH_4$, in the following concentrations:

$AlCl_3$	400 g/L or 3 *M*
$LiAlH_4$	15 g/L or 0.4 *M*

Various additives have been recommended to reduce grain size and treeing, but a periodic reverse (PR) current cycle is said to be at least as efficacious.

Anhydrous conditions are required. Since the lithium aluminum hydride will destroy any traces of water present in the ether, no special precautions are required in making up the bath; but anhydrous conditions must be maintained by the use of inert atmospheres such as dry nitrogen or argon. This requirement, of course, entails the use of a closed system. A dry box or glove box is suitable for small installations; larger ones call for fairly elaborate constructions.*

The hydride bath depends for its conductivity on the reaction of aluminum chloride with lithium aluminum hydride:

$$LiAlH_4 + 3AlCl_3 \rightarrow 4AlHCl_2 + LiCl$$

The major carrier of current is probably etherated Li^+, but little or no theoretical investigation of the bath has been reported. Other ionic species probably present are $AlCl_4^-$ and $AlHCl_3^-$.

Although soluble aluminum anodes are used in the bath, there is some question whether they dissolve in a form that replenishes the metal in plateable form. The original workers reported that although aluminum dissolved, it was in a form from which it could not be redeposited; consequently the life of the bath was limited to the aluminum originally present. Later workers did not confirm this finding, reporting that with proper solution circulation, anodic replenishment was successful.

Temperature of operation is from about room temperature to 60°C; higher temperature favors better conductivity, but causes considerably higher rates of evaporation. Since the bath has relatively high resistance, ohmic heating occurs

*See, e.g., A. E. Buschow and C. H. Esola, *Plating,* **55,** 931 (1968).

and cooling may be required. Cathode and anode current densities average about 200 A/m^2.*

Fused-Salt Electrolysis

REFRACTORY METALS

The refractory metals (so-called from their relatively high melting points) are in Groups IVB, VB, and VIB of the periodic system: titanium, zirconium, and hafnium (IVB); vanadium, niobium, and tantalum (VB); and molybdenum and tungsten (VIB). None can be deposited in the pure state from aqueous solutions; chromium is the only member of any of these three groups that can be so deposited, and then only with some difficulty, as the low cathode efficiency and rather unusual operating conditions demonstrate. Attempts to deposit them from organic electrolytes have likewise been ineffective. Molybdenum and tungsten alloys with the iron-cobalt-nickel group have been deposited, and cobalt-tungsten alloys, especially, have some attractive possibilities.

The reasons for their failure to deposit from aqueous solutions, however, are probably not the same for all these refractory metals. The electrode potentials of the Group IVB metals titanium, zirconium, and hafnium are probably too negative to be attained in the presence of depositable hydrogen; but for the others, insofar as their electrode potentials have been determined (mostly by indirect methods), their deposition should be possible. Failure is usually attributed to kinetic factors: the extreme stability of their complexes both with the solvent and with other ligands, to the extent that not enough energy can be supplied to break these bonds. In the case of tungsten in particular and of the rest to a lesser extent, much research has been published concerning their deposition from water solution, and success has been claimed. None of these claims, however, has been independently verified, and it is now generally accepted that none of the metals in this group can, in fact, be deposited from solutions of their salts in water. The reason for the relatively easy deposition of alloys of tungsten and molybdenum with metals of the iron group is not readily apparent.

Some properties of these metals (except vanadium and hafnium, which have been little studied as electrodeposits) are listed in Table 15-4.

All can be deposited from molten-salt electrolytes, but usually the deposits are dendritic or powdery rather than coherent electroplates, useful for the recovery of the metals perhaps but not for electroplating purposes.

A process for the deposition of coherent deposits of all these metals except titanium, in the form of useful, adherent plates, has been reported. It is based on the electrolysis of molten fluorides; and for the four metals tantalum, niobium,

*For details see W. B. Harding, "Modern Electroplating," cited in the Bibliography and references cited therein.

Table 15-4 Properties of Some Refractory Metals

	Ti	*Zr*	*Nb*	*Ta*	*Mo*	*W*
Atomic number	22	40	41	73	42	74
Atomic weight	47.9	91.22	92.9064	180.948	95.94	183.85
Density, g/cm^3	4.507	6.45	8.57	16.6	10.2	19.3
Melting point, °C	1668	1852	2467	2996	2610	3410
Boiling point, °C	3535	3580	5127	5425	4800	≈5000
Electrical resistivity, $\mu\Omega$-cm, 20°C	47.8	44.1	13.1	12.4	5.17	5.5
Electrical conductivity, % IACS	3	—	12	13	34	—
Electronic configuration	(Ar) $3d^2 4s^2$	(Kr) $4d^2 5s^2$	(Kr) $4d^4 5s^1$	(Xe) $5d^3 6s^2$	(Kr) $4d^5 5s^1$	(Xe) $5d^4 6s^2$
Crystal structure	hcp	hcp	bcc	bcc	bcc	bcc
Tensile strength, MPa*	234	170	245	250–500	165	1480
Electrode potential, E^0, V						
$M^{n+} + ne^- \rightarrow M$	−1.63 $n = 2$	—	≈ −1.1 $n = 3$	−1.12 $n = 5$	≈ −0.02 $n = 3$	$WO_3 + 6H^+ + 6e \rightarrow W + 3H_2O$ −0.09

*Figures are not strictly comparable since conditions of samples were different.

molybdenum, and tungsten the process was carried to a production scale and for some little time was practiced commercially.

The plates are dense and coherent, about 99.98 percent pure metal. Appearance and smoothness are said to be comparable to those of deposits from a Watts nickel solution at equivalent thicknesses; thicker deposits, up to 6.5 mm, have been produced with only moderate roughness.

The optimum conditions vary from metal to metal. A typical bath composition consists of a fluoride of the refractory metal, for example, ZrF_4 or K_2TaF_7, in a mixture of alkali fluorides, such as the eutectic mixture of potassium, lithium, and sodium fluorides; the container is a nickel or graphite crucible. An inert atmosphere of argon is required (many of these metals easily form nitrides, so that nitrogen is not an ''inert'' gas in this connection). Temperature ranges between 700 and 850°C, and current densities range between 50 and 1250 A/m^2.

For satisfactory results, the atmosphere must be essentially free of air and moisture; impurities in the salts should be low. Each metal must be in its appropriate valence state: 3+ for Mo, Cr, and V; 4+ for Nb, Zr, and Hf; 4.5+ for W; and 5+ for Ta. Anodes are of the metal being deposited; anode and cathode efficiencies are about 100 percent.*

*See S. Senderoff, ''Modern Electroplating,'' cited in the Bibliography.

16 Alloy Plating

Most cast or wrought metals are used in the form of their alloys rather than in the pure state; almost the only large-scale exception is copper, which is most used for its electrical conductivity. This is degraded by any alloying additions, so that for electrical purposes the purer the metal, the better. In other cases, properties of alloys can be varied over a wider range than those of pure metals; metallurgists can, by judicious alloying, provide metallic materials that combine mechanical and physical properties tailored to the projected use almost at will.

In contrast, most electrodeposits are almost pure metals, if we neglect the very small percentages of codeposited elements arising from addition agents, such as sulfur and carbon in bright nickel, selenium in some bright copper deposits, and the like. This situation is mainly the result of the greater difficulty in controlling alloy-plating processes, as will appear. Another reason is that by varying the conditions of electrodeposition, electrodeposited metals can be obtained in a number of physical conditions not requiring alloying. Thus nickel, depending on the bath used and the operating conditions, can be hard or soft, stressed or stress-free, etc.

Nevertheless, electrodeposited alloys are probably worthy of more practical consideration than they have received, since their possibilities are almost limitless. Metallurgically produced alloys, of course, are not limited to binary types: in fact very many practical alloys consist of three, four, or even more elements. For electroplating purposes, however, the difficulties in control multiply almost exponentially as the number of alloy constituents increases over two; only one ternary alloy system (copper-tin-zinc) has been deposited on a practical scale,* and quaternary and higher-order alloys have not been attempted at all. For all practical purposes, electrodeposition of alloys is confined to binary systems.

In order to warrant consideration, electrodeposited alloys must possess suffi-

*The lead-tin-copper alloys sometimes used for bearings is another, but the copper constituent is dripped into what is essentially a lead-tin bath.

cient advantages to outweigh the additional costs and control problems inherent in the deposition processes. Such advantages may include better physical properties, better corrosion resistance, decorative appeal not obtainable otherwise, magnetic properties, ability to be heat-treated, ability to substitute for more expensive metals, and others.

Some examples of these advantages may be cited: the yellow color of brass is desirable because solid brass is highly prized for its appearance; no single metal deposit can duplicate this color. Brass is also the preferred coating for steel wire to enhance its adhesion to rubber in automotive tires.

Bronze (copper-tin) can simulate the color of gold for decorative purposes when the use of gold itself is not warranted for reasons of cost. Bronze is also somewhat superior to copper in corrosion resistance as an undercoat for nickel-chromium deposits, and it is harder than copper, thus does not so easily cut through in buffing.

Some base-metal alloys can take the place of precious metals in particular applications; the saving in metal cost can more than compensate for the additional complications in the plating process. The use of yellow bronze to simulate gold has been mentioned; the tin-nickel alloy can substitute for gold in some (by no means all) electronic applications.

Nickel-phosphorus and nickel-cobalt alloys have magnetic properties that render them useful as thin films in computer applications.

Nickel-iron alloys save money by substituting the cheap metal iron for the more expensive nickel in some uses.

Cobalt-tungsten alloys show promise of improvements over hard chromium, since they retain their hardness well at elevated temperatures, and the plating process both is more efficient and has better throwing power than chromium plating. Little use, however, has resulted yet from these advantages.

Alloying additions to gold deposits have several purposes: decreasing costs and modifying color and physical properties such as hardness and wear resistance.

Tin-lead alloys are superior to either metal alone in corrosion resistance (low-tin alloys) and in solderability (high-tin alloys).

The number of alloy electrodeposits that have been investigated on a beaker scale is almost certainly well in the hundreds; see Fig. 16-1. Most of these studies have been theoretical investigations concerned with cathode potentials, the structure of the deposits, the effect of operating variables on the composition of the deposits, and similar data. But from this large mass of research—some of it of high quality, some not so high—only a few alloy-plating systems have evolved into fairly widespread practical utilization:

Gold alloys, already considered in Chap. 13

Tin-lead; see Chap. 13

Brass, or copper-zinc; bronze, or copper-tin; tin-nickel; these three will be discussed in this chapter.

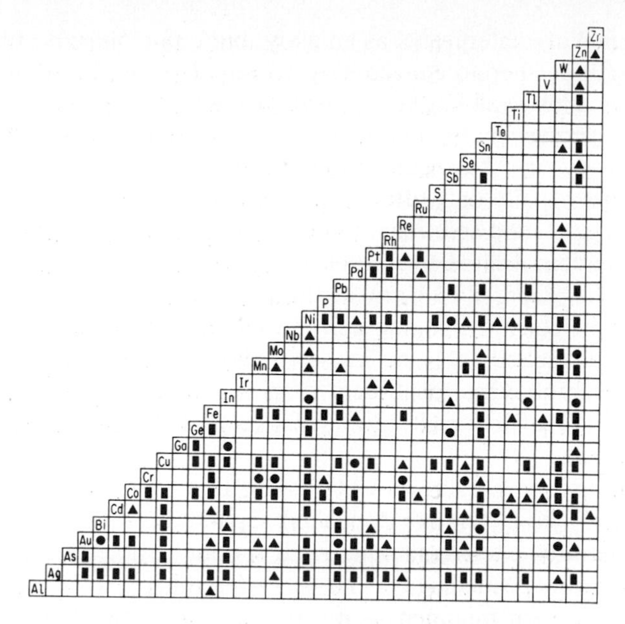

Fig. 16-1 Binary alloys reported up to 1970. [From A. Krohn and C. W. Bohn, *Plating,* **58**, 237 (1971). *Reproduced by permission.*]

A few nickel and cobalt alloys with phosphorus and some other elements are used to some extent in computer technology, but for the most part are produced by methods other than electrodeposition.

Nickel-phosphorus and nickel-boron alloys, produced not by electrodeposition but by the process of "autocatalytic" deposition, are considered in Chap. 17.

An exhaustive treatise on alloy deposition is available*; see also C. L. Faust, "Modern Electroplating," cited in the Bibliography.

General Discussion

THE NATURE OF ALLOYS

There are several available definitions of the word *alloy;* as good as any is this: An *alloy* is a substance that has metallic properties and is composed of two or more chemical elements at least one of which is a metal. This definition must not be taken too literally: a mere mixture of iron powder and copper powder

*A. Brenner, "Electrodeposition of Alloys, Principles and Practice," 2 vols., Academic Press, New York, 1963.

accordingly might be interpreted as an alloy, but common sense tells us that it is not an alloy. It is therefore necessary to stipulate further that the elements composing the alloy shall not be distinguishable by the unaided eye.

Alloys may, however, be inhomogeneous, i.e., composed of more than one phase, and many are; but such inhomogeneity is observable only under the microscope and usually only after appropriate chemical treatment. Alloys may be of many types: solid solutions, eutectic mixtures, intermetallic compounds, and even more complicated structures combining all of these.

Alloys may be prepared by many methods; the most common is simply to melt one of the major constituents and add the minor metals in such form that they dissolve or are melted into the mass. "Master alloys" containing a larger proportion of the alloying constituent than is wanted in the final alloy may be used. In the case of low-melting alloys like tin-lead, the molten metals may simply be mixed.

Alloys can also be prepared by diffusion; this route is fairly common with electrodeposits. The metals are deposited separately, then heat is applied, and the metals interdiffuse to form an alloy. (It should be noted that this method for forming an electrodeposited alloy is not considered here, since it is not truly alloy plating.) Actually heat treatment is not required: two metals placed in intimate contact with each other may form an alloy at the interface, given sufficient time. The time, however, may be years before significant mixing takes place.

Alloys may also be electrodeposited. Under proper conditions, two or more metals can codeposit on a cathode, forming an alloy whose phase structure and properties are usually (but by no means always) comparable to those of a similar alloy produced by thermal methods.

Alloys may be homogeneous (single phase) or heterogeneous (consisting of two or more phases). Single-phase alloys may be solid solutions or intermetallic compounds; multiphase alloys may be mixtures of solid solutions, mixtures of intermetallic compounds, or mixtures of virtually unalloyed elements. In the last-named, metals considered to have no mutual solubility are usually at least minimally soluble, so that their alloys consist of mixtures of extremely dilute solid solutions of each metal in the other.

Metals that are closely related chemically, such as copper and silver, lead and tin, or zinc and cadmium, nevertheless may have very small mutual solubility, and their alloys consist of virtually unalloyed elements in a very fine state of subdivision.

ALLOY PLATING: GENERAL

There are two principal reasons why alloy plating is not as widely used as the known utility of alloys in general would lead one to expect: (1) relatively few electroplated alloys have shown any marked advantages over pure metal deposits; (2) alloy plating processes are much more critical in control than corresponding processes for depositing pure metals.

Lack of Significant Advantages

Many alloys are useful because they are heat-treatable to develop desired properties of hardness, strength, and so on. By means of heat treatment and subsequent cooling or quenching, alloying constituents may be precipitated in metastable forms, particular phases may be stabilized, inhomogeneous alloys may be homogenized, etc. With relatively thin electrodeposits, many of these treatments are not available, since not only the deposit but the whole article must be so treated, and the heat treatment appropriate to the deposit may often be just the wrong treatment for the bulk material. Thus most (not all) electrodeposited alloys must be accepted just as deposited.

Difficulty in Control

This is perhaps even more of a stumbling block than the above. Not only must two or more metals be codeposited, but their ratio in the deposited alloy must be maintained within acceptable limits. It is not enough to codeposit, say, copper and zinc; if a 70/30 alloy is wanted, a 50/50 alloy will be unacceptable. Alloy deposits will change composition with changes in solution composition, current density, temperature, pH, and other variables. Control of operating conditions and composition, therefore, is much more critical than with deposition of a single metal. True, with a pure metal conditions must be monitored and maintained, but at least there is no concern about the composition of the deposit.

The usual way of maintaining the composition of a plating solution fairly constant is by the use of soluble anodes of the metal being deposited; in this way, about as much metal is dissolved at the anode as is deposited at the cathode, and the metal concentration of the bath remains fairly constant. Admittedly this system of anodic replenishment is often imperfect: cathode and anode efficiencies may not balance, so that adjustments are required based on chemical analyses. But the problem is complicated in alloy plating.

Few alloys are single-phase; and when two or more phases are present in the anode, it is only by luck that they are equally soluble under the conditions of electrolysis. Thus one phase may dissolve while another remains passive; the passive material may be loosened as the surrounding phase dissolves away and may simply fall into the bath as anode particles. Some unusual alloys, such as tin-nickel, cannot be prepared by thermal methods, and there is thus no way of making anodes of the same composition and phase structure as the deposit.

Many thermally prepared alloys of excellent properties are composed not of two, but of three or more metals. The difficulties in alloy electrodeposition, however, increase geometrically with the number of constituents, and ternary alloys become extremely difficult to control in production. Alloys of higher order are practically unknown in practical electroplating, even though common enough in metallurgy.

Alloy plating has been a fertile field for researchers, who are content to confine themselves to 100- or 250-mL beakers, to produce a few deposits under given sets of conditions, report the results, and then discard the bath and make up a fresh one. The number of combinations of the plateable metals in the form of binary alloys is very large—the total is calculable but depends on the assumptions made—and the possible plating solutions from which such alloys can be deposited is also very large. The number of publications resulting from such work is correspondingly great. Such publications are numerous in the literature, and many are valuable contributions to our knowledge of anode and cathode processes, but many are not; and very few have had any relationship to practical electroplating. In spite of the large number of studies reported, very few alloy-plating systems have made their way from laboratory to plant, and fewer still have found a permanent place in the repertory of the practical electroplater. The reasons for this have already been cited.

FUNDAMENTAL CONSIDERATIONS IN ALLOY PLATING

Basically the problem of codepositing two (or more) metals in the form of an alloy does not differ from that of depositing a single metal. However, in practice finding conditions for the deposition of an alloy in the form of an adherent, coherent, and useful deposit is not so simple.

Two conditions are basic for usefully codepositing two metals: (1) at least one of the metals must be capable of being independently deposited alone; (2) their deposition potentials must be fairly close together. This second condition follows from the fact that the more noble metal deposits preferentially, often to the exclusion of the less noble metal; in fact, this behavior is basic to the process of electrorefining. An electrode can have only one potential at a time; for two reactions to take place simultaneously at an electrode—in this case, for two metals to codeposit—they must take place at the same potential. Thus for simultaneous deposition of two metals in any useful form, conditions must be arranged so that the more electronegative (less noble) potential of the less noble metal can be reached without the use of an excessive current density; the potentials of the two metals must be brought close together if they are not already so.

In this discussion of alloy plating, we shall confine our attention to binary alloys (alloys composed of two metals or one metal plus a nonmetal), for reasons already stated.

The emf series, or table of standard electrode potentials, (Table 2-1) is a rough guide for deciding whether two metals can be codeposited from simple salt solutions. It must be emphasized that the series is only a rough guide, not an infallible one. The numbers apply only to very specific conditions: metals in equilibrium with solutions of their simple ions at unit activity. Under actual deposition conditions, equilibrium is not attained, and the deposition potentials are always somewhat more negative than the standard potentials because of polarization.

The standard potentials thus represent thermodynamic conditions and do not take into account kinetic factors. As we have already noted, at first glance the table of standard electrode potentials would lead us to predict that no metal more negative than hydrogen could be deposited from aqueous solution, because hydrogen would deposit preferentially. But hydrogen actually deposits at a much more negative potential because of polarization (hydrogen overvoltage in this case), and consequently the potentials of some metals can be reached in aqueous solution without the discharge of hydrogen, or with only partial discharge. The most electronegative metal that can be deposited from water solution is manganese, with a standard potential of −1.18 V.

The opposite wrong conclusion may also be drawn from the table of standard potentials. The metals vanadium, molybdenum, germanium, and tungsten have electrode potentials between −0.253 and −0.09 V. Cobalt, to take one example, has a potential of −0.277 V and is easily deposited from water solution. Yet none of the four mentioned metals can be so deposited in the pure state. This failure to deposit (and many attempts have been made and reported) must then be due to some other factor than the thermodynamics of the reactions. Such situations are fairly common in chemistry, where not every thermodynamically possible reaction does in fact take place. Failure is usually attributed to kinetic or steric factors, but this may be simply an acknowledgment of ignorance.

In spite of these inconsistencies, the table of electrode potentials is not useless for prediction of alloy-plating possibilities. Metals close together in the table are in general easier to codeposit than metals that are far apart. Examples of easily depositable pairs (from simple acid solutions) that follow prediction from the table are lead-tin, copper-bismuth, nickel-cobalt, and nickel-iron; in each pair the potentials are less than 100 mV apart. There are some exceptions to this generalization also: zinc and nickel can be deposited from a simple salt bath even though their potentials are 500 mV apart, and silver and palladium do not readily codeposit even though their potentials are only about 200 mV apart.

How close together must the potentials be to offer good possibilities of codeposition? This question cannot be answered categorically, but in general the metals should have standard potentials within about 200 mV of each other to be considered "close."

THE NERNST EQUATION IN ALLOY PLATING

We have seen that a metal will deposit when the cathode potential reaches that at which deposition can take place, which is in turn governed primarily by the Nernst equation, modified by polarization:

$$E = E^0 + (RT/n\mathfrak{F}) \ln a + P \qquad (16\text{-}1)$$

or

$$E = E^0 + (0.059/n) \log a + P \qquad \text{at } 25°\text{C} \qquad (16\text{-}1a)$$

where E^0 is the standard electrode potential of the metal, the second term takes account of the thermodynamic concentration of the metal ions in solution (called the activity, a), and P is a catch-all term for the nonequilibrium factors in the equation subsumed under the heading of polarization.

Equation (16-1) or (16-1*a*) represents the usual form of the Nernst equation for a metal electrode; it is written in this way because interest normally focuses on a pure metal in contact with a solution of its ions. In considering alloys, however, the equation must be given in its complete form:

$$E = E^0 + (RT/n\mathfrak{F}) \ln a_{M^{n+}}/a_M \tag{16-2}$$

where (see p. 17fn) the activity of the oxidized form (the metal ions) is in the numerator of the logarithmic term and that of the reduced form—here the activity of the metal in the alloy—is in the denominator. This complete form of the equation is required because the activities of the metals in an alloy may have a major effect on the conditions for their codeposition. The denominator is often neglected because a_M is taken to be unity in the case of pure metals, but this cannot be assumed here.

That cathodic reaction will take place which requires the least negative potential, as defined by Eq. (16-2). For example, in a solution containing both copper ions, Cu^{++} and hydrogen ions, H^+, the deposition of copper will normally take place before hydrogen can be evolved, because the evolution of hydrogen requires a much more negative potential than the deposition of copper. In the case of zinc, the metal can be deposited on most substrates because, although the standard potential of zinc is considerably more negative than that of hydrogen, hydrogen overvoltage on most substrates is high enough (P in the equation) to allow the deposition potential of zinc to be reached.

If two cathodic processes require almost the same potential for their occurrence, they may take place simultaneously at about the same rates. For two different cathodic processes to take place simultaneously, we have

$$E_1 = E_1^0 + RT/n\mathfrak{F} \ln a_1 + P_1 \approx E_2 = E_2^0 + RT/n\mathfrak{F} \ln a_2 + P_2 \tag{16-3}$$

For alloy deposition to occur, therefore, the situation breaks down to three possibilities: (1) E_1^0 must be about equal to E_2^0 so that minor adjustments in the activities of the metals (a) will allow E_1 to equal E_2; or (2) if E_1^0 does not equal or approach E_2^0, large differences in a must be provided for; or (3) P_1 and P_2 must be far enough apart to equalize the dynamic deposition potentials. In most cases of alloy deposition, either (1) or (2) provides the mechanism.

If the standard electrode potentials of the two metals concerned are close together and they can be plated satisfactorily from the same type of solution, we have the simplest case in alloy plating, exemplified by nickel-cobalt and tin-lead alloys. The standard potentials of nickel and cobalt are 27 mV apart, and those of tin and lead about 10 mV apart. The more noble of the two pairs are nickel and lead, but in both cases the potentials are close enough so that codeposition is easily accomplished from simple salt solutions: for tin and lead from fluoborates, which are generally regarded as uncomplexed; and for nickel-cobalt, from sulfate-chloride solutions in which complexing, if it takes place at all, is weak.

There are, however, relatively few cases of metals sufficiently close together in the emf series that their alloys can be deposited from simple solutions of their salts. Therefore in order to deposit most alloys, it is necessary to adjust their potentials in some way so as to more or less equalize them. This can be done in two ways, as we have stated above; only one is really practical.

First, static potentials can be equalized* by lowering the concentration of the more noble metal in the solution. This method is relatively ineffective. In the case of monovalent metals, the Nernst equation predicts that a tenfold change in the concentration will change the potential by only 59 mV; for a divalent metal, by about 30 mV, etc. Thus if the potentials are, say, 1 V (or 1000 mV) apart, the potential of the more noble metal must be changed by about 800 mV or more; this requires that the more noble metal must be at an activity about 800/59 or 10^{13} (since this is a logarithmic function) less than that of the less noble one in the case of univalent metals, or 800/30 or 10^{27} less than that of the other for divalent metals. But the use of such a dilute solution of the more noble metal is entirely impractical. A 10^{-27} *M* solution is beyond the limits of most analytical techniques; as soon as a few atoms of the metal have been deposited, the cathode film would be entirely stripped of metal ions, and controlling the composition of such a dilute bath would be entirely impossible.†

The other approach to equalizing the static potentials, and thus indirectly the deposition potentials, of two metals is to regulate the activity (or activities) by complex formation. Complex ion baths often have better throwing power than simple ion baths; they may yield finer-grained, smoother, and brighter deposits. and they prevent the immersion deposition of a more noble metal upon a less noble one, as we have already seen in discussion of single metal-plating baths. These advantages hold for alloy plating also; to them is added the essential one so that electrode potentials can be more easily equalized by the employment of appropriate complexing agents.

Brenner (op. cit.) divides complex types of alloy-plating baths into two groups: (1) "single" complex baths, in which the complexing agent is the same for both metals; (2) "mixed" complex baths, in which either one metal is complexed and the other remains in the form of a simple salt, or the complexing agents for the two metals are different. There is no fundamental difference between these two types, but type 2 is rather easier in control and formulation than type 1.

For the practical cases of alloy deposition to be discussed in this chapter, or which have been mentioned previously, the systems are of all three types.

In the tin-lead alloy system, both metals are in the form of their simple

*By "equalize" we mean here not actually making the potentials exactly equal, but bringing them close enough together for codeposition.

†In this discussion we neglect the difference between activity and concentration; this seems justified because: (1) the treatment is not strictly quantitative; (2) since two metals are involved, the activities appear as a ratio a_1/a_2 or f_1c_1/f_2c_2, and to a first approximation $f_1 \approx f_2$ in most cases; and (3) the activity appears in a logarithmic term so that minor differences between a and c have little effect on the answer.

We are also neglecting the denominator in Eq. (16-2), the activities of the metals themselves in the alloy. These are not usually known, and for most cases this approximation is justified, at least for practical purposes.

uncomplexed ions; their deposition potentials differ by only about 10 mV. Nickel-iron and nickel-cobalt also fall in this group.

The copper-zinc alloy brass is of type 1: both metals are complexed by cyanide ion, but the copper-cyanide complex is much more stable than the zinc-cyanide complex, and cyanide concentration has much more effect on the copper potential than on the zinc potential; furthermore, the zinc is also complexed by hydroxyl ion, which forms no copper complex.

For the copper-tin alloy bronze, the two metals are complexed by different ligands: copper by the cyanide ion, which forms no tin complex, and tin by the hydroxyl ion, which forms no copper complex.

For the tin-nickel alloy, the situation is not entirely clear; probably fluoride complexes of both metals are involved, but there is some evidence that the two metals are bound in a multinuclear complex containing both tin and nickel.

The use of complexing agents is the most important method of equalizing the potentials of the two metals in alloy plating. In solutions of complex ions, the potentials of the metals in the complex are shifted to more negative (less noble) values; the extent of the shift depends on the strength of the complex, i.e., the stability constant or dissociation constant of the complex ion formed (see Chap. 8). As in the plating of single metals, the most useful complex former is cyanide ion, because this ion forms complexes with so many metals and the complexes vary greatly in their strength or formation constants. This shift to more negative values tends to crowd the electrode potentials closer together.

A SAMPLE CALCULATION

While it is possible to calculate the change in potential caused by the complexing agent, from a knowledge of the stability constant of the complex, such calculations can yield results only suggestive of the possibility of alloy deposition, and not truly quantitative results. The following illustrative calculation should be considered with this caveat in mind.

If two metals are to be codeposited, their ions must be in an electrolyte in which their deposition potentials are the same or nearly so. The static electrode potentials are given by the Nernst equation, which for the case of alloy deposition is given by Eq. (16-3); for E_1 and E_2 we may read E_{d1} and E_{d2}, where d denotes deposition potential.

In order to calculate the separate deposition potentials E_{d1} and E_{d2}, a knowledge of both the activities and the P values for both metals is required, under varying conditions of temperature, current density, pH, concentration, and other factors. Such calculation is normally not possible in a quantitative sense, but it is possible to determine whether under a set of conditions alloy plating looks promising.

For example, the E^0 values for copper and zinc are far apart in the emf series; on the face of it, codeposition does not appear possible, at least from solutions of simple salts of the two metals. But by the use of complexing ions, it can be shown that the copper-ion concentration can be made so much smaller than the zinc-ion concentration that brass plating should be at least attempted.*

We assume that the copper-cyanide complex is $Cu(CN)_3^{--}$; its dissociation constant is 5.6×10^{-28}, but (since this will appear in a logarithmic term) little error is introduced by using 10^{-28} for simplicity. Similarly, the zinc cyanide complex is taken to be $Zn(CN)_4^{--}$ for which the dissociation constant is 10^{-17}. Then

$$\frac{(Cu^+)(CN^-)^3}{[Cu(CN)_3^{--}]} = 10^{-28} \tag{16-4}$$

$$\frac{(Zn^{++})(CN^-)^4}{[Zn(CN)_4^{--}]} = 10^{-17} \tag{16-5}$$

Let $A = (Cu^+)$ and $B = (Zn^{++})$; then in Eq. (16-4) (CN^-) will be 3A. In a 0.5 M solution of the cuprocyanide ion $Cu(CN)_3^{--}$, its concentration will be $0.5 - A$, but here A can be neglected as being so much smaller than 0.5. Then

$$(A)(3A)^3 = 0.5 \times 10^{-28} \tag{16-6}$$

or $9A^4 = 0.5 \times 10^{-28}$ and $A \approx 0.5 \times 10^{-7}$ or $(Cu^+) = 0.5 \times 10^{-7}$ in the copper solution, and $(CN^-) = 3A = 1.5 \times 10^{-7}$ M.

A similar calculation gives (B) $(4B)^4 = 10^{-17}$, or $B = (Zn^{++}) \approx 3 \times 10^{-4}$ and $(CN^-) = 4(Zn^{++}) = 12 \times 10^{-4}$.

On mixing the two solutions, the CN^- concentration must be the same, and it will be about that in the zinc solution, which is much higher than that in the copper solution. Then the Cu^+ concentration is approximately

$$(Cu^+)(12 \times 10^{-4}) = 10^{-28} \qquad \text{or} \qquad (Cu^+) = 3 \times 10^{-20}$$

Inserting these values into Eq. (16-3), we have for copper

$$E = 0.52 + 0.059 \log 3 \times 10^{-20} = 0.52 - 1.15 = -0.63 \text{ V}$$

and for zinc

$$E = -0.76 + 0.059/2 \log 3 \times 10^{-4} = -0.97 \text{ V}$$

This calculation shows that the deposition potentials of copper and zinc have been brought closer together by means of complexing both ions with the cyanide ion. This is not, of course, the whole story: zinc also forms a hydroxyl complex; the factor P has not been accounted for; and furthermore, it is likely that deposition takes place not from the very small amount of free metal ion present as a result of the dissociation of the complexes, but directly from the complexes themselves. Thus the only conclusion we can draw from this calculation is that codeposition of copper and zinc might be possible, since their potentials are within reasonable distance of each other. Complicating factors, making exact

*We should admit in all honesty that such calculations are usually made *after* we already know that the alloy can be plated.

calculation impossible or very difficult, have been mentioned in part. They include: (1) deposition takes place, in all probability, not from any free Cu^+ and Zn^{++} ions that might be present, but directly from the complexes $Cu(CN)_3^{--}$ and $Zn(OH)_4^{--}$, so that the potentials of $Cu^+ \rightleftharpoons Cu^0$ and $Zn^{++} \rightleftharpoons Zn^0$ are not really directly involved; (2) the assumption that the copper complex is $Cu(CN)_3^{--}$ is probably only partially true; (3) the metal being deposited is neither copper nor zinc but brass, an alloy of the two—its potential is needed for meaningful insertion into the equations. In many cases the energy released by alloy formation may aid in deposition, and this is not taken into account in the simple calculations suggested here.

An extreme case of a situation where formation of the alloy itself aids in driving the reaction forward is the formation of amalgams in deposition into a mercury cathode: sodium metal, which cannot be deposited from aqueous solution, can nevertheless be deposited into a mercury cathode; in fact, this is the basis of the mercury cell for the production of sodium hydroxide. The high overvoltage of hydrogen on a mercury surface helps to explain this deposition, but still it would not be possible without the additional driving force supplied by the energy of formation of the sodium-mercury alloy or amalgam.

Hence such calculations cannot give quantitative predictions, but the exercise does show that there is some hope of codepositing zinc and copper from complex ion baths. This expectation is, of course, borne out by the process of brass plating.

To summarize: two metals may probably be codeposited from their simple salt baths if their standard potentials are close together. If they are far apart (say, more than about 200 mV), either one or both probably must be complexed in order to bring their deposition potentials closer together.

REPLENISHMENT OF THE METAL CONTENT OF ALLOY-PLATING BATHS

The great majority of laboratory investigations of alloy-plating systems have been conducted on a small beaker scale, and the workers have been interested only in cathodic reactions: the nature of the alloy produced, or its structure and composition, using platinum or other inert anode and without regard to bath life. A few deposition experiments are conducted, and the bath is then discarded and a new one made up. This type of investigation is satisfactory as a preliminary step toward the development of an alloy-plating system, but stops far short of practical utility. For regular production, some way must be found to maintain the bath in day-to-day operation, and this entails some means of replenishing the metal content of the bath as alloy is deposited at the cathode. The two metals must be added to the bath in the same proportion as they are deposited, and in roughly the same amounts. Several means of replenishing the metal content of alloy-plating baths have been used, and all are to some extent satisfactory.

1. Alloy anodes of the same composition as the deposit. This is the ideal method, and is strictly comparable to processes for single metal plating. It is not always applicable, however. Alloys may be inhomogeneous, and anodes of the alloy may therefore not corrode satisfactorily in the bath. One phase may be more soluble than another: solution of this phase may simply cause the less soluble phase to sludge off as solid particles in the bath. Or the thermally prepared alloy may not have the same phase structure as the electrodeposited one, and its anodic behavior may not be comparable to its cathodic behavior.

2. Inert anodes with chemical replenishment. This system is usually not practical because of the buildup of extraneous ions necessarily added along with the metals, but it is used in some gold alloy–plating systems.

3. Anodic replenishment of one metal, chemical replenishment of the other. This system works well if the metal being replenished chemically is a relatively minor constituent of the deposited alloy. In red bronze (10 to 15 percent tin, balance copper) plating, copper anodes may be employed and the tin replenished chemically, because chemical replenishment accounts for only 10 to 15 percent of the total metal plated; this amount can usually be tolerated owing to drag-out and other losses.

   In tin-nickel plating, nickel anodes are used and the tin is replenished chemically by additions of stannous chloride. Although this would seem to entail a buildup of chloride ion in the solution, that does not in fact occur, and it must be assumed that chlorine is evolved at the anodes, unnoticed because of the brisk ventilation required in any case.

4. Separate anodes of the two metals, hung on the same bus bar. Such systems are perhaps possible, but seldom practical. Unless the metals are very close together in the emf series, it is unlikely that they will dissolve at the same potential. One metal is almost sure to act as an inert anode.

5. Separate anodes of the two metals, hung on separate bus bars with separate controls. This system is quite workable, but somewhat unwieldy. The amount of current to each metal is under close control, and the potentials of the two metals can be regulated independently of each other; but the setup is complicated.

6. Alternate use of the two metals as anode. One metal is hung on the anode bus; after a certain time interval, these anodes are removed and anodes of the other metal are put in their place. The composi-

tion of the solution is permitted to oscillate around a mean; some variation is usually tolerable, and the method is entirely practical but quite unhandy, since it entails excessive handling of the anodes.

INFLUENCE OF VARIABLES IN ALLOY PLATING

Independent variables in alloy plating, as in all plating, are current density, temperature, agitation, pH, and concentrations of bath constituents. Normally, these must be more tightly controlled in alloy plating than in single metal plating, because it is likely that a change in any one will affect one metal more than the other and thus change the composition of the alloy. Exact effects can be determined only by experiment, because there are as yet too many unknowns in alloy plating to permit generalizations; but the following usually, though not always, are true.

Current Density

Increase in current density normally tends to increase the proportion of the less noble metal in the alloy deposit. Some addition agents may reverse this generalization, by having more effect on the polarization behavior of one metal than on the other. The extent of the change is normally greater in simple salt solutions than in complex solutions, and greatest when the codepositing metals are in complex ions with a common ligand than when the anions of the complexes are different.

Agitation

Increase in agitation usually increases the proportion of the more noble metal in the deposit, thus offsetting the effect of an increase in current density. By mechanically bringing fresh solution to the cathode face and decreasing the thickness of the cathode film, agitation offsets the normal tendency for more rapid depletion of the more noble metal in the cathode film. This effect is usually less pronounced when the metals are associated with complex ions than when they are in the form of simple ions, and more pronounced when the two metals are associated with the same anion than when they are complexed by different anions.

Temperature

Increase in temperature usually increases the proportion of the more noble metal in the deposit; it has about the same effect as agitation. However, the effect is more complicated, since temperature change may also alter the degree of dissociation of complexes and have various effects on polarization factors.

pH

This variable is more important in regulating the physical properties of the deposit than its composition; its effect is likely to be specific to particular alloy systems.

Bath Composition

The concentrations of the two metals in alloy-plating baths have a direct effect on the composition of the deposit, but not necessarily to the same degree: the higher the ratio M_1/M_2 in the bath, the higher the ratio in the deposit, but the ratios may not change to the same extent.

OTHER CONSIDERATIONS IN ALLOY PLATING

Current Efficiency

The current efficiency in alloy plating may be higher or lower than that for each single metal under comparable conditions. In some cases, where it has been possible to plate a metal like tungsten as an alloy but not singly, its plating efficiency has been raised from zero to a net positive number.

Because (rather surprisingly) several publications have used an incorrect method of reporting current efficiency in alloy plating, it may be well to show how to calculate the electrochemical equivalent of an alloy:

$$W_a = \frac{W_1 \cdot W_2}{f_1 W_2 + f_2 W_1} \tag{16-7}$$

where W_a is the electrochemical equivalent of the alloy, W_1 and W_2 are the electrochemical equivalents of the two metals (in grams per coulomb or other comparable units), and f_1 and f_2 are the fractions by weight of the corresponding metals in the alloy: $f_1 + f_2 = 1$.

Throwing Power

Since current density can change the composition of the deposit, as noted above, the composition of the alloy deposit may not be uniform over a shaped cathode on which the current density varies. Thus for alloy plating, in addition to the usual throwing power concerned with weight or thickness of metal deposited over a cathode, we are concerned with what may be called *composition throwing power*. This may be deduced from a knowledge of the effect of current density on plate composition, or may be determined experimentally by analyzing deposits at various areas of a shaped cathode. For any practical alloy-plating system, either one of these criteria must be met: (1) the composition throwing power must be fairly good or (2) the composition of the deposited alloy must not be critical.

Brass

Brass, an alloy of copper and zinc, has been plated for well over 100 years; the proportions are normally about 70 percent copper, 30 percent zinc. Although many types of solution have been investigated, the only practical plating bath has been, and remains, the cyanide solution, the principles of which have already been discussed.

Most brass is plated for decorative purposes, for articles such as furniture, builders' hardware, and handbag frames. Brass deposits over aluminum, zinc die castings, steel tubing, and stampings substitute for solid brass where weight or cost is a factor. Much brass plating is a flash deposit over bright nickel, the nickel being used to provide brightness and the brass just thick enough to give the desired color.

Although arsenic (see below) can be used as a brightener in decorative brass plating, most brass deposits are smooth and mat, without leveling characteristics. The parts may be brushed, tumbled, or relieved by buffing for decorative effects.

Functional uses for brass plating include its use for rubber adhesion to steel; also it has been used for its lubricating properties, on such items as steel shell casings and aircraft engine parts.

The thickness of brass deposits varies from a mere flash up to as much as 25 μm; most deposits are in the lower end of this range. Most decorative brass deposits are protected by a clear lacquer.

So-called white brass, containing 30 percent copper and 70 percent zinc, has been used as a substitute for bright nickel on toys, tubular furniture, and interior automotive hardware and trim. The Ford Motor Company has used white brass between layers of copper and bright nickel, followed by chromium, on some automotive exterior trim.

Bath compositions reported in the literature vary over a wide range; these ranges are subsumed in the compositions listed in Table 16-1. There is no "standard" brass-plating bath, but most operating baths fall within the ranges noted.

The cyanides of copper and zinc furnish the required metal content. Excess cyanide, above that needed to dissolve the metal cyanides, is called *free cyanide;* it is arbitrarily defined as that in excess of the amount required to form the complexes $Cu(CN)_3^{--}$ and $Zn(CN)_4^{--}$. This free cyanide is required to aid in anode corrosion and to form a reserve of cyanide ions to prevent the precipitation of insoluble metal cyanides.

As free cyanide increases, the cathode efficiency decreases and the composition of the deposit changes; although high free cyanide would appear to lead to a lower copper content of the deposit, there is some disagreement on this score, with some workers reporting the opposite effect.

Sodium carbonate acts as a buffer; it is sometimes added as such, but in any case will form in any operating bath as a result of carbon dioxide absorption and

Table 16-1 Brass-plating Solutions

	"Regular"		High speed		White brass		"Bronze"*	
	g/L	*Molarity*	*g/L*	*Molarity*	*g/L*	*Molarity*	*g/L*	*Molarity*
Copper cyanide, CuCN	25–50	0.28–0.56	60–100	0.67–1.1	16–20	0.18–0.22	30–55	0.33–0.61
Zinc cyanide, $Zn(CN)_2$	10–30	0.085–0.26	5–30	0.043–0.26	34–40	0.29–0.34	2–4	0.017–0.034
Total sodium cyanide, NaCN	45–90	0.92–1.8	90–150	1.8–3.1	52–60	1.1–1.2	38–67	0.78–1.4
"Free" sodium cyanide, NaCN	8–35	0.16–0.71	5–45	0.1–0.92	4.5–6.5	0.09–0.13	2.25–4.5	0.05–0.09
Sodium hydroxide, NaOH	—	—	5–75†	0.13–1.9	30–37.5	0.75–0.94	—	—
Sodium carbonate, Na_2CO_3	30	0.28	0–15	0–0.14	37.5	0.35	0–30	0–0.28
Sodium sulfide, Na_2S	—	—	—	—	0.25	0.003	—	—
Rochelle salts, $KNaC_4H_4O_6 \cdot 4H_2O$	—	—	—	—	1.5–2.2	0.005–0.008	15–45	0.05–0.16
By Analysis:								
Copper as metal, Cu	18–35	0.28–0.55	43–71	0.68–1.1	11–14	0.17–0.22	21–39	0.33–0.60
Zinc as metal, Zn	5.6–17	0.088–0.27	2.8–17	0.044–0.27	19–22	0.30–0.35	1.1–2.2	0.017–0.035
pH	10.3–11		—		—		10.3	
Temperature, °C	25–55		—		21–29		25–60	
Cathode current density, A/m^2	30–150		—		400		20–200	
Percent Cu in anodes	70–80		70–75		35		92–95	

*Refers to color of deposit—true bronze is copper-tin.

†May use potassium hydroxide.

anodic oxidation of cyanide. Sodium hydroxide is added to some baths; it increases the conductivity. Since zinc also forms a hydroxyl complex, it tends to affect the zinc content of the deposit; increase in pH increases the zinc content, by transferring some zinc from the cyanide to the hydroxyl complex, from which it is more readily deposited.

Ammonium ion is used to control the color of the deposit; although not shown in the table, ammonia is usually added to brass baths in small amounts, and since it evaporates at the operating pH and temperature, it must be replaced on a regular basis. Some ammonia is formed as a result of cyanide decomposition.

Arsenic in small amounts (7.5 to 30 mg/L) is added as a "brightener" and to correct for copper-red deposits. Excess arsenic should be avoided: the deposit will become white and the anodes encrusted. Occasionally nickel is used for the same purpose, and phenol has been used to decrease the zinc content of the deposit.

Arsenic may be added by dissolving about 240 g As_2O_3 in 1 L of 50% sodium hydroxide solution; about 15 to 30 mL of this solution is added to each liter of plating bath. Arsenic is not used in "high-speed" baths, for which proprietary brighteners are available.

Since most brass plating is done for decorative purposes, the color of the deposit is more important than its composition. And, surprisingly, off-color red deposits can indicate either an excess or a deficiency of either metal; i.e., "red" deposits may indicate either too much or too little copper. The skill and experience of the operator are paramount in controlling decorative brass plating, since there are no simple rules.

For example, although most brass plating is done at relatively low current densities of about 50 A/m², increasing the cathode current density may either increase or decrease the zinc content of the deposit; the relative nobility of the two metals varies with changes in the bath composition. In general, at very low current densities (10 A/m²) the deposit is high in copper; as the current density increases, the copper content passes through a minimum at about 50 A/m² (for most bath formulations) and then gradually increases as the current density is increased to 500 A/m². If the free cyanide is increased, the effect of current density is less evident.

Each specific bath formulation requires its own temperature of operation. Increase in temperature normally causes an increase in copper content of the deposit: about 2 percent copper for each 1°C rise in temperature.

Control of pH is normally required; ordinary glass electrodes cannot be used, but type "E" electrodes are satisfactory. Raising the pH will correct "copper-red" deposits; this may be done with either sodium hydroxide or ammonia, though the latter is not suitable above pH 11.5. Over pH 11.5 either sodium or potassium hydroxide is used for pH control. Lowering of pH is accomplished with acid salts such as sodium bicarbonate: caution is necessary to avoid release of poisonous hydrocyanic acid gas.

Most brass baths operate with a cathode efficiency of about 75 percent; this may be increased by use of agitation, operating with as low a free cyanide as practical, maintaining a high metal concentration, and raising the temperature.

Anode efficiency may be increased by agitation, PR current, increase in temperature, and increase in free cyanide; decreasing the metal content; and using high-copper or pure copper anodes. Anode efficiencies of up to 95 percent are obtainable under ideal conditions.

Throwing power is improved by increasing free cyanide, using low metal content, high current density, and low pH. Obviously many of these remedies are contradictory, and compromises must be made.

In general anodes are of the same composition as the deposit, although use of copper anodes has been recommended. In any case, anode metal should be as pure as possible, and some metals, notably tin, antimony, nickel, arsenic, lead, and iron, as far as possible should be absent or present only in traces for good anode corrosion.

Bronze

In general tin alloys are plated rather easily; in part this is due to the ease with which alkaline stannate solutions can be mixed with solutions of the alkaline cyanide complexes of other metals. Thus many of the tin alloy–plating baths are of type 2 (see p. 371), in which the complexing agents for the two metals are different. Among the tin alloys reported have been those with copper, zinc, silver, gold, and cadmium, all plated from cyanide-stannate solutions; and a few, principally tin-nickel and tin-lead, for which the plating solution is of a different type.

Tin-lead alloys were already discussed in Chap. 13.

Tin-nickel alloys are discussed later. Of the stannate-cyanide solutions, the one that has enjoyed greatest acceptance in the industry is bronze, or copper-tin, usually containing 8 to 15 percent tin; the much higher-tin alloy speculum (45 percent tin) has had some success in decorative applications but, at least in the United States, has not been widely used.

Tin and copper can be codeposited from stannate-cyanide solutions in all proportions, depending on solution composition. Principal emphasis has been placed on the alloys in the 8 to 15 percent tin range, known as red bronze. At about 8 to 10 percent tin, the alloy is almost as red as copper; when the tin content rises to about 15 percent, the color is golden yellow; and at 20 percent or more tin the deposit is almost white. The following discussion is limited to the 8 to 15 percent range.

Bronze plating has been in limited use for some years, particularly as a stopoff in the nitriding of steel gears. The nickel shortage of the early 1950s prompted intensive search for nickel substitutes; white brass received considerable impetus from this development, as has been mentioned; and a further result was the finding that bronze, under substandard thicknesses of nickel or even no nickel at all—i.e., directly under chromium—was superior to copper in corrosion protec-

Table 16-2 Bronze-plating Solution

	g/L	*Molarity*
Potassium stannate, $K_2Sn(OH)_6$	60	0.20
Potassium hydroxide, KOH	7.5	0.13
Copper cyanide, CuCN	40	0.45
Potassium cyanide, KCN	90	1.38
Wetting agent*	2.5†	—
Addition agent*	5% vol.	—
Brightener*,‡	1.5–2.5†	
By analysis:		
Tin as metal, Sn	19–26	0.16–0.22
Free potassium hydroxide	4–11	0.07–0.20
Copper as metal, Cu	26–34	0.41–0.54
Free potassium cyanide	30–37.5	0.46–0.58
Temperature, °C	60–70	
Cathode current density, A/m²	200–1000	

*Proprietary.

†mL/L.

‡Optional.

tion of steel substrates. Bronze plating was, however, not widely adopted for this purpose, partly because nickel returned to good supply before convincing test results could be widely publicized. But the practical applicability of bronze deposits was demonstrated on a plant scale, and the deposit was adopted principally for its decorative color, in which it can resemble gold at a small fraction of the cost, and for some other uses such as builders' hardware and radio and television chassis. It is more expensive than copper, but superior to it in corrosion resistance, hardness, and resistance to cutting through in buffing.

A typical bronze-plating solution has the composition and operating conditions listed in Table 16-2. Without a brightener, deposits are mat but smooth; proprietary brighteners are available.

In this solution the copper is complexed by the cyanide ion and the tin by hydroxyl ion; thus the activities of copper and tin can be controlled independently. High free alkali inhibits deposition of tin and raises the copper content of the deposit, and high free cyanide has corresponding effects on the copper, raising the tin content.

Increasing temperature tends to favor tin deposition. Variations in metal content of the bath have direct corresponding effects on the composition of the deposit, but free alkali and free cyanide are more important control factors than metal content.

How closely the deposit composition must be controlled depends on the application. As an undercoat for other metals, or as a nitriding stopoff, deposit composition is not critical; for decorative applications, color matching is of greater significance. For gold-colored deposits, the temperature should be closely controlled in the 68–71°C range, and the free cyanide and tin contents should be near the high end of the range given in the table.

Although bronze anodes can be used, the current density at which they dissolve efficiently is rather low. More flexibility in operation, with no sacrifice in economy, is attained by using copper anodes and replenishing the tin content by regular additions of potassium stannate, preferably controlled by use of an ampere-hour meter. Dual anode circuits may also be used; in this case the tin anodes must be filmed, as for stannate tin plating.

A proprietary process for bronze plating, based on copper and potassium cyanide plus stannous sulfate or chloride, with sodium or potassium pyrophosphate to complex the tin, plus Rochelle salts, is available.

Tin-zinc–alloy plating (70 to 80 percent tin) has been promoted and used as a partial replacement for cadmium; the solderability of the deposit is excellent (in spite of the high zinc content), and the deposit appears to resist formation of gray tin (see Tin Plating). The corrosion resistance of the deposit is somewhat intermediate between those of cadmium and zinc, depending on exposure conditions. Typical bath compositions are listed in Table 16-3. With the increasing pressure to discontinue or reduce the use of cadmium plating, tin-zinc may become more prominent than it has been so far.

Table 16-3 Tin-Zinc Alloy Plating Solution

	Still tanks		*Barrels*	
	g/L	*Molarity*	*g/L*	*Molarity*
Potassium stannate, $K_2Sn(OH)_6$	120	0.40	95	0.32
Zinc cyanide, $Zn(CN)_2$	11.3	0.10	15	0.13
Potassium cyanide, KCN	30	0.46	34	0.52
Potassium hydroxide, KOH	7.5	0.13	11.3	0.20
By analysis:				
Tin as metal, Sn	38–53	0.32–0.45	32–40	0.27–0.34
Zinc as metal, Zn	4.5–7.5	0.07–0.11	6.8–10	0.10–0.15
Free potassium hydroxide	5–8.3	0.09–0.15	8.3–11.3	0.15–0.20
Total potassium cyanide*	38–53	0.58–0.81	41–60	0.63–0.92
Temperature, °C	63–67			
Cathode current density, A/m^2	100–800			
Anode current density,† A/m^2	150–250			
Anode composition	80% Sn, 20% Zn			

*Total KCN is higher than that added because it includes the cyanide combined with zinc in $Zn(CN)_4^{--}$.

†Anodes must be filmed as in stannate tin plating (see Tin Plating).

Tin-Nickel

Although stannate-cyanide baths for plating tin-nickel alloys have been investigated, the practical process for plating this alloy is based on acid fluoride solutions. The alloy contains 65 percent tin and 35 percent nickel; this corresponds closely to the intermetallic compound SnNi, and x-ray examination confirms this structure. This phase does not appear on the thermal equilibrium diagram of the tin-nickel system, and it has been prepared only by electrodeposition.

The deposit is almost inert to a wide variety of reagents; when free of pores, the deposit provides excellent protection from corrosion in outdoor atmospheres, both industrial and marine. For application over a steel substrate, a copper undercoat—as little as a mere flash plate—substantially improves its performance.

The composition and operating conditions for the tin-nickel–plating solution are listed in Table 16-4. It has been suggested that about 50 g/L of ammonium chloride be added to a fresh solution to regulate the internal stress; once the bath has been worked for a while, further additions are unnecessary.

Additives and brighteners have been proposed, but are not in general use.

The deposit is fairly bright, but not comparable to a fully bright nickel; it has a slightly pink tinge, unlike nickel or chromium, and is thought by many to be more pleasing than these finishes. It has been used on indoor appliances such as toasters and percolators. The deposit is solderable, though not as easily as tin or tin-lead; solderability can be improved by a very thin flash coating of gold.*

Tin-nickel plate is used to some extent as an etch resist in the production of printed circuits.

The composition of the deposit remains constant at 65/35 tin-nickel in spite of fairly wide fluctuations in bath composition and operating conditions. This constancy has been attributed to the existence in the solution of a binuclear complex containing tin and nickel in the same ion, in the 1:1 atomic ratio, from which the deposition takes place. But the exact mechanism of the action of the tin-nickel process has not been entirely elucidated.

In operating the bath, the most important variables are the pH and the fluorine content. The latter should be at least equal to the total tin content, to ensure that enough fluorine is present to complex all the tin as SnF_4^{--} and SnF_6^{--}; the latter results from oxidation of stannous to stannic tin as the bath ages. In SnF_4^{--}, $(F^-) = 0.64$ g/g Sn, and in SnF_6^{--}, $(F^-) = 0.97$ g/g Sn. Thus if $(F^-) = (Sn)$, enough F^- is present to fulfill the above condition with a margin of safety.

Although anodes of tin-nickel alloy can be used, they are both difficult to cast and brittle; the NiSn phase cannot be prepared thermally, and such anodes will

*See, for example, M. Antler, M. Feder, C. F. Hornig, and J. Bohland, *Plating and Surface Finishing*, **63**(7), 30 (1976).

Table 16-4 Tin-Nickel–plating Solution

	g/L	*Molarity*
Stannous chloride, $SnCl_2$	49	0.26
(or $SnCl_2 \cdot 2H_2O$)	(58)	0.26
Nickel chloride, $NiCl_2 \cdot 6H_2O$	300	1.26
Ammonium bifluoride, NH_4HF_2	56	1.0
Ammonium hydroxide, NH_4OH	to pH 2.0–2.5	
By analysis:		
Stannous tin, Sn^{++}	26–37.5	0.22–0.32
Nickel, Ni	60–83	1.0–1.4
Total fluorine*	34–45	1.8–2.4
Temperature, °C	65–71	
Cathode current density, A/m^2	100–300	
Anode current density, A/m^2	to 500	

*Total fluorine should be at least as high as total tin: Sn(II) + Sn(IV); see text.

be mixtures of Ni_3Sn_4 and Ni_3Sn_2. If separate anodes of tin and nickel are used, the former are subject to intergranular corrosion leading to their gradual disintegration. The most practical system for replenishment of the bath is the use of nickel anodes, with additions of stannous chloride on a scheduled basis, preferably using an ampere-hour meter; the nickel anodes should be bagged with nylon. Such replenishment would at first sight appear to entail a buildup of chloride ion in the bath, but in practice this effect has not been experienced. It is probable that chlorine is evolved at the anodes; since good ventilation is required in operation, such evolution goes unnoticed.

pH is controlled by ammonium hydroxide (to raise) or hydrofluoric acid or ammonium bifluoride (to lower). Fluorine deficiency is corrected by either of these latter compounds.

The bath is a solvent for most metals, so that work dropped into the tank should be recovered promptly; impurities are best removed by dummying. Hull cell tests are useful in control.

The solution contains large amounts of fluoride ion at high temperature and low pH; ventilation must be efficient and workers suitably protected by resistant clothing and rubber gloves, and equipment must be corrosion-resistant. Coils and heaters may be of graphite, nickel, or heavily nickel-plated steel or copper. Note that hydrofluoric acid presents a particularly insidious hazard when its fumes are inhaled or the liquid comes into contact with the skin.

Although tin-nickel may be plated directly on steel, the outstanding corrosion resistance is realized only if an undercoat of copper or bronze is used. ASTM Standard B 605 specifies thicknesses for various service conditions and offers further details.

Tin-nickel is thermally stable up to about 500°C*; above this temperature it disproportionates into Ni_3Sn_2 and Ni_3Sn_4, with a consequent increase in volume which may cause spalling. Any high-temperature application must be viewed with this in mind.

Fluoborate baths have also been proposed for plating the alloy, but so far as is known formulations have not been published.

Tin-Cobalt

Tin-cobalt alloys, sometimes containing a small amount of a third metal and comparable in many ways to tin-nickel, have been promoted as a replacement for decorative chromium deposits. Advantages claimed include higher efficiency, color resembling that of chromium, better throwing power and covering power, and ease of barrel plating, known to be a problem with chromium. At the time of writing, literature available had been either promotional or only in patents, so that no evaluation of these claims can be considered here.†

*There is some disagreement on this maximum temperature; lower figures have been reported.

†Some technical data are given by J. Hyner, *Plating and Surface Finishing,* **64** (2), 32 (1977).

PART THREE

RELATED PROCESSES

17
Autocatalytic ("Electroless") Plating

There are several methods of applying a metallic coating to a substrate which do not require the application of an outside source of electric current, but which nevertheless belong in the repertory of the electroplater. Some methods, including vacuum metalizing, hot dipping, hard facing, metal spraying, and siliconizing or chromizing, are outside our scope; but three techniques are closely enough related to electroplating to be included. These are contact plating, immersion plating, and autocatalytic, or "electroless," plating.

Contact plating merely replaces the outside source of current with an internal galvanic couple, which provides the required flow of electrons. It is an old art, seldom used any more. For example, pieces of work to be plated may be piled alternately with pieces of zinc in a solution containing a tin salt; the galvanic couple set up between the zinc sheets and the work deposits a coating of tin on the latter. Or the work may be hung on aluminum wires in a stannate tin bath, with comparable effect. This technique will not be discussed further; it is rather obvious and in any case obsolescent.

In immersion plating (also called *cementation* or *displacement deposition*) the work is less noble than the metal in solution; the dissolved metal deposits on the work. As already discussed, most such deposits are of no practical utility, being nonadherent, powdery, or both; but some tin immersion processes are in use, as are a few processes for "gold wash." (Coincidentally, many, if not most, contact and immersion processes are applicable to tin; the explanation for this situation is not obvious.)

The third technique, and the present subject, is autocatalytic or electroless plating, in which the deposited metal is reduced from its ionic state in solution by means of a chemical reducing agent rather than by an electric current.

There exists considerable confusion between electroless and immersion plating; since the increase in popularity of the electroless method, many workers, unfortunately including many whose papers appear in the literature, have used the word *electroless* to include processes properly classified as immersion or cementation reactions.* The distinction between the two methods is a real one that should be observed.

Brenner and Riddell invented ''electrodeless'' nickel plating in 1946, more or less serendipitously during an investigation of the effect of chemical additives in nickel plating. One of the additives tried was sodium hypophosphite, which yielded a cathode efficiency of more than 100 percent; this led to the correct conclusion that chemical reduction was involved, and further investigation uncovered the original process. The practical utility of such a process was quickly appreciated, and interest in and investigations of chemical reductive plating have continued almost without letup since that time. There are now several practical techniques for electroless (the word soon lost the ''-de-'') plating, suitable not only for nickel but for several other metals as well.

More recently, the name *autocatalytic* has been preferred, perhaps because of the misuse of the term *electroless,* as mentioned above. The words are interchangeable and will be so used in this chapter. *Autocatalytic plating* may be defined as ''deposition of a metallic coating by a controlled chemical reduction that is catalyzed by the metal or alloy being deposited'' (ASTM B 374).

The process provides a continuous buildup of a metal or alloy coating on a suitable substrate by simple immersion in an appropriate aqueous solution; a chemical reducing agent in the solution supplies the electrons for the reaction $M^{n+} + n\ e^{-} \rightarrow M^0$, but the reaction takes place only on a ''catalytic'' surface. Electroless plating thus differs in both its mechanism and its results from:

1. Immersion or displacement plating, which requires no reducing agent in the solution. The electrons are furnished by the substrate itself, and immersion deposition ceases as soon as the substrate is completely covered by the coating, whereas electroless plating knows no limit to the thickness of deposits obtainable.
2. Homogeneous chemical reduction processes such as silvering, where deposition occurs indiscriminately over all objects in contact with the solution, and often in the body of the solution itself.

Since autocatalytic plating takes place only on a catalytic surface, it is evident that once deposition is initiated, the metal deposited must itself be catalytic if deposition is to continue. Not all metals, therefore, are capable of being plated autocatalytically. Electroless deposition progresses essentially linearly with time, similar to electroplating at constant current density; there is, at least theoretically, no limit to the thickness of deposits that can be produced.

The chemical reducing agents used in electroless plating are much more expensive sources of electrons than the electric current; where conventional electroplating techniques are suitable, electroless plating cannot compete, nor is it at present applicable to some metals that can be easily electroplated, e.g., tin

*Is this due to a desire to use ''in'' words?

and chromium. But electroless plating possesses a unique combination of characteristics that render it useful in many applications:

1. The throwing power is essentially perfect; deposits are laid down on any surface to which the solution has free access, with no excessive buildup on projections or edges.
2. Deposits are often less porous than electrodeposits.
3. Power supplies, electric contacts, bus bars, and electrical measuring instruments are not required.
4. Deposits can be produced on nonconductors (with appropriate pretreatment).
5. Some deposits have unique chemical, mechanical, or magnetic properties.

Electroless plating has been applied to produce deposits of nickel, cobalt, palladium, platinum, copper, gold, silver, and some alloys containing one or more of these metals (reports of electroless chromium deposition have not been confirmed). Reducing agents used have included hypophosphite, formaldehyde, hydrazine, borohydrides, amine boranes, and some derivatives of these compounds. The original electroless plating process used sodium hypophosphite to produce deposits of nickel, and this process (or modifications of it) is still perhaps the most important of the electroless plating processes. The deposits are not pure nickel, but alloys of nickel with 3 to 15 percent phosphorus.

Nickel

Hypophosphite-based Baths

Sodium hypophosphite was the reductant first used by the inventors of electroless nickel plating, and it remains the most important. The composition of these baths has been modified in the ensuing 30 years but not radically changed. The essential ingredients are a nickel salt, sodium hypophosphite as reducing agent, and a compound (usually a salt of an organic acid) which serves as both a buffer and a mild complexing agent for nickel. Improvements have consisted in addition of materials that increase the rate of deposition and improve the stability of the bath. The last-named is an important factor, since one of the principal sources of difficulty in electroless plating is "catastrophic" or spontaneous decomposition, which is the tendency, under some conditions, for the reducing action to take place not only on the catalytic surface but throughout the solution, leading to the precipitation of all the metal contained in the form of a powder, or precipitation on the walls of the containing vessel.

No more than one g-atom of nickel is produced per mole of hydrogen gas evolved from the reaction, which may be formulated as:

$$\underset{\text{hypophosphite ion}}{2H_2PO_2^-} + 2H_2O + Ni^{++} \rightarrow Ni^0 + H_2 + 4H^+ + \underset{\text{phosphite ion}}{2HPO_3^{--}} \qquad (17\text{-}1)$$

The mechanism of this reaction has been the subject of considerable study and some disagreement; but at present the most likely route seems to be Lukes'* hydride mechanism, in which the actual reducing agent is a hydride ion formed by catalytic decomposition of the hypophosphite ion:

$$H_2PO_2^- + H_2O \xrightarrow{\text{catalytic surface}} HPO_3^{--} + 2H^+ + H^- \qquad (17\text{-}2)$$

The hydride ion H^- then reacts with Ni^{++} (or a hydrolyzed nickel ion adsorbed on the surface, such as $NiOH^+$):

$$2H^- + Ni^{++} \rightarrow Ni^0 + H_2 \qquad (17\text{-}3)$$

Hydride ion may also be consumed by reaction with hydrogen ion:

$$H^- + H^+ \rightarrow H_2 \qquad (17\text{-}4)$$

This reaction reduces the efficiency of hypophosphite usage.

According to Eq. (17-1), 2 mol of sodium hypophosphite should produce 1 g-atom of nickel; in practice only about 0.7 g-atom of nickel is usually produced. Since sodium hypophosphite is expensive, efficiency in its utilization is a principal object of process research; alkaline baths are somewhat higher in this respect than acid baths.

Electroless nickel deposits from hypophosphite baths are not pure nickel; they contain from 3 to about 15 percent phosphorus, depending on bath composition and operating conditions. The phosphorus is present (after heat treatment) as nickel phosphide, Ni_3P. The reactions for phosphide formation have not been entirely elucidated; it has been proposed that phosphorus derives from an intermediate of short life, such as metaphosphorous acid, HPO_2, which can be formed along with the hydride ion [Eq. (17-2)]. Then

$$2PO_2^- + 6H^- + 4H_2O \rightarrow 2P + 3H_2 + 8OH^- \qquad (17\text{-}5)$$

Such a mechanism is at least consistent with the observation that the phosphorus content of the deposit decreases with increasing pH of the bath; see Fig. 17-1.

Other mechanisms than the hydride theory have been proposed for electroless nickel plating using hypophosphite; but at present this route seems most widely accepted.

*R. M. Lukes, *Plating*, **51**, 969 (1964).

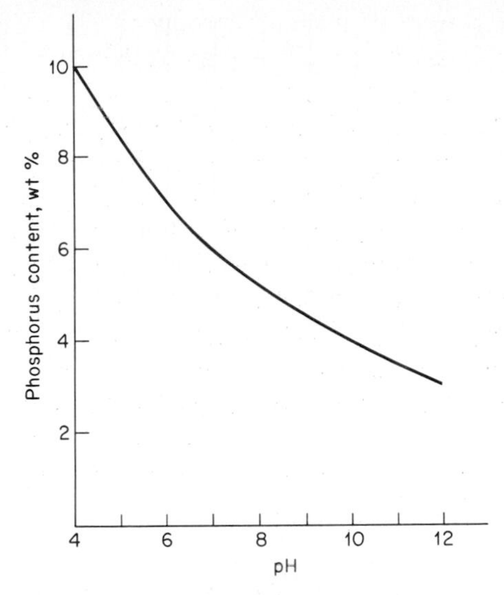

Fig. 17-1 Effect of pH on phosphorus content, electroless nickel.

Bath Compositions

Two distinctly different types of electroless nickel baths are in use; acid baths at pH 4 to 7 and ammoniacal baths at pH 8 to 11. Commercially the acid baths have been the more important.

Table 17-1 lists typical formulations for acid and alkaline baths (baths A and D) as well as two of the many modifications of acid solutions which have been proposed (baths B and C). In bath A, the hydroxyacetate (glycolate) serves several functions: as a buffer, as a complexing agent for nickel, and as a rate promoter. In the alkaline bath, the citrate and ammonium ions provide the complexing action required to prevent precipitation of nickel hydroxide.

The formulas listed as baths B and C contain various additives for stabilization and increasing the rate of deposition. Baths containing fluoride ion are suitable for plating on aluminum and titanium.

In addition to these formulations, many suppliers offer proprietary baths, which may or may not be essentially similar to those in the table.

EFFECTS OF VARIABLES AND ADDITIVES

pH

From acid baths, the deposition rate decreases with pH (Fig. 17-2). It is usually not practical to operate at pH below 4, where the rate decreases much more sharply than between 4 and 6. Conversely, operation at pH above 6 carries the

Table 17-1 Electroless Nickel-plating Solutions, Hypophosphite Type

Bath:	*A (acid)*		*B (acid)*		*C (acid)*		*D (alkaline)*	
	g/L	M	*g/L*	M	*g/L*	M	*g/L*	M
Nickel chloride, $NiCl_2 \cdot 6H_2O$	30	0.125	—	—	—	—	30	0.125
Nickel sulfate, $NiSO_4 \cdot 6H_2O$	—	—	21	0.08	12	0.045	—	—
Sodium hypophosphite, $NaH_2PO_2 \cdot H_2O$	10	0.094	24	0.23	22	0.21	10	0.094
Sodium hydroxyacetate, $HOCH_2COONa$	50	0.51	—	—	—	—	—	—
Acetic acid, CH_3COOH	—	—	—	—	9.4	0.16	—	—
Lactic acid, $CH_3CHOHCOOH$	—	—	27	0.30	—	—	—	—
Molybdic oxide, MoO_3	—	—	—	—	0.009	—	—	—
Propionic acid, CH_3CH_2COOH	—	—	2.2	0.03	—	—	—	—
Lead acetate, $Pb(C_2H_3O_2)_2$	—	—	0.001	—	—	—	—	—
1,3-diisopropyl thiourea, $(C_3H_7NH)_2CS$	—	—	—	—	0.004	—	—	—
Sodium citrate, $Na_3C_6H_5O_7 \cdot 2H_2O$	—	—	—	—	—	—	84	0.29
Ammonium chloride, NH_4Cl	—	—	—	—	—	—	50	0.93
pH	4–6		4.6		5.5		8–10	
Temperature, °C	90		95		82		90	
Plating rate, μm/h	15		20		—		6	

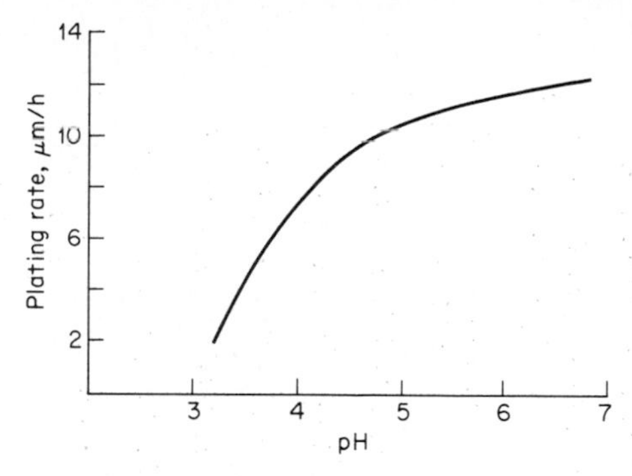

Fig. 17-2 Effect of pH on plating rate, acid electroless nickel.

danger of precipitation of sparingly soluble compounds; the most suitable pH thus is usually between 4.2 and 5. Equations (17-1) and (17-2) show that hydrogen ions are products of the plating reaction; in order to stabilize the pH in operation, therefore, buffers such as hydroxyacetate, acetate, citrate, succinate, and lactate are included in the formulations of the baths. Even so, periodic additions of alkali will be necessary to neutralize the acid formed in operation. A 10% sodium hydroxide or sodium carbonate solution should be added slowly with stirring to the well-cooled bath; ammonia is also sometimes used.

Alkaline electroless nickel baths are not so sensitive to pH within the normal range; it is usual to maintain pH by periodic addition of ammonia to replace that evaporated as well as to neutralize the hydrogen ions formed in the deposition reaction.

Phosphite

As shown by Eqs. (17-1) and (17-2), phosphite is a normal product of the deposition reaction; this may cause precipitation of nickel phosphite, which is only sparingly soluble. Such precipitation removes nickel from the bath; in addition, it may lead to inferior deposits and may trigger spontaneous decomposition. Complexing agents for nickel ion, such as hydroxyacetate, lactate, or glycine, tend to inhibit this precipitation. However, on continued use, phosphite-ion concentration continues to increase, and sooner or later it must be removed or the bath must be discarded. Ammoniacal baths are somewhat more tolerant of phosphite ion than are acid baths.

Temperature

Operating temperature is one of the most important variables affecting the rate of deposition, which increases nearly exponentially with temperature; see Fig. 17-3.

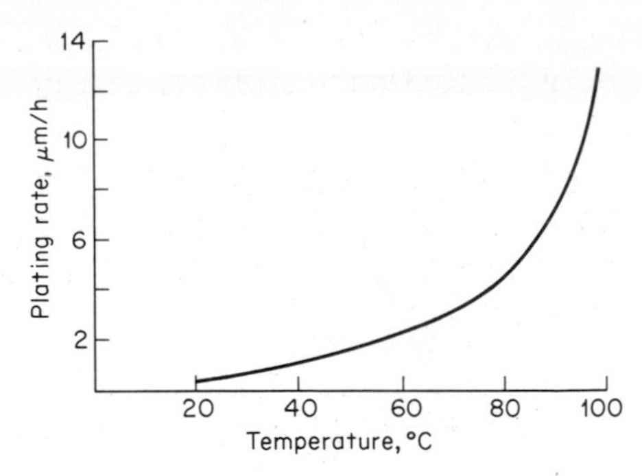

Fig. 17-3 Effect of temperature on plating rate, electroless nickel.

Most baths, therefore, are used at as high a temperature as practical in order to take advantage of the higher rates of deposition.

Baths must be heated in such a way that local overheating is avoided; overheating may cause spontaneous decomposition. Tanks jacketed with hot water or steam are advantageous. A few baths, mostly proprietary, may be operated near room temperature; usually they are used for application of thin deposits on activated nonconductors, to be further plated electrolytically.

Exaltants

In acid electroless nickel baths, some anions have the property of increasing the deposition rates; these have been called *exaltants*. Many of them are identical to the complexing agents already mentioned: lactate, hydroxyacetate (or glycolate), and succinate; other exaltants include propionate, acetate, glycine, malonate, and fluoride. Their mode of action is still uncertain, although some theories have been proposed.

Stabilizers

Although electroless nickel baths may be operated for considerable periods, there comes a time when catalytic nuclei form throughout the bath and cause rapid and "catastrophic" or spontaneous decomposition: a black nickel-phosphorus precipitate forms throughout the bath. This tendency is greatest with the more rapidly acting baths; it may be triggered by precipitation of insoluble phosphites or hydroxides, or by particulate matter from the atmosphere such as dust. Baths may become unstable if too large an area of work is used per volume of solution; it has been suggested that about 125 cm^2/L is appropriate.

Since spontaneous decomposition is presumably promoted by an excess of catalytic surface in solution, the addition of certain catalytic poisons may overcome the tendency for decomposition. Organic and inorganic thio compounds, and some heavy-metal cations, act in this way. For example, 1 to 5 mg/L of thiosulfate, 5 to 10 mg/L ethyl xanthate, 1 to 5 mg/L lead ions, or 2 to 10 mg/L stannous ions can be used to stabilize electroless nickel acid baths. Other compounds found to have this effect are molybdenum trioxide, arsenic, and thiourea and its derivatives; some examples are shown in Table 17-1. Since the stabilizers also decrease the catalytic activity, they must not be present in too high a concentration or else the bath will no longer function. When their concentrations are in the optimum range, apparently they adsorb preferentially on the minute nuclei of particulate matter in preference to the catalytic surface on which plating is taking place; but excess of stabilizer will adsorb on any surface.

Most studies of stabilizers have been carried out on acid electroless nickel baths; little information is available on their action in alkaline solutions.

Brighteners

Electroless nickel deposits are semibright to mat as deposited; many of the brighteners used for bright nickel plating also act to brighten electroless nickel deposits.

Agitation

Agitation is desirable in electroless nickel plating to aid in dislodging gas bubbles, replenish solution in holes and crevices, and decrease the concentration gradient in the layer next to the work (analogous to the "cathode layer" in electroplating). Air, work rod, or solution agitation may be used. Agitation also may increase deposition rate.

Substrates

Autocatalytic deposition takes place only on "catalytic" substrates. These include nickel, cobalt, steel, rhodium, and palladium; some more active metals such as aluminum and beryllium are also receptive, perhaps because the initial deposit is nickel produced by chemical displacement.

Noncatalytic metals such as copper, brass, and silver can be made receptive by touching them with a part that is actively plating in the electroless bath, or by momentarily making them cathodic using a source of direct current. Another method is to immerse the metal for a short time in a dilute acidic solution of palladium chloride, followed by thorough rinsing (necessary because otherwise the palladium chloride might cause decomposition of the electroless bath); the palladium chloride solution may contain about 0.1 g/L $PdCl_2$ and 0.2 mL/L HCl. The thin immersion deposit of palladium is sufficient to initiate the nickel plating; once this is formed, the nickel itself is, of course, autocatalytic.

These procedures are not successful with some metals including lead, cadmium, bismuth, and tin. They may be plated with electroless nickel by first applying a copper or nickel strike electrolytically.

The most commonly used substrate for electroless nickel coatings is steel;

coated steel parts are normally baked at about 260°C for 1 to 4 h to improve adhesion and relieve possible hydrogen embrittlement. Aluminum and magnesium benefit from electroless nickel plating, which confers a hard and corrosion-resistant coating on these rather soft metals. Although it is possible to plate aluminum directly, it is usually given a zincate treatment before electroless plating. Plated aluminum should be heated for 1 h at 190°C to improve adhesion, if this can be done without deleterious effects on the substrate; plated high-strength aluminum alloys may be heated at 140°C for 2 h or more.

Titanium alloys are electroless nickel-plated to prevent wear and galling; adhesion is improved by heat treatment at 425°C or more, and for best adhesion diffusion bonding is accomplished by heating in vacuo for 4 h above 680°C.

Stainless steels may be given a Wood's nickel strike before electroless plating for better adhesion.

Electroless nickel is widely used for plating on nonconductors, considered in more detail in Chap. 19. For this application low-temperature baths are preferred, usually of the alkaline type. Only enough electroless nickel is deposited, after activation with palladium, to permit building up the thickness by conventional electroplating.

Composition and Properties of Deposits

As previously mentioned, nickel deposits from hypophosphite electroless baths are not pure nickel, but contain up to 15 percent phosphorus. Deposits as plated are lamellar, consisting of supersaturated solid solutions of phosphorus in nickel. On heat treatment at 400°C, an intermetallic phase Ni_3P is precipitated, and the grain size increases. Deposits containing about 7 percent phosphorus consist of 50 percent (vol.) Ni_3P and 50 percent (vol.) Ni. Deposits lower in phosphorus consist of Ni_3P in a nickel matrix, and higher-P deposits just the reverse (Ni in a Ni_3P matrix); therefore there are often abrupt changes in properties of the deposits at around the 7 percent level.

Phosphorus Content

The phosphorus content of deposits is affected by bath pH: decrease in pH results in higher P content (see Fig. 17-1). Most acid baths produce deposits containing about 7 to 10 percent P, and most alkaline baths about 5 to 7 percent. Increase in hypophosphite content raises the P content of the deposits at a given pH. The effect of temperature varies with pH; increase in temperature of operation may either raise or lower the P content of the deposit, depending on the pH and composition of the bath.

Hardness and Wear Resistance

Electroless nickel deposits are harder and more abrasion-resistant than electroplated nickel; this property has accounted for a large proportion of the uses of the

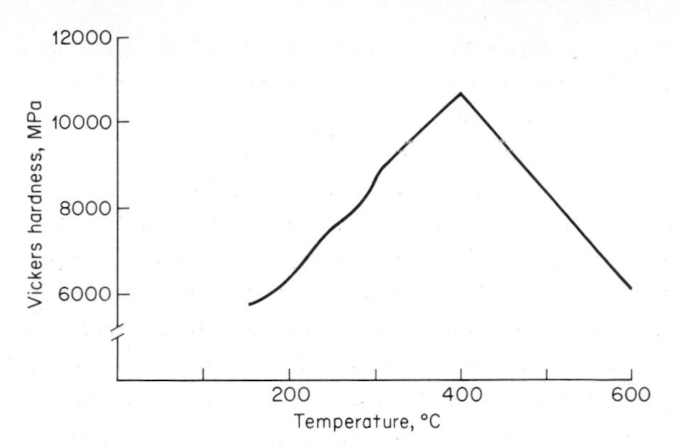

Fig. 17-4 Effect of heat treatment on hardness, electroless nickel.

deposit: on hydraulic cylinders, rotating shafts, pistons, gears, thread guides, cutting tools, and dies. Electroless nickel deposits as plated have a Vickers hardness of about 5000 to 6000 MPa (500 to 600 kg/mm^2); this can be increased to about 10 000 MPa (1000 kg/mm^2) by heat treating for 1 h at 400°C (see Fig. 17-4).*

Internal Stress

High stress is undesirable, since it may cause peeling, cracking, or blistering of the deposits. Internal stress increases with increase in bath pH or decrease in P content. The difference in coefficient of linear expansion between the deposit and the substrate affects the stress in the deposits. More specialized texts should be consulted for details.

Corrosion Prevention

Electroless nickel deposits tend to be lower in porosity than electrodeposited nickel coatings, and therefore are generally superior, thickness for thickness, to the latter. The protection afforded steel substrates is somewhat improved by heat treating for 3 h at 650°C or above, to provide a diffusion layer. Deposits from alkaline baths tend to be less protective than those from acid baths.

Deposit Uniformity

Since current distribution is of no concern in electroless plating, deposits can be much more uniform on shaped parts and in holes and recesses than electrolytically produced coatings. However, the solution still must have access to the

*See also E. Johnson and F. Ogburn, *Surface Technology,* **4,** 161 (1976); and Safranek, cited in the Bibliography.

substrate surface; therefore efficient agitation may be required, especially for parts having small-diameter holes or for the inside of tubing.

Magnetic Properties

Magnetic properties of electroless nickel deposits are often of interest. Electroless nickel deposits from acid baths usually are not ferromagnetic; deposits from alkaline baths, containing less phosphorus, are slightly magnetic. Heat treatment increases magnetism. Magnetic gages can be used for measuring the thickness of as-plated electroless nickel deposits on steel, with little error, by assuming the deposits to be nonmagnetic.

APPARATUS

Owing to the inherent properties of electroless nickel baths—as well as electroless baths for other metals—the choice of tanks and auxiliary equipment is important; they must be such that the deposit does not form on the walls of the container, heating coils, pumps, filters, etc.

Vessels for containing electroless nickel baths may be of glass, plastics, or passivated stainless steel. They must be resistant to nitric acid, since this reagent must be used occasionally to dissolve any nickel deposit which eventually forms on them in spite of precautions. Polyethylene bags have been used as disposable tank linings. Anodized titanium has also been suggested. Some workers recommend setting up the plating tanks in pairs, so that the solution can be transferred, with filtration, from one to the other at the end of every day or shift; the empty tank is filled with nitric acid solution to dissolve residual nickel. This system permits immediate transfer from one tank to the other if incipient spontaneous decomposition threatens, owing to the presence of catalytic particles.

Unless the operation is strictly batch, with discard of the solution after a short operational life, electroless plating baths should be continuously filtered and replenished. A typical process scheme is shown in Fig. 17-5. The plating solution is cooled to about 60°C before replenishment and adjustment of pH and then filtered and reheated before it is returned to the plating tank. Even continuously replenished solutions, however, must eventually be dumped because of buildup of by-products.

NICKEL-BORON

Reducing agents other than sodium hypophosphite are suitable for producing electroless nickel (and other metal) deposits. Among them are borohydrides, amine boranes, and hydrazine. The boron-containing reductants yield deposits containing 0.3 to 10 percent boron. These deposits are harder as plated and more resistant to oxidation than Ni-P deposits. They are also heat-treatable to yield even greater hardness and wear resistance. Some applications for Ni-B deposits include knife blades, pump housings and impellers, glass molds, and polymer

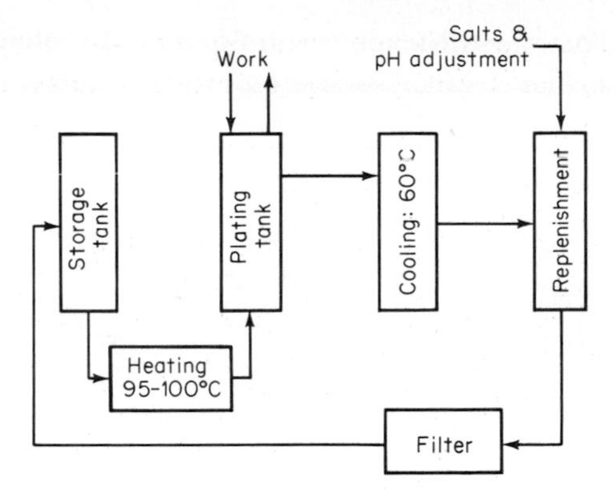

Fig. 17-5 Schematic diagram of electroless plating setup.

embossing tools. Ni-B deposits are also useful for their solderability and electrical properties, and they have been promoted as partial substitutes for gold in electrical and electronic applications.

Borohydride Baths

Sodium borohydride, $NaBH_4$, is a powerful reducing agent; theoretically one mole can reduce four equivalents of metal ions, vs. one for hypophosphite. Borohydrides decompose rapidly unless the solution is highly alkaline, and electroless baths using this agent are usually at pH above 11. Complexing agents must be used, of course, to keep nickel in solution; these include tartrate, citrate, ammonia, ethylenediamine, triethylenetetramine, EDTA, and succinate.

Stabilizers are required to inhibit spontaneous decomposition of the bath with the production of nickel and nickel-boron powders. Stabilizers include organic divalent sulfur compounds such as thiodiglycolic acid and acetylenedithiosalicylic acid; lead salts may also be used.

The deposition reaction is probably

$$2Ni^{++} + NaBH_4 + 2H_2O \rightarrow 2Ni^0 + 2H_2 + 4H^+ + NaBO_2 \qquad (17\text{-}6)$$

Thus 1 mol of borohydride should produce 2 g-atoms of Ni and at least 2 mol of hydrogen, if the utilization of borohydride is 100 percent efficient. The production of nickel boride may be represented as

$$4Ni^{++} + 2NaBH_4 + 6NaOH \rightarrow 2Ni_2B + 6H_2O + 8Na^+ + H_2 \qquad (17\text{-}7)$$

A typical formulation is listed in Table 17-2.

Baths are used in a batch operation, with replenishment of alkaline borohydride solution until 60 to 80 percent of the nickel has been deposited; then the bath must be discarded. Replenished baths sooner or later contain so much sodium metaborate, $NaBO_2$, that deposition rates are unacceptably low. An

Table 17-2 Electroless Nickel-Boron Solution, Borohydride Type

	g/L	*Molarity*
Nickel chloride, $NiCl_2 \cdot 6H_2O$	30	0.13
Sodium hydroxide, NaOH	40	1.0
Ethylene diamine, 70%, $(CH_2)_2(NH_2)_2$	86 mL	0.90
Sodium borohydride, $NaBH_4$	0.6	0.016
Thallium(I) nitrate $TlNO_3$	0.07	—
pH	13–14	
Temperature, °C	90	

aqueous solution of magnesium chloride may be used to precipitate magnesium metaborate.

Amine Borane Baths

Amine boranes are addition compounds of amines with boron hydride, of general formula $R_3N—BH_3$; R may be a hydrogen, alkyl, or aryl group. Dimethylamine borane (DMAB) is the most commonly used of this class; it is $(CH_3)_2HN—BH_3$. Other analogous compounds have been used. The deposits from these baths contain boron, similar to borohydride baths. At present, amine borane–based baths are somewhat more practical than the borohydride type; but since both enjoy only limited use, this situation is subject to change. A typical amine borane bath is shown in Table 17-3.

Both sodium borohydride and DMAB are considerably more expensive than sodium hypophosphite; this has inhibited commercial use of these baths. Nevertheless nickel-boron deposits have some desirable properties, as mentioned, and increased commercial acceptance is quite possible.

Table 17-3 Electroless Nickel-Boron Solution, Amine Borane Type

	g/L	*Molarity*
Nickel chloride, $NiCl_2 \cdot 6H_2O$	30	0.13
Sodium glycolate, $HOCH_2COONa$	15	0.15
Dimethylamine borane, $(CH_3)_2NHBH_3$	1.0	0.017
Lead acetate, $Pb(OOCCH_3)_2$	0.02	—
pH	5–6	
Temperature, °C	65	

Hydrazine Baths

Though not widely used, hydrazine (N_2H_4) as a reducing agent in electroless nickel has the advantage that the deposit is almost pure nickel instead of a phosphorus or boron alloy. Deposits are 97 to 99+ percent nickel; the balance consists of some oxygen and nitrogen, with other trace elements.

Two hydrazine-based baths are listed in Table 17-4.

Other Metals

Nickel remains preeminent among the metals deposited by chemical reductive, autocatalytic, or electroless methods. Nevertheless the same principles have been applied to the deposition of several other metals, which have had somewhat more limited but still significant practical application.

Cobalt

Cobalt is interesting principally for its magnetic properties in communications and computer applications: switching and memory devices. Cobalt is usually deposited from hypophosphite baths, analogous to the electroless nickel baths already discussed, and the deposits are cobalt-phosphorus alloys. A typical formulation is shown in Table 17-5.

The principles are much the same as for nickel. Cobalt, however, is deposited only from alkaline hypophosphite baths; deposition rates are somewhat slower than for nickel, and the tendency for spontaneous decomposition is somewhat less.

Amine boranes can be used as reducing agents, in a manner analogous to

Table 17-4 Electroless Nickel Solutions, Hydrazine Type

	Bath E		*Bath F*	
	g/L	*Molarity*	*g/L*	*Molarity*
Nickel chloride, $NiCl_2 \cdot 6H_2O$	4.8	0.02	—	—
Nickel acetate, $Ni(C_2H_3O_2)_2$	—	—	60	0.34
Hydrazine, N_2H_4	32	1.0	100 mL	3.1
Sodium tartrate, $Na_2C_4H_4O_6 \cdot 2H_2O$	4.6	0.02	—	—
Glycolic acid, $HOCH_2COOH$	—	—	60	0.79
Tetrasodium EDTA $(NaO_2C_2H_2)_4(NCH_2)_2$	—	—	25	0.066
pH	10.0		11.0	
Temperature, °C	95		90	

Table 17-5 Electroless Cobalt-Phosphorus Solution

	g/L	*Molarity*
Cobalt sulfate, $CoSO_4{\cdot}7H_2O$	32	0.11
Ammonium chloride, NH_4Cl	50	0.93
Sodium citrate, $Na_3C_6H_5O_7{\cdot}2H_2O$	84	0.29
Sodium hypophosphite, $NaH_2PO_2{\cdot}H_2O$	20	0.19
pH	9–10 with $NH_3(aq)$	
Temperature, °C	95	

nickel, to produce cobalt-boron alloys, but the process is not widely practiced. Table 17-6 shows the use of DMAB for electroless cobalt-boron deposition. Unlike hypophosphite cobalt baths, DMAB can be used under acidic conditions; deposition rates are somewhat higher, and the baths can be operated at room temperature.

Copper

The principal use for electroless copper deposits is as a part of the sequence for plating on nonconductors; it is widely used in the metalizing of printed-circuit boards and for plating on plastics (see Chap. 19), as the starting coat for subsequent electroplating. There is considerable controversy about whether electroless copper or electroless nickel is superior for these applications; both are used widely. As with nickel, many proprietary formulations are available.

Most commercial electroless copper baths employ formaldehyde as the chemical reducing agent. The reducing power of formaldehyde increases with pH; most baths are operated at pH above 11. The alkalinity is provided by sodium hydroxide. Other ingredients include a copper(II) salt, usually copper(II) sulfate; a complexing or chelating agent to hold the copper in solution; and various stabilizers and other additives. The most common chelating agent is tartrate ion, but several others are used, including EDTA, amines and amine derivatives, and glycolic acid.

The stoichiometry of the reaction is as follows:

$$Cu^{++} + 2HCHO + 4OH^- \rightarrow Cu^0 + H_2 + 2H_2O + 2HCO_2^- \qquad (17\text{-}8)$$

indicating that 2 mol of formaldehyde and 4 mol of alkali are required for each g-atom of copper deposited. In practice, more formaldehyde and alkali are consumed, indicating that side reactions take place; one of these is probably the disproportionation of formaldehyde with alkali to methanol and formate ion:

$$2HCHO + OH^- \rightarrow \underset{\text{methanol}}{CH_3OH} + \underset{\text{formate}}{HCOO^-} \qquad (17\text{-}9)$$

The mechanism of copper deposition is believed to be similar to that for nickel, involving a hydride ion originating from the formaldehyde:

$$2H^- + Cu^{++} \rightarrow H_2 + Cu^0 \qquad (17\text{-}10)$$

Electroless copper baths are not so intrinsically stable as electroless nickel baths; one of the principal problems in the development of electroless copper plating was spontaneous decomposition, and many of the baths had rather short operating life. The decomposition probably arises from a competing reaction, which is noncatalytic (does not require a catalytic surface) and takes place in the bulk of the solution:

$$2Cu^{++} + HCHO + 2OH^- \rightarrow Cu_2O + HCOO^- + 3H_2O \qquad (17\text{-}11)$$

The cuprous oxide (Cu_2O) particles can disproportionate:

$$Cu_2O \rightarrow Cu^0 + CuO \qquad (17\text{-}12)$$

forming catalytic copper powder in the bulk of the solution; this causes general bath decomposition. Several means of inhibiting or controlling this undesirable side reaction have been proposed. Simply bubbling air or oxygen through the bath is effective, as is vigorous agitation, which tends to maintain air saturation in the bath. Small amounts of oxidizing agents such as chromate are also effective.

Various catalytic poisons stabilize electroless copper baths, but they must be present in sufficiently small amounts to enable plating to proceed at practical rates. Many such stabilizers have been proposed, including mercaptobenzothiazole (MBT), thiourea, cyanide ion, vanadium pentoxide, methyl butynol, selenium compounds, and others.

Particulate matter (dust, etc.) entering the bath from the air can also serve as nuclei for copper deposition and thus decomposition of the bath; continuous filtration of particles larger than about 2 μm is recommended, preferably using replaceable polypropylene-wound cartridge filters.

Table 17-6 Electroless Cobalt-Boron Solution

	g/L	*Molarity*
Cobalt sulfate, $CoSO_4 \cdot 7H_2O$	25	0.089
Sodium succinate, $(CH_2COONa)_2$	25	0.15
Sodium sulfate, Na_2SO_4	15	0.11
Dimethylamine borane, $(CH_3)_2NHBH_3$	4	0.069
pH	5	
Temperature, °C	70	

Table 17-7 Electroless Copper Solutions

Bath:	*G*		*H*		*J*		*K*		*L*	
	g/L	M	*g/L*	M	*g/L*	M	*g/L*	M	*g/L*	M
Copper(II) sulfate, $CuSO_4 \cdot 5H_2O$	3.6	0.014	5	0.02	30	0.12	30	0.12	13	0.052
Sodium potassium tartrate, $KNaC_4H_4O_6 \cdot 4H_2O$	25	0.089	25	0.089	100	0.35	140	0.50	66	0.23
Sodium hydroxide, NaOH	3.8	0.10	7	0.18	50	1.25	42	1.1	20	0.5
Sodium carbonate, Na_2CO_3	—	—	—	—	32	0.30	25	0.24	—	—
"Versene T"	—	—	—	—	—	—	17	—	—	—
Mercaptobenzothiazole, MBT*	—	—	—	—	—	—	—	—	0.013	—
Formaldehyde, HCHO, 37%†	10	0.12	10	0.12	29	0.36	167	2.1	38‡	0.47
Temperature, °C	22		20		25		25		25	
Plating rate, μm/h	0.5		0.75		2.5		20		2.5	

*Added as 10g/L MBT in 0.2 *M* NaOH: $C_6H_4N{=}C(SH)S$ (ring)

†mL/L.

‡12.5 percent methanol as preservative.

Saubestre* reviewed the various methods for stabilizing electroless copper baths. He concluded that air-agitated filtered baths should contain 40 to 250 g/L methanol and chelating agents for copper(I) ion, preferably compounds containing nitrogen and sulfur in a ring structure.

Typical formulations for electroless copper plating are listed in Table 17-7. Bath G is a dilute bath that may be maintained almost indefinitely by replenishment of the constituents, continuous filtration, and lowering of pH with sulfuric acid during idle periods. Bath H is somewhat more concentrated; it may also be used for continuous operation. The area of work in the bath should be limited to 250 cm^2/L of solution, to avoid premature spontaneous decomposition. The rate of plating is pH-dependent, with a maximum at about pH 12.8. Both baths G and H are rather slow in rate of deposition, but are sufficiently inexpensive to be used for batch operation and then discarded.

Baths J, K, and L work rather more rapidly; bath L contains a little mercaptobenzothiazole to stabilize it. Thiourea, 0.1 mg/L, has been used similarly.

Vessels for containing electroless copper baths are usually of plastic such as polyethylene or polypropylene. A spare tank should be available for transfer of solution in case copper should deposit on the tank.

Careful control and maintenance of proper concentration and pH are important in operation of electroless copper baths.

Electroless copper will deposit spontaneously on copper, gold, silver, platinum, iron, cobalt, nickel, palladium, or rhodium. The most important use of electroless copper in effect uses palladium as the substrate: in the preparation of nonconductors for electroplating and in circuit board manufacture, the sensitiz-

*E. B. Saubestre, *Plating,* **59,** 563 (1972).

Table 17-8 Electroless Palladium Solutions

	Bath M		*Bath N*	
	g/L	*Molarity*	*g/L*	*Molarity*
Palladium(II) chloride, $PdCl_2$	2	0.01	3.9	0.02
Ammonium hydroxide, NH_4OH, 27%	160*	1.2	350*	2.7
Ammonium chloride, NH_4Cl	26	0.49	—	—
Disodium EDTA, $(NaO_2C_2H_2)_2(HO_2C_2H_2)_2(NCH_2)_2$	—	—	34	0.1
Sodium hypophosphite, $NaH_2PO_2 \cdot H_2O$	10	0.094	—	—
Hydrazine, N_2H_4	—	—	0.3	0.01
Temperature, °C	50		80	

*mL/L.

Table 17-9 Electroless Gold Solution

	g/L	Molarity
Potassium gold cyanide, $KAu(CN)_2$	5.8	0.02
Potassium cyanide, KCN	13	0.20
Potassium hydroxide, KOH	11.2	0.20
Potassium borohydride, KBH_4	21.6	0.40
Temperature, °C	75	

ing of the substrate is accomplished by deposition of a very thin deposit of palladium.

Palladium

Palladium can be plated autocatalytically using either hydrazine or hypophosphite as reducing agent. Formulations are listed in Table 17-8.

Palladium deposits have been applied for electric contacts and connectors, especially in the communications industry; it may substitute for gold in some applications.

Gold

Most so-called electroless gold-plating formulations are, in fact, immersion or displacement plating baths, but at least one truly autocatalytic process uses the formulation listed in Table 17-9.

Silver

A DMAB bath for silver has the formula listed in Table 17-10.

Table 17-10 Electroless Silver Solution

	g/L	Molarity
Silver cyanide, $AgCN$	1.34	0.01
Sodium cyanide, $NaCN$	1.49	0.03
Sodium hydroxide, $NaOH$	0.75	0.02
Dimethylamine borane, $(CH_3)_2NHBH_3$	2.0	0.035
Thiourea, $CS(NH_2)_2$	0.0003	—
Temperature, °C	55	

Other Metals

Reports of the electroless or autocatalytic deposition of several other metals have appeared from time to time. Many are based on the misuse of the word *electroless* already discussed, and some are simply unverifiable (at least up to now) by independent workers. It is believed that the metals already mentioned (plus perhaps other metals of the platinum group) exhaust the scope of electroless plating at this time.

18 Immersion Plating

Immersion plating is the deposition of a metallic coating on a substrate by chemical replacement from a solution of a salt of the coating metal. Thus it differs from autocatalytic plating in not requiring a chemical reducing agent to reduce the metal ions to metal, and from "contact" plating in not requiring a galvanic couple to generate an internal flow of electrons. The reducing agent is the substrate metal itself, and the coating consists of atoms of the metal in solution, reduced from their ionic state by the substrate.

Immersion methods are useful in many industrial processes, but most are not concerned with the quality of the coating. A widely used method of recovering the last traces of copper from leaching solutions, mill tailings, and the like is to "cement out" the copper by using scrap iron:

$$Fe + Cu^{++} \rightarrow Cu + Fe^{++}$$

Here the aim is recovery of the copper, using the much cheaper metal iron to accomplish it; and the physical form in which the copper is recovered is of little moment. Similarly, zinc dust is used to recover cadmium during the zinc refining cycle:

$$Zn + Cd^{++} \rightarrow Cd + Zn^{++}$$

Zinc dust is also used to purify electroplating solutions from contamination by unwanted metals and to recover precious metals from waste solutions. Again, the physical form of the metal deposited is not of interest.

When the metal ions in solution precipitate as a metallic coating on the substrate in a useful form—as a coherent and adherent deposit—the process is immersion plating. It is a technique of limited application, but where applicable it has several advantages.

Like autocatalytic plating, it requires no electric circuitry or source of power, and its throwing power is essentially perfect. Unlike autocatalytic plating, it is an inexpensive technique, requiring only a properly formulated solution of metal salts with no requirement for relatively expensive chemical reducing agents, nor for various additives for stabilization and other purposes; nor is the careful control necessary with autocatalytic plating required.

Advantages of immersion plating are, thus, simplicity, minor capital expense, and the ability to deposit in recesses and on the inside of tubing. Immersion tin coatings, for example, have been used on the inside of copper tubing to inhibit "green stain" in water supplies.

On the other hand, immersion plating is applicable to relatively few situations, and the thickness of deposits obtainable is usually extremely limited, because as soon as the substrate metal is completely, or almost completely, covered by the deposit, the reaction stops or slows down to the rate at which substrate metal is available through pores or discontinuities in the coating. In those few cases where immersion plating is suitable, its advantages are evident. For practical purposes, these processes are concerned with only two coating metals: tin and its alloys with copper, and gold.

IMMERSION TINNING

Tin may be plated by immersion on three substrates: copper and its alloys, mild steel, and aluminum alloys. The coating thicknesses of tin on copper and steel are in no way comparable to those obtainable by conventional electroplating; a typical thickness for tin on copper is about 3 μm in 1 h.

On Copper and Copper Alloys

Tin is immersion-plated on copper and its alloys brass and bronze for several purposes: on brass pins for coloring and for preventing staining of textiles by the green corrosion products of copper; on some electrical parts for color, to distinguish between one conductor and another; and on copper conductors in printed circuitry and allied arts as an aid to soldering. In the last-named use, there is considerable controversy over whether the relatively thin deposits yielded by this method are sufficiently thick to be of more than temporary benefit. In any case, some producers do use such methods.

Copper is normally more noble than tin, and thus it would not be expected that the reaction

$$Cu + Sn^{++} \rightarrow Sn + Cu^{++}$$

would take place spontaneously. However, the electrode potential of copper can be made much more negative by the incorporation of a complexing agent for copper in the tinning solution. Cyanide ion is used in some formulations; thiourea also forms stable complexes with copper and is an ingredient of some solutions. A few of the older formulas appear to rely on the disproportionation of tin(II) in alkaline solution—stannite ion $Sn(OH)_4^{--}$—to tin and stannate, so that half the tin is deposited on the work. Little theoretical investigation of immersion-tinning mechanisms has been reported.

Table 18-1 Immersion Tinning Formulas for Copper*

Bath	*A*		*B*		*C*		*D*		*E*		*F*	
	g/L	*Molarity*	*g/L*	*Molarity*	*g/L*	*Molarity*	*g/L*	*Molarity*	*g/L*	*Molarity*	*g/L*	*Molarity*
Stannous chloride, $SnCl_2$	4	0.02	8–16	0.04–0.08	4	0.02	—	—	5	0.026	20	0.11
(or $SnCl_2 \cdot 2H_2O$)	4.8	0.02	9.5–19	0.04–0.08	4.8	0.02	—	—	6	0.026	24	0.11
Potassium stannate, $K_2Sn(OH)_6$	—	—	—	—	—	—	60	0.20	—	—	—	—
Sodium hydroxide, NaOH	5.6	0.14	—	—	—	—	—	—	—	—	—	—
Potassium hydroxide, KOH	—	—	—	—	—	—	7.5	0.13	—	—	—	—
Sodium cyanide, NaCN	50	1.0	—	—	—	—	—	—	—	—	—	—
Potassium cyanide, KCN	—	—	—	—	—	—	120	1.8	—	—	—	—
Thiourea, $CS(NH_2)_2$	—	—	80–90	1.1–1.2	50	0.65	—	—	80	1.1	75	1.0
Sulfuric acid, H_2SO_4	—	—	—	—	20	0.2	—	—	—	—	—	—
Hydrochloric acid, HCl (conc.)†	—	—	10–20	0.1–0.2	—	—	—	—	—	—	50	0.5
Citric acid, $C_6H_8O_7$	—	—	—	—	—	—	—	—	16	0.083	—	—
Sodium hypophospite, $NaH_2PO_2 \cdot H_2O$	—	—	—	—	—	—	—	—	—	—	16	0.15
Wetting agent	—	—	—	—	—	—	—	—	—	—	1	—
Temperature, °C	20		50–boil		25		20–65		—		—	
Time, min	1–2		5		5–30		2–20		—		—	

*Note patent caveat in Preface.

†mL/L.

Some typical formulas are listed in Table 18-1; in addition, several proprietary compounds are available. The bath shown as formula F in the table represents a definite improvement in speed over most of the others; although it contains sodium hypophosphite, the mechanism of its action is believed not to be similar to that of electroless plating baths; its function is not understood with certainty.

On Steel

Steel wire is often immersion-plated with tin, or more usually tin-copper alloys, for coloring the wire for such uses as bobby pins, paper clips, and similar items. The deposit also serves as a drawing lubricant. The process is known as "liquor finishing." The color of the copper-tin alloy deposit is varied by controlling the percentages of copper and tin salts in the solution. Formulations are listed in Table 18-2.

On Aluminum Alloys

Most immersion deposits of tin on aluminum are nonadherent; in fact, the hydrogen generated in the reaction often blows the deposit off the substrate during the plating operation. However, the aluminum alloys used for automotive pistons, containing significant quantities of silicon, are satisfactorily plated with substantial thicknesses of tin—considerably thicker than those usually resulting from immersion processes—by simple immersion in a solution of potassium stannate (or sodium stannate). The tin coating acts as a lubricant during the running-in period, preventing scoring of the cylinder walls by the abrasive aluminum oxide. Most aluminum-alloy pistons for internal-combustion engines are so tinned.

The parts are cleaned and rinsed, then dipped in 1:4 nitric acid for about 20 s; after rinsing, they are immersion-tinned in alkali stannate solution, about 45 to 70 g/L, for 3 to 4 min at 50–75°C. Cold and hot rinses complete the process. Coatings are mat white and considerably thicker than usual immersion coatings: up to 5 μm or more.

Table 18-2 Liquor Finishing Solutions

	"Bronze"		*Tin (white finish)*	
	g/L	*Molarity*	*g/L*	*Molarity*
Stannous sulfate, $SnSO_4$	7.5	0.035	0.8–2.5	0.004–0.012
Copper(II) sulfate, $CuSO_4 \cdot 5H_2O$	7.5	0.03	—	—
Sulfuric acid, H_2SO_4	10–30	0.1–0.3	5–15	0.05–0.15
Temperature, °C	20		90–100	
Time, min	≈5		5–20	

The tinning reaction may be formulated as

$$4Al + 3K_2Sn(OH)_6 \rightarrow 3Sn + 4KAlO_2 + 2KOH + 8H_2O$$

As work is processed, the tin content of the bath decreases and the free alkali (+ aluminate ion, which reacts much like free alkali) increases. Free alkali should be kept below about 10 g/L. This may be accomplished by additions of acetic acid, and tin content may be replenished by additions of alkali stannate. As the tin content decreases, the activity of the bath may be maintained by raising the temperature. Baths must be dumped sooner or later, since this process of neutralization and replenishment cannot be carried on indefinitely.

Proprietary processes are available that are said to be adaptable to most aluminum alloys, as well as to commercially pure aluminum.

IMMERSION GOLD

The so-called gold-dip process consists in immersing the work, usually of brass, into a suitable gold solution for a short time. The zinc and copper comprising the brass go into solution and are replaced by gold. Such coatings are very thin, perhaps not more than 0.025 μm thick. Considering their thinness, these coatings are relatively pore-free; this has been attributed to the fact that the volume of gold deposited is about twice that of the copper and zinc displaced. These coatings are suitable only for inexpensive and perishable articles, principally for sales appeal since the word *gold* can be used in describing them (but not "gold-electroplated").

A typical formulation is shown in Table 18-3. This is a very old art, and many "secret" formulas have been promoted, but the one listed in the table is said to be as satisfactory as any.

Other Metals

In porcelain enameling, a "nickel dip" is used on steel as a means of securing adhesion of the ground coat; a dilute (7 to 15 g/L) solution of nickel sulfate is used

Table 18-3 Gold-dip Solution

	g/L	*Molarity*
Gold(I) cyanide, AuCN	2.4	0.01
(gold as metal, Au)	2.1	0.01
Potassium cyanide, KCN	12	0.18
Temperature, °C	80	

at pH 3 to 4 and at about 70°C. Although this is immersion plating by definition, it is practiced not by electroplaters but in the ceramic industry.

The zincate process for plating on aluminum involves the deposition of a thin film of zinc on the aluminum substrate, as already described in Chap. 5. This process and the analogous stannate immersion process are also immersion plating, but the finish so applied is only an intermediate step and is never used as a final coating.

19
Plating on Nonconductors

The problem of electroplating on nonconductors is essentially to provide a more or less adherent conductive coating on the substrate by nonelectrolytic means, so that conventional electroplating methods can be applied just as to any metallic substrate. The reasons for desiring to apply an electrodeposit to a nonconductor range from purely "artistic" or sentimental (such as the bronzing of baby shoes) to decorative, to entirely functional, as in printed circuits. In any case, electroplating on nonconductors has been practiced about as long as electroplating itself, but the art, or technology, underwent a fundamental change in 1960 or thereabouts.

Saubestre* accordingly divides the technology of plating on nonconductors into two distinct periods, with 1960 somewhat arbitrarily serving as the dividing line. The processes used, the results obtained, and the purposes for which nonconductors were, in general, coated with metals differed considerably in the two periods.

In the earlier period, dating from the beginning of the plating art itself in the 1830s or 1840s to about 1960, nonconductors were plated principally for artistic or decorative purposes, and the results could be characterized as follows. Adhesion of the metal to the substrate was poor, so that parts had to be completely encapsulated. Deposits had to be thick enough to ensure that they would not come off in normal use. Such adhesion as was achieved was obtained by mechanical interlocking, in turn obtained by purely mechanical means such as roughening the surface to be coated by sanding, vapor blasting, or (in the case of glass) by chemical etching. Much hand work was involved in applying the conductive coating. The nature of the substrate was unimportant, since the means used to coat it were applicable to materials as diverse as leaves, baby shoes, glass, plastics, wood, plaster, and so on: the chemistry of the substrate played no part in the success of the process.

*In "Modern Electroplating," cited in the Bibliography.

By contrast, plating on plastics in the modern or post-1960 era has the following characteristics: Relatively good adhesion is obtained, so that complete encapsulation is unnecessary and thinner electrodeposits can be applied; this adhesion is obtained by means of specially formulated chemical conditioners, and through both mechanical and chemical bonding (although the exact nature of the bond is still a matter of controversy). Handwork can be almost entirely eliminated, and the chemistry of the substrate plays an important part in the process, with the corollary that only certain substrates are capable of coating by these more modern procedures. These substrates are primarily specially formulated acrylonitrile-butadiene-styrene (ABS) copolymers and polypropylene, although means for applying comparable procedures to other plastics such as polysulfones and polyamides are in various stages of development.

The preceding discussion should not be construed to imply that the older art is completely obsolete; baby shoes and other articles of sentimental or artistic value are still being metalized by the older techniques. When it is desired, for any reason, to electroplate on nonconductors not adaptable to the newer processes, the old methods must be used.

In addition to electroplating as a method of metalizing nonconductors, other techniques are available: vacuum metalizing, cathode sputtering, and metal spraying. But these are not within the scope of this book.

To electroplate on a nonconducting medium, it is necessary that the surface of that medium be made conductive in some way; once this is accomplished, the step of electroplating on the surface does not differ substantially from any other type of electroplating. Thus perhaps a more descriptive title for this chapter would be "The Preparation of Nonconductors for Electroplating."

CLASSICAL METHODS

In these methods for rendering the surface conductive, the principal aim was to achieve some sort of mechanical adhesion by roughening the substrate—if it did not have a rough surface already—followed by application of a conducting substance which would permit application of an electrolytic deposit. Adhesion was minimal, and total encapsulation of the substrate was necessary. The first step was roughening, usually by mechanical means and occasionally by chemical treatment. Next came sealing, if the substrate was porous such as wood; this was done with molten wax, shellac, varnish, or lacquer. After the sealant had hardened, the next and most critical step was application of a conductive coating. Several procedures could be used.

Bronzing

A finely divided metallic powder, usually copper, in a suitable binder such as varnish, is applied to the surface with a brush. When the surface is completely covered, the copper is immersion-plated with silver from a silver cyanide solution.

Graphiting

To some surfaces graphite adheres readily: wax, rubber, and some polymers. Finely divided graphite is applied to the surface with a brush; if no binder is used, the graphite is rubbed to a high luster with cotton. In some cases the surface is then sprinkled with iron powder, and an immersion deposit of copper is applied from a copper sulfate solution.

Metallic Paints

These are typically silver dispersed in a flux; this is applied and the piece is fired at 400–800°C to produce an adherent conductive coating. The high temperature obviously limits this technique to ceramics and glasses. Modifications, using oven-baked or air-dried suspensions, yield results similar to bronzing. Various organometallic compounds have been synthesized which are applied to the substrate and subsequently decomposed by heat, leaving a metallic coating.

Metalizing

A metallic coating is produced by chemical means; the usual coating is silver, produced by mixing two solutions: (1) an ammoniacal solution of silver nitrate, and (2) a reducing solution such as formaldehyde or hydrazine. The solutions are preferably sprayed on the part from a two-nozzle gun so that they mix as they hit the part. This procedure differs from electroless plating in that the reduction takes place throughout the solution and on all objects in contact with it, not just on "catalytic" surfaces. This technique is mentioned briefly in Chap. 20 (Electroforming) as a means of rendering nonmetallic mandrels conducting.

After rendering the surface conductive by any one of these techniques, the next step is electroplating. The conductivity of the surface thus prepared is still not comparable to that of massive metals, so that electroplating must be started at low current densities until a fair thickness of electrodeposit is built up. The usual first plate is a copper strike from a copper sulfate solution ($CuSO_4 \cdot 5H_2O$, 70 to 100 g/L; H_2SO_4, 2 mL/L; pH 2 to 2.5). About 7 to 12 μm of copper is deposited from this solution, after which a standard copper sulfate solution can be used to yield the desired final thickness. Overplates of other metals can, of course, be added if desired. Any buffing or mechanical finishing required must be done carefully to prevent overheating of the part, which might cause the poorly adherent plate to pop off.

THE MODERN ERA OF PLATING ON PLASTICS

The results of plating on nonconductors by the techniques mentioned so far were satisfactory enough for purely decorative and artistic purposes, as well as for electroforming in which adhesion to the substrate is actually detrimental. Several developments, both in electrodeposition and in plastics technology, took place about 1960, which ushered in a new era in plating on nonconductors, specifically on certain synthetic plastics. These processes are still being improved and developed, but even in their present state the process of "plating on plastics" has assumed an important place in the technology of electrodeposition. Plated plastics compete directly with plated zinc die castings for many applications, in particular for automotive trim and appliances. Although more expensive than zinc on a weight basis, plastics are so much lighter than zinc that total costs may be comparable, and the trend to weight reduction in automobiles, spurred by the energy crisis, will no doubt increase the use of both functional and decorative plastic components. The technology of plating on plastics differs in almost all respects from the procedures so far discussed; it depends on the following developments, among others.

Plateable Plastics

The producers of ABS resins developed copolymer compositions and molding techniques especially adapted to the conditioning and surface treatments required for the subsequent plating operation. Special grades of ABS became available; later the techniques were modified to include grades of polypropylene and some other polymers among the plateable plastics.

Chemical Conditioning

The need for mechanical roughening of the surface, to provide some minimal adhesion of the electroplate, was eliminated by the development of chemical conditioners which both roughened the surface and altered its chemical composition so that the subsequent electroless deposit adhered to the plastic by both mechanical and chemical bonding. (Whether the bond is mainly mechanical or chemical, or a little of both, is a matter of some controversy, which cannot be entered into here. It must suffice to say that the bond is sufficiently strong so that complete encapsulation of the part by the electrodeposit is no longer necessary, nor are excessively thick electrodeposits. The adhesion generally is not comparable to that obtained in conventional electroplating on metallic substrates, but is still sufficient for many purposes.)

Autocatalytic Plating

Electroless copper baths appeared on the market which were not so subject to rapid decomposition, and which could be replenished rather than discarded after a short period of use. Electroless nickel baths also appeared which were more adaptable to the total process of plating on plastics. Most of these developments were, and are, proprietary.

Bright Acid Copper Electroplating Baths

The development of bright acid copper baths which both yielded brilliant deposits and had good leveling characteristics made possible the use of much thinner copper deposits than were previously required and eliminated the need for mechanical polishing; now deposits only 7 to 15 μm thick suffice for satisfactory results.

These innovations made possible the development of integrated systems, fully or partially automated, for producing plated plastic parts which were readily accepted by many segments of industry. Early troubles included lack of reliable control resulting in high reject rates, but these proved to be the usual startup problems suffered by most new processes. The technique is now well established, and further improvements continue to be reported.

What Is a Plateable Plastic?

At present by no means all synthetic plastics are suitable for the processes which have been developed for preparing plastics for electroplating. Acrylonitrile-butadiene-styrene (ABS) and polypropylene have been studied most thoroughly and remain the most widely used; producers of other synthetic plastics such as polysulfones, polyamides, and polycarbonates have, with varying success, introduced variations in the process to make it adaptable to their products. Not only the chemical composition but also the shape and size of the parts affect their plateability. As the size and complexity of the parts increase, so do the problems in plating them: sharp variations in cross section, ribs, corners, and holes all pose problems. In general, a uniform and deep surface porosity must develop on the substrate during the preplating, or "conditioning," cycle.

Conditioning of the surface should result in a uniform etching which will be receptive to the steps that follow. The molding process is a very important part of producing a plastic component that will accept an adherent plate; thus, as in other branches of metal finishing, the treatment of the parts even before they reach the finishing department is a vital element in the total process. Molding techniques are not within our present scope, but complete cooperation and mutual understanding between the producer of the plastic parts and the finisher are an indispensable requirement.

Pores developed on etched ABS are believed to result from removal of butadiene spheres, leaving voids which aid in mechanical adhesion. Suppliers of plating-grade polypropylene often add materials such as titanium dioxide or talc which appear to provide active sites for nucleation and subsequent deposition of metal.

Design of plastic parts to be plated is important, but the same guidelines applicable to metallic parts also apply in general to plastics: convex surfaces are easier to plate than concave ones, large flat surfaces should be crowned, edges should be rounded, narrow and closely spaced slots and holes are difficult to plate, and blind holes should be avoided, as should deep, V-shaped grooves.

THE PLATING CYCLE

Cleaning

Cleaning generally is not an important element in preparing plastics for plating, except that fluorocarbon mold-release compounds should not have been used. Removal of other mold-release agents, or fingerprints or other minor soils, may be accomplished by mild alkaline cleaning, followed by an acid dip to remove alkaline residues.

Solvent Treatment

This is required if the plastic is not readily wetted by the conditioner that follows. The solvent serves to alter the surface of the plastic by making it readily wettable by the conditioner; some swelling of the plastic usually accompanies this treatment. The solvent should have some chemical action on the polymer, so as to render it hydrophilic (readily wettable by aqueous media) but not enough to cause gross degradation of the chemical, physical, and mechanical properties of the plastic. ABS and polypropylene usually do not require any solvent treatment, since they are already hydrophilic; but some other less commonly used plastics such as polycarbonates, polyvinyl chloride, polysulfones, and many others generally do. Which solvent to use will be suggested by the supplier of the plastic.

Conditioning

Conditioning provides some interlocking roughness on the surface, to obviate the need for mechanical treatments and to provide sites for chemical bonding of metals to be applied later. For ABS and polypropylene, these conditioners generally contain chromic and sulfuric acids, with or without other additives. Other plastics may be treated with alkaline oxidizing solutions.

Four of the more commonly suggested conditioners for ABS and polypropylene are:

1. Chromic acid, CrO_3, 75 g/L; sulfuric acid, H_2SO_4, 250 mL/L
2. Potassium dichromate, $K_2Cr_2O_7$, 90 g/L; sulfuric acid, 600 mL/L
3. Chromic acid, 50 g/L; sulfuric acid, 100 mL/L; hydrofluoric acid, HF, 100 mL/L
4. Chromic acid up to 900 g/L or more; no sulfuric acid

These solutions are used for 1 to 5 min at 20–35°C. Underconditioning results in incomplete wetting, leading to "skip plating" (as the name implies, skip plating is manifested by unplated areas on an otherwise satisfactory product). Overconditioning may result in generalized degradation of the surface, leading to poor adhesion of the plate and possible blistering.

Various other conditioning formulas have been suggested, and (as with this whole field) many proprietary formulations are marketed.

Preparation of the Catalytic Surface

This step is the critical one in plating on plastics, and most practitioners utilize proprietary and patented processes. (The general caveat concerning patented processes, mentioned in the Preface, is particularly relevant in this chapter, and the discussion of sensitization-nucleation processes and formulations should be read with this in mind.)

A certain amount of confusion exists concerning the terminology used for the steps of rendering the plastic surface receptive to a metallic coating. The definitions adopted by ASTM for these steps are as follows (ASTM B 374):

> Sensitization: the adsorption of a reducing agent, often a stannous compound, on the surface.
>
> Nucleation: the preplating step in which a catalytic material, often a palladium or gold compound, is adsorbed on the surface. The catalyst is not necessarily in its final form.
>
> Postnucleation: the step where, if necessary, the catalyst is converted to its final form. This is the final step prior to electroless plating.

After the surface of the plastic has been conditioned by any suitable method, as already discussed, it must be rendered ''catalytic'' so that it will receive an autocatalytic or electroless deposit (Chap. 17). This may be accomplished in several ways.

Sensitizing-Nucleating

This is a two-step process; in the first, or sensitizing, step, a reducing agent is adsorbed on the surface of the plastic. The usual sensitizer, or reducing agent, is stannous chloride, although titanium(III) chloride has been used. The stannous chloride solution may typically consist of 20 g/L $SnCl_2$ plus 40 mL/L hydrochloric acid. The solution is used at 20–25°C, and immersion time is 1 to 3 min. The free acid concentration of the sensitizing bath must be maintained by periodic additions of acid to prevent hydrolysis of the tin salt. Rinsing following this step must be thorough, since if any unadsorbed tin(II) ions are dragged over into the nucleating solution, the latter will be decomposed.

The next step is ''nucleation,'' which renders the plastic surface catalytic for electroless nickel or copper plating. Nucleating agents are typically precious-metal salts, palladium being most common. The sensitizer adsorbed on the surface is readily oxidized and reduces the catalytic metal salt to metal, depositing it on the surface at discrete sites in the metallic state. A typical formula for a nucleating solution is: palladium chloride, $PdCl_2$, 0.25 g/L [or equivalent tetrachloropalladic(II) acid, H_2PdCl_4]; hydrochloric acid, 2.5 mL/L.

The stannous ions adsorbed on the surface react with the palladium ions in solution to form palladium metal:

$$Sn^{++} + Pd^{++} \rightarrow Sn^{4+} + Pd$$

The solution is used at 20–40°C for 30 to 60 s. Thorough rinsing must again follow, since drag-in of palladium chloride into the electroless plating bath would decompose the latter.

Sensitizing-Nucleation/Solubilizing (One-step Process)

More recently, the sensitizing and nucleating steps have been combined into one solution, and most users, except the smaller shops, appear to have adopted this method, which was first applied to the processing of printed circuits but is now in general use. In this method, the sensitizer and nucleating agent are combined into one solution; this is followed by a solubilizing solution which removes the excess sensitizer. The process has obvious advantages in requiring fewer steps and in being less sensitive to insufficient rinsing. A typical formula is:

Stannous chloride, $SnCl_2$	2 g/L
Palladium(II) chloride, $PdCl_2$	0.2 g/L
Hydrochloric acid, HCl	10 mL/L

The palladium chloride is reduced by tin(II) to colloidal palladium; this colloid is, in turn, stabilized by the presence of hydrolyzed tin(IV) formed by the oxidation of the stannous ions. In some formulations sodium stannate is deliberately added as an additional stabilizer.

The bath is used at 20–40°C for about 1 min. The palladium particles, adsorbed on the plastic surface, are, however, surrounded by tin(IV) ions (or colloidal particles of stannic oxide hydrate); these must be removed, so that the palladium particles will be exposed to the electroless plating solution. In addition, if any tin is left on the surface and electroless copper plating follows, an intermediate layer of tin-copper alloy may be formed. This layer may be brittle and weaken the bond between the plastic and the copper deposit.

The solubilizing solution (called an *accelerator*) must dissolve enough of the protective layer of tin ions, without dissolving so much palladium as to destroy the catalytic activity of the surface; mixtures of fluoboric and oxalic acids are typical of solubilizing solutions.

When the surface of the plastic has been rendered catalytic by either of these methods, or variations of them, it is now ready for deposition of electroless copper or nickel, to be followed by conventional electroplating. Electroless plating has already been discussed (Chap. 17). Since only the surface of the plastic is conductive, and the electroless deposit is quite thin, the electrical conductivity of the part is not comparable to that of metallic articles, and electroplating must be started at relatively low current densities to avoid burning at contact points. Otherwise electroplating is more or less conventional.

TESTS FOR PLATED PLASTICS

Conformance and performance tests for plated plastics are specialized, and therefore are mentioned here rather than in the general chapter on testing. The tests most usually applied are the peel, or Jacquet, test and the thermal cycling test. In a rough way, these tests measure the degree of adhesion of the metal coating to the plastic, but they cannot be regarded as true adhesion tests. The tests are merely outlined here; for details the applicable ASTM specifications should be consulted.

Jacquet or Peel Test (ASTM B 533)

"The so-called adhesion of a metal electrodeposit to a plastic substrate, as measured by the peel test, is not a measure of the attachment of that film to the substrate, but rather a measure of the strength of the weakest layer in the metal-substrate combination." In effect, this means that failure often occurs not at the metal-plastic interface, but in the plastic itself. In spite of this limitation, the test is useful as a quantitative measure of the degree to which the plated plastic may be expected to survive in service, which answers the question, "Is the deposit likely to come off?" Parallel marks are scribed on a flat surface of the plated plastic article, 25 mm apart, and the metallic coating is gripped in a standard fitting and peeled off at a 90-degree angle at a constant rate of 25 mm/min. The force required to peel the coating is measured directly; results are reported in pounds per inch. (Unfortunately, at this time the industry has not adopted the metric system, and peel strengths are reported in conventional units. While metrication can be achieved by reporting the results as kg/25 mm, this cannot be reduced further to kg/mm because the width of the test strip affects the measurement.) A minimum value of 5lb/in. (2.27 kg/25 mm) is considered passable, but values up to 22 lb/in. (10 kg/25 mm) or more are common. (As noted, it is not permissible to translate 10 kg/25 mm to 0.4 kg/mm.) For details of the apparatus and method, consult the cited ASTM specification.

Thermal Cycling Test

Because the coefficients of thermal expansion of plastics and of metals differ widely, the degree to which a plated plastic part will survive alternate heating and cooling without failure is an indirect measure of the quality of the adhesion and of the probable performance of the part in service. The plated part is heated to 60°C for parts destined for mild service conditions or up to 85°C for parts designed for severe service; they are maintained at the high temperature for 1 h, allowed to cool to room temperature, then refrigerated to −30 or −40°C (for mild and severe service, respectively) for 1 h. The progression heat-room temperature-cool-room temperature constitutes one complete cycle; the number of cycles a part should survive without blistering is subject to agreement. Details of this test are given in ASTM B 533.

The CASS test (Chap. 3) is also used as a measure of the quality of the plating.

Stress

There is no agreement on tests for determining whether a plastic part has been properly annealed before conditioning. Several tests are used, but have not been standardized. Immersion in glacial acetic acid, and in various solvents, will often reveal, by cracking of the plastic, that internal stresses have been introduced into the part by the molding process; these will result in unsatisfactory plating.

Appearance

The defects to which plated plastics are subject are those common to all plated articles; but in addition several are peculiar to these items. A guide to the evaluation of the appearance of plated plastic surfaces is available as ASTM B 532. Most of the defects arise from the plastic itself; a few are a result of poor plating practice.

Thickness

Many of the standard methods of thickness measurement, considered in Chap. 23, are applicable to measuring the thickness of metallic coatings on nonconductors; see ASTM B 554. These methods include the microscopic, coulometric, beta backscatter, and eddy current. Specifications of thickness for various service conditions, as well as further test methods, are offered in ASTM B 604.

ADHESION OF METALS TO PLASTICS

The mechanism by which metals adhere to plastic substrates, after the preplating treatments already discussed, remains a subject of some controversy. According to some, the adhesion is mainly mechanical: i.e., the metallic coating simply interlocks with undercuts, shallow pits, and channels in the plastic; these provide anchoring points for the metallic coating. On the other hand, proponents of the chemical theory maintain that the conditioning treatment, involving the use of oxidizing agents, alters the chemistry of the plastic surface to provide true chemical bonding between it and the metal, through the presence of oxide ions on the surface.

This subject is active in the literature, and it cannot be said that either theory has been entirely substantiated. It seems most likely that the adhesion is due to a combination of both factors.

There is also considerable disagreement as to whether the palladium nucleating solution involves colloidal phenomena or true complex formation. No definitive statements can be made at this writing, and current literature should be consulted.

20 Electroforming

Electroforming is defined as the production or reproduction of articles by electrodeposition upon a mandrel or mold that is subsequently separated from the electrodeposit. (In exceptional cases, the mandrel or mold may remain with the finished article.) A mandrel is a form used as a cathode in electroforming—a mold or matrix.

The history of electroforming as a means of producing manufactured articles is practically parallel to that of electroplating itself; one of the earliest reports on the subject, by Jacobi of the Academy of Sciences in St. Petersburg, Russia, told of the reproduction of articles by ''galvanoplasty'' in 1838. Interest in the process has continued to the present, but more recent times have seen an increase in its industrial as contrasted with the purely artistic possibilities, stimulated by the more complicated shapes of some required parts and improvements in the solutions available. Electroforming has taken its place now alongside other standard methods of producing metal articles; it possesses both advantages and disadvantages compared with these more conventional techniques.

During the early period of electroforming (also called electrotyping, and earlier ''galvanoplasty''), the ability of the process to produce exact copies was applied mainly to the reproduction of art objects. Its earliest functional applications were in the manufacture of electrotype printing plates and the preparation of phonograph record stampers. The accuracy of reproduction required in these applications is critical—especially with the advent of the microgroove long-playing record—and illustrates the capabilities of electroforming as a precision tool. More recently many other applications of electroforming have become important, some of which employ similar abilities of precise reproduction; others take advantage of some other capabilities of electroforming such as the ability to produce complicated shapes, exemplified by radar waveguides, Pitot tubes, and Venturi meters.

Applications

Some of these applications of electroforming may be illustrated:

Manufacture of duplicating plates: electrotypes, phonograph record masters and stampers, embossing plates

Thin-walled sections: foil, sheet, hypodermic needles, fine-mesh screen, seamless tubing

Precision parts: molds and dies for rubber and plastics, and for dentures

Parts difficult or impossible to make by other techniques: radar waveguides, Pitot tubes, Venturi meters, surface roughness gages, fountain-pen caps, musical instruments.

ADVANTAGES AND DISADVANTAGES OF ELECTROFORMING

Advantages

1. The metallurgical properties of some electrodeposited metals can be controlled over wide ranges by choice of a suitable metal and by adjusting plating bath compositions and operating conditions. Properties can sometimes be realized which cannot be duplicated in massive metal.
2. Parts can be produced in quantity, with very great dimensional accuracy, limited only by the accuracy possible in machining the mandrel. This can be in the order of 2.5 μm. All parts produced from the mandrel will be dimensionally identical, provided that the mandrel material is properly chosen.
3. Reproduction of fine detail is unmatched by any other method of mass production. Perhaps the best illustration of this is the production of masters and stampers for microgroove phonograph records, where modulations necessary for hi-fi are about 13 nm (13×10^{-6} mm). In spite of the extremely small tolerance, the master can be faithfully reproduced by electroforming even to the third generation of copies. Similarly, a buffed, highly polished, or deliberately roughened surface (as for surface roughness standards) can be faithfully reproduced for several generations.
4. The size of articles that can be electroformed is limited only by the size of the facilities available. Electroformed articles that have been produced range from hypodermic needles, foil 2.5 μm thick, and 400-mesh screen, to articles of complicated geometry weighing several hundred kilograms, to tolerances of less than 25 μm.

5. Some shapes that cannot be made by any other method—at least without excessive amounts of machining and scrap losses—can be easily electroformed. Examples are radar waveguides with two right-angle bends, with the interior made to close tolerances and with a high surface finish.

6. Metal sandwiches can be built up: radar waveguides of copper, with an interior lining of silver.

7. Electroforming is equally adaptable to production of single pieces and to large production runs. For the latter, it is possible to start with a master which may be unique and irreplaceable, such as an original phonograph record master recording, and proceed to several generations of alternate positive and negative reproductions. When sufficient positives have been made, only negatives are produced which are used for the final stamping operation.

Disadvantages

1. Electroforming is a basically expensive method, and is chosen only when alternative methods are impossible or even more expensive.

2. The production rate of individual pieces is slow, sometimes measured in days, though thin-walled sections sometimes may be produced in an hour or so. But precision machining operations are also slow and require the constant attention of skilled artisans, and by manifolding, the total production of electroformed articles may in some cases be comparable to other methods.

3. There are definite design limitations. Sharp angles and deep and narrow recesses cause problems; and sudden changes in cross section or wall thickness must be avoided unless postplating machining can be done.

4. The excellent reproduction of surface detail entails its own penalties: minute imperfections, scratches, pits, nodules, etc., in the mandrel will be duplicated in all the pieces made from it.

5. Many metals are deposited in a state of stress, which may cause problems when the deposit is separated from the mandrel.

In brief, electroforming should be considered as a production tool when: (1) the difficulty and cost of other means of production are excessive; (2) unusual physical properties are required in the part; (3) extremely close dimensional tolerances are required, especially on internal dimensions; and (4) very fine reproduction of detail is required.

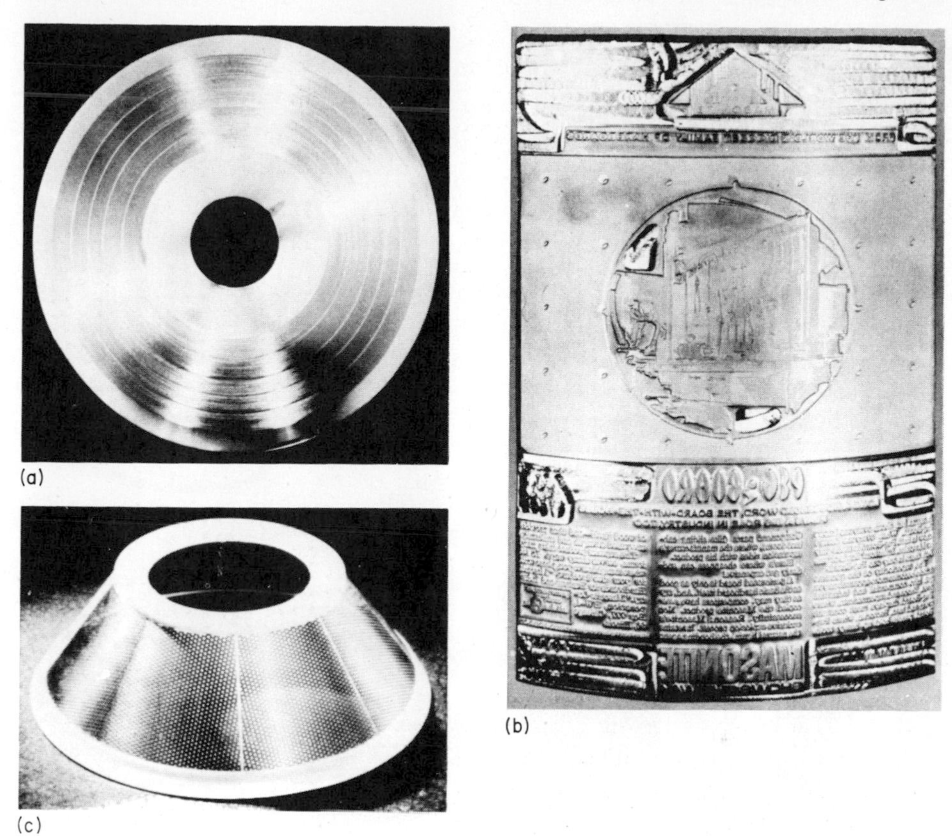

Fig. 20-1 Functional electroforms: (*a*) phonograph record stamper; (*b*) electrotype printing plate; (*c*) screens and grids.

Typical items that are, or have been, produced by electroforming are illustrated in Figs. 20-1 and 20-2.

PRINCIPLES

The principles of electroforming do not differ from those of conventional electroplating. However, because the physical properties of the deposit are of paramount importance; because the deposits are usually much thicker and plating times are often measured in hours or even days rather than minutes, and finally because design considerations are much more important than in usual electroplating practice, electroforming is far more demanding of good practice in all aspects of the operation. Minor errors and oversights and some departures from the best practice that might still result in acceptable work in decorative or protective plating cannot be tolerated, and will almost surely cause rejection of

Fig. 20-2 Electroformed consumer items: (*a*) switch plates; (*b*) coffee percolator; (*c*) salt and pepper shakers; (*d*) hollow ware.

electroforms. Therefore the requirements for acceptable electroforming are, as stated, not different from those for electroplating—only much more stringent. In this chapter accordingly we shall not attempt a complete set of directions for the electroformer, for which specialized texts should be consulted or for which no written directions can take the place of practical experience.* We shall merely outline the salient principles upon which electroforming practice is based.

The principal difference between electroplating and electroforming is that, in the former the aim is a deposit with the best possible adhesion to the substrate, whereas in the latter the adhesion should be minimal—just sufficient so that the

*Excellent guidance is available in: ASTM B 431, Recommended Practice for Processing of Mandrels for Electroforming; ASTM B 450, Recommended Practice for Engineering Design of Electroformed Articles; and ASTM B 503, Recommended Practice for the Use of Copper and Nickel Electroplating Solutions for Electroforming.

deposit remains attached to the substrate during the operation, but is easily separated afterward. It follows that highly critical aspects of electroforming are the type, design, and materials of the mandrels used for the production of the electroform.

Mandrels

Mandrels are of two types: permanent and expendable. As the names imply, permanent mandrels are used over and over, while expendable mandrels are used once and are destroyed in the process of separating them from the electroform.

Permanent mandrels are usually metallic, although plastics (after suitable metalization to render them conductive) can be used. The metals most commonly used are those which do not easily accept an adherent deposit without special treatment: stainless steel, chromium-plated metals, and nickel or nickel-plated metals.

Expendable mandrels are ordinarily of waxes and plastics, which may be softened by heat for removal from the electroform; low-melting or fusible metals, which can be melted away; soluble metals which can be dissolved by reagents that do not attack the electroform; and plaster, which can be broken away.

Permanent Mandrels

Stainless steels can be machined to close tolerances and are stable in dimensions on continued use. High-precision nickel tubes, surface roughness standards, screens, and small orifices are commonly electroformed on stainless steel mandrels.

Electroplated metals such as chromium and nickel, on steel, copper, or brass substrates, combine the advantages of the easy workability of the substrate with the lack of adhesion of electrodeposits on the chromium or nickel. Consumer items such as trays, pitchers, bowls, pen caps, and decanters are examples of electroforms produced on this type of mandrel.

Copper and brass, though they must be specially treated to enable subsequent separation of the electroform, are preferred by engravers and other artisans when chasing or photoengraving procedures are required to produce the form.

Nickel mandrels are used when the mandrel itself must be electroformed; this is a less expensive operation than producing multiple mandrels by working, engraving, or hand finishing. Phonograph record stampers perhaps represent the most typical example of this type of mandrel.

Plastics—including epoxy resins, vinyl plastisols, and rigid vinyl sheets—are used for producing surface textures and intricate shapes. They must be metalized, of course, before the electroforming operation. They are most useful when close dimensional tolerances are not required.

Mandrels for some low-production items can be made of plaster or wood; these materials must be sealed to prevent absorption of liquid from the plating bath: liquid resin materials such as acid-resistant varnishes may be used. This type of mandrel may be permanent or expendable.

Expendable Mandrels

Fusible metals can be cast into molds of brass, bronze, copper, nickel, or steel when fine surface detail and smooth finish are required. When these are not necessary, plaster, plastics, rubber, or wood molds can be used. Fusible metals are commonly alloys of tin, lead, bismuth, and cadmium in various proportions. The electroform is separated by melting out the fusible metal, preferably in silicone-based oils having flash points considerably higher than the melting point of the metal.

Soluble metals are those that can be removed chemically; aluminum and its alloys constitute the principal example. Aluminum is soluble in caustic alkali solutions, which do not attack most electroformed metals such as nickel and copper. Aluminum alloys can be forged, machined, ground, and lapped to close surface and dimensional tolerances.

DESIGN OF MANDRELS

In electroforming, the side of the electroform next to the mandrel (the cathode) will reproduce exactly the finish and smoothness of the mandrel; the side away from the mandrel, the solution side, is not under such close control and may tend to be somewhat rough or nodular. Mandrels are designed to produce the desired pattern and texture on either the inside or the outside surface of the part. When the outside surface of the part is the significant area, the internal surface of the mandrel upon which it is formed must be designed carefully. When the inside surface of a part is important, the external surface of the mandrel is the one requiring the special care. In the former case the mandrel is called a positive mandrel; in the latter, a negative mandrel.

Certain design features are common to all mandrels, but may be more important for negative mandrels, because it is easier to work on external than on internal surfaces of the finished electroform.

Sharp internal angles should be avoided.

When two deposits of electroformed metal are used, care must be exercised to ensure adhesion of the second deposit to the first.

Deep grooves should be avoided.

Corners should have a radius of at least 0.8 mm.

A part intended for production on permanent mandrels must be designed with sufficient draft or taper to permit withdrawal of the mandrel without damaging

Table 20-1 Preparation of Nickel Mandrels

1. Cathodic alkaline clean: trisodium phosphate, 23 g/L; sodium tripolyphosphate, 23 g/L, 80–83°C, 10 to 25 s, 500 A/m²
2. Rinse, 70–80°C
3. Cathodic cyanide: sodium cyanide, 45 g/L, room temperature, 10 s, 200 A/m²
4. Rinse, 70–80°C
5. Acid dip, 5% (vol.) sulfuric acid, 5 s
6. Rinse, cold water
7. Dichromate treatment: sodium dichromate, 3.8 g/L, 30 to 60 s
8. Rinse, cold water
9. Electroform

either the electroform or the mandrel. If this is not possible, expendable mandrels must be used.

Mandrels must be carefully cleaned to obtain uniform coverage and an electroform free of surface imperfections. Permanent mandrels must also be passivated to enable separation of electroform from the mandrel. Nonmetallic mandrels must be metalized to yield an electrically conductive surface.

Various preplating cycles are appropriate for the preparation of mandrels before electroforming. These are summarized in the tables that follow, but also see ASTM B 431 already cited.

Table 20-1 lists the steps for preparing nickel mandrels.

Stainless steel and chromium have passive surfaces to which deposits do not ordinarily adhere, and passivation can usually be omitted. See Table 20-2.

Aluminum cannot be plated directly with nickel, or other metals; the immersion zincate procedure is generally used, similar to that described in Chap. 5, and then copper-plated.

Fusible metal mandrels are treated as shown in Table 20-3.

Table 20-2 Preparation of Stainless Steel Mandrels

1. Degrease (if necessary) chlorinated solvent
2. Water rinse
3. Cathodic clean: trisodium phosphate, 45 g/L, 60–82°C, 1 min, 500 A/m²
4. Rinse, 38°C. If water break occurs, repeat (3)
5. Electroform

Table 20-3 Preparation of Fusible Metal Mandrels

1. Degrease (if necessary) chlorinated solvent
2. Water rinse
3. Cathodic clean: 45g/L trisodium phosphate, 60°C, 30 to 60 s, 6 V
4. Water rinse
5. Anodic sulfuric acid: 3% sulfuric acid, 27°C, 5 to 10 s, 6 V
6. Water rinse
7. Electroform

Nonmetallic mandrels may be metalized by use of the silvering procedure; see Table 20-4.

METALS FOR ELECTROFORMING

Although about 18 metals can be easily deposited from aqueous solution, only three are readily adaptable to electroforming: copper, nickel, and iron. To a somewhat lesser extent, silver, gold, and cobalt have been used, and nickel-cobalt alloys appear to possess some advantages. For exotic purposes, aluminum mirrors, deposited from nonaqueous solutions (see Chap. 15), have been electroformed. Zinc, cadmium, tin, and lead have physical properties not suitable for electroformed articles; chromium is too brittle, though it is used to face electro-

Table 20-4 Preparation of Nonmetallic Mandrels

1. Mild alkaline clean
2. Spray rinse
3. Sensitize: 2.5 to 10 g/L stannous chloride + 40 mL/L 35% (conc.) hydrochloric acid; spray on mandrel until continuous film is obtained
4. Spray rinse
5. Apply silver film: double-nozzle gun;
   Solution (1): 55 to 60 g/L silver nitrate + 60 to 75 mL/L 28% (conc.) ammonium hydroxide
   Solution (2): 100 g/L dextrose or 65 mL/L 40% (vol.) formaldehyde
   Solutions (1) and (2) are sprayed from twin nozzles so that they mix on the surface of the mandrel. Proprietary silvering solutions available. *Note:* Do not store solution (1) as explosive compounds may be formed.
6. After rinsing, electroform

Table 20-5 Typical Stress Values of Electroformed Metals

Metal	*Bath*	*Stress, MPa**
Copper	cyanide	34–83
	sulfate	2.8–14
	fluoborate	5–21
Nickel	Watts	110–207
	Watts + H_2O_2	275+
	all-chloride	207–310
	fluoborate	96–172
	fluoborate + H_2O_2	96–172
	sulfamate	3.4–55
	sulfamate + chloride	55–110
	sulfate	110–138
Iron	chloride	90–138
Silver	cyanide	14–27.6
Gold	cyanide	compressive

*Tensile.

forms of other metals. Little has been done with the platinum group metals, and it is doubtful that their properties, as deposited, would lend themselves to this technique. The crown used in the investiture of Prince Charles of England as Prince of Wales was made by electroforming in gold.

The choice of metal for electroforming is dictated by the properties required in the finished article. Table 20-5 summarizes the properties obtainable from the most commonly electroformed metals; more details are given in later tables.

Copper is fairly inexpensive and sufficiently strong for many purposes; it deposits at a high rate from acid solutions in a relatively low-stressed state. When required for corrosion resistance, the bulk of the form can be made of copper and overplated with nickel.

Nickel is the most widely used metal for electroforming; its physical properties can be regulated over a wide range by control of the plating variables; and it has been the subject of more study than any other plating metal, so that more is known about the effects of variables, addition agents, and bath compositions on the properties of the deposits.

Iron is cheap and strong; but (as discussed in Chap. 14) most deposits, except those from the hot and very corrosive chloride bath, tend to be brittle. The deposit is not corrosion-resistant, and usually requires protection by overplating with nickel or chromium. Its principal use has been for printing plates at the Bureau of Printing and Engraving.

Table 20-6 Nickel Electroforming Solutions

	Concentrations, g/L		
	Sulfamate	*Watts*	*Fluoborate*
Nickel sulfamate, $Ni(SO_3NH_2)_2$	375	—	—
Nickel fluoborate, $Ni(BF_4)_2$	—	—	300
Nickel sulfate, $NiSO_4 \cdot 6H_2O$	—	330	—
Nickel chloride, $NiCl_2 \cdot 6H_2O$	6	45	—
Boric acid, H_3BO_3	30	38	30
Temperature, °C	50	60	60
pH (electrometric)	4.0	4.0	4.0
Current density, A/m^2	220–650	270–1080	430–1000

PLATING SOLUTIONS FOR ELECTROFORMING

Plating solutions for electroforming do not differ in principle from ordinary plating baths for decorative and protective purposes. But because the physical properties of electroforms are usually of much more concern than those of ordinary electrodeposits and because electroforms are usually much thicker than ordinary electroplates, the composition and operating conditions of electroforming tend to be more critical than those for the thinner deposits used in standard electroplating. Many addition agents, added for brightness, leveling, and other purposes, are not allowable in electroforming solutions because they introduce trace contaminants into the deposits which have deleterious effects on the properties of the electroforms. And because thicker deposits are involved, speed of plating is of more importance; plating times, long in any case, should be minimized.

Nickel is the most commonly electroformed metal. The most widely used baths are the Watts, sulfamate, and fluoborate solutions, having the compositions and operating characteristics listed in Table 20-6. The principal physical properties obtainable from these three baths are listed in Table 20-7. Effects of variables, as summarized in ASTM B 503, are listed in Tables 20-8 and 20-9.

Copper is usually plated from sulfate or fluoborate solutions, as shown in

Table 20-7 Typical Mechanical Properties of Nickel, Baths of Table 20-6

	Sulfamate	*Watts*	*Fluoborate*
Hardness, DPN	190	140–160	185
Tensile strength, MPa	745	352	514
Elongation in 25 mm, %	15	30	17
Internal stress, MPa	10–69	124	110–180

Table 20-10, having physical properties summarized in Table 20-11; effects of variables are summarized in Tables 20-12 and 20-13.

STRESS IN DEPOSITS

Internal stress in deposits may present a problem to the electroformer. High stress may cause the deposit to adhere so firmly to the mandrel that parting becomes difficult or impossible. After removal of the electroform, high stress may cause distortion of the piece. Stress may cause other problems also, such as curling up at the edges of the electroform or even exfoliation, since by its very nature electroforming is carried out with minimal adhesion to the substrate.

Copper and silver pose few problems in this respect, but nickel is deposited from most ordinary baths in a condition of moderate to high tensile stress, and electroforming solutions for nickel must be formulated with this in mind. Stress is

Table 20-8 Variables Which Affect Mechanical Properties of the Deposit—Watts Solution

Property	*Operational*	*Solution composition*
Tensile strength	Relatively independent of plating solution temperature within range suggested.	Increases with increasing nickel content.
	Relatively independent of changes in cathode current density.	Increases with increasing chloride content.
	Relatively independent of pH variation within range suggested.	
Elongation	Increases with temperature to 55°C followed by slight decrease at higher temperature.	Decreases with increasing nickel content.
	Relatively independent of pH variation within range suggested.	
Hardness	Decreases with temperature rise to 55°C but increases with higher temperatures.	Increases with increasing nickel content.
	Decreases significantly with increasing cathode current density to 540 A/m^2. At higher current densities the hardness increases with increasing current density.	Increases with increasing chloride content.
Internal stress	Relatively independent of plating solution temperature.	Increases slightly with increasing nickel content.
	Decreases slightly, then increases with increasing cathode current density.	Increases markedly with increasing chloride content.
	Relatively independent of pH variation within range suggested.	

Table 20-9 Variables Which Affect Mechanical Properties of the Deposit—Nickel Sulfamate Solution

Property	*Operational*	*Solution composition*
Tensile strength	Decreases with increasing temperature to 49°C, then increases slowly with further temperature increase.	Decreases slightly with increasing nickel content.
	Increases with increasing pH.	
	Decreases with increasing current density.	
Elongation	Decreases as the temperature varies in either direction from 43°C.	Increases slightly with increasing nickel content.
	Decreases with increasing pH.	Increases slightly with increasing chloride content.
	Increases moderately with increasing current density.	
Hardness	Increases with increasing temperature within operating range suggested.	Decreases slightly with increasing concentration of nickel ion.
	Increases with increasing solution pH.	Decreases slightly with increasing chloride content.
	Reaches a minimum at about 1300 A/m².	
Internal stress	Decreases with increasing solution temperature.	Relatively independent of variation in nickel-ion content within range suggested.
	Reaches a minimum at pH 4.0–4.2.	Increases significantly with increasing chloride content.
	Increases with increasing current density.	

Table 20-10 Copper Electroforming Solutions

	Concentrations, g/L	
	Sulfate	*Fluoborate*
Copper(II) sulfate, $CuSO_4 \cdot 5H_2O$	210–240	——
Copper(II) fluoborate, $Cu(BF_4)_2$	——	225–450
Sulfuric acid, H_2SO_4	52–75	——
Fluoboric acid, HBF_4	——	to pH 0.15–1.5
Temperature, °C	21–32	21–54
Current density, A/m²	100–1000	800–4400

Table 20-11 Typical Mechanical Properties of Copper, Baths of Table 20-10

	Sulfate	*Fluoborate*
Tensile strength, MPa	205–380	140–345
Elongation, %	15–25	5–25
Hardness, Vickers, 100-g load	45–70	40–80
Internal stress, MPa	0–10	0–105

influenced both by the type of bath used and by addition agents, if used. Without addition agents, the sulfamate bath, free of chlorides, yields the lowest stress, followed by sulfamate-chloride, fluoborate, Watts, and all-chloride baths. In general, increased chloride content increases stress in all types of bath. Most wetting agents have little effect; but when hydrogen peroxide is used as an antipit, stress may be raised except in the fluoborate solution, where it appears to have little effect.

Table 20-12 Variables Which Affect Mechanical Properties of the Deposit—Acid Copper Sulfate Solution

Property	*Operational*	*Solution composition*
Tensile strength	Decreases slightly with increasing solution temperature.	Relatively independent of changes in copper sulfate concentration within the range suggested.
	Increases significantly with increase in cathode cd.	Relatively independent of changes in sulfuric acid concentration within the range suggested.
Elongation	Decreases with increasing solution temperature.	High acid concentrations, particularly with low copper sulfate concentration, tend to reduce elongation slightly.
	Increases slightly with increasing cathode cd.	
Hardness	Decreases slightly with increasing solution temperature.	Relatively independent of copper sulfate concentration.
	Relatively independent of change in cathode cd.	Increases slightly with increasing acid concentration.
Internal stress	Increases with increasing solution temperature.	Relatively independent of copper sulfate concentration.
	Increases with increasing cathode cd.	Decreases very slightly with increasing acid concentration.

Table 20-13 Variables Which Affect Mechanical Properties of the Deposit—Copper Fluoborate Solution

Property	*Operational*	*Solution composition*
Tensile strength	Increases with increasing solution temperature.	Increases with copper fluoborate concentration
	Increases with increasing cathode current density.	Relatively unaffected by fluoboric acid concentration.
Elongation	Increases with increasing solution temperature.	Increases with increasing copper fluoborate concentration.
	Decreases with increasing cathode current density.	Relatively unaffected by fluoboric acid concentration.
Hardness	Decreases with increasing solution temperature.	Decreases with increasing copper fluoborate concentration.
	Increases with increasing cathode current density.	Relatively unaffected by fluoboric acid concentration.

Some compounds such as saccharin or *p*-toluenesulfonamide can reduce stress to zero or even cause it to become compressive. However, these and other stress reducers introduce sulfur into the deposit, which may affect high-temperature properties.

PARTING

Permanent Mandrels

These have been treated to minimize adhesion between the mandrel and the electroform; parting is accomplished by one of the following methods.

Impact: a sudden blow with a hammer

Gradual force, applied by a hydraulic ram, a wheel-puller, or similar device

Heating, say in a hot oil bath, either to melt or to soften a parting compound or to take advantage of the difference in thermal expansion between the electroform and the mandrel

Cooling, as in a mixture of Dry Ice (solid carbon dioxide) and acetone, which uses the same principle as heating

Prying with a sharp tool, used carefully on flat pieces such as phonograph record stampers.

Expendable Mandrels

Zinc alloys can be dissolved in hydrochloric acid. Aluminum alloys are dissolved in strong, hot sodium hydroxide solutions. Low-melting alloys are melted and shaken out; the alloy may be re-used. If "tinning" occurs (i.e., the electroform retains a coating of the low-melting alloy), the coating may be dissolved with strong nitric acid (if the electroform is nickel). Plastics may be softened by heat so that the mandrel may be withdrawn, after which the electroform is cleaned with a solvent. Wax may be melted out and the electroform cleaned with a solvent, as for plastics.

Wood and plaster may be mechanically broken away from the form.

Backing

For some applications, such as in electrotyping and the forming of molds, the electroform itself is not of sufficient bulk or thickness, and must be backed up with some other material; this may be a cast metal, an additional electrodeposit, or thermosetting resins.

21 Conversion Coatings

A conversion coating is a coating produced by chemical or electrochemical treatment of a metallic surface that gives a superficial layer containing a compound of the metal—for example, chromate coatings on zinc and cadmium and oxide coatings on steel. Some anodic coatings constitute exceptions; see Chap. 22.

In a sense, almost all metals exposed to the atmosphere have on their surfaces "conversion coatings" formed by the chemical reaction of constituents of the atmosphere with the metal: oxide or hydrated oxide films such as rust on iron, sulfide tarnish films on copper and silver, basic carbonates and sulfates that constitute the patina on copper, etc. The term *conversion coating,* however, is for practical purposes restricted to coatings deliberately formed by the metal finisher for a specific purpose by means of a specific and controlled chemical or electrochemical process.

Among the more widely used conversion-coating processes are chromate conversion coatings, phosphating, black oxide finishes for iron and copper alloys, and various coloring processes for copper, brass, bronze, and other metals. Only the first two will be considered here; and, because such processes are almost invariably proprietary, or in a few cases disclosed in texts of the "home workshop recipe" type, complete discussion will not be attempted. Both chromating and phosphating are very useful adjuncts to metal finishing, but formulations and operating directions must be obtained from vendors.

CHROMATE CONVERSION COATINGS

Chromate conversion coatings can be applied to electrodeposited zinc, cadmium, silver, copper, brass, and tin, as well as to aluminum, magnesium, zinc die-cast alloys, hot-dip galvanized parts, and some other metals, usually by simple immersion in aqueous solution although some electrochemical processes are available.

The primary purpose of chromate finishes on zinc and cadmium is to retard the formation of white corrosion products upon exposure to stagnant water, moist atmospheres, or stagnant environments containing organic vapors, such as may emanate from certain plastics, paints, and other organic materials. Chromate finishes will not prevent the growth of metallic filaments, commonly known as whiskers.*

Coatings generally contain oxides of the basis metal and trivalent and hexavalent chromium in varying proportions, except that colorless coatings contain little or no hexavalent chromium. They may be produced by either chemical or electrochemical processes containing hexavalent chromium compounds with one or more of certain anions which act as activators, film formers, or both. There is evidence that over an extended period, chromate coatings undergo some chemical changes even under ordinary conditions. These changes increase with increase in temperature. At temperatures above about 65°C these changes take place fairly rapidly, converting the soluble hexavalent chromium into an insoluble compound and thereby reducing its protective value under salt spray and humid conditions. Colorless or light iridescent coatings appear to be less sensitive to elevated temperatures than are heavy chromate coatings.

The quality of the chromate film depends to a large extent on the chemical purity and the physical condition of the basis surface to which it is applied. In order to produce an acceptable coating, it is essential that the surface be properly cleaned and free of heavy metallic impurities such as lead, copper, and contamination such as brightener occlusions and oxides, which interfere with the chromating reaction.

The thickness of the coating to be chromated should be not less than 4.0 μm, and the thickness requirement on the coating and chromated finish should apply *after* the chromate treatment. The color and luster produced by a given treatment will depend to some extent on the surface condition of the metal to which it is applied and may vary from part to part, or even on one single part.

Chromate conversion coatings on zinc and cadmium deposits offer several advantages: the brilliance of the plated surfaces is improved by an action that polishes or smooths the surface. The solutions remove haze or other surface films that interfere with specular reflection.

Of equal importance is the greater resistance to finger staining imparted by the treatment. The improvement in corrosion protection may be substantial: the time to appearance of "white rust" (the white corrosion products of zinc) in salt spray is significantly increased, and the shelf life of plated articles stored under mild conditions is enhanced.

Chromate conversion coatings on zinc are of three general types: clear, iridescent, and colored. The last offers maximum protection and is often called out in military specifications, both for its corrosion resistance and because the olive drab color matches much other military hardware. Olive drab conversion coatings are also used on washing machine parts for resistance to soaps and

*ASTM B 201: Testing Chromate Coatings on Zinc and Cadmium Surfaces.

detergents, on under-the-hood automotive parts, and some refrigerator parts. Black chromate conversion coatings are also available.

Chromate conversion coatings also accept dyes. Such dyed coatings have applications in cameras (black), household items in which the color of brass, say, can be matched, and for similar applications; and for color coding of industrial items such as screws, nuts, and bolts to differentiate between various thread counts on otherwise similar-appearing items. Color coding has assumed increased importance as industry begins the process of metrication. For example, screws produced to U.S. customary specifications are given one color, those produced to metric specifications, a different one. The film is dyed immediately following rinsing of the chromate; after the film has set, by drying and aging, it cannot be dyed.

Chromate conversion coatings are also applied as a base for paints and other organic finishes. The coating provides "anchor points" to which the paint film can adhere; it also retards the spread of corrosion if the paint is scratched or damaged.

The chromate conversion coating is formed by a chemical reaction between the metal surface and the hexavalent chromium in the solution: the metal is oxidized, and the Cr(VI) is reduced to Cr(III). During the course of the reaction, the pH at the metal-liquid interface rises; this causes the Cr(III) compounds to precipitate on the surface in the form of a gel, which entraps some of the Cr(VI) from the solution.

Most chromate conversion solutions contain wetting agents, which help to make the reaction more uniform, reduce drag-out, and prevent staining during transfer to the first rinse.

A typical analysis of a chromate conversion coating on zinc is shown in Table 21-1.

Table 21-1 Typical Analysis of a Chromate Film on Zinc

	Percent	
Cr(VI)	8.7	
Equivalent to CrO_4^{--}		19.4
SO_4^{--}	3.3	
Cr(III)	28.2	
Equivalent to Cr_2O_3		41.8
Zn^{++}	2.1	
Na	0.3	
Water (at 110°C)	19.0	
Unaccounted for (oxygen or water bound at 110°C)	14.1	

Table 21-2 Cycle for Chromating Electrodeposited Coatings

1. Electroplate	6. Chromate
2. Cold-water rinse	7. Cold-water rinse
3. Cold-water rinse	8. Cold-water rinse
4. Neutralize	9. Hot-water rinse
5. Cold-water rinse	10. Dry

The quality of the electrodeposit, such as zinc, is very important in determining the quality of the conversion coating. A smooth and fine-grained deposit, free of microscopic hills and valleys, will produce a satisfactory coating; the deposit also should be relatively pure metal.

The process of chromating removes some metal; therefore it is imperative that the electrodeposit to be chromated not be too thin. The minimum thickness is about 4 μm, since as much as 1.3 μm may be removed in the chromating process.

For satisfactory chromate conversion coatings on zinc and cadmium deposits, the plating solution, and therefore the deposit itself, should be free of contaminant metals; especially deleterious are copper, chromate (or chromium in other forms), and lead.

Typical cycles for electrodeposited coatings of zinc, cadmium, silver, or copper to be chromated are shown in Table 21-2. Cycles for massive metals and for galvanized zinc coatings are somewhat modified, as shown in Table 21-3.

Control of Chromating Solutions

Time, temperature, pH, and concentration are the principal factors determining the success of the chromating operation. Time is generally about 10 to 30 s; temperature of the solution ranges from 24 to 35°C, and the pH from somewhat less than zero to about 2.8. The concentration is as specified by the vendor, and it is controlled by analysis for hexavalent and trivalent chromium and for dissolved metals.

The lower end of the pH range generally results in the maximum amount of chemical polishing and in clear coatings; at the upper end of the range, colored coatings are produced which have the greatest corrosion resistance. High temperature tends to yield accelerated stripping of the deposit and iridescent, cloudy, and loose deposits; low temperature (below 21°C) gives thin, milky, and dull coatings.

Chromate films, as mentioned, are amorphous gels, and consequently they are sensitive to heat. Excessive heat dehydrates the gel and may affect its corrosion performance.

Table 21-3 Cycle for Chromating Massive Metals (Al, Mg, Zn Alloys)

1. Soak clean	6. Chromate
2. Cold-water rinse	7. Cold-water rinse
3. Activate or deoxidize	8. Hot-water rinse
4. Cold-water rinse	9. Dry
5. Cold-water rinse	

Testing Chromate Coatings on Zinc and Cadmium Surfaces*

Procedures for evaluating the protective value of chemical and electrochemical conversion coatings produced by chromate treatments of zinc and cadmium surfaces are summarized here from the ASTM document cited.

This protective value is usually determined by salt spray test and by determining whether the coating possesses adequate abrasion resistance. Tests are applicable to chromate coatings of the colorless (both one- and two-dip), iridescent yellow or bronze, olive drab, black, colorless anodic, yellow, or black anodic types,† and of the dyed variety, when applied to surfaces of electrodeposited zinc, mechanically deposited zinc,‡ hot-dipped zinc, rolled zinc, electrodeposited cadmium, mechanically deposited cadmium,‡ or zinc die castings (colorless coatings are also referred to as clear-bright or blue-bright coatings).

Time to failure depends on the type of coating tested. Some expected protective values (time to white corrosion products in the neutral 5 percent salt spray test) are listed in Table 21-4; it should be emphasized that these are not necessarily endpoint requirements but typical values. Usually, failure is defined as the first appearance of white corrosion products on significant surfaces visible to the unaided eye at normal reading distance, except that presence of white corrosion products at sharp edges and at junctions between dissimilar metals should not be considered failure. Occasionally, the first appearance of red rust may be regarded as failure.

Before subjecting a chromate coating to test, it must be aged at room temperature in a clean environment for at least 24 h after the chromating treatment. The test surfaces must be free of fingerprints and other extraneous stains and must not be cleaned except by gentle wiping with a clean, soft, dry cloth to remove loose surface particles. Oily or greasy surfaces should not be tested, nor is the use of solvents for degreasing recommended.

*ASTM B 201, loc. cit,

†Some proprietary chromating processes are electrolytic, the part being made anodic in the solution.

‡By the process known as Peen Plating,™ not otherwise discussed in this text.

Abrasion Resistance

To determine whether the coating is adherent, nonpowdery, and abrasion-resistant, rub the chromated surface with a gritless, soft gum eraser (Artgum) for 2–3 s by hand (about 10 strokes) using normal pressure (about 70 kPa) and a stroke about 50 mm long. The chromate coating should not be removed or worn through to the underlying metal.

Test for Colorless (Clear) Coatings

Determine the presence of a colorless (clear) coating by placing a drop of lead acetate testing solution on the surface (50 g hydrated lead acetate in 1 L of water, adjusted to pH 7.3–7.7 by acetic acid). Allow the drop to remain on the surface for 5 s. Remove the testing solution by blotting gently, taking care not to disturb any deposit that may have formed. A dark deposit or black stain indicates *lack* of an acceptable chromate coating.* For comparison, treat an untreated surface similarly; on such a surface a black spot forms almost immediately.

Chromate coatings are widely specified by the military and other government procuring agencies; some of their specifications are listed in Table 21-5.

CHROMATE FINISHES FOR ALUMINUM

This process is the subject of ASTM B 449, Recommended Practice for Chromate Treatments on Aluminum, summarized here. It covers producing, on commercial aluminum and aluminum alloys, chromate conversion coatings by

*This test is to some extent controversial, and one should be alert for revisions.

Table 21-4 Expected Protection of Chromated Zinc Coatings in 5% Neutral Salt Spray

Type of coating	*Expected minimum hours to white corrosion of zinc*
One-dip colorless (clear bright)	12
Two-dip colorless (clear bright)	24
Black dip	48
Anodic colorless	48
Anodic black	96
Iridescent yellow or bronze	96
Anodic yellow	150
Olive drab	150

Table 21-5 Some U.S. Government Specifications for Chromate Conversion Coatings

Zinc plate	QQ-Z-325a
Cadmium plate	QQ-P-416b
Silver plate	QQ-S-365a
General	MIL-STD-171B
Aluminum	MIL-C-5541A
Magnesium	MIL-M-3171C

Note: This list is not complete; see also Cross Index of standards, in ASTM annual "Book of Standards," Part 9.

chemical methods from aqueous acidic solutions containing hexavalent chromium plus acid radicals and film-forming chemicals. (Distinguish this from chromic acid anodizing, Chap. 22.)

Color and color uniformity will vary somewhat between one alloy and another and from a polished surface to an etched surface. Iridescence and variations in color intensity from one area of the surface to another are normal and not objectionable.

Coatings are usually applied by immersion or spray, but application by brush or swab is also used, mainly for touchup. Roll coating is used for continuous strip processing.

Chromate conversion coatings on aluminum are used to retard and prevent corrosion, as a base for paint, and as a protective surface having lower electrical contact resistance than anodized coatings. They can be used to change the emissivity and absorption properties of the metal and can be dyed for identification purposes.

The coatings generally provide protection in proportion to film thickness. They provide good protection against marine and humid environments. When coatings are dried at room temperature, hexavalent chromium compounds, loosely bonded in the film structure, are slowly soluble on exposure to aqueous media and will heal minor abrased or scratched areas.

The hexavalent chromium compounds in chromate conversion films become insoluble in aqueous media when they are heated above 71°C, thus eliminating the self-healing characteristics and gradually reducing the protective value of the film as the temperature and time at temperature increase. Corrosion protection is reduced to a greater degree in heavy than in thin coatings. The paint-base properties of the coatings are not adversely affected by the short elevated drying times normally used in production.

The aluminum surface should be free of all foreign substances and soils—grease, oils, paints, cleaning compounds, welding fluxes, and heat-treat and other oxides. For cycle details, see ASTM B 449.

Classification of Chromate Coatings in Aluminum

Class 1: Maximum corrosion resistance and for use unpainted. Color ranges from yellow to brown and coating weight from 320 to 1100 mg/m^2.

Class 2: General purpose for corrosion resistance and as a paint base. Color from iridescent to yellow, coating weight from 110 to 380 mg/m^2

Class 3: Colorless films for decorative purposes, and yellow to colorless for low electrical resistance; thinner and less corrosion resistant than class 2.

All three classes may be used as a paint base, but class 2 is preferred. The performance of the coating depends on the environment of use; in general it is in proportion to the thickness of the coating. Tests of chromate coatings on aluminum are detailed in ASTM B 449.

PHOSPHATE COATINGS

Another widely used type of conversion coating is the phosphate coating, used primarily as a base for subsequent painting but also for other purposes such as for lubrication during drawing or shaping and providing improved corrosion protection. Like chromate conversion-coating processes, phosphate coatings are invariably produced from proprietary or patented formulations and processes; therefore they need not be discussed in detail.

When a metallic surface such as iron is exposed to the corrosive environment of an acid, it is dissolved, and if insoluble corrosion products are formed, the latter may, under appropriate conditions, deposit on the metal surface in the form of a coating. Phosphoric acid has the required properties. The iron phosphates formed in the corrosion reaction are deposited on the iron surface in the form of amorphous or crystalline iron phosphates, which tend to protect the surface from further attack, and, if in the proper form, serve as an excellent bonding surface for paint or other organic coatings. In addition, the phosphoric acid may contain salts of other metals, such as zinc and manganese, and the coating will be mixtures of iron phosphate with phosphates of these metals.

Most metallic surfaces do not form good bases for paint films; a phosphate coating provides mechanical "keying" points for paint adhesion. It also converts the metallic surface into a poor conductor, so that underfilm corrosion is retarded: it provides an insulating barrier to the flow of corrosion currents.

The three types of phosphate coatings in general use are iron phosphates, zinc phosphates, and manganese phosphates. The simplest is iron phosphating, in which the substrate metal itself supplies the cations for formation of the phosphate film. Iron and zinc phosphating may be accomplished by power-spray or immersion techniques; manganese phosphating is accomplished by immersion only.

The largest application for phosphate coatings is as a substrate for paint bonding, These coatings substantially improve the adhesion, impact resistance, and flexibility of paint films, and inhibit underfilm rusting if the paint film is scratched or scored.

Iron phosphating is used principally for protecting stampings for cabinets, metal furniture, shelving, and siding; the corrosion resistance provided is minimal; its only use is as a substrate for painting.

Zinc phosphating is used to prepare automobile and truck bodies and appliances before painting. It is also used in cold-forming operations in conjunction with drawing and forming lubricants to extend the life of dies; it imparts excellent corrosion resistance.

Manganese phosphating is used on frictional and bearing surfaces such as piston rings, valve tappets, cam shafts, gears, and the like as a protection against wear and seizing of the surfaces during break-in periods. The coating acts to prevent metal-to-metal contact and also serves as a reservoir for oil films. Manganese phosphate coatings are not used as a paint base.

Iron phosphating solutions consist essentially of alkali or ammonium acid phosphates plus additives for accelerating the reaction such as oxidizing agents: nitrites or chlorates. The stoichiometry (much simplified) may be represented as

$$12Fe + 8NaH_2PO_4 + 10H_2O + 7O_2 \rightarrow \underbrace{2Fe_3(PO_4)_2 \cdot 8H_2O + 2Fe_3O_4}_{\text{coating}} + 4Na_2HPO_4$$

The pH of the solution rises during the reaction, as indicated by the conversion of monosodium acid phosphate to the disodium salt.

The mildly acidic monosodium dihydrogen phosphate acts to dissolve the iron, forming the soluble ferrous phosphate $Fe(H_2PO_4)_2$; some hydrogen is evolved at the cathodic sites on the surface. Formation of the coating itself, a mixture of ferrous phosphate and magnetite, Fe_3O_4, is a secondary reaction. Even when oxidizing agents are present, the iron present in the phosphate is in the Fe(II) state, owing to the hydrogen evolution mentioned. The most effective pH for iron phosphating is between 4.5 and 5.3.

Other competing reactions occur, including the formation of the very insoluble ferric phosphate, $FePO_4$, which precipitates from the bath as a waste product.

The reactions occurring in zinc phosphating and manganese phosphating are more complicated. In the former, a zinc phosphate is present in solution at the outset; the bath generally consists of an aqueous complex mixture of primary zinc phosphate, $Zn(H_2PO_4)_2$, free phosphoric acid, and oxidizing agents. Somewhat simplified, the reactions are

$$Fe + 2H_3PO_4 \rightarrow Fe(H_2PO_4)_2 + H_2$$

$$3Zn(H_2PO_4)_2 \rightleftharpoons Zn_3(PO_4)_2 + 4H_3PO_4$$

$$2Zn(H_2PO_4)_2 + Fe(H_2PO_4)_2 + 4H_2O \rightarrow \underbrace{Zn_2Fe(PO_4)_2 \cdot 4H_2O}_{\text{coating}} + 4H_3PO_4$$

$$3Zn(H_2PO_4)_2 + 4H_2O \rightarrow \underbrace{Zn_3(PO_4)_2 \cdot 4H_2O}_{\text{coating}} + 4H_3PO_4$$

When iron articles are immersed in a zinc phosphating solution, the hydrogen ions from the phosphoric acid pickle the surface; iron dissolves with the liberation of hydrogen at local cathodic sites. The decrease in the hydrogen ion concentration shifts the equilibrium to the right, the solubility product of the tertiary zinc phosphate is exceeded, and it begins to precipitate on the metal surface. The zinc phosphate coating formed consists of a thin layer of zinc phosphophyllite, $Zn_2Fe(PO_4)_3 \cdot 4H_2O$, next to the metal surface, with an overlay of hopeite, $Zn_3(PO_4)_2 \cdot 4H_2O$.

Zinc phosphating is best accomplished at a pH range of 3.1 to 3.4. Ferric phosphate precipitates as a waste product and must be periodically removed.

As in iron phosphating, oxidizing agents accelerate the formation of the coating and also serve to precipitate dissolved iron as sludge and act as metal cleaners by oxidizing residual soils. Commonly employed oxidizing agents are, as with iron phosphating, sodium nitrite and sodium chlorate. Other ions are often employed in zinc phosphating solutions, usually to modify the crystal structure of the coating; these include nickel, calcium, and fluoride ions.

Phosphate coatings, if they are to function satisfactorily, must consist of dense, tightly packed, fine crystals; the coating must be uniform and thin, free of powder, and flexible, and must possess satisfactory corrosion resistance. These factors are attained by proper cleaning and other pretreatment cycles, and by proper formulation and operation of the phosphating baths.

Several other conversion-coating processes are in use, including the formation of black oxide coatings on steel and various processes for coloring copper, brass, and other metals. Like the other conversion-coating methods mentioned in this chapter, these processes are invariably proprietary (even though patents may have expired, few finishers attempt to formulate their own solutions), and directions for use must be obtained from vendors.

22 Anodizing Aluminum

Anodizing is an electrolytic process in which a metal is made the anode in a suitable ele‹ .rolyte so that when an electric current is passed through the electrolyte, the metal surface is converted to a form of its oxide that has useful decorative, protective, or other desirable properties. The electrolyte provides oxygen ions that react with metal ions to form the oxide, and hydrogen is released at the metal or carbon cathode. Depending upon the solvent action of the selected electrolyte on the anodic oxide, the operating conditions employed, and voltage/current relationships, the metal anode continues to be consumed and converted to an oxide coating which progresses inward. The last-formed oxide is adjacent to the metal-coating interface.

Anodizing differs from electroplating in two significant respects. In electroplating, the work is made the cathode, and metallic coatings are deposited on the work. In anodizing, the work is made the anode, and its surface is converted to a form of its oxide that is integral with the metal substrate.

Anodizing processes have been developed for many metals. However, those used with aluminum are of the greatest commercial significance. Magnesium is anodized for improved resistance to corrosion and abrasion by procedures similar to those used with aluminum. Zinc can be electrochemically treated as an anode in a complex proprietary aqueous electrolyte developed under the auspices of the International Lead-Zinc Research Organization (ILZRO). Although this process is commercially referred to as "anodizing," it does not produce an oxide coating. Rather, via high-voltage spark discharge, a fritted semifused surface that enhances resistance to abrasion and corrosion is formed. Other metals including copper, cadmium, silver, and steel usually are anodized to achieve decorative effects which often are fugitive unless they are protected by an organic overcoating.[1]

Note: This chapter is by D. Jack George, Aluminum Company of America, Alcoa Laboratories, Alcoa Center, PA 15069.

*Superior numbers refer to references cited at the end of this chapter, p. 478.

THEORY OF ANODIC OXIDE FORMATION

The mechanism of anodic oxidation is complex, and some aspects are not completely understood. In accordance with Faraday's law, 1 gram-equivalent of pure aluminum (8.9938 g) reacts electrochemically when 96 500 C of electricity is passed through the aluminum anode. However, not all of this aluminum appears as aluminum oxide in the coating. A significant ratio can be calculated between the weight of coating and the metallic aluminum or aluminum alloy removed. If all the aluminum were converted to oxide, this "coating ratio" would be 1.89 ($Al_2O_3/2Al$). With the porous and adsorbent type of coating, this ratio is significantly lower, seldom exceeding about 1.60. The observed coating ratios for several electrolytes and coating times are given in Table 22-1. The coating ratio is lowered significantly by an increase in the sulfuric acid content or the temperature of the electrolyte. The ratio is also lower with aluminum alloys than with the pure metal. A lowering of the coating ratio indicates, in general, an increase in porosity and decrease in abrasion resistance. It is known, moreover, that the coating is not all aluminum oxide, but contains chemically bonded and adsorbed substances such as sulfate and water from the electrolyte.

Based on studies conducted with a sulfuric acid electrolyte, it has been suggested that even under the most favorable conditions, anodic coating weight is only 61 percent of the theoretical weight calculated from Faraday's law. Considering the fact that as much as 13 percent of the coating is a form of sulfate, the divergence from the theoretical amount of aluminum oxide is even greater.[2]

Table 22-1 Relation between Weight of Anodic Coating and Aluminum* Removed

Electrolyte	*Current density, A/m²*	*Time of anodic oxidation (min)*	*Ratio of weight of coating to metal removed*
Sulfuric acid (15% solution at 25°C)	130	15	1.41
	130	30	1.33
	130	60	1.19
Sulfuric acid, 6% plus Oxalic acid, 6%, at 21.1°C	130	30	1.47
Sulfuric acid, 6% plus Oxalic acid, 6%, at 32.2°C	130	30	1.32
Chromic acid (3% solution at 40°C)	30	60	0.87
Oxalic acid (3% solution at 57.8 to 30°C)	130	15	1.31
	130	30	1.27
	130	60	0.95

*Aluminum, 99.8 percent purity, used for tests.

Anodic coatings formed on aluminum in an electrolyte that has little or no capacity to dissolve the oxide are called *barrier* coatings. They are essentially nonporous and have a limited thickness proportional to the applied voltage (1.3–1.4 nm/V). The barrier thickness represents the distance through which ions penetrate the layer of oxide under the influence of the applied potential. Once the limiting barrier thickness has been reached, it becomes an effective barrier to further ionic and electron movement; current flow drops to a very low leakage value, and oxide formation ceases.

When electrolytes that have appreciable solvent action on the oxide are employed, the barrier layer does not reach its limiting thickness, and current continues to flow. This results in the development of a "porous" oxide structure. Porous coatings may be several tens of micrometers thick. However, a thin layer of barrier oxide always remains at the metal-oxide interface.

Figure 22-1 represents a model of a porous type of anodic coating as envisioned by Keller, Hunter, and Robinson[3] and illustrates typical dimensional relationships of barrier to porous oxide as formed in a phosphoric acid electrolyte. This concept has been confirmed essentially by others with only minor modifications in pore configuration and dimensions.

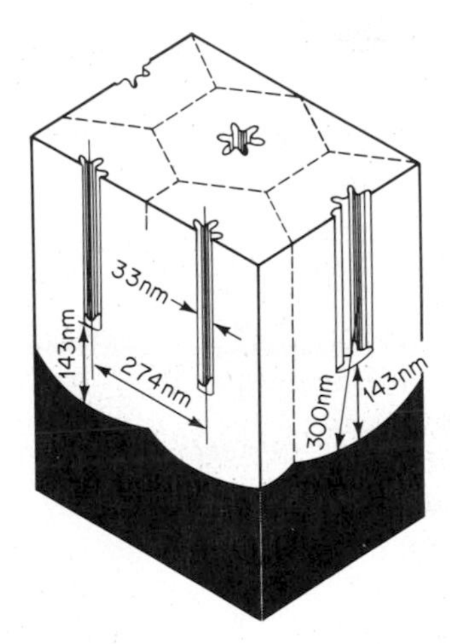

Fig. 22-1 Structure of 120-V phosphoric acid coating, constructed on cross section of cell base pattern. Dimensions of pore, cell, cell wall, barrier, and radius of curvature are shown in nanometers. Magnification: 65 000 ×.[3]

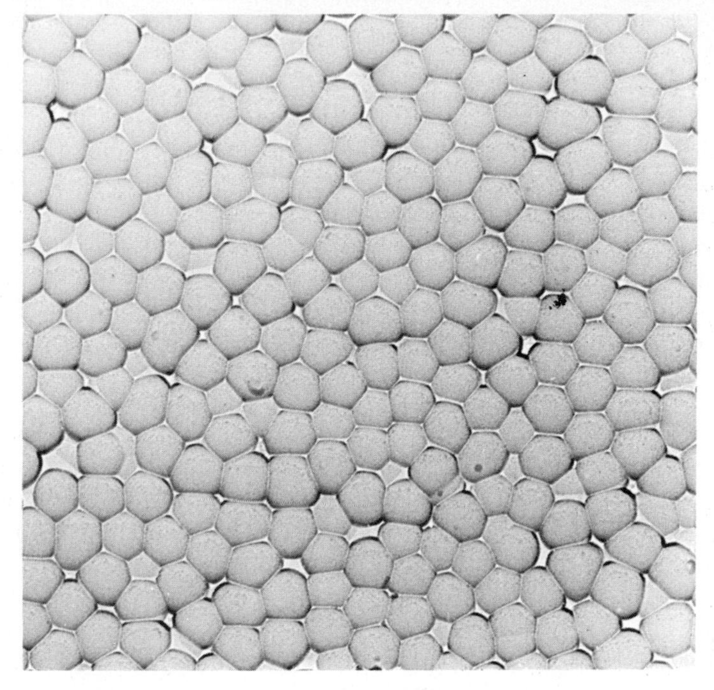

Fig. 22-2 Underside of an anodic oxide coating showing cell base pattern. Magnification: 80 000 ×. *(Courtesy H. A. Burns, Alcoa Laboratories.)*

MORPHOLOGY OF ANODIC COATINGS

Advances in the science of microscopy and especially the availability of the electron microscope have enabled researchers to describe more clearly the structural features of anodic coatings.

Electron micrographs of the underside of an anodized oxide coating reveal the presence of close-packed cells of essentially amorphous oxide (Fig. 22-2). There can be billions of cells per cm^2, their size being a function of anodizing voltage (Fig. 22-3). As can be seen in Fig. 22-4, which depicts a surface and cross-sectional view, each cell has a single pore. Pore size is influenced by a number of factors, including type of electrolyte, temperature, and voltage/current relationships.[4] Pores extend downward to the barrier oxide (Fig. 22-5). The structures of anodic coatings formed in phosphoric, sulfuric, chromic, and oxalic acids differ only in pore and cell dimensions.

ANODIZING IN SULFURIC ACID

Many electrolytes have been suggested for anodizing aluminum,[5] some of which might be classified as general purpose, whereas others are intended to achieve

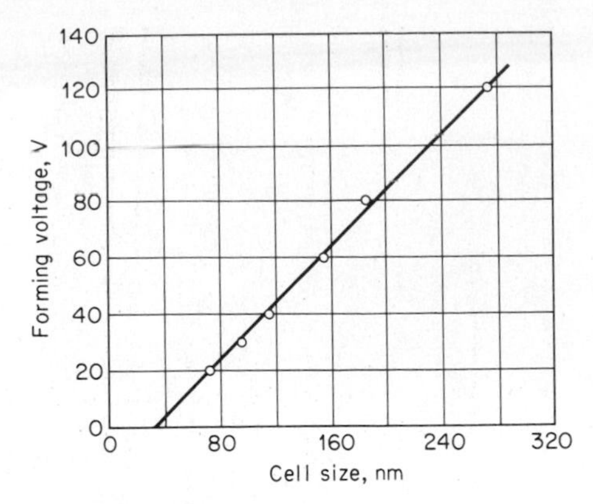

Fig. 22-3 Relation between cell size and forming voltage in 4 percent phosphoric acid electrolyte.[3]

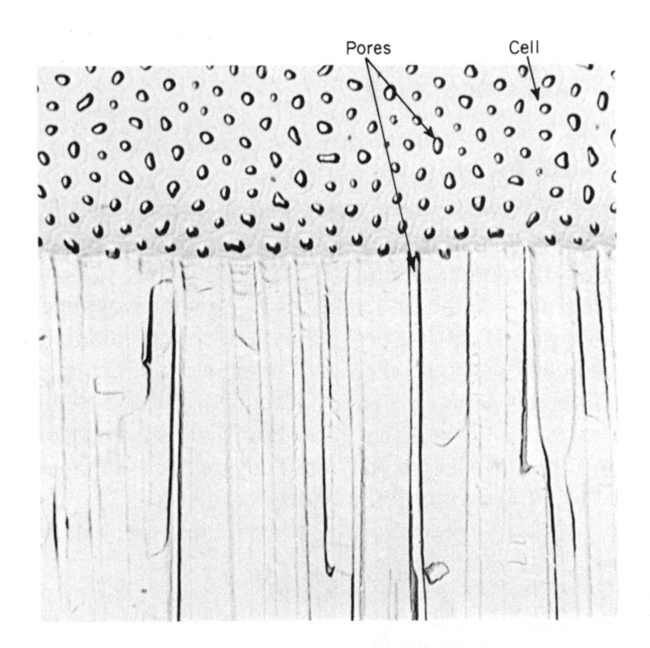

Fig. 22-4 Surface and cross section of anodic oxide coating showing cells and pores. Magnification: 60 000 × . *(Courtesy R. A. Burns, Alcoa Laboratories.)*

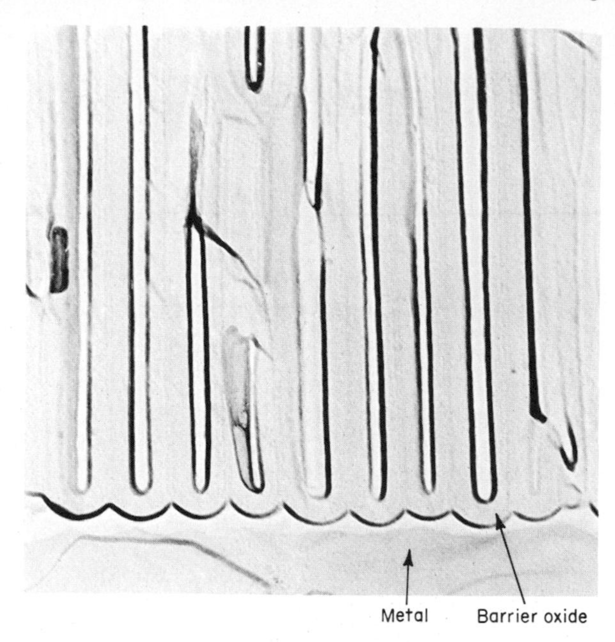

Fig. 22-5 Cross section of anodic oxide coating showing pores extending to barrier oxide. Magnification: 80 000 ×. *(Courtesy R. A. Burns, Alcoa Laboratories.)*

some specific objective. The processes most widely used employ sulfuric acid in water at concentrations of 12 to 25% (wt.) This type of electrolyte is relatively inexpensive and easy to control, and it results in coatings with a wide range of aesthetic or functional properties.

General-purpose or "conventional" sulfuric acid anodic coatings applied for decorative and protective purposes range in thickness from 2.5 to 30 μm. Typically, they are produced at a temperature of 21°C, a current density of 130 to 260 A/m², and a voltage range of 12 to 22 V.

The effects of time of oxidation and temperature of the electrolyte on the thickness of the oxide coating are illustrated in Fig. 22-6. To obtain these data, 1100 alloy sheet was anodized in 15 percent sulfuric acid electrolyte at a current density of 130 A/m² and at several temperatures. With the electrolyte at a temperature of 21°C, the increase in thickness of the coating is linearly proportional to the time of oxidation over a period of about an hour or longer. If the temperature of the electrolyte is increased, the coating thickness obtained for a given oxidation period decreases, and the linear portion of the curve is shortened (Fig. 22-6). This has led to the selection of a temperature of about 21°C with the 15 percent sulfuric acid electrolyte for the production of general-purpose coatings.

The oxide coating occupies a greater volume than the aluminum metal from

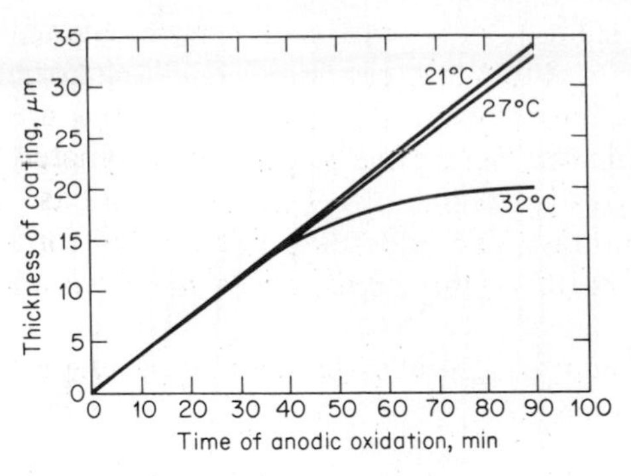

Fig. 22-6 Alloy 1100: effect of anodizing time on anodic coating thickness at three temperatures; 15 percent sulfuric acid, 130 A/m^2.

which it is formed; therefore, when there is no appreciable solvent action upon the oxide coating, the thickness of the section increases. This increase in thickness is about one-third the thickness of the coating. For example, a cylinder with an oxide coating 24 μm in thickness would be increased 8 μm per surface, or 16 μm in diameter. However, when the solvent action of the electrolyte is greater, the increase will be smaller or there may even be a decrease in the diameter of the piece. These changes are ordinarily small, but where close tolerances are specified for parts that must fit together, they must be considered.

The weight of the oxide coating is a more sensitive measure of the solvent action of the electrolyte than is the thickness of the coating. Oxide, for example, may be dissolved from within the pores, reducing the weight without appreciably decreasing the thickness of the coating. The effect of time of oxidation on weight of coating is illustrated by the data of Fig. 22-7, which were obtained at three different electrolyte temperatures. In comparing Figs. 22-6 and 22-7, it can be seen that temperature has little effect on coating thickness, whereas coating weight is lowered substantially as temperature is increased. This accounts for the increase in porosity and loss of abrasion resistance as the temperature of the electrolyte increases. With long oxidation periods, for instance, 1 to 2 h, and with other conditions such as high temperature and low current density, there may even be a net loss in weight of an article after oxidation.

OTHER ANODIZING ELECTROLYTES

Dozens of electrolytes have been developed for producing special effects or characteristics on aluminum. Those that have attained some commercial significance follow.

Chromic acid anodic coatings are opaque, are limited to a maximum thickness of about 10 μm, and are rarely used for decorative purposes. Typically, they are formed in a 3–10% solution of chromic acid at 40°C. Standard practice is to raise the voltage slowly to 40 V and continue anodizing for about 30 min at this constant voltage. Chromic acid anodizing processes are among the earliest developed for aluminum. Before the advent of chromate chemical conversion-coating processes, this type of coating was used as a base for paints, especially for military applications. They continue to be used for this purpose and for anodizing complex parts where complete rinsing of the anodizing electrolyte is very difficult.

Phosphoric acid anodizing processes have been suggested as a pretreatment for electroplating, since this type of anodic coating is rather porous and provides a substrate for mechanical locking of the electroplate.

Oxalic acid electrolytes produce yellow coatings that are somewhat harder than conventional sulfuric anodic coatings. Because this process generally is more expensive than sulfuric acid processes, it is not used extensively in the United States. Oxalic acid–sulfuric acid mixtures are employed for producing hard anodic coatings, although low-temperature hard anodizing processes based on sulfuric acid alone are competitive.

Sulfonated organic acids in combination with sulfuric acid are employed to develop so-called integrally colored anodic coatings on controlled alloys. Shades of bronze, gold, gray, and black have found wide acceptance for architectural applications. These coatings are harder and denser than conventional sulfuric acid coatings, and the color develops as a result of the alloy and temper employed together with highly controlled anodizing procedures. Details on these processes are proprietary.*

Boric acid electrolytes, often with additions of borax, are popular for produc-

*Duranodic process: Aluminum Company of America; Kalcolor process: Kaiser Chemical and Aluminum Company.

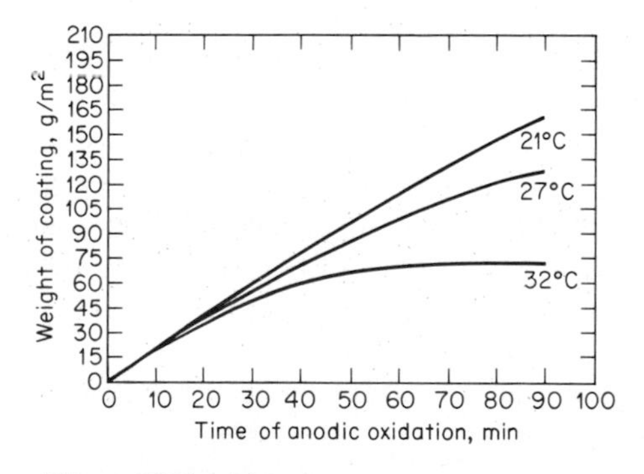

Fig. 22-7 Aluminum sheet (1100): effect of electrolyte temperature and time of oxidation on oxide coating weight; 15 percent sulfuric acid, 130 A/m^2.

ing thin barrier oxide coatings for electrical capacitors. Citrates and tartrates also are employed for this purpose.

EFFECT OF ALLOY COMPOSITION

Since anodic oxidation or anodizing involves conversion of the aluminum surface into an oxide coating, the alloy and its metallurgical structure have important effects on the characteristics of this coating. Not only does the alloy composition have a pronounced effect on the density of the anodic coating, but it also affects its appearance. Alloy constituents or impurities may impart coloration to the coating. Sometimes they do not cause coloration; instead, their presence makes coatings look opaque rather than transparent. The appearance of such coatings is dull and lacks a metallic luster.

Wrought alloys represented by sheet, plate, forgings, wire, rod, bar, and extrusions generally are more suitable for anodizing than cast alloys. Nonuniform appearance after anodizing is often experienced with castings owing to large grains, surface porosity, segregation of alloy constituents, dross inclusions, or flow lines. Also, anodic coatings on cast products can be less dense and less protective, especially if the alloy contains appreciable amounts of copper.

Pure aluminum develops the most transparent anodic coatings. With the exception of magnesium, most major alloy constituents reduce the transparency of the anodic coating. Additionally, certain constituents can impart a discrete color to the coating: e.g., silicon turns it gray; chromium or copper, gold; and manganese, tan to brown.

Primary producers of aluminum offer specialty alloys, or they specially control the production of conventional alloys for applications where uniform appearance and a specific aesthetic effect are important. For competitive reasons, composition, temper, and fabrication control often are not published. However, guidance on alloy selection can be obtained readily.

The characteristics of some of the more popular aluminum alloys when anodized in a sulfuric acid electrolyte are shown in Table 22-2.

COLORING ANODIC COATINGS

In addition to so-called integral-color anodic coatings, where the color results from the metallurgical characteristics of the aluminum alloy and specially developed anodizing procedures, most aluminum alloys can be anodized in general-purpose electrolytes and subsequently colored. Porous-type anodic coatings can be colored with organic dyes, certain inorganic pigments, and electrolytically deposited metals.

Organic Dyes

After anodizing and thorough cold-water rinsing, parts are immersed in a heated (65°C) aqueous or organic solvent solution containing several grams per liter of

Table 22-2 Anodizing Characteristics of Some Aluminum Alloys Anodized in a Conventional Sulfuric Acid Electrolyte

Alloy	*Type and form*	*Characteristics*
5252 5457 5657	High-purity Al alloyed with Mg in sheet form	Luster is below that of super-purity base Al alloys, but they respond well to bright-anodizing, have good mechanical properties, and are economical for applications such as automotive and appliance trim. Alloying constituents do not tint coating.
6063 6463	Al-Mg_2Si extrusion alloys	These heat-treated alloys combine strength and good response to anodizing. Heat treatment dissolves Mg_2Si, improving strength and luster after anodizing. Applications include automotive, appliance and architectural trim.
7005 7046	Al-Zn-Mg alloys in sheet, plate, or extrusions	Give bright to diffuse luster, depending upon base purity. Age naturally at room temperature to give high strengths, or may be heat-treated.
1100 5005	Sheet and plate	Alloy 1100 is commercially pure Al having some Si and Fe which diminish luster of anodized finish. Presence of Si can give coating a gray tint. Alloy 5005 has some Mg for modest strengthening.
3003	Al-Mn alloy in sheet or plate	Has excellent formability, and though widely used for anodic finishing, has less luster than 1100 or 5005 because of Mn. Since MnO_2 is brown, the thicker anodic coatings have a tan or brownish tint.
5052	Al-Mg-Cr alloy in sheet or plate	Structural alloy widely used for welded parts. Presence of Cr gives thicker anodic coatings a yellowish tint.
6061 7075	Al-Mg-Si-Cu-Cr and Al-Zn-Mg-Cu-Cr in extrusions, sheet, and plate	Heat-treated alloys have lower luster than 5052 because Al purity is lower. High-strength alloys containing Cr and Cu give the thicker anodic coatings a yellowish tint.
2011 2014 2024	High-Cu alloys of Al in extrusions, sheet, and plate	Heat-treated, high-strength alloys. Give lower-density anodic coatings because most of the Cu dissolves out during anodizing. Thicker coatings show yellowish tint.

Table 22-3 Mineral Pigmentation Processes for Anodic Coatings on Aluminum[6]

First solution	*Second solution*	*Mineral pigment*	*Color of anodic coating*
Ferric ammonium oxalate	None	Iron oxide	Gold
Chromium oxalate	None	Chromium oxide	Green
Lead acetate	Ammonium hydrosulfide	Lead sulfide	Brown to black
Cobalt acetate	Ammonium hydrosulfide	Cobalt sulfide	Black
Cobalt acetate	Potassium permanganate	Cobalt oxide	Bronze
Potassium ferrocyanide	Ferric nitrate	Ferric ferrocyanide	Prussian blue

dye. Dye concentration and pH control vary with the particular dye, and recommendations are available from dyestuff manufacturers. Immersion times of 5 to 15 min are typical. After dyeing, parts are rinsed in cold water and sealed.

Mineral Pigments

Impregnation of anodic coatings with mineral pigments involves precipitation of insoluble materials such as metal oxides, sulfides, and ferrocyanides in the oxide pores. Sometimes this is a two-step process. Examples of several processes are shown in Table 22-3.[6] Generally, colored anodic coatings achieved via organic dyestuffs are not as resistant to fading as those resulting from mineral pigmentation processes. On the other hand, the former are much easier to control and of much greater commercial value as a finishing system. They are especially suitable to multicolor applications since they can be resisted preferentially and bleached to allow repeated dyeings in alternate colors.

Electrolytic Deposition

After anodizing in sulfuric acid, metals can be electrolytically deposited in the oxide pores to achieve shades of gray, bronze, gold, black, and red. Proprietary processes* have been developed based on coloring electrolytes containing nickel, cobalt, tin, selenium, tellurium, vanadium, cadmium, copper, iron, magnesium, lead, and calcium although those employing nickel, tin, or cobalt are of

*Anolok: Alcan International Ltd.; Eurocolor: Pechiney, Ugine Kuhlmann; Colinal: Alusuisse, Aluminium Ltd.

greatest commercial significance. Essentially, the technology of electrolytically depositing metal into the oxide coating can be likened to electroplating. Many colored anodic coatings of this type have good resistance to heat and fading and have been exploited primarily for architectural applications.

SEALING ANODIC COATINGS

The utility and performance of anodic coatings on aluminum often depend upon the type and quality of postanodizing treatment employed. The term *sealing* generally denotes a treatment which renders the coating nonabsorptive or introduces into the coating a material that enhances or modifies the characteristics of the anodic coating. Sealing usually involves subjecting the anodic coating to a hot aqueous environment which causes hydration of the coating. With certain aqueous solutions, components of the sealant are absorbed by the coating. Other solutions permit precipitation of materials into the coating by hydrolysis of specific metal compounds.

When anodic coatings are subjected to pure water at elevated temperatures, the water reacts with the surface of the aluminum oxide to form boehmite:

$$Al_2O_3 + H_2O \xrightarrow{\Delta} 2AlOOH$$

The process of sealing involves a dissolution of oxide and reprecipitation of voluminous hydroxide inside the pore. Prolonged sealing causes aging and densification of the precipitate.

Some aqueous sealants contain metal salts, the oxides or hydroxides of which may be coprecipitated with the aluminum hydroxide. With such sealants, benefits accrue from this reaction as well as from the hydration. Dichromate sealants are useful because they combine the attributes of hydration with the corrosion-inhibiting characteristics of chromate ions.

The resistance of anodic coatings to staining and corrosion depends upon the absorptivity of the coating. Therefore, sealing treatments usually are employed immediately after anodizing. Improperly sealed coatings can account for pitting attack, color change, and mottling. Highest dielectric values also require sealing of the coating.

Choice of sealant depends upon the kind of environment to which the anodic coating will be subjected and any special performance requirements. Also to be considered are any postsealing surface treatments, e.g., painting or adhesive bonding. It is possible to add materials to a sealant to accomplish specific objectives, and it is practical to employ a dual sealing treatment in order to capitalize on the peculiar advantages of each.

TYPES OF SEALANTS AND APPLICATIONS

Water

The most widely used sealant employs water, although only high-quality water is effective. Distilled or deionized boiling water—low in solids and free of phosphates, silicates, fluorides, and chlorides—is required. Mixed-bed ion-exchange systems or a two-column type charged with a strong-base anion and a strong-acid cation exchange resin are satisfactory for furnishing sealing water. In some finishing lines, continuous ion exchange of the sealing water is used to remove deleterious ions dragged in by racks or anodized items.

The presence of trace amounts of phosphates and silicates (5 and 10 ppm, respectively) significantly retards hydration of the coating. Chlorides and fluorides also will prevent adequate sealing and even cause pitting attack of the anodic coating.

Sealing temperature is important and should be 98–100°C; lower temperatures require much longer sealing times. The time required for formation of boehmite approaches infinity at temperatures below about 65°C.

To preclude undue attack of the coating and to permit proper hydration, the sealing water should be maintained at a pH of 5.5 to 6.5. Buffers, such as sodium acetate, are sometimes used to facilitate pH control.

Sealing time should not be shorter than 10 min for thin (2.5-μm) decorative coatings and can be as long as 60 min for thicker (25-μm) coatings. Extension of sealing time ensures more complete hydration, but longer times do not entirely compensate for poor control of other sealing variables, such as temperature, pH, and water purity.

Surfactants and dispersing agents may be added to the water to minimize the formation of a powdery smudge on the work. Care must be exercised that they are of the type that does not adversely affect the adhesion of paints or other applied coatings used subsequently. Phosphates also are effective for minimizing smudge, but they retard hydration significantly.

A variation of water sealing is steam sealing. The prime advantage of this technique is that contaminant-free moisture is ensured. The major disadvantage is that equipment costs are much higher than for tank sealing using boiling water. Compared with conventional water-sealing methods, equivalent sealing is claimed to be faster with steam, the sealing rate increasing with increasing steam temperature.

Nickel Acetate

It is believed that sealing in solutions of nickel acetate is effected by hydrolysis, resulting in precipitation of colloidal nickel hydroxide in the pores of the coating. The coating is not colored by this reaction because the finely divided nickel hydroxide is almost colorless. Concurrent with the precipitation process, a reaction of aluminum oxide with water to form boehmite occurs, as in plain water sealing. In the case of certain dyed coatings, the nickel ion may combine with the dye molecule to form a metal complex or a less soluble nickel salt of the dye. This reaction may cause a color change depending upon the particular dyestuff involved. In general, however, nickel acetate sealing minimizes leaching of the dye during the sealing treatment and improves lightfastness. Nickel acetate–sealed chromic acid anodic coatings are often used to mask or "resist" areas on parts to be "hard-anodized."

The addition of 1 to 5 g/L of nickel acetate in pure water yields a very popular sealant, especially for dyed anodic coatings. The solution is maintained at a pH of 5.2 to 5.5 and a boiling temperature. Immersion times of 3 to 10 min are typical.

Pure water—free from phosphates, silicates, halides, and heavy-metal ions other than nickel—is required, just as with water sealing. Nickel acetate solutions are somewhat less sensitive to the deleterious effects of these contaminants. Nonetheless, it is important to prevent accumulation of these ions and to regularly discard solutions which have become contaminated.

Other Metal Salts

In addition to nickel acetate, salts of aluminum, cobalt, zinc, copper, lead, and chromium have been used to seal anodic coatings. The choice among these is based upon the utility and cost of the chemicals employed and the compound formed as a result of the reaction between the particular dyestuff and the metal ion. It is also possible to combine two or more salts in a single sealant. Certain organic dyestuff–metal ion combinations are more stable than others. Usually, the preferred sealant is noted by the manufacturer of the particular dyestuff or mineral pigment.

Nickel sulfate, nickel nitrate, and alkali metal salts such as sodium, potassium, or calcium sulfate or chloride have been used for sealing but are not of commercial significance in the United States. Sodium molybdate has also been used by itself or in combination with other metal salts or with dichromates.

Dichromates

For improved resistance to saline environments, anodic coatings are sealed in a 5% solution of sodium or potassium dichromate. Such solutions are operated at boiling temperature; immersion time is usually 15 min. At a pH of 5.0 to 6.0, maximum chromate-ion sorption and hydration of the coating occur, assuming the absence of interfering ions such as phosphates and sulfates. Resulting coatings are yellow. However, if the concentration is lowered to 0.5 g/L, only a trace of color is apparent.

Dichromate-sealed coatings are not so resistant to staining by aqueous materials as those sealed in water or metal-salt seals. For example, some trace of stain is usually apparent after testing in accordance with the ASTM B 136 dye stain test. Neither do dichromate-sealed coatings provide the maximum dielectric values. For this reason they are not preferred as treatments when the anodic coating must withstand breakdown by impressed voltages, such as might be encountered in electrical applications. Also, they are not as effective as nickel acetate seals for rendering the coating resistant to breakdown during subsequent anodizing treatments.

Potassium or sodium dichromate is sometimes added to sealing water, when the chloride content of the sealer exceeds 30 ppm and causes surface smudge. A concentration of 0.01 g/L is effective and will not color the anodic coating.

Silicate

Alkali-metal silicate solutions have been used to seal anodic coatings. However, they are not especially popular commercially. The mechanism of such treatments is not clear, but, in addition to hydration, it is probable that aluminum silicates are formed.

Sodium or potassium silicate solutions are usually employed, sodium silicate being the more popular. In the case of sodium silicate, the preferred ratio of Na_2O to SiO_2 is 1:3; with potassium silicate, a ratio of 1:4 is recommended. A boiling 5% solution and an immersion time of 30 min are typical, although sealing times of less than 1 min may be adequate.

Whereas trace amounts of silicate (10 ppm) in a water sealant appear to retard hydration of the coating, larger concentrations provide a film of ''water-glass'' with its characteristic resistance to alkalis and certain environments.

Organic Materials

Waxes are often used to seal anodic coatings, especially when good ''nonsticking'' or release characteristics are required. Examples are ice-cube trays and threaded-screw machine products.

Treatments such as these do not provide maximum resistance to corrosion and weathering. They should be considered as special treatments to be used for release properties or lubricity in applications where maximum resistance to corrosion is not of prime importance. Often they are used after other sealing treatments to impart slipperiness.

Soaps, molybdenum disulfide, and dispersions of polytetrafluoroethylene (e.g., Teflon*) are also employed to decrease the coefficient of friction of the anodized finish. The physical dimensions of some organic polymers are so much larger than the typical pore diameter of an anodic coating that it is doubtful that impregnation of the coating can occur. Thus, treatments of this type are not seals per se, and the resultant surface characteristics may be somewhat transitory.

Lacquers are often applied to anodic coatings to improve resistance to staining and weathering and to increase dielectric properties. Usually, they are used after an aqueous sealing treatment, but sometimes are applied directly after anodizing. Therefore, they are called *seals,* although this is perhaps a loose generalization. Acrylics or modified acrylics are popular and are available for dip or spray application. Often, such treatments are used to prevent staining and attack of the anodic coating by alkaline building products used during construction.

Vapor Techniques

Anodic coatings may be impregnated with certain organic materials in the vapor phase. The sealing material is placed in a closed container with the anodic

*Trademark of E. I. du Pont de Nemours & Company.

coating and heated to temperatures ranging from 93 to 260°C. In the case of certain resin-forming substances, the monomer is vaporized, adsorbed by the coating, and polymerized, thus forming a solid resin in the pores.

Waxlike materials can also be applied by vapor phase treatments. Stearic acid, Carbowax,* and paraffin are examples of materials which can be vaporized. When such materials condense, the pores of the anodic coating are plugged with the solid material. Vapor sealing is not important commercially, primarily because of the rather cumbersome procedures involved. Its prime utility would be for applications where aqueous sealing systems or hydrated anodic coatings might be undesirable.

Dual Seals

Two-step sealing systems are sometimes used to capitalize on the special attributes of each system. For example, a brief (1- to 2-min) nickel acetate seal can be followed by a longer (10- to 15-min) treatment in boiling pure water. This sequence is employed with dyed coatings to minimize leaching of the color, limit the formation of surface smudge associated with nickel acetate, and maximize the degree of sealing. Separate nickel acetate and dichromate treatments also are common, especially when resistance to saline exposure is paramount.

Insofar as lacquering treatments are sometimes called *seals,* another dual treatment would involve water or metal salt sealing before lacquering. This combination is used to provide maximum resistance to alkaline building products.

General

Whereas sealing treatments are required for many good reasons, certain coating characteristics are adversely affected by certain sealants. Resistance to abrasion, for example, is lowered by as much as 10 percent when a pure boiling-water seal is employed (measured by Taber Abraser instrument). Also, the temperature at which an anodic coating will craze is seal-dependent. Seemingly, the more complete the seal, the lower the crazing temperature. Fatigue failure also may be hastened by sealing.

For these and other reasons, the choice of a sealant should be made carefully. Table 22-4 provides general guidelines.

EQUIPMENT

Anodic coatings are produced by batch, bulk, and coil techniques on manual or automatic equipment similar to that employed for electroplating. Irrespective of the technique employed, certain basic considerations must be observed.

*Trademark of Union Carbide Corp. for polyethylene glycols and related compounds.

Table 22-4 Sealing Systems for Anodic Coatings

Type	*Composition*	*Time (min)*	*Temperature*	*pH*	*Remarks*
Water	Pure (deionized)	10–60	Boiling	5.5–6.5	Most widely used system
Steam	Pure	5–30		Neutral	
Nickel acetate	1–5 g/L	3–30	Boiling	5.2–5.5	Recommended for most dyed coatings
Dichromate	50 g/L	15	Boiling	5.0–6.0	Imparts yellow color, excellent resistance to saline environments
Silicate	50 g/L	30	Boiling	Neutral	Ratio of Na_2O to SiO_2 is 1:3
Wax	Proprietary	10–15	Boiling	6.5–7.0	Not suitable for exterior use; good nonstick characteristics
Lacquer	Acrylics or modified acrylics	—	—	—	Spray or dip
Teflon	TFE or FEP	—	—	—	Application techniques are proprietary. Intended to provide low coefficient of friction

Electrical Contact

Anodic coatings are dielectric; therefore, the original positive contact with the aluminum surface must be maintained throughout the entire anodizing cycle. Otherwise, the electrical insulating character of the initially formed anodic oxide effectively impedes flow of current to the work.

Racks

Aluminum or commercially pure titanium are the only practical materials for racking individual parts or for bulk anodizing. When aluminum is used for racking, the anodic oxide coating must be removed from contact areas after each anodizing cycle, usually by etching in caustic soda. For coil anodizing, electrical contact can be made via copper or brass contact rolls that are located outside of the anodizing tank. This arrangement requires good maintenance of the contact rolls to avoid arcing and localized overheating of the aluminum web. Liquid contact techniques eliminate arcing problems associated with contact rolls and permit use of higher anodizing current densities. Essentially, this approach relies on dual anodizing cells where the aluminum web is made the cathode in the first cell and the anode in the second cell.

Cooling and Agitation

During anodizing, electrical energy is converted to heat which must be removed to maintain the selected electrolyte operating temperature. This is accomplished by air or mechanical agitation of the electrolyte and the use of cooling coils or an external heat exchanger. The refrigeration requirement is determined by calculating the wattage input of the largest load to be anodized.

Tank Linings

Type 316 stainless steel, antimonial lead, or tellurium lead linings are satisfactory and can be used as the cathode. Inert linings of rubber, plastic, or glass can also be used provided suitable metallic cathodes in the form of coils or auxiliary electrodes are provided.

Power Supply

Motor generators or rectifiers can be used to provide the required direct current. Solid-state silicon rectifiers featuring constant current and voltage control provide a reliable and versatile power source. For most anodizing, a power supply of 24-V capacity is adequate, although up to 100 V may be required for hard anodizing and certain integral-color processes. Copper anode bars are preferred for conducting the current from the power supply to the rack.

Fume Removal

An exhaust system for removal of fume or spray, constructed of corrosion-resistant material, is required. Capacity calculation should be based on an air flow of about 5.0 m^3 per 0.1 m^2 of solution surface.

NOMENCLATURE

The Aluminum Association Designation System for Aluminum Finishes provides a convenient way to classify anodic finishes as well as chemical and mechanical treatments for aluminum (Table 22-5). Anodized finishes are designated by the letter A followed by a two-digit numeral, the first denoting class of coating and the second denoting type of coating.

Since the introduction of this system in 1964, it has gained in popularity. However, the Alumilite* designation system, introduced in 1931, continues to be used commercially.†

Another important designation system relates to military applications (Spec MIL-A-8625) which classifies chromic acid and sulfuric acid anodic coatings as Types I and II, respectively, and hard coatings as Class III.

APPLICATIONS AND PROPERTIES

Reasons for anodizing are to: (1) increase corrosion resistance, (2) increase paint adhesion, (3) permit subsequent plating, (4) improve decorative appearance, (5) provide electrical insulation, (6) permit application of photographic and lithographic emulsions, (7) increase emissivity, (8) increase abrasion resistance, and (9) detect surface flaws.[7] Even a casual review of the many attributes of anodic coatings and the markets they serve illustrates clearly the tremendous contribution anodizing has made to the growth of the aluminum industry.

Corrosion Resistance

Aluminum, in its natural form, has a high inherent resistance to corrosion owing to an ever-present thin oxide film. Thicker, controlled anodic coatings enhance this characteristic so that aluminum is widely accepted for outdoor applications including marine hardware, automotive trim, architectural curtain walls, windows, and storefronts.

*Trademark name Aluminum Company of America.

†Assistance in converting Alumilite designations to Aluminum Association designations is available from Aluminum Company of America.

Paint Adhesion

In addition to providing an excellent substrate from the standpoint of adhesion, anodic coatings provide an inert and abrasion-resistant barrier to attack of the metal should the organic overcoating be damaged or partially removed. Sealed anodic coatings must be free from surface films associated with some sealing procedures, especially those containing surfactants. Also, final rinsing of sealed coatings requires high-quality water to avoid deposition of contaminants harmful to adhesion of organic coatings. Painted anodic coatings are particularly valuable for critical aerospace and military hardware. Examples are torpedoes and helicopter blades.

Plating Substrate

The natural oxide on aluminum prevents it from being electroplated directly. However, porous, discontinuous anodic coatings, such as can be formed on aluminum in a phosphoric acid electrolyte, are sufficiently electrically conductive to provide a suitable substrate for electroplating, This type of anodic coating is characterized by a relatively irregular profile that enhances mechanical locking or keying of the electrodeposit to the anodic coating.

Decorative Appearance

Proper choice of alloy, preanodic finishing system and anodizing procedure enables aluminum to exhibit an almost endless variety of surface appearances. Anodized products can be bright and specular, dull and diffuse, directional or nondirectional in texture, any color or combination of colors; and anodized finishes can be combined with organic coatings to provide a multitude of aesthetic effects. Jewelry, sporting goods, appliance trim, building components, cooking utensils, hardware, decorative plaques, and nameplates illustrate successful use of this attribute.

Electrical Insulation

Anodic coatings can withstand as much as 40-V/μm breakdown voltage, and they do not char at elevated temperatures. They are used for transformer windings, electronic cabinetry, and many high-temperature applications. Special barrier-type anodic coatings are the foundation of the aluminum electrolytic capacitor industry.

Photographic and Lithographic Substrates

When porous anodic coatings are impregnated with light-sensitive materials (silver halides), the anodized sheet or plate can function like photographic film.

Table 22-5 Designation System for Anodic Coatings (*Adapted from The Aluminum Association Designation System for Aluminum Finishes*)

Type of finish	*Designation**	*Description*	*Examples of methods of finishing*
General	A10	Unspecified	
	A11	Preparation for other applied coatings	2.5-μm anodic coating produced in 15% H_2SO_4 at 21 $\pm$1°C at 130 A/m^2 for 10 min.
	A12	Chromic acid anodic coatings	To be specified.
	A13	Hard, wear- and abrasion-resistant coatings	To be specified.
	A1X	Other	To be specified.
Protective and decorative (coatings less than 10 μm thick)	A21	Clear coating	Coating thickness to be specified. 15% H_2SO_4 used at 21 $\pm$1°C at 130 A/m^2.
	A211	Clear coating	Coating thickness—2.5 μm minimum. Coating weight—6 g/m^2 minimum.
	A212	Clear coating	Coating thickness—5.0 μm minimum. Coating weight—12 g/m^2 minimum.
	A213	Clear coating	Coating thickness—7.5 μm minimum. Coating weight—18 g/m^2 minimum.
	A22	Coating with integral color	Coating thickness to be specified. Color dependent on alloy and process methods.
	A221	Coating with integral color	Coating thickness—2.5 μm minimum. Coating weight—6 g/m^2 minimum.
	A222	Coating with integral color	Coating thickness—5.0 μm minimum. Coating weight—12 g/m^2 minimum.
	A223	Coating with integral color	Coating thickness—7.5 μm minimum. Coating weight—18 g/m^2 minimum.
	A23	Coating with impregnated color	Coating thickness to be specified. 15% H_2SO_4 used at 21 $\pm$1°C at 130 A/m^2, followed by dyeing with organic or inorganic colors.
	A231	Coating with impregnated color	Coating thickness—2.5 μm minimum. Coating weight—6 g/m^2 minimum.

*The complete designation must be preceded by AA—signifying Aluminum Association.

When anodic coatings are used for lithographic plates, the anodic coating exhibits excellent hydrophilic qualities and further protects image areas from wear.

Emissivity and Reflectivity

These terms usually describe surface characteristics in the infrared and visible regions of the spectrum of energy, respectively. Emissivity describes a material's

Type of finish	*Designation**	*Description*	*Examples of methods of finishing*
	A232	Coating with impregnated color	Coating thickness—5.0 μm minimum. Coating weight—12 g/m^2 minimum.
	A233	Coating with impregnated color	Coating thickness—7.5 μm minimum. Coating weight—18 g/m^2 minimum.
	A24	Coating with electrolytically deposited color	Coating thickness to be specified. 15% H_2SO_4 at 21 $\pm$1°C at 130 A/m^2 followed by electrolytic deposition of inorganic pigment in the coating.
	A2X	Other	To be specified.
Architectural Class II† (10 to 17 μm coating)	A31	Clear coating	15% H_2SO_4 used at 21 $\pm$1°C at 130 A/m^2 for 30 min, or equivalent.
	A32	Coating with integral color	Color dependent on alloy and anodic process.
	A33	Coating with impregnated color	15% H_2SO_4 used at 21 $\pm$1°C at 130 A/m^2 for 30 min, followed by dyeing with organic or inorganic colors.
	A34	Coating with electrolytically deposited color	15% H_2SO_4 at 21 $\pm$1°C at 130 A/m^2 for 30 min, followed by electrolytic deposition of inorganic pigment in the coating.
	A3X	Other	To be specified.
Architectural Class I† (17 μm and greater coating)	A41	Clear coating	15% H_2SO_4 used at 21 $\pm$1°C at 130 A/m^2 for 60 min, or equivalent.
	A42	Coating with integral color	Color dependent on alloy and anodic process.
	A43	Coating with impregnated color	15% H_2SO_4 used at 21 $\pm$1°C at 130 A/m^2 for 60 min, followed by dyeing with organic or inorganic colors, or equivalent.
	A44	Coating with electrolytically deposited color	15% H_2SO_4 at 21 $\pm$1°C 130 A/m^2 for 60 min, followed by electrolytic deposition of inorganic pigment in the coating.
	A4X	Other	To be specified.

†Aluminum Association Standards for Anodized Architectural Aluminum.

capability of giving up or reradiating absorbed heat to its surroundings. Reflectivity is that portion of incident energy that is not transmitted or adsorbed by the material.

Since emissivity and reflectivity are surface phenomena, the surface finish, including the type and thickness of anodic coating employed, can be selected to achieve the desired effect. For maximum emissivity values (0.8–0.9) in the far infrared region (9.3 μm), sulfuric acid anodic coatings of about 25 μm are optimum. In the far infrared region, thickness of the anodic coating, not its color, is the controlling factor. However, at shorter wavelengths, color of the coating influences emissivity. Thinner coatings, and coatings formed in other electrolytes such as oxalic, chromic, and phosphoric acids, have lower values.

Highest reflectivity for visible radiation (ca. 84 percent) is achieved with thin, sulfuric acid anodic coatings (2.5–5.0 μm) formed on highly polished, brightened, high-purity aluminum.

Anodic coatings with high emissivity are used in aerospace applications where, under vacuum conditions, heat dissipation is controlled by radiation. Many applications are also found in the electronics and machinery and equipment fields where aluminum products function as heat sinks.

Abrasion Resistance

Anodic coatings on aluminum characteristically are hard and resistant to abrasion. Still, the term *hard anodic coating* usually is reserved for a special class of extra-thick and extra-hard anodic coatings used chiefly for their wear-resistant characteristics. Normally, appearance and, to a lesser extent, resistance to corrosion are secondary attributes. There are a number of commercial hard anodizing systems identified by trade names such as Alumilite, Hardas, Martin, and Sanford.

Hard anodic coatings are usually applied in a sulfuric acid electrolyte operated at relatively low temperatures (−4 to +10°C) and high current densities (250 to 400 A/m^2). Coatings are usually 25 to 75 μm thick.

Hard-anodizing most aluminum alloys causes each surface to grow by an amount equivalent to about one-half of the anodic coating thickness produced.

Excellent resistance to abrasion relates to the hardness of the aluminum oxide coating. However, a hardness value, as measured by a conventional indentation test (e.g., Vickers, Rockwell, Brinell, Knoop), contributes little to any assessment of the wearing quality of an anodized article. These tests point-load the hard anodic coating, forcing it into the much softer aluminum substrate. Thus, the value obtained is influenced by the mechanical properties of the particular alloy and temper from which the coating was formed.

The Taber Abraser test* is used often for determining relative resistance to rubbing wear; it requires flat, 10.2-cm square panels. Weighted abrasive wheels ride on the rotating test panel. Various wheels and different weights can be used, and test results are reported in different ways. The results of one study using type CS-17 wheels and 1000-g loads appear in Figure 22-8. Here the weight loss was divided by the density of the material being tested to develop a "wear index." These data show a 25-μm-thick hard anodic coating on alloy 6061-T6 to perform almost the same as a case-hardened steel and much better than carbon steel or Type 304 stainless steel. Compared with uncoated alloy 6061-T6 material, the

*Teledyne Taber, North Tonawanda, New York.

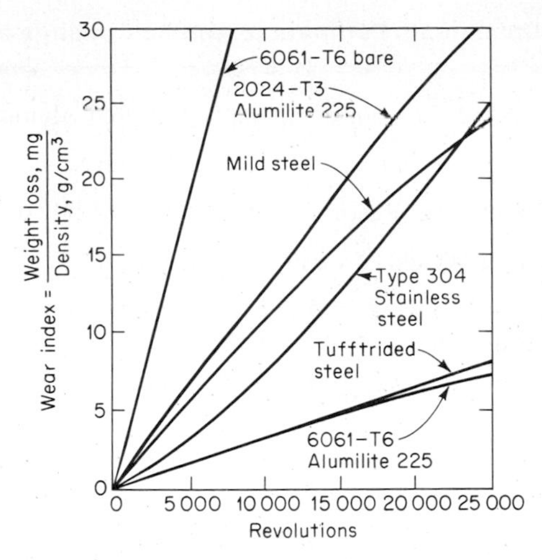

Fig. 22-8 Taber Abraser test results on various coatings and materials.

anodized specimen was about 13 times more resistant to abrasion. Although most anodic coatings are sealed after anodizing to enhance resistance to staining and corrosion, hard anodic coatings are normally not sealed because sealing can reduce abrasion resistance 10 to 20 percent.

Despite the fact that they are not sealed, these coatings perform well in saline environments. Compared to conventional anodic coatings which usually must withstand a few hundred hours of salt spray exposure without failure, hard anodic coatings withstand months of salt spray exposure per ASTM B117 with no signs of failure.

Large quantities of aluminum are hard-anodized for hundreds of applications such as gears, pistons, fan blades, gun scopes, guide tracks, missile components, and fuel nozzles.

Surface Analysis

Since anodic coatings reproduce the surface from which they are formed, anodizing, especially in chromic acid, is a useful tool to make minute surface flaws more visible. This technique is used also to study metallurgical characteristics of aluminum substrates.

TESTING AND EVALUATION

The American Society for Testing and Materials, and particularly its Committee B.08 on Electrodeposited Metallic Coatings and Related Finishes, has developed

Table 22-6 ASTM Documents Pertinent to Anodic Coatings on Aluminum

B110:	Test for Dielectric Strength of Anodically Coated Aluminum
B117:	Salt Spray (Fog) Testing
B136:	Measurement of Stain Resistance of Anodic Coatings on Aluminum
B137:	Measurement of Weight of Coating on Anodically Coated Aluminum
B244:	Measurement of Thickness of Anodic Coatings on Aluminum with Eddy-Current Instruments
B287:	Acetic Acid–Salt Spray (Fog) Testing
B368:	Copper-Accelerated Acetic Acid–Salt Spray (Fog) Testing (CASS Test)
B457:	Measurement of Impedance of Anodic Coatings on Aluminum
B487:	Measurement of Metal and Oxide Coating Thicknesses by Microscopical Examination of a Cross Section
B529:	Measurement of Coating Thicknesses by the Eddy-Current Test Method; Nonconductive Coatings on Nonmagnetic Basis Metals
B538:	FACT (Ford Anodized Aluminum Corrosion Test) Testing
B580:	Specification for Anodic Oxide Coatings on Aluminum
B588:	Measurement of Thickness of Transparent or Opaque Coatings by Double-Beam Interference Microscope Technique
B602:	Sampling Procedures for Inspection of Electrodeposited Metallic Coatings and Related Finishes
D2244:	Standard Method for Instrumental Evaluation of Color Differences of Opaque Materials
E167:	Standard Recommended Practice for Goniophotometry of Reflecting Objects and Materials
E429:	Measurement and Calculation of Reflecting Characteristics of Metallic Surfaces Using Integrating Sphere Instruments
E430:	Measurement of Gloss of High-Gloss Metal Surfaces Using Abridged Goniophotometer or Goniophotometer
G23:	Recommended Practice for Operating Light- and Water-Exposure Apparatus (Carbon-Arc Type) for Exposure of Nonmetallic Materials
G25:	Standard Recommended Practice for Operating Enclosed Carbon-Arc Type Apparatus for Light Exposure of Nonmetallic Materials

a number of standards, specifications, test methods, and recommended practices that are useful to those interested in anodic coatings on aluminum (Table 22-6).

Table 22-7 ISO Documents Pertinent to Anodic Coatings on Aluminum

ISO 2376:	Anodization (Anodic Oxidation) of Aluminium and Its Alloys—Insulation Check by Measurement of Breakdown Potential
ISO 2767:	Surface Treatments of Metals—Anodic Oxidation of Aluminium and Its Alloys—Specular Reflectance at 45°—Total Reflectance—Image Clarity
ISO 2931:	Anodizing of Aluminium and Its Alloys—Assessment of Quality of Sealed Anodic Oxide Coatings by Measurement of Admittance or Impedance
ISO 2932:	Anodizing of Aluminium and Its Alloys—Assessment of Sealing Quality by Measurement of the Loss of Mass after Immersion in Acid Solution
ISO 3210:	Anodizing of Aluminium and Its Alloys—Assessment of Sealing Quality by Measurement of the Loss of Mass after Immersion in Phosphoric-Chromic Acid Solution
ISO 3211:	Anodizing of Aluminium and Its Alloys—Assessment of Resistance of Anodic Coatings to Cracking by Deformation

International Organization for Standardization, TC79, Sub. 2 on Anodized Aluminum, has adopted additional standards that are particularly important for international trade (Table 22-7).

GENERAL

Although no attempt has been made to discuss mechanical, chemical, and electrochemical treatments applied before anodizing, it must be realized that these finishing systems are inextricably part of the technology essential to the use of anodizing systems. Figure 22-9 shows typical flow diagrams employed for

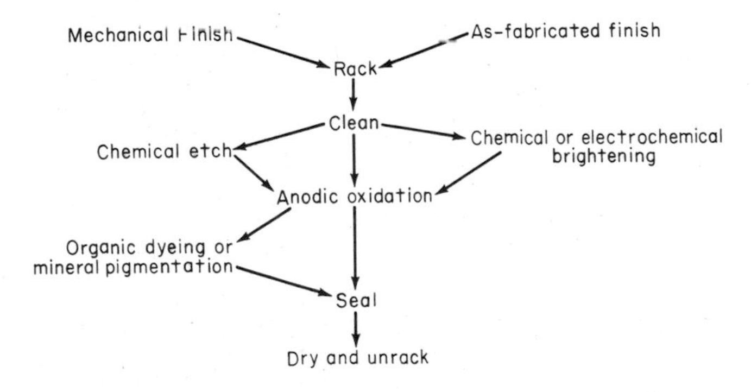

Fig. 22-9 Flow diagrams showing typical processing sequences for various anodized finishes. (Water rinse after each wet process is understood.)

various anodized finishes. Also, it must be understood that the scope of this chapter does not permit reference to the hundreds of important technical documents that have been written over the past 118 years on the anodic oxidation of aluminum.

REFERENCES

1. N. P. Fedotev and S. Ya. Grilikhes, "Electropolishing, Anodizing, and Electrolytic Pickling of Metals," pp. 258–268, Robert Draper, Ltd., Teddington, England, 1959.
2. G. H. Kissin, B. E. Deal, and R. V. Paulson, "The Finishing of Aluminum," chap. 2, p. 23, Reinhold Publishing Corp., New York, 1963.
3. F. Keller, M. S. Hunter, and D. K. Robinson, *J. Electrochem. Soc.,* **100,** 411 (1953).
4. G. C. Wood, "Oxides and Oxide Films," vol 2, pp. 167–279, Marcel Dekker, Inc., 1973.
5. S. Wernick and R. Pinner, "The Surface Treatment and Finishing of Aluminium and Its Alloys," 4th ed., vol. 1, chap. 6, Robert Draper, Ltd., Teddington, England, 1972.
6. K. R. Van Horn (ed.), "Aluminum," vol. 3, p. 662, American Society for Metals, Metals Park, Ohio, 1967.
7. T. Lyman, ed., "Metals Handbook," 8th ed., vol. 2, pp. 620–627, American Society for Metals, Metals Park, Ohio, 1964.

PART FOUR
SPECIAL TOPICS

23 Specifications and Tests

Specifications

A specification is "a concise statement of a set of requirements to be satisfied by a product, material, or a process indicating, whenever appropriate, the procedure by which it may be determined whether the requirements are satisfied. As far as possible, it is desirable that the requirements be expressed numerically in terms of appropriate units together with their limits."*

Specifications are useful to all three parties to a transaction. For purchasers, they are a means of ensuring that the items bought will be of acceptable quality, and that, regardless of where the material is bought, it will meet the requirements. For manufacturers or vendors, the specifications are a means of indicating exactly what is required, of controlling the quality of their output, and ensuring them that they are on equal terms with their competitors who must meet the same requirements. For the ultimate consumer, the public, specifications can be an assurance that the goods offered for sale have met certain minimum standards deemed appropriate for the articles involved.

Specifications are issued by government agencies, by private firms, and by national standardizing bodies such as the American Society for Testing and Materials (ASTM), the American National Standards Institute (ANSI), and specialized industry societies like the Society of Automotive Engineers (SAE) and many others in the United States; by the British Standards Institute (BSI) in Britain, Deutsche Industrienormalen (DIN) in Germany, etc.; and by international organizations such as the International Standards Organization (ISO).

Thousands of specifications are issued by various departments of the U.S. government. Most are either federal specifications, meant to be used by all government agencies, or military (MIL) specifications issued by the Department

*ASTM.

of Defense. Other countries, of course, have corresponding specification-writing agencies, their official standing differing from country to country.

Among consensus bodies, the most important in the United States is ASTM, but many other more specialized organizations also issue specifications related to their interests. Specifications issued by standardizing bodies like ASTM represent a consensus of suppliers, purchasers, and "general interest"—the last including experts in the field who have no immediate responsibility for either purchasing or supplying materials. These consensus standards may or may not be accepted by individual purchasers and suppliers; i.e., a firm may either write its own specifications or may state that the material supplied shall conform to ASTM Designation——. The ISO is attempting to write consensus standards that will be accepted internationally.

TYPES OF SPECIFICATIONS

Specifications are of three general types: process, product, and performance. Process specifications are generally internal within a company: they spell out in great detail the process to be used to manufacture an article, for the guidance of the people on the shop floor who will actually do the work. Suppliers to that firm may also have to meet such a process specification.

The second type is the product specification: it details the dimensions, materials, and properties of the article concerned.

The third type specifies performance: it details the tests that must be passed by the finished article, or components of it, in order that it be considered satisfactory for the intended use. Unfortunately, performance tests (such as accelerated corrosion tests) are not yet well enough developed for industry to rely completely on this type of specification, and it is usually necessary to place reliance on product specifications, often supplemented by performance requirements.

METAL-FINISHING SPECIFICATIONS

In the early days of electroplating, few specifications were in use. Most finishing was for decorative purposes, and it usually sufficed if the product had the required appearance. By experience, the electroplater learned how long to leave the work in the plating tank in order to ensure that the result would be satisfactory to the customer. As metal finishing became more and more an integral part of the manufacturing operation, it became increasingly subject to purchasers' specifications: the deposit must meet minimum thickness requirements, pass certain corrosion tests, possess satisfactory solderability, be sufficiently adherent to the substrate, and satisfy other requirements appropriate to the projected use of the article.

Consensus standards for many industries are published by ASTM. The work of ASTM is done by Technical Committees; that having jurisdiction over electroplated and related finishes is Committee B-8. In addition to specifications for plated coatings, B-8 also issues "Recommended Practices," which do not have the force of requirements but are nevertheless (as many chapters in this book have shown) sources of excellent advice, and Standard Methods of Test, some of which will be discussed in this chapter. Specifications related to electroplated finishes are published annually by ASTM (presently in Part 9 of the "Annual Book of Standards") and are recommended reading; an appendix to Part 9 correlates ASTM specifications with federal and military specifications. Like all ASTM specifications, B-8 documents are continually reviewed and revised when necessary. Users of ASTM—or in fact any—specifications should always use the latest available revision.

Specifications for electroplated coatings may call out several attributes.

Thickness

Thickness of the deposit is usually one of the most important requirements, and a minimum thickness is normally required. Sometimes several thicknesses are named, each one appropriate for a given type of service, which may be characterized as mild, severe, etc. If this is done, the meaning of these words must be illustrated: *mild* may mean indoors under dry conditions, *severe* may mean outdoors in all weathers, etc. Since coatings fail most often at their thinnest points, minimum thickness is usually specified, and often the method used to determine it is also spelled out, or alternative methods may be referred to. Sometimes average thickness may be called for instead of, or in addition to, the minimum.

Coverage

The areas that are to be plated are designated; they are usually *significant surfaces*, which are defined as those visible either directly or by reflection, or those whose corrosion products will be visible. Or they may be defined as those areas that can be touched with a sphere 19 mm in diameter; this recognizes the fact that it may not be possible to apply the minimum thickness inside holes or at bases of angles. Alternatively, the significant surfaces may be indicated on the drawing of the part.

Adhesion

Most specifications require that the coating be adherent and may instruct how this requirement shall be determined.

Corrosion Prevention

Many specifications also require that the coating pass a corrosion test; if so, the number of hours in test before significant corrosion occurs will be stated.

Appearance and Surface Finish

Requirements that the surface be free of certain imperfections are usual; if the coating must be "bright" or "satin" or in some other condition, unfortunately this requirement must be left to the inspector's judgment, since no standard tests are available.

Other properties of the coating may be subject to specification also, e.g., contact resistance, solderability, conductivity, hardness, porosity, smoothness, stress, and ductility.

Quality of Substrate

It is usual to specify that the basis metal be free of defects that will be detrimental to the coating; in effect this means that the finisher should not attempt to plate parts that are not satisfactory as received and cannot be suitably treated (e.g., buffed) to put them in such condition.

Supplementary Requirements

These may include a requirement for heat treatment to relieve hydrogen embrittlement, the application of conversion coatings, or other treatments.

Specifications may include purely business matters: who is to pay for tests, conditions for acceptance or rejection, sampling procedures, and similar matters.

Specifications for the *supplies* used in electroplating are issued by the government, but so far they have not been the subject of ASTM standards. Such materials include, for the most part, the anodes and salts used in plating processes. Normally, supplies bought from reputable vendors will meet the needs of the plater for purity and suitability. Private companies may, of course, insist on specifications of their own for such materials.

Tests

In order to determine conformance with specifications, it became necessary to develop tests, so that today a large number of tests are available, along with suitable instruments for performing some of them. Many of these tests are sufficiently standard to be the subject of "consensus" specifications issued by ASTM, and for details of such tests the appropriate ASTM documents cited below should be consulted. Other tests are not so well advanced, or there is no general agreement on their applicability or reliability; in such cases each purchaser may prefer particular tests, and the vendor (the electroplater) has no choice but to meet them.

It must also be borne in mind that the process of electroplating may alter the properties of the substrate, usually unfavorably. The presence of a coating may have significant effects on the fatigue properties of the basis metal (especially true of chromium deposits), or the plating process (including cleaning and pickling) may cause hydrogen embrittlement of the basis steel.

COATING THICKNESS

Tests for the thickness of deposits are probably the most widely used of all tests for coatings, both metallic and organic. Other things being equal, the performance of plated coatings improves about linearly with their thickness, and thicknesses are almost always specified in specifications for plated finishes. Usually the minimum thickness is specified; sometimes average thickness is called out. The difference between minimum and average thickness can be large; it reflects the difficulty in producing absolutely uniform coatings on shaped parts. Minimum thickness is usually specified at significant surfaces. The usual practice is to specify minimum thickness on that significant surface of the plated article which is known to have received the thinnest coating.

Many methods for measuring coating thickness are available; they may be classified as (1) destructive, (2) semidestructive, and (3) nondestructive. Although the names are self-explanatory, it may be added that destructive tests require that the part to be examined be damaged in such a way that it is not usable; semidestructive tests destroy the deposit being tested, but not the part, which may be stripped and replated.

The most widely used methods for testing thickness are in the latter two categories; the principal destructive test, microscopic examination, is time-consuming and has been shown not to be as accurate as once believed.

Nondestructive methods make use of instruments especially developed for the purpose; although they rely on different principles for their operation, these principles are similar: the interposition of a coating between the basis metal and the instrument changes some property which can be measured by the instrument, and the magnitude of this change depends on the thickness of the coating.

Most methods employ commercially available instruments or gages: magnetic, eddy current, beta-backscatter, x-ray fluorescence, or coulometer. All instruments, and noninstrumental methods, have some limitations, and no one of them can be singled out as the most satisfactory in all situations.

None of the available methods is capable of an accuracy better than about ±5 percent, and even this only under the most rigorously controlled conditions of use. This is not a serious limitation, because for almost all uses of electrodeposited coatings better accuracy is unnecessary: in fact, it would have little meaning because the substrate and the coating are usually rough enough so that the exact thickness of the coating is difficult or impossible to define. The difference between precision (or reproducibility) and accuracy must be recalled. One operator, or one laboratory, can often duplicate its own measurements with great precision; but many round-robin tests among different laboratories, all measuring the same part, have clearly shown that operator idiosyncracy is a large factor in every method of measuring thickness.

Fig. 23-1 Magne-gage.

THICKNESS GAGES

Several thickness gages are commercially available; they measure an average thickness over a small area which depends on the design of the instrument. This area ranges from about 1 mm^2 to several cm^2. All gages require calibration against reference standards of known thickness, usually supplied by the manufacturer.

No gage is without possibility of error; errors or erratic readings may arise from several causes: foreign material on the surface of the coating or on the part of the gage head or probe that comes in contact with the coating, improper positioning of the gage head or probe, surface roughness of the coating or the substrate, using the probe too near an edge of the specimen, and curvature of the surface too great for the design of the probe. Readings may be sensitive to the composition and structure of the coating and substrate, and to residual magnetism in the latter. If the substrate is too thin, erroneous readings may result.

These limitations and sources of error are, or should be, explained in the literature supplied by the manufacturer of the instrument. Sometimes suitable calibration can eliminate these errors, e.g., calibrating the instrument with a standard having the same structure, such as curvature.

The range of coating thicknesses measurable by such gages is very wide; it depends in individual cases on the design of the gage and the coating-substrate combination. Gold coatings as thin as 0.7 μm and copper coatings as thick as 2.5 mm can be measured.

Magnetic Gages (ASTM B 499, B 530)

These depend on the magnetic properties of the plated surface. If the deposit, the substrate, or both are magnetic, the magnetism measured is a function of the thickness of the deposit. Some instruments can be calibrated directly in thickness units.

The earliest magnetic gage was developed by Brenner, and is available as the "Magne-gage." Various modifications using the same basic principle are available; the smallest is the size of a pencil and operates with a spring mechanism. The larger ones incorporate electronic circuits and require alternating current or battery power. Most are used to measure nonmagnetic coatings on steel, though with proper calibration nickel coatings can also be measured. Gages of this type measure, in effect, the decrease in magnetism caused by the interposition between the magnetic substrate and the magnet of the gage of a nonmagnetic deposit; usually the function measured is the force required to pull the magnet away from the specimen. See Fig. 23-1.

Some thickness gages of the magnetic type depend on the reluctance of the magnetic circuit formed when the two poles of a magnet are placed on the plated surface; the reluctance of the circuit depends on the coating thickness if the coating or the substrate is magnetic. Changes in the reluctance can be measured mechanically or electronically.

All magnetic gages are sensitive to the magnetic condition and properties of the test specimen, to surface curvature, to roughness, and the like. Very soft coatings such as lead require special handling lest they be indented by the probe; probes also have a tendency to stick to such coatings.

Eddy-current Gages

When a coil carrying a high-frequency current is positioned on a specimen, the electromagnetic field of the coil induces eddy currents in the specimen. The frequency of the coil current is so chosen that the eddy currents penetrate the deposit and extend into the substrate. The magnetic field of the eddy current changes the impedance of the probe coil, and this change is measured by an appropriate circuit. If the electrical conductivity of the coating differs from that of the substrate, the magnitude of the eddy currents and coil impedance will vary with the thickness of the deposit.

Several eddy-current gages are commercially available. They have the advantages of not requiring a nonmagnetic coating on a magnetic substrate, or vice versa, and are applicable to nonconductive coatings, as are magnetic gages. On the other hand, each deposit-substrate combination requires a separate calibration, and the gages are sensitive to the electrical properties of the coatings

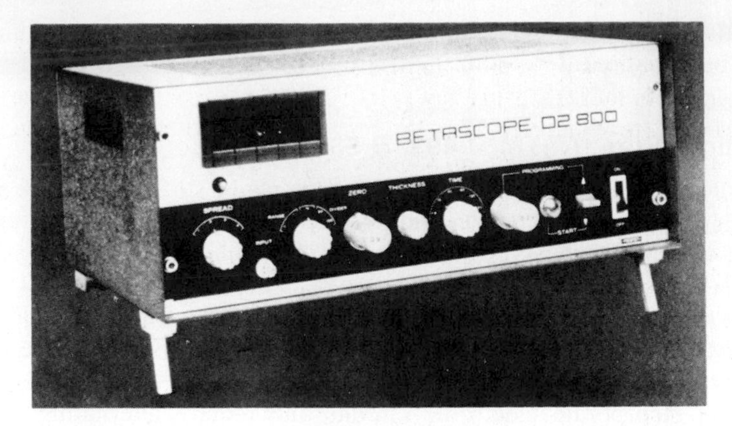

Fig. 23-2 Beta-backscatter thickness tester.

resulting from variations in plating techniques. ASTM B 529 covers the use of eddy-current gages to measure nonconductive coatings on nonmagnetic basis metals, and ASTM B 244 relates this specifically to anodic coatings on aluminum. Conversion coatings are too thin to be measured by this technique.

Beta-backscatter Gages (ASTM B 567)

The measurement of coating thickness by the beta-backscatter principle is based on the different degrees to which beta particles are scattered when they encounter substances of different atomic numbers. This makes it possible to determine the thickness of a coating, provided that a sufficient difference in atomic number exists between the coating and its substrate. This separation should be at least 15 percent of the atomic number of the substrate material.* The coating surface is irradiated by beta rays emitted from a small sealed radioisotope source placed close to an aperture in a mask against which the specimen is firmly held. This aperture defines the measuring area. Some of the radiation is absorbed or transformed within the specimen, but a certain amount will be backscattered toward the source. The proportion of radiation backscattered is related to factors such as the atomic numbers of the coating and substrate materials, the area of the specimen exposed to the radiation, the thickness or mass per unit area of the coating, and the energy of the radiation. If all the other variables are fixed, the intensity of the backscattered radiation is a function of the thickness or mass per unit area of the coating. The intensity of the backscattered radiation is measured by a radiation detector placed behind or near the sealed source. The exact

*Thus nickel (atomic number 28) or copper (atomic number 29) on iron (atomic number 26) is not a suitable combination for this type of gage, but gold (atomic number 79) on copper is readily measured, etc.

relationship between the backscatter intensity and the corresponding coating thickness must be established by the use of standards.*

These gages, in fact, measure not thickness, but weight (or mass) of coating per unit area of the specimen. These numbers can, of course, be translated into thickness units, and gages are so calibrated. For most metals this causes no difficulties, but there has been considerable disagreement regarding the relationship of thickness to weight per unit area in the case of gold coatings, since the density of electrodeposited gold can vary with plating conditions. The prime advantages of these gages are simplicity and rapidity in use and ease of reading. They have proved especially valuable for measuring thin coatings of the precious metals and have been used to measure coatings over small areas, as little as 2 mm in diameter. Some models have been designed for continuous measurement of moving strip. See Fig. 23-2.

Electrical Conductance Gages

The conductance of a plated coating varies with its thickness, and this principle has been employed in the construction of a gage for measuring the thickness of silver coatings on the inside of a stainless steel waveguide.

Thermoelectric Gages

These gages depend on the voltage that develops between hot and cold junctions of dissimilar metals. The probe of the gage has an electrically heated prong that heats a small zone of the specimen; this establishes a temperature gradient along a portion of the deposit-substrate interface. The resulting thermoelectric potential is measured between the heated prong and a cold contact by means of an appropriate circuit; it is a function of the coating thickness. A gage of this type is available in England.

Radiation Measurement Methods

(See also Beta-backscatter Gages, above). Coatings and substrates are subjected to x-ray, gamma-ray, electron, and neutron radiations; reflected, transmitted, or induced radiations are measured. Except for x-ray fluorescence methods, these techniques have not found wide acceptance—not because of any inherent faults, but because simpler and less exotic methods seem sufficient for most practical purposes. X-ray fluorescence methods are suitable for rapid measurement of most coating-substrate combinations, with good precision. They must be calibrated with standard thickness samples. Principal limitations are the high cost of equipment and the fact that in general they are limited to thin coatings, not much

*This paragraph (except the footnote) is quoted from ASTM B 567-72, Measurement of Coating Thickness by the Beta Backscatter Principle.

over 25 μm. The method is sensitive to weight per unit area rather than directly to thickness, similarly to beta backscatter.

These methods, because of their high cost, generally are limited to high-production and repetitive operations. See ASTM B 568.

CHEMICAL METHODS FOR LOCAL THICKNESS

These methods are older than the instrumental ones already considered, and have (except for the coulometric) more or less gone out of favor. Most depend on the time required for a reagent to penetrate the coating and expose the substrate.

Spot Tests

A drop of reagent is applied to the coating, and the time it takes to expose the basis metal is measured. For many years such tests were the only ones available for measuring the thin decorative chromium coatings, as described in ASTM B 556; but now they have been essentially superseded by the coulometric method, which is capable of much greater accuracy.

Dropping and Jet Tests

Drops, or jets, of chemical reagent are allowed to fall on the specimen, and the number of drops, or the time, taken to expose the substrate is measured. Equipment is commercially available. Such tests are useful when, for whatever reason, instrumental methods cannot be used; but at best these tests do not have high precision or accuracy. See ASTM B 555.

Coulometric or Anodic Solution

This method, which according to many workers is capable of the greatest accuracy of any of the standard thickness-measuring techniques, is described in ASTM B 504. The thickness of the coating is determined by measuring the quantity of electricity (coulombs) required to dissolve the coating anodically from a known and accurately defined area. A small metal cell is commonly employed, which is filled with an appropriate electrolyte chosen for the substrate-coating combination being tested. The test specimen serves as the bottom of the cell, and an insulating gasket between the cell and the specimen defines the test area, about 10 mm^2. With the specimen as anode and the cell as cathode, a constant direct current is passed through the cell until the coating has dissolved, when a sudden change in voltage occurs. This change can be made to trigger an appropriate electronic circuit and shut the instrument off. The thickness of the coating may be calculated from the number of coulombs passed, the area, the electrochemical equivalent of the coating metal, the anodic current efficiency, and the density of the coating.

Commercial instruments can be calibrated directly in thickness; the instrument has a dial which automatically adjusts for the various factors mentioned.

MEASUREMENT OF LENGTH

These types of measurements are physical like the instrumental, but differ in that length measurements are made directly.

Microscopic

In this method the coating thickness is measured directly, by a magnified image of a cross section of the coating. Standard metallographic methods are used for mounting, polishing, and etching the specimen. A filar micrometer ocular or image-splitting eyepiece is used to measure the thickness; sometimes the image is projected on a ground-glass plate.

The microscopic method is direct, though destructive, and at first sight would seem to be the most accurate. It has often been used as a referee method, but recent work has thrown considerable doubt on its claim to this distinction. The method is slow and completely destructive and requires skill and experience. Measurements are necessarily made over a very small and localized area, and without considerable replication over several parts of a specimen—which adds to the time required—cannot be relied on for checking general thickness variations over the surface of a part. Operator idiosyncracy and error have been shown to detract from its accuracy, and agreement among laboratories is often poor. See ASTM B 487. Properly used, the microscopic method is no doubt reliable, but few routine laboratories, apparently, meet this requirement.

The accuracy of the microscopic method is considered to be about ± 1 μm or ± 5 percent, whichever is greater, when the cross section is perpendicular to the coating. Taper sectioning could give improved accuracy but is seldom used.

Chord Method

If a coating on a circular surface is just cut through with a file, the width of the cut gives a measure of the thickness of the deposit; see Fig. 23-3. The method also can be used in reverse: the deposit on a flat surface is just cut through by means of a circular wheel. One of the earliest methods (often called the Mesle chord, after its inventor) of measuring coating thickness, it is seldom used any more.

Micrometer

If a measurement is made at the same spot on an article before and after plating, or before and after stripping the coating, a direct measurement of the coating thickness is obtained. Alternatively, the basis metal can be stripped from the coating and the latter measured directly. This method is of little use for thin

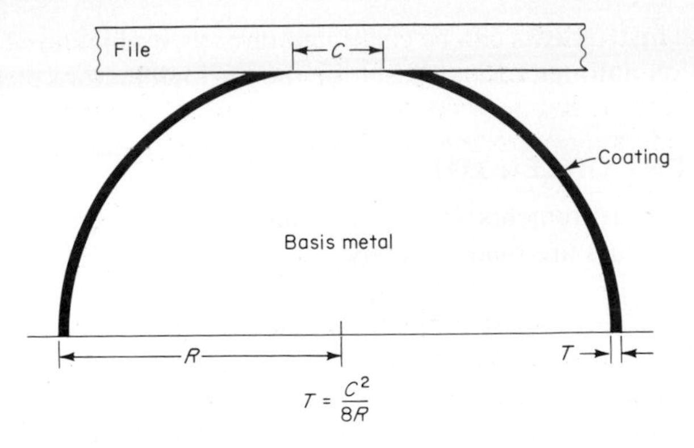

Fig. 23-3 Chord method for determining coating thickness.

coatings, but possesses fair accuracy for heavy deposits such as may be applied in hard chromium plating or electroforming. Dial-gage micrometers can be used similarly.

Interference Microscope

See ASTM B 588. The interference fringes formed between an optical flat and the surface of the plated specimen are observed with a microscope; the fringe pattern appears as alternate parallel light and dark bands. Any discontinuity in the plated surface will displace the fringe pattern, and the displacement is a measure of the difference in level of the surface of the specimen. For details see the ASTM method cited.

AVERAGE THICKNESS

Several methods are available for determining average, rather than local, thickness, as calculated from the weight and surface area of the coating:

$$\text{Thickness} = \frac{\text{weight}}{\text{density} \times \text{area}}$$

The density is assumed to be the handbook value; this assumption may not be strictly true for many electrodeposits, but it is usually within acceptable limits of error. The area of the specimen is determined by conventional methods. The weight of the coating may be determined by several techniques.

Chemical Analysis

The entire test specimen is dissolved in a suitable reagent, and the amount of coating metal is determined by analysis of the solution. The method is general, and its accuracy is limited only by that of available analytical methods. Sources of error may include overlooking some minor constituent of the coating metal, such as cobalt in nickel deposits, or a metal present in both the coating and the substrate, such as a tin coating on steel containing traces of tin.

Stripping

Frequently it is possible to strip a deposit chemically without attacking the substrate. Then the specimen of known area may be weighed before and after stripping to obtain the coating weight. In a variation, the coating may be stripped and the stripping solution analyzed for the coating metal; such methods have been used routinely for obtaining the weight of tin on tinplate, since a simple titration suffices for the analysis and the buret is calibrated directly in coating weight. Other variations of this principle have been developed for specific deposit-substrate combinations.

Several other methods, including making use of the difference in density between the coating and the substrate, and measuring the heat evolved when the coating is dissolved, have been described; they are not used widely but may have their individual applications. The calorimetric method is especially useful for small mass-produced parts such as are typically barrel-plated.

STANDARD THICKNESS SAMPLES

Thickness gages depend for their use on calibration by means of samples having known coating thicknesses. Such standards usually are available from suppliers of the instrument, and the National Bureau of Standards has made standard thickness samples available. If for special reasons a laboratory must prepare its own standards, plating conditions must be rigorously set to produce uniform coatings, at least on those parts of the cathode that are to be used as standard samples.

Laboratories may choose to plate sheets larger than the standard being prepared and then cut the standard from the center of the sheet where the deposit is likely to be most uniform. However, now it is necessary to measure the thickness of this cut-out portion by some independent method, and all such methods are destructive. Therefore many cathodes must be prepared under identical conditions, and it is assumed that those destroyed by the measurement truly represent the usable standards.

Metal or plastic foils or sheets of known and uniform thickness have been used. They are placed over a specimen of the uncoated substrate, and a gage reading is taken just as for a plated deposit. The method is practical, especially for nonmagnetic coatings on steel. The principal source of error is that the separate foil may not be in sufficiently intimate contact with the substrate during the calibration.

Since coating thickness must almost always be determined, we have considered the available techniques in some detail. In addition, the ASTM standards cited should be consulted for further information. Other tests for plated coatings are not so universally employed, and the discussion of them will be briefer.

ADHESION

The adhesion of an electrodeposited coating to its substrate is as important as its thickness. Unfortunately, adhesion tests are either only qualitative or quantitative but not suited to routine testing and appropriate for research purposes only. For quality control, therefore, adhesion is usually tested only to the extent that it proves good enough for the intended use; i.e., it is a go–no go test.

Qualitative Tests

Adhesion tests suitable for routine use are the subject of an ASTM standard, B 571, summarized here. The interpretation of the results of adhesion tests is a matter of some controversy; if possible, more than one test should be conducted. The end use of the coated article or its method of fabrication may suggest the most appropriate test: e.g., an article that is to be subsequently formed might be tested by a draw or bend test; an article that is to be soldered might be tested by a heat-quench test. Many tests are limited in their application to specific coatings, thickness ranges, or other properties.

"Perfect" adhesion obtains if the bond between the coating and the substrate is greater than the strength of either one. Good plating practice will usually yield "perfect" adhesion, so that in effect adhesion tests are tests of good plating practice. For many purposes, the objective of the adhesion test is to detect any adhesion less than "perfect." Thus any means are used to try to separate the coating from the substrate: most adhesion tests involve some sort of physical abuse of the specimen such as prying, pulling, hammering, bending, heating, beating, sawing, grinding, scribing, chiseling, or a combination of these. If the coating peels, flakes, or lifts from the substrate, the adhesion is not perfect.

If the size and shape of the item to be tested do not permit the use of one of these tests, a test panel may be used; it must be of the same material as the specimen and be plated along with the parts in question. Obviously, all adhesion tests are destructive; if the parts are very valuable, the use of test panels may be necessary.

Bend Test. Bend the part with the coated surface away, over a mandrel until its two legs are parallel; the diameter of the mandrel should be 4 times the thickness of the sample. Examine the deformed area under low magnification, say, 4X, for peeling or flaking of the coating from the substrate. If the coating fractures or blisters, a sharp blade may be used to attempt to lift off the coating. Brittle coatings may crack under this test, but cracks per se are not evidence of poor adhesion unless the coating can be peeled with a sharp instrument.

Burnishing Test. Rub a coated area, about 5 cm^2, with a smooth-ended tool for about 15 s. The pressure is sufficient to burnish the coating but not dig into it. Blisters, lifting, or peeling should not develop. The test is not suitable for thick coatings.

Chisel-knife Test. Use a sharp cold chisel to penetrate the coating, or at a coating-substrate interface exposed by sectioning the specimen. If it is possible to remove the deposit, the adhesion is not satisfactory. This is not applicable to soft or thin coatings.

Draw Test. Form a sample about 60 mm in diameter into a flanged cap about 38 mm in diameter, to a depth up to 18 mm, by use of a set of adjustable dies in a punch press. The adhesion of the coating may be observed directly or further evaluated by the chisel-knife test. Evaluation is uncertain because the test also involves the ductility of the coating and the substrate.

File Test. Saw off a piece of the coated specimen and inspect it for detachment at the deposit-substrate interface. Apply a coarse mill file across the sawed edge from the substrate toward the coating so as to raise it, using an angle of about 45 degrees to the coating surface. This test is not suitable for soft or thin coatings.

Grind-saw Test. Hold the coated article against a rough emery wheel so that the wheel cuts jerkily from the substrate toward the deposit; a hack saw may be used. This is not suitable for thin or soft coatings.

Heat-quench Test. The article is heated in an oven to a temperature prescribed for the coating-substrate combination (Table 23-1); then the specimen is quenched in water at room temperature. Note that the heating may actually improve adhesion by diffusion alloying, or it may form a brittle alloy layer at the interface; these effects limit the applicability of the test.

Impact Test. In essentials, this test consists merely of hammering the specimen severely to see whether blistering or exfoliation occurs.

Peel Test. A strip of steel or brass is bonded to the specimen by solder or a suitable adhesive; at an angle of 90 degrees the strip is pulled off the specimen. Failure at the deposit-substrate interface evidences poor adhesion. (See also Quantitative Tests, below).

Push Test. Drill a blind hole 7.5 mm in diameter from the underside of the specimen until the point of the drill tip comes within about 1.5 mm of the deposit-substrate interface on the opposite side. Support the specimen on a ring about 25 mm in diameter and apply steady pressure over the blind hole, using a hardened steel punch 6 mm in diameter, until a button sample is pushed out. Exfoliation or

Table 23-1 Temperature Test Guide

	Coating material				
Substrate	*Chromium, nickel nickel + chromium, copper, temperature, °C*	*Tin, temperature, °C*	*Lead, tin-lead, temperature, °C*	*Zinc, temperature, °C*	*Gold and silver, temperature, °C*
Steel	250	150	150	150	250
Zinc alloys	150	150	150	150	150
Copper and copper alloys	250	150	150	150	250
Aluminum and aluminum alloys	220	150	150	150	220

peeling of the coating in the button or crater area is evidence of poor adhesion. This is not suitable for soft, thin, or very ductile deposits.

Scribe-grind Test. Scribe two or more parallel lines or a rectangular grid pattern on the article using a hardened steel tool; the distance between the scribed lines should be about 10 times the nominal thickness of the coating, with a minimum of 0.4 mm. Cut through the coating to the substrate in a single stroke. If any portion of the coating between the lines breaks away from the substrate, adhesion is inadequate.

If not explicitly stated, in all of these tests peeling, flaking, blistering, or exfoliation of the coating is evidence of poor adhesion.

Suggestions regarding tests suitable for various substrate-coating combinations are summarized in Table 23-2 (from ASTM B 571).

QUANTITATIVE TESTS

Most quantitative tests for adhesion—i.e., tests that attempt to express the force necessary to separate the coating from the substrate in numerical terms—are not suitable for routine use but are research tools. An exception is the peel test for electroplated plastics; see Chap. 19. But, as was stated, this test may measure not the adhesion of the metal to the plastic, but rather the weakest point in the combination, which is usually within the plastic itself.

Two direct tests of adhesion have found fairly wide acceptance for research purposes; both require specially prepared specimens and are therefore not adapted to control of a plating process or as acceptance tests.

In the Ollard test, all but the end of a rod is stopped off; the end is plated with a fairly thick deposit, which is allowed to grow laterally over the stopoff. After plating it is machined so that the deposit has a lip extending radially from the rod. The underside of the lip is in the same plane as the original flat end of the rod. The rod is then pushed or pulled through a close-fitting cylinder; the lip of the deposit is engaged by the cylinder, and the deposit is pulled off. The force necessary to do this can be measured; see Fig. 23-4. Various versions of this test have been published, all substantially equivalent.

Direct-pull tests depend on being able in some way to grip the coating over a definite area, so that it may be pulled off directly, by a spring balance or in a tensile machine. This has been done in several ways, all subject to some objection. (1) A plug or rod is soldered or brazed to the coating. This necessitates heating, which may increase or decrease the adhesion to be measured. (2) A grip is attached to the coating with an adhesive; naturally the strength of the adhesive limits the applicability of this method, although newer developments in adhesives have raised the limit. (3) A mushroom-shaped nodule of metal (cobalt is usually used) is electroformed on the coating; see Fig. 23-5. Although this method is applicable generally, it is very time-consuming and hardly suitable for routine use.

Table 23-2 Adhesion Tests Appropriate for Various Coatings

	Coating material									
Adhesion test	*Cadmium*	*Chromium*	*Copper*	*Lead and lead-tin alloy*	*Nickel*	*Nickel and chromium*	*Silver*	*Tin and tin-lead alloy*	*Zinc*	*Gold*
Bend			*		*	*				
Burnish	*		*	*	*	*	*	*	*	*
Chisel		*	*		*		*			
Draw	*	*	*	*	*	*	*	*	*	*
File			*		*	*				
Grind and saw		*			*	*				
Heat/quench		*	*	*	*	*	*	*		*
Impact		*	*		*	*		*		
Peel	*		*	*	*		*	*	*	*
Push		*			*	*				
Scribe	*		*		*		*	*	*	*

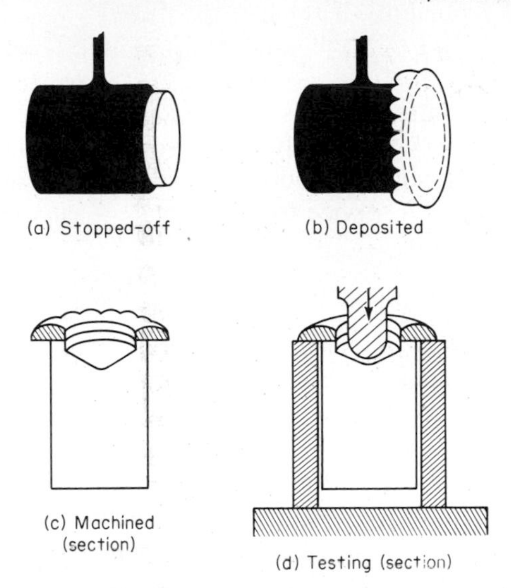

Fig. 23-4 Ollard adhesion test.

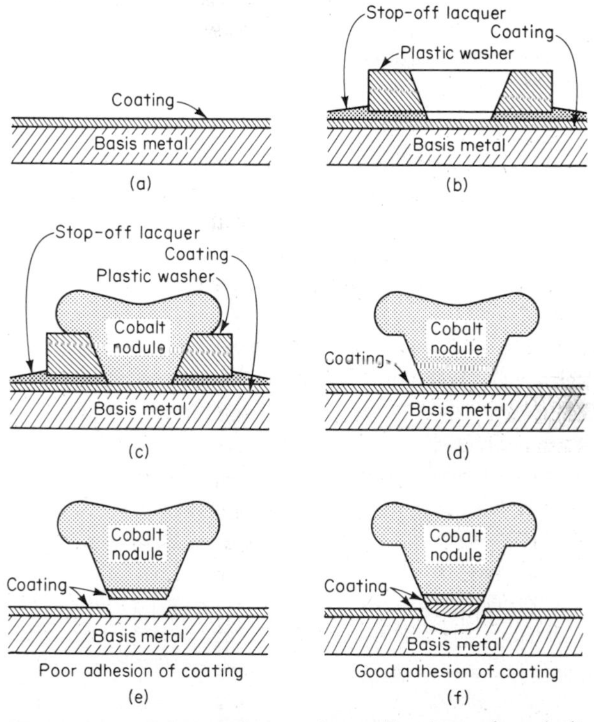

Fig. 23-5 Direct-pull test for adhesion: A cobalt nodule is electroformed on the coating surface and pulled off.

CORROSION

The subject of corrosion testing was discussed in Chap. 3. We repeat here the warning that actual service is the only final test of corrosion resistance, and that accelerated tests, while necessary both for research and control purposes, must be shown to correlate with service to be of value.

POROSITY

Porosity tests are sometimes considered a subclass of corrosion tests, since a porous coating will presumably permit access of corrodents to the substrate. This is important when the coating is more noble than the substrate, as in copper-nickel-chromium on steel, tin on steel, and similar combinations. The presence of pores in electrodeposited coatings may, however, be important in themselves: in many electronic applications, exposure of substrate metal through pores in a gold coating may be deleterious irrespective of their implications for corrosion resistance.

Some of the accelerated corrosion tests may be considered to be porosity tests. Chemical reagents may be used to indicate the presence of pores by reacting with the basis metal to form a colored product. For example, potassium ferricyanide reacts with iron ions to produce an intense blue color. Gelatin-coated paper dipped in salt solution may be placed on a plated steel surface to be tested; iron ions from the exposed substrate are absorbed into the gel. When the paper is then immersed into a potassium ferricyanide solution, blue spots are observed at the sites of the pores. This is known as the Ferroxyl test, one of the earliest devised for porosity testing.

Electrographic tests are similar, but make use of a current to drive the ions of the substrate metal into the test paper or gel. The part is made anodic, and the paper or gel is placed between it and a conforming cathode. Colored spots appear at pore sites.

The color-forming reaction is chosen in accordance with the nature of the substrate metal: ferricyanide for iron, dimethylglyoxime for nickel, etc.

A simple hot-water test has been used for tinplate on steel: rust spots appear at the sites of pores.

In some tests, the substrate metal dissolving through pores is determined by chemical analysis.

Porosity tests have been criticized because it is never certain whether they reveal preexisting pores or actually cause the formation of new ones. This question is open.

APPEARANCE

For decorative deposits appearance is obviously an important property, but it is usually evaluated subjectively. Many attempts have been made to quantify the measurement of "brightness" or specular reflection; although this can be done, the results do not necessarily correlate with the subjective notion of satisfactory appearance, and it is the latter that determines acceptability. Probably the human eye is sensitive to more properties that go to make up what is recognized as brightness than can be quantitatively measured.

Smoothness of the surface is one factor in bright appearance, although by no means the only one; but it can be measured, and several instruments are commercially available for measuring surface roughness. These instruments either record a magnified profile of the surface or give a meter reading of an average size of surface irregularities. A stylus is drawn over the surface, and its vertical movements are converted into electrical impulses which are recorded. For soft coatings such as tin and lead, such instruments are unreliable because the stylus itself can change the surface.

SOLDERABILITY

Solderability is the ease with which a surface, such as an electrodeposit, is wetted by solder.* The property is important, especially in the electronics industry in which soldered joints must be produced quickly and reliably. Usually this must be accomplished without the use of corrosive fluxes whose residues would be deleterious to the functioning of the circuits, often miniature. Solderability is also an important consideration in the use of tinplate to be used in the manufacture of side-seamed tin cans. Many tests have been devised for this property called solderability; none has emerged as a universal standard. It is well to remember that such tests in reality measure the properties of the surface-solder-flux combination, not that of the surface alone. A surface easily soldered with the aid of a corrosive flux like zinc-ammonium chloride might fail if used in conjunction with a mild flux like rosin in alcohol. Also there are many types of solder, and the test should be conducted with the same type of solder and flux that are to be used in production.

A critical summary of the various solderability tests is offered by Long.† Commercial instruments are available for conducting some of these tests, which can be divided into classes according to the principle involved.

Joining by soldering involves the penetration and filling of a clearance between two components with solder, to provide a joint having integrity, electrical continuity, or heat transfer in the finished assembly. Solder joints form liquid- and gas-tight seals; they provide some mechanical strength. The solder must wet and penetrate mating surfaces at a temperature suitable to the soldering action, usually below 425°C. Fluxes are normally required to aid in this action by

*The term *solder* subsumes a wide variety of alloys, the purpose of which is to join two surfaces either electrically or mechanically. Most are tin-lead alloys, but some incorporate other metals such as silver or antimony. That most often used in the electronics industry approximates the eutectic of the tin-lead system at about 62 percent tin.

†J. B. Long, in R. Sard, H. Leidheiser, Jr., and F. Ogburn (eds.), "Properties of Electrodeposits, Their Measurement and Significance," The Electrochemical Society, Princeton, N.J., 1975.

dissolving surface oxides or other films both on the deposit and on the solder itself. Most tests are of one of the following types:

1. Capillary penetration or capillary rise
2. Spreading (area of spread)
3. Dip
4. Time of wetting

Depending on the soldering operation contemplated, one type of test may be more relevant than another; see Long (loc. cit.) for a useful summary of the tests that are most appropriate in each application.

Capillary Penetration Tests

These tests are used in quality control in tinplate manufacture. Essentially, a sheet of tinplate is folded over itself; under controlled conditions of temperature and fluxing, the extent to which solder wets the internal surface of the fold is a measure of the solderability of the tinplate. It is also possible to measure the capillary rise of molten solder between two wires twisted together, and other modifications have been devised. These tests are simple to perform and require no complicated equipment; on the other hand, their reliability is questionable except for certain specific applications.

Area of Spread

A fixed volume of solder is placed, along with a suitable flux, on the surface to be tested; this is heated in a controlled manner, causing the solder to melt and spread over the surface. The area of spread is a measure of the wetting properties of the solder-flux-surface combination. This area can be measured in various ways, and by experience in production a minimum can be set below which the surface is considered unsatisfactory for the application. The test can also be used in research, to compare various surfaces such as electrodeposits. In a variation, the measurement of the area of spread is replaced by measurement of the height of the solder blob after spreading: the lower the height, the more area the solder has covered. From this measurement a "spread factor" can be calculated and again related to the minimum acceptable for the application.

In still another variation, the contact angle between the solder blob and the plane surface can be used as a measure of "spreadability."

Area-of-spread tests, especially that which measures the height of the blob of solder (the Pessel test), are easy and rapid and require little equipment. But in some cases, the solder, instead of spreading *over* the surface of the electrodeposit, cuts through it and spreads *under* the surface, between the deposit and the substrate. Interpretation in such a case is questionable.

Dip Tests

These are among the easiest to apply: the test specimen is simply immersed in a pot of molten solder, and the degree of coverage is noted. The test has several modifications, some of which are specified by military agencies and consensus bodies such as the Institute of Printed Circuits. Interpretation is usually simply by visual examination. Thus the tests are difficult to quantify, but they are useful in production control and acceptance testing as go–no go tests.

Time-of-Wetting Tests

Several means have been devised for determining the length of time required for a solder to completely wet a specimen surface. In one such test, specimens are rotated through a pot of molten solder at decreasing speed, so that their residence time is progressively increased. Thus the minimum time required before the specimen is completely wetted by the solder can be determined.

The globule test, mentioned in ASTM B 579, is a time-of-wetting test in which the specimen of wire or component lead, previously fluxed, is lowered horizontally into a molten globule of solder, which is thereby cut in two. The time in seconds for the solder to flow around the wire and unite above it is a measure of solderability. Commercial machines are available that incorporate this test.

These time-of-wetting tests can be automated, and they have the advantage of measuring an important property: not only the ultimate spreadability of the solder, but the time required to attain it, which is important in highly automated electronic assembly processes. Long (loc. cit.) believes that the time-of-wetting test is, on the whole, the most versatile and reliable of those available.

Many solderability specifications incorporate a requirement that the article retain this property for a period of time; in such case the specimen may be aged, either naturally or in an accelerated atmosphere such as of high temperature and high humidity for stated periods. For instance, the specimens may be suspended over boiling water in a closed vessel or kept in a humid room at elevated temperature.

Mechanical Properties

For many applications of electroplating, the mechanical properties of the deposits are not of primary concern. In individual cases, however, one or more of them may be of paramount importance. Mechanical properties of metals include hardness, residual stress, tensile strength, ductility, yield strength, modulus of elasticity, fatigue behavior, and abrasion resistance. Whether one or more of these properties should be tested depends on the application contemplated.

HARDNESS

Hardness is the most frequently measured mechanical propery of electrodeposits, but it is not truly a single, definable property of a material. Although the concept is readily comprehended, its measurement is subject to several uncertainties. It usually connotes the ease with which a material is penetrated by another material that is "harder." But the act of penetration itself involves changes in several fundamental structural features of the material being tested: atomic bonds are broken (the cohesive strength of the material is exceeded); as the metal is pushed aside by the penetrator, its atoms must slide past one another, and this involves internal friction. As the metal is penetrated, it may work-harden to some extent. As the penetrator enters the sample, the material must flow past it; therefore the coefficient of friction between the metal and the penetrator enters the equation. Hardness, then, as measured by penetration tests, results from several properties, and the final result depends on the method of measurement.

Several hardness-testing techniques are described; most applicable to electrodeposits is the indentation method.

A mark, or indentation is made with a penetrator, called an indentor, which is applied to the specimen under a known load. The size of the mark is measured, and a simple formula involving this and the test load is used to calculate a hardness number. The principal variables in indentation testing are the shape of the indentor and the method of applying the load. Many designs are possible, and correspondingly numerous machines are marketed. The only ones suitable for electrodeposits are the Vickers and the Knoop. Because many electrodeposits, particularly those whose hardnesses are of interest, are very hard, diamond is the only material suitable for the indentor.

Modern machines use only pyramid-shaped indentors. The Vickers indentor is a symmetrical pyramid with an included angle between the faces of 136 degrees. It makes a square mark on the sample (Fig. 23-6*a*). The Knoop indentor is a nonsymmetrical pyramid with included angles between corners of 130 and 172.5 degrees. It makes a diamond-shaped mark of which only the long diagonal is measured (Fig. 23-6*b*). For the Vickers indentor, both diagonals are measured, and the average is used in calculations.

Manufacturers of the machines usually supply tables from which hardness numbers can be read off for combinations of indentation size and load. The Vickers hardness is calculated as follows.

$$H_V = \frac{1.854L}{D^2}$$

where H_V = Vickers hardness number
L = load in kg
D = diagonal of the indentation in mm

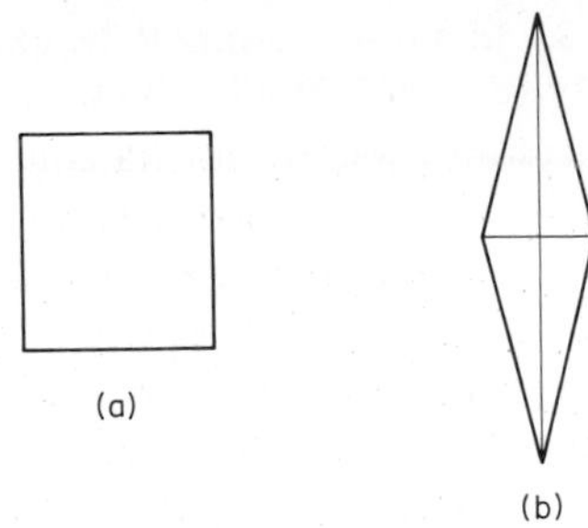

Fig. 23-6 (*a*) Vickers indentor; (*b*) Knoop indentor.

The dimensions of the Vickers hardness number, therefore, are in kilogram-force per square millimeter (kgf/mm^2); to convert this to the SI unit pascal (Pa), 1 kgf/mm^2 = 9.8 MPa (megapascal or 10^6 pascal).

For the Knoop indentor, the formula is

$$H_K = \frac{L}{0.070D^2}$$

where H_K = Knoop hardness number
L = load in kg
D = *long* diagonal of the indentation in mm; dimensions are the same as for the Vickers number

(The Vickers and Knoop tests are examples of a larger class known generically as diamond pyramid hardness, or DPH, tests. Another common hardness test for metals is the Brinell, in which a hardened steel ball of known diameter is impressed by a known load and the depth of the impression is measured. This test, yielding the Brinell hardness number, or BHN, is not well adapted to electrodeposits.)

Choice between the Vickers and Knoop tests is determined by several factors; the most important is the thinness of most electrodeposits. This poses the danger that one will measure not the hardness of the deposit but that of the substrate, or at least that the latter will influence the result. This is known as the *anvil effect.* To eliminate this effect, the depth of the indentation must be no more than one-tenth of the thickness of the coating being tested, and preferably no more than one-fifteenth. For a given load, the depth of the Knoop indentation is only about one-seventh that for the Vickers, making the Knoop the choice in most cases. Minimum deposit thicknesses for avoiding the anvil effect in the Knoop test are shown in Table 23-3. Loads of less than 25 g are undesirable because results tend to be irreproducible. For this reason many decorative coatings are too thin to be measured by surface indentation, and it is necessary to prepare cross sections by metallographic grinding and polishing and test the cross section rather than the

Table 23-3 Minimum Deposit Thickness Necessary to Avoid Anvil Effect, μm

	Load, g		
H_K	*25*	*100*	*200*
100	28	56	76
300	15	30	46
500	13	25	36
700	10	20	30
900	10	18	25
1100	8	18	23

surface. Again the Knoop indentor is preferable because the indentation is narrow.

On the other hand, the Vickers indentor is less sensitive to errors arising from the elastic properties of the material under test. As the load is decreased, the Knoop indentor will give erroneous results sooner than the Vickers.

In reporting and specifying hardness numbers, it follows from what has been said that the indentor and load must be specified. Conversions between the various hardness numbers, using different indentors and loads, are not reliable. In short, the hardness number is valid only for the actual conditions and indentors used to determine it.

There are many sources of error in hardness testing, including poor sample preparation, porosity, inclusions, and cold working of the sample during metallographic sectioning. Indentors may become distorted in use.

The most reliable measures of hardness are those made on cross sections of the deposit rather than on the surface; for details consult, for example, ASTM B 578, Standard Method for Measurement of Microhardness of Electroplated Coatings.

RESIDUAL (OR INTERNAL) STRESS

Residual stress can be important in many applications of electroplating. It can cause distortion of the workpiece or cracking of the deposit. Although its causes are not completely understood, certainly distortion of the atomic lattice is a factor. If the atoms of the electrodeposit are closer together than their normal spacing, the result is a deposit with residual tensile stress, and the deposit tends to pull on the substrate. On the other hand, if the atoms in the deposit are farther apart than normal, a compressive internal stress will result, and the deposit will push on the substrate. When the substrate is massive and does not distort under these stresses, the deposit itself may crack to relieve the stress. Even if cracking does not occur, the "locked-in" stresses may contribute to early failure via stress-corrosion cracking.

Although many methods have been suggested for measuring internal or residual stress in electrodeposits, two have emerged as most useful, or at least most widely used: the rigid strip and the Spiral Contractometer.

In the rigid-strip technique, a thin basis-metal strip is mounted in a fixture that holds it rigid while a deposit is applied. When deposition is complete, the sample is taken from the fixture and allowed to distort freely. The radius of curvature of the strip is measured, and from this the residual stress is calculated. This calculation is subject to some uncertainty, but a widely accepted formula is

$$S = \frac{2Et^2d}{NL^2}$$

where S = residual stress
E = elastic modulus of the test strip
t = thickness of the test strip
d = deflection due to plating
N = thickness of the deposit
L = length of the test section

More convenient, especially for routine use, is the Spiral Contractometer, a commercially available instrument (Fig. 23-7). The instrument is hung in the plating tank from the cathode rod. It has a helical strip of metal, one end rigidly anchored and the other free to move; as it moves, it actuates a pointer on the instrument dial. A compressive stress causes the helical cathode to unwind, and a tensile stress winds it tighter. The instrument can be readied for re-use by stripping the deposit after each test. The reading on the dial is converted to residual stress by a suitable factor furnished with the instrument.

Results with the Spiral Contractometer do not always check those of the rigid strip, and furthermore they may not be reproducible from laboratory to laboratory; they are useful, however, for comparative purposes within a single establishment and as a method for maintaining a process under set control limits. The significance of the absolute values obtained with the instrument is questionable.

STRENGTH AND DUCTILITY

Cracking of deposits may be a serious limitation to their usefulness; the stronger and more ductile a deposit, the less likely it is to crack in service. Although measurements of strength and ductility of massive metals are well-established techniques in metallurgy, thin electrodeposits pose special problems.

Ductility (the ability of a material to deform plastically without fracturing) is often expressed as a percent elongation in a tensile test, and it is measured at the same time as tensile strength in testing massive metals. But ductility is not independent of the thickness of the metal if it is less than 0.25 mm, which most electrodeposits are. Thus although thick electroformed foils can be prepared for measuring the strength and ductility of plated metals, the results may not apply to the more usual thin deposits of 25 μm or less.

Fig. 23-7 Spiral contractometer.

Strength is not similarly dependent on thickness, and if desired, a thin foil can be machined into the standard tensile specimen shape for insertion in the conventional tensile tester as used by metallurgists. This method is at best poorly adapted to thin foils, of the thickness of usual electrodeposits, and ductility is usually measured by other techniques.

Several mandrel tests have been devised. A deposit on a sheet substrate is bent (as a composite) around mandrels of successively smaller diameter until cracking of the deposit is observed; a suitable set of mandrels has diameters ranging from about 3 to about 50 mm. The percent elongation (a measure of ductility) may then be calculated from

$$E(\%) = \frac{100t}{d + t}$$

where d = diameter of the mandrel
t = thickness of the composite (not the electrodeposit)

Other mandrel tests have been reported, based on similar principles.

The micrometer bend test (Chrysler ductility rating) is based on a different principle. Here a strip of the deposit alone, 6.35 mm wide, is bent into a U shape and placed between the anvils of an ordinary machinist's micrometer. The micrometer screw is closed down on the sample until it fails. The elongation is

$$E(\%) = \frac{100t}{d - t}$$

where d = distance between the jaws at fracture
t = thickness of the specimen

See also ASTM B 490.

A vise test may also be used as a qualitative measure of ductility; it consists simply of placing a sample of the deposit between the jaws of a special vise and bending it back and forth through 180 degrees until it breaks; the number of bends is an indication of ductility.

A method of testing both strength and ductility of thin-sheet specimens is the hydraulic bulge test, developed for thin electrodeposits by Read and his coworkers. Figure 23-8*a* is a schematic of the test. The specimen in the form of a disk is clasped between a bottom platen through which oil can be pumped and an upper platen with a circular opening whose diameter is accurately known. The oil under pressure deforms the specimen into a bulge, as shown in Fig. 23-8*b*. The oil

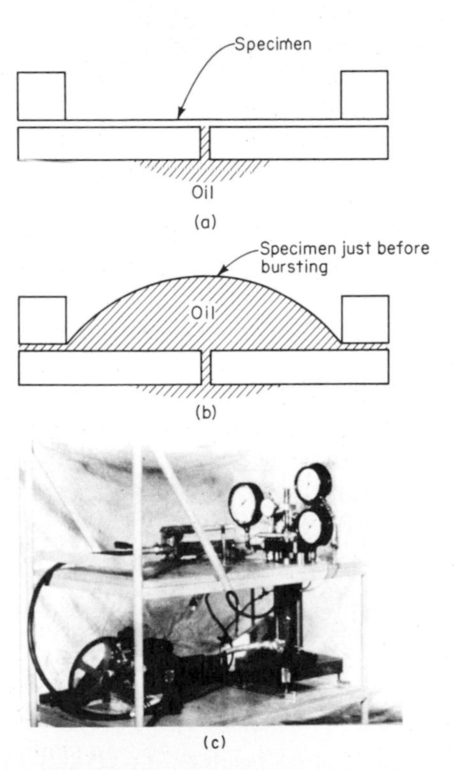

Fig. 23-8 Hydraulic bulge test: (*a*) diagram of specimen before test; (*b*) specimen after test; (*c*) hydraulic bulge tester.

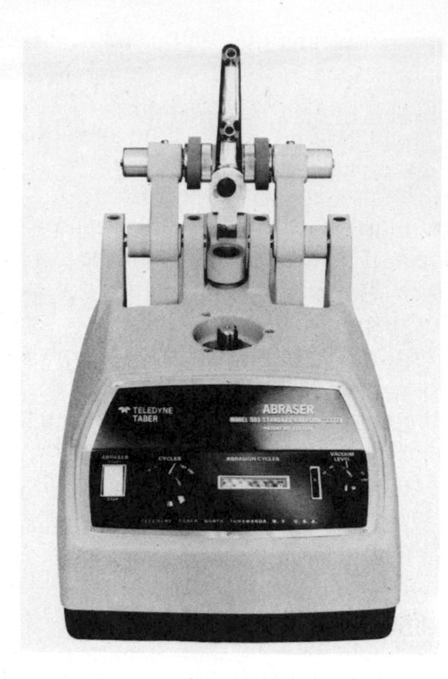

Fig. 23-9 A model of the Taber Abraser. *(Teledyne Taber, North Tonawanda, N.Y.)*

pressure and the height of the bulge at fracture are measured. The tensile strength is

$$S = \frac{PR}{2t}$$

where P = oil pressure
R = radius of curvature of the bulge
t = thickness of the specimen

Calculation of ductility is more complicated and is usually accomplished by means of a graph. Unfortunately, commercial instruments for performing the bulge test are not available, but a good machine shop should be able to build one from standard components.*

ABRASION RESISTANCE

Abrasion resistance is often confused with hardness, and it is true that many hard materials are also resistant to the loss of thickness caused by wear and rubbing.

*For details see T. A. Prater and H. J. Read, *Plating*, **36** 1221 (1949); *ibid*, **37**, 380 (1950); *ibid*, **38**, 142 (1951).

But this correlation is by no means universal, and if the application involves actual wear (as in bearing surfaces), abrasion resistance itself should be tested. No standard test is universally accepted, but the Taber Abraser is widely used (Fig. 23-9). In this instrument the test panel is rotated while two abrasive wheels held by weighted arms produce a rubbing action. An annular ring of wear is produced on the specimen by the abrasive wheels, and the degree of wear can be observed visually or measured by weight loss. Various types of wheels are available, and the load weight can be varied.

In its present state of development, this test for abrasion resistance is not an infallible indication of service behavior, and for reliable prediction the deposit-substrate combination should be tested under actual conditions of use.

The tests mentioned are those most often performed on electrodeposited coatings; in particular, tests for thickness (or weight per unit area) are almost universally necessary for both process control and acceptance testing; and corrosion tests are common. Other tests mentioned are performed less frequently. In addition, special circumstances may render other tests necessary: coefficient of friction, electrical and heat conductivity, contact resistance, magnetic properties, emissivity, and others as relevant to particular end uses. For the most part such tests are not specific to electrodeposits, and standard methods are used.*

Sometimes the properties of the basis metal are altered by the plating process. In particular, hydrogen embrittlement may be caused, and fatigue limit may be changed. Tests for hydrogen embrittlement usually are carefully controlled ductility tests. The Lawrence hydrogen gage is available for measuring uptake of hydrogen; it is favored by some aerospace industries. Fatigue properties are measured by standard metallurgical techniques.

Tests for the properties of anodic coatings on aluminum are cited in Chap. 22.

*A critical review of tests mentioned here, and others, is R. Sard, H. Leidheiser, Jr., and F. Ogburn (eds.), "Properties of Electrodeposits, Their Measurement and Significance," The Electrochemical Society, Princeton, N.J., 1975.

24

Control and Analysis

In the previous chapter we examined the tests that may be required to determine the acceptability of the products of the finisher's efforts; in the final analysis, these tests also test the finisher's success in controlling these processes. If the plated or otherwise finished parts are acceptable, by any appropriate standards, then the cleaning, rinsing, plating, and postplating cycles also must have been acceptable. But there are many in-process tests that can, and should, be applied to ensure that matters are under control and to render it more likely that the finished product will be satisfactory. Finding that a number of parts coming off the plating line are rejects is a rather late, and expensive, method of determining that something is out of control in the solutions used.

The plater, therefore, can more easily and cheaply ensure the success of the operation by conducting various in-process tests that will signal incipient trouble which can often be corrected before much harm is done. Such tests as the water-break test, already considered under Cleaning (Chap. 5), can indicate that the cleaning cycle is operating properly. We are concerned here with control of the plating solutions themselves.

All plating solutions operate satisfactorily within a given range of current densities, temperatures, pH's, and solution compositions; outside these ranges the results will be unacceptable. These ranges are identified in the individual chapters or in the vendor's literature if the process is proprietary. In addition to the determination of these variables, there exist some plating tests that are often most helpful in determining the condition of the solution.

SAMPLING

In order to analyze a solution or to perform other small-scale tests upon it, a representative sample must be taken; otherwise the analytical results and any other tests will be meaningless. Mere grab samples of solution taken from a plating tank are useless unless it is certain that the solution in the tank is perfectly homogeneous.

It is a common mistake to adjust the volume or level of solution in the tank and then take a sample without properly agitating to ensure this homogeneity. The water added will lie above the denser plating solution. A similar mistake is to add chemicals and take a sample without being sure that they have been evenly distributed throughout the solution. Many operators do not realize how thorough and prolonged agitation must be to ensure that the solution has the same composition throughout; this is especially true of large tanks, and when solid chemicals have been added. (It is usually poor practice, in any case, to replenish solutions by adding solid chemicals; they should be dissolved, either in water or in a small sample of the plating bath, in a container separate from the plating tank. The resulting solution is then added. This does not avoid the necessity for good mixing after additions, but it renders the mixing easier.)

Except when the plating tank has a very small volume, it is good practice to adopt a standard procedure for all baths. The sample should be taken by "thieving": a glass or plastic tube, about 10 to 15 mm in diameter, is inserted through the solution with the top end uncovered. As the tube passes through the bath, it fills with solution. When the tube reaches the bottom and the level in the tube is the same as the level of the bath, cover the top of the tube tightly with the thumb, withdraw the tube, and transfer the liquid quickly to a sample container, which must be clean and dry or previously rinsed out with the solution being tested. Repeat this operation at a few locations in the plating tank; the resulting sample is then mixed and carefully identified. The temperature of the solution should be recorded, so that, if necessary, the sample can be reheated to the bath temperature for any tests.

Many properties of the solution may be measured. Usually pH and chemical composition are the most important; conductivity, surface tension, and specific gravity are other inherent properties that may be tested. In addition, various plating tests may be performed, such as Hull cell, bent cathode, throwing power, electrode efficiencies, and others that indicate the plating characteristics of the bath.

pH AND ITS MEASUREMENT

Most baths operate best within certain limits of pH, but there are exceptions. Highly alkaline solutions such as stannate tin and highly acidic ones such as fluoborate and chromic acid solutions operate at a pH outside the usual ranges of measurement, and where measurement of pH is insensitive. The alkalinity of stannate baths is determined directly, by a titration for free hydroxyl ion; and the acidity of fluoborate baths is found similarly by titrating for free hydrogen ion. For most other baths, measurement and control of pH are important.

Meaning of pH

Probably the most important and most frequently measured property of plating solutions is their pH. A brief description of this concept follows; for more thorough and precise discussion, texts on electrochemistry should be consulted.

Pure water ionizes to a slight extent, according to

$$H_2O \rightleftharpoons H^+ + OH^- \tag{24-1}$$

(More precisely, $H_2O + H_2O \rightleftharpoons H_3O^+ + OH^-$, but the discussion is not affected by this approximation.) As with all such equilibria, the equilibrium constant of water may be written as a dissociation constant:

$$K_{diss} = \frac{a_{H^+} \cdot a_{OH^-}}{a_{H_2O}} \qquad (a = \text{activity}) \tag{24-2}$$

But for reasonably dilute solutions, a_{H_2O} is relatively constant and for practical purposes may be included in the total constant, yielding

$$K_w = a_{H^+} \cdot a_{OH^-} \tag{24-3}$$

where K_w is called the ion product of water. At 25°C, K_w = very nearly 10^{-14}. It follows that at 25°C

$$a_{H^+} = a_{OH^-} = \sqrt{K_w} = \sqrt{10^{-14}} = 10^{-7} \tag{24-4}$$

Even if a_{H^+} and a_{OH^-} are not equal, it is still true that

$$a_{H^+} \cdot a_{OH^-} = 10^{-14} \tag{24-5}$$

K_w varies with temperature; the variation is about 5 percent per °C. K_w is little affected by the presence of solutes, and only in very concentrated solutions need this variation be taken into account.

Acids and Bases

There are various definitions of acids and bases, and for theoretical considerations the Lewis concept of electron-pair donors and acceptors is probably the most significant. For aqueous solutions, including plating solutions, the Brönsted definition, however, is adequate: acids are substances that donate hydrogen ions (protons), and bases are substances that accept hydrogen ions (acid = proton donor; base = proton acceptor). Thus hydrogen chloride, which is a nonelectrolyte in many solvents, in water ionizes according to

$$HCl + H_2O \rightleftharpoons H_3O^+ + Cl^- \tag{24-6}$$

Here, in the Brönsted sense, HCl is an acid and H_2O is a base. Equally, H_3O^+ is an acid and Cl^- is a base, though a very weak one since the equilibrium is far to the right; we have a system of *conjugate acids and bases*. In the case of ammonia

$$NH_3 + H_2O \rightleftharpoons NH_4^+ + OH^-$$

Water is an acid, since it donates hydrogen ions to the ammonia, and ammonia is a base since it accepts them; equally, NH_4^+ is an acid and OH^- is a base.

It appears that water itself is either an acid or a base, depending on the substance with which it reacts; therefore it is an *ampholyte*. This double function of water is shown in

$$H_2O + H_2O \rightleftharpoons H_3O^+ + OH^- \qquad (24\text{-}7)$$

Since HCl reacts with water practically completely, a 0.01 N solution of HCl will have a hydrogen ion activity of 10^{-2}; therefore the activity of OH^- will be 10^{-12} [Eq. (24-5)]. In 0.01 N NaOH (a strong base, completely dissociated into Na^+ and OH^- ions) the OH^- activity is 10^{-2}, so that $a_{H^+} = 10^{-12}$ (at 25°C). Thus an acid solution contains a preponderance of hydrogen ions and a small concentration of hydroxyl ions; an alkaline (or basic) solution contains the reverse.

In 1909 Sørenson suggested the pH scale, which is merely a logarithmic method of expressing the hydrogen ion activity of a solution; it avoids the continual use of negative exponents. pH was defined as the negative logarithm of the activity of the hydrogen ion (to the base 10).

$$\text{pH} = -\log_{10} a_{H^+} \qquad (24\text{-}8)$$

Although the original definition of pH—the negative logarithm of the hydrogen ion activity—is useful in practice, it encounters theoretical difficulties and cannot any longer be accepted literally. This is because it is impossible to *measure* the activity of an individual ion, and the definition therefore is, in a sense, without meaning; that is, it is not "operational." For an aqueous solution pH is now defined, arbitrarily but operationally and reproducibly, by the Bates-Guggenheim convention:

$$\text{pH}_x = \text{pH}_s + \frac{E_x - E_s}{2.3026RT/\mathfrak{F}}$$

where R, T, and $\mathfrak{F}$ have their usual meanings; the pH_x of the unknown medium is calculated from that of an accepted standard (pH_s) and the measured difference in the emf's (E) of the electrode combination, when the standard solution is removed from the cell and replaced by the unknown.

In spite of these difficulties, the definition of pH as the "negative logarithm of the hydrogen ion activity" is a useful concept, if the user bears in mind its limitations. For more complete discussion, see almost any modern text on physical chemistry or electrochemistry, especially R. G. Bates, "Determination of pH, Theory and Practice," Wiley, New York, 1964.

In the following discussion this concept of pH will be assumed. Also, we shall often equate activities with concentrations; this introduces an error that is usually not very serious in solutions less concentrated than about 0.1 N.

Thus in pure water or in neutral solutions, where $a_{H^+} = a_{OH^-} = 10^{-7}$, pH = 7; if the solution is acid, that is, has an excess of hydrogen ions over hydroxyl ions, the pH is less than 7; and if the solution is alkaline, the pH is greater than 7. (Less frequently the function pOH may be encountered. It has comparable meaning: the negative logarithm of the hydroxyl-ion activity.)

At any temperature, pH + pOH = pK_w at that temperature, so that if K_w is known, a pH determination also determines the activity of both hydrogen and hydroxyl ions.

The pH scale is logarithmic; an *increase* of 1 unit in pH denotes a *tenfold decrease* in hydrogen-ion activity (or to a sufficiently close approximation the respective concentrations).

For practical purposes, in aqueous solution, the pH scale is useful in the range from 0 to 14. Beyond these limits activity coefficients are sufficiently different from unity to render the meaning of pH somewhat ambiguous; other considerations, such as ion association, also complicate matters.

Neutral solutions have a pH of 7 at 25°C; solutions of pH between about 4 and 6 are considered mildly acidic, pH below 3 strongly acidic. Solutions of pH between 8 and 10 are mildly alkaline, above 10 strongly alkaline. These are, of course, relative terms.

The fact that K_w changes with temperature renders it important in determining pH that compensation be made for this. pH meters (see below) usually incorporate means for such compensation.

For strong acids and bases, calculation of pH from their concentrations is relatively easy. For weak acids or alkalis such as acetic acid and ammonia, the dissociation constants of the solute must be taken into account.

pH and Plating

The control of pH of plating solutions is necessary to maintain the acidity or alkalinity which has been determined to produce the best results; appearance, stress, leveling, electrode efficiencies, and other properties may depend on operation at the set pH.

One reason for the importance of pH in plating is that most metal hydroxides are insoluble, and, depending on their solubility products, they will precipitate at various pH's. The pH in the cathode film is usually higher than that in the bulk of the solution, but it is the latter which is usually measured.

Exact calculation of the pH at which this precipitation will occur usually is not practical because metal ion concentrations in plating baths are so high that equating concentration to activity would be highly inaccurate, and activity coefficients are not known for these concentrated solutions.

Too low a pH may cause hydrogen evolution and consequent loss of efficiency; in particular baths, pH also has specific effects on hardness, stress, and other deposit properties.

Usually, high accuracy is not required in measuring pH for plating baths. Most practical plating processes work well in a range of pH units at least 0.5 unit wide, and a precision of ±0.2 pH units is adequate for control. However, the method of measurement should be consistent: if the best pH for operation has been determined by a particular method, that method should be used for control, since, without needlessly complicated adjustments, different measurement techniques can give differing results.

The methods for determining pH are colorimetric and electrometric; either is satisfactory. With proper instrumentation and care, the colorimetric method can be just as accurate as the electrometric, but for daily shop use the electrometric methods are considered to be the more accurate.

Electrometric

Electrometric methods depend on the measurement of the potential of an electrode in contact with a solution containing hydrogen ions, Fundamentally, the method depends on the use of the hydrogen electrode: a platinized platinum electrode bathed in hydrogen gas, at which the reaction $H_2 \leftrightarrows 2H^+ + 2e^-$ is reversible when catalyzed by platinum black (a platinized platinum electrode is a sheet of platinum coated with platinum black, which in turn is very finely divided platinum metal). The potential of this electrode is considered the primary standard, and by definition is zero at all temperatures when the activity of the hydrogen ion is unity. Since single electrode potentials cannot be measured by any known technique, what is actually measured is the potential between the hydrogen electrode and a secondary electrode such as the calomel or silver–silver chloride.

Except for some types of demanding research work, the hydrogen electrode is seldom used, since the setup is inconvenient and requires too many precautions. Thus for daily shop or laboratory use, indicator electrodes are used that are simpler, more rugged, and sufficiently accurate for all practical purposes. These secondary electrodes include the quinhydrone, antimony, and glass; the last-named has superseded all others in commercial instruments and is the only one requiring discussion here.

The most common electrode system in technical use is the glass-saturated calomel pair; most commercial instruments have both electrodes mounted in the same holder. The potential of the glass indicator electrode depends linearly on the pH of the solution in which it is immersed; the change of potential follows the Nernst equation, being about 59 mV per pH unit at 25°C. The measuring circuit is essentially a potentiometer or vacuum-tube voltmeter, but the scale may be made to read directly in pH units. Most commercial instruments have double scales and can be used as either millivoltmeters or pH meters.

At the interface of a thin glass membrane and a solution in contact with it, an emf is set up which depends on the pH of the solution. Thus if a glass bulb contains a solution of constant pH and a suitable reference electrode, the emf of the glass electrode is a measure of the hydrogen-ion concentration of the solution. In practice, the emf of the glass electrode is measured against a calomel reference electrode; because of the high resistance of the glass a sensitive potential-measuring device such as a vacuum-tube voltmeter (VTVM) or potentiometer circuit is used.

The procedure for using a pH meter is to initially standardize the instrument emf with a buffer solution of known pH. The change of potential occurring when the glass electrode system is immersed in the test solution is then measured; it yields the difference in pH between the buffer and the test solutions. A buffer solution is one that resists change in pH on addition of small quantities of acid or alkali. Many such solutions are known, for all ranges of pH, and their compositions are available in most standard handbooks.* Many such solutions are available ready to use from laboratory supply houses.

Modern electrometric pH meters are highly automated and require little attention; the glass electrode and the reference electrode are contained in one simple glass vessel, which is readily inserted into the solution to be measured. The meter is provided with calibrations that can compensate for temperature changes; and standard-pH buffer solutions can be purchased. The emf in millivolts is converted directly into pH units by the scale of the meter; some meters have digital readout. The necessary power for activating the potentiometer circuit may be either from the mains or from self-contained batteries.

That these commercial meters require little attention does not mean they require *no* attention. The instructions of the manufacturer for storing the electrode, for compensating for temperature, and for other maintenance functions should be followed, or incorrect readings will be obtained.

Glass electrodes may be used in contact with most solutions, including oxidizing, reducing, and unbuffered solutions. In contact with solutions high in sodium ion above pH 9 or so, the special "E-type" electrode should be used. In general, glass electrodes are not suitable for use in fluoride-containing solutions because of the chemical reaction between fluorides and glass.

Standard commercial instruments are equipped with a glass electrode and a reference calomel electrode, both contained in one bulb which is kept immersed in distilled water until needed. The bulb is immersed in a known buffer solution (preferably of pH near to that of the test solution) frequently for adjustment, and a temperature correction is made by adjusting a dial. Then the bulb is immersed in the unknown solution and a reading is taken, either on a meter or by digital readout. After use the bulb should be rinsed off with a jet of distilled water and replaced in the beaker of distilled water.

Colorimetric Methods

Colorimetric methods of pH measurement are, in theory, at least as precise as those with the glass electrode, but in practice are likely to be somewhat less so because the precautions required are usually not practical for daily shop use. They depend on the color of certain organic compounds, called indicators, which change color with change in pH of the solution. Indicators are organic compounds having acidic or basic character, which undergo the equilibrium

$$\mathrm{HIn} \rightleftharpoons \mathrm{H^+} + \mathrm{In^-}$$

This equilibrium has its own dissociation constant:

*For example, "Handbook of Chemistry and Physics," "Lange's Handbook of Chemistry," and most textbooks on electrochemistry.

$$K = \frac{[H^+][In^-]}{[HIn]}$$

If the acid form of the indicator, HIn, has a color different from that of the basic form, In^-, the color will depend on the relative amounts of the two forms present, which in turn depends on $[H^+]$, as the equation shows. The color therefore becomes an indication of the hydrogen-ion concentration.

Very accurate estimation of pH can be made with comparison colorimeters, but these instruments are not common; most colorimetric determinations are made with pH papers, which are widely available. An indicator will change color at a pH which depends on its dissociation constant; indicators are available that cover the whole pH range. Papers may cover a wide range or a narrow one; both types are useful. The wide-range papers give an indication of the approximate pH; after this, a narrow-range paper can be used for more precise determination.

pH papers are particularly useful for routine control because of their ease in use. A strip is simply dipped into the test solution, and the color of the strip is compared with a standard color chart on the container. This sort of determination can be carried out right at the tank, without taking a sample and using relatively unskilled labor.* Colorimetric and electrometric pH determinations often differ by about 0.1 pH unit; in the directions for proper pH operation, the method of determination should be specified.

Colorimetric methods are liable to error in strongly oxidizing or reducing solutions or others that react with the organic dye. They may be also unreliable in strongly colored solutions such as nickel-plating baths (although this color may be compensated for in instrumental colorimetric methods).

In adjusting the pH of the bath, once it has been determined, it is best to titrate a small sample, say 1 L, with acid or alkali, until the proper pH is reached. Simple calculation then gives the amount of acid or alkali to be added to the operating bath. Usually a little less than that calculated should be first added and another determination made, to avoid overshooting.

Other Measurements

In addition to pH determinations, other inherent properties of plating solutions are somewhat less frequently of interest; they include specific gravity, surface tension, and conductivity. (Chemical analysis is considered later.)

SPECIFIC GRAVITY

For most plating solutions, specific gravity is not especially useful as a measure of their concentration or composition, since too many variables are involved and

*Operators should be tested for color-blindness!

the sensitivity of the test is too low. In a few cases, however, a simple specific-gravity test with a hydrometer can yield useful information.

The specific gravity is a fair measure of the concentration of chromic acid–plating baths, of simple copper sulfate–sulfuric acid baths, and of chromate anodizing solutions. Tables are widely available relating specific gravity to concentration of active constituents. The operator must be sure, however, that impurities that may contribute to the specific gravity of the solution do not falsify the results; hydrometer measurements are thus useful in routine control but are no substitute for occasional analysis by more accurate methods. Specific gravity (sp. gr.) determinations can also be used to detect stratification of plating baths, by comparing the sp. gr. of the solution at different levels.

Although more precise methods (such as the Westphal balance and the pycnometer) are available for density determinations, simple hydrometers are precise enough for the purposes of the metal finisher. The hydrometer should be of sufficiently narrow range that significant results can be obtained; and, as with all such measurements, readings should be taken at eye level to avoid parallax. A known volume of solution may be weighed on a balance if a hydrometer is not available. Many hydrometers are calibrated in degrees Baumé, but this scale is obsolescent and those reading in sp. gr. are preferable. The equation relating degrees Baumé to sp. gr. is, for liquids heavier than water,

$$\text{Sp. gr.} = \frac{145}{145 - \text{Baumé}}$$

This may be the occasion to note the difference between specific gravity and density, two terms that are often confused even in the scientific literature. Specific gravity is a dimensionless quantity, being the ratio of the density of the sample to that of water at the temperature of its maximum density. Density has the dimensions of mass per unit volume, and if expressed in grams per cubic centimeter, it is numerically identical to specific gravity, within any error significant in practical work.

SURFACE TENSION

The surface tension of a liquid is defined as the inward force acting on the surface of the liquid due to the attraction of the molecules below the surface, or the property, due to molecular forces, existing in the surface film of all liquids which tends to contract the volume into a form with the least surface area; the particles in the surface film are inwardly attracted, thus resulting in tension. Its dimensions are dynes per centimeter (dyn/cm), or in SI units newtons per centimeter (N/cm); 1 dyn = 10^{-5} N. Surface tension is to a great extent a measure of how well a liquid will spread over a surface: the lower, the better. Thus measurements of surface tension became important to the plater when wetting agents began to be used in plating solutions as anti-pitting agents. The concentration of such wetting agents (surfactants) is not easily determined by standard analytical techniques, and surface tension measurements are used for control of such additives. Surfactants are used also in cleaning solutions, and similar considerations apply.

The surface tension of water is relatively high, about 73 dyn/cm at room temperature, decreasing to about 65 dyn/cm at 70°C. (For comparison, ethyl alcohol has a surface tension of about 22 dyn/cm at room temperature.) The addition of most inorganic salts to water does not change the surface tension very much, whereas the addition of very small amounts of "surfactants" (detergents, surface-active agents) can reduce it considerably—to as little as 25–35 dyn/cm. Several ways of measuring surface tension of solutions are available.

Capillary Rise

This method is comparable to the similar method of measuring the solderability of surfaces, which also depends to some extent on surface tension. When the end of a capillary tube is placed in a liquid, a film of the liquid rises in the tube; this column of liquid is supported by the surface tension acting along the circumference of the tube. The extent of the rise is a measure of the surface tension.

$$\gamma = \tfrac{1}{2} hrdg$$

where γ = surface tension, dyn/cm
h = height of rise, cm
r = radius of the capillary, cm
d = density of the liquid, g/cm^3
g = the gravitational constant, cm/s^2 (about 980 to sufficient precision at or near sea level)

This method is seldom used directly by platers, but an instrument based on its principles is available, the Sur-Ten Meter.

Drop Weight

The weight of a drop of liquid that will remain suspended from the end of a glass tip is a measure of surface tension. Practical results can be obtained by counting the number of drops that fall from an orifice of known diameter for a given volume of liquid. By knowing this figure for pure water at the same temperature, the surface tension of the test liquid can be calculated. This is the principle of the stalagmometer, which is essentially a glass capillary tube having a small bore (0.5 mm or less), with a bulb blown in it to form a reservoir and the bottom ground flat. After thorough cleaning, the instrument is filled with distilled water and the number of drops is counted as the liquid falls between two marks on the tube. The operation is repeated with the test liquid. Then

$$\gamma = \frac{W_x \gamma_W \times \text{sp. gr. of unknown}}{\text{no. of drops of unknown}}$$

where γ = the surface tension of the test or unknown liquid
W = number of drops of water
γ_W = surface tension of water (73 at room temperature)

Torsion Balance

When a ring of fine wire is immersed in a liquid, it is held up by the forces of surface tension; the extent of the force preventing the ring from sinking is a measure of the surface tension. This principle is used in the DuNuoy Tensiometer, which employs a platinum-alloy wire ring and a sensitive balance. The method is probably the most accurate of those mentioned, but on the other hand requires the most expensive instrumentation and more exacting care; the additional precision is probably not needed in practical electroplating.

Nevertheless all surface tension measurements must be made with care if they are to be significant. The Sur-Ten meter, the stalagmometer, and the ring of the Tensiometer must be thoroughly clean; the last-named must be burned off after each test. Traces of surfactants from cleaning solutions can have large effects on the results and render them meaningless.

Surface tension measurements are most often made on nickel-plating solutions containing surfactants as anti-pits; the vendor's instructions will specify the proper values. Such measurements may also be valuable in control of cleaning solutions.

(Surface tensions are, in fact, interfacial tensions; we have assumed in the discussion above that the interface is that of the liquid in question with the air above it. More precisely, the air is saturated with water vapor, and other impurities may be present; these considerations are not significant for present purposes.)

CONDUCTIVITY

Control and measurement of conductivity is of little value for plating solutions, but is significant for control of rinsewaters and is sometimes specified in electronics specifications as a measure of residual salts. For greatest accuracy in measurements of conductivity, extreme precaution is required, and textbooks on experimental electrochemistry should be consulted. For control of rinsewaters, self-contained conductivity bridges are available which are direct-reading and sufficiently accurate for routine control. They may be made to activate solenoid valves that control the flow of rinsewater, so that when the conductivity gets too high (too high a salt concentration in the rinse), the flow of water is increased; and when it falls low enough, the water flow is shut off or decreased to save water.

Plating Tests

In addition to tests such as pH determination, chemical analysis, and the others already discussed, which give information on the composition of the solution, there are several tests that tell more directly the plating characteristics of the bath; properly interpreted, they can be of great aid in informing the electroplater of necessary adjustments, i.e., trouble-shooting. The most important and most widely used of these is the Hull cell test.

HULL CELL

The Hull cell, described and patented by R. O. Hull in 1939, is a trapezoidal box of nonconducting material (see Fig. 8-7); an anode is laid against the right-angle side and a 100 × 70-mm cathode is laid against the sloping side, connected to a current source with alligator clips. When a current is passed through the solution sample contained in the cell, the current density along the sloping cathode varies in a known manner, so that the character of the plate at a range of current densities is determined in one experiment. This current density variation is plotted in Fig. 24-1. Similar plots are provided by the manufacturer along with the purchase of the cell. The standard Hull cell contains 267 mL, a volume selected because 2 g of material added to the sample corresponds to a 1-oz/gal

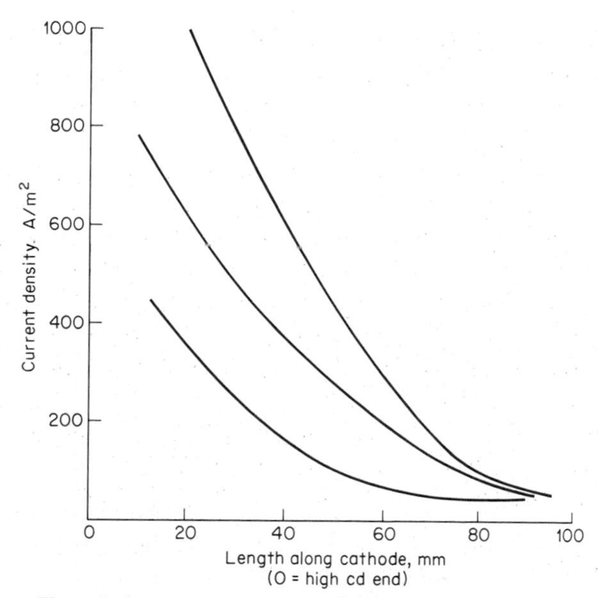

Fig. 24-1 Current variation over Hull cell cathode.

(7.5-g/L) addition to the main plating bath. The current density at any point on the cathode is approximately represented by the formula

$$cd\ (A/m^2) = 100I(5.102 - 5.24 \log L)$$

where L = the length (in centimeters) along the cathode (between 6.4 and 82.5 mm)

I = amperes

In practice, one simply lays the cathode along the graph (Fig. 24-1) or special scales supplied by the vendor, to determine the current density.

Hull cells are usually of polymethyl methacrylate, which allows observation of the cathode during plating. Ceramic material is also used. Some cells have provision for heating and for agitation; movement of a stirring rod across the face of the cathode can simulate the type of agitation used in practice. In any case, conditions in the plating bath should be simulated as much as possible in the Hull cell test for significant results.

Cathodes of the proper size are supplied by makers of the cells. They are usually of zinc-plated sheet steel, from which the zinc deposit is stripped just before use by immersion in hydrochloric acid (wiping off any smut), or of polished brass, protected by a strippable lacquer. Nickel-plated panels are useful for testing chromium-plating solutions. Naturally platers can cut their own panels from sheet metal if they wish.

Current used in the cell varies from 1 to 3 A, and time from 2 to 10 min, depending on the type of solution being tested. The current source should preferably provide current with no more than 5 percent ripple.

The appearance of the Hull cell cathode after plating can offer considerable information about the condition of the solution; a typical cathode may appear somewhat as in Fig. 24-2, which is illustrative only and does not look like any particular cathode.

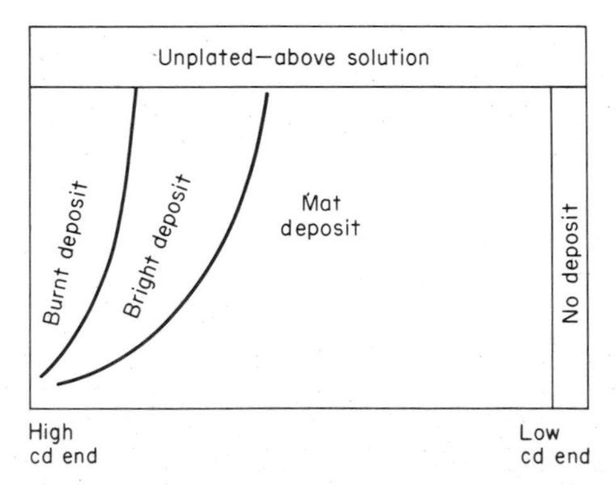

Fig. 24-2 A typical Hull cell cathode (illustrative only).

The bright range, coverage, appearance of the plate, and other factors, as influenced by current density, all appear from one Hull cell test. Need for purification by carbon treatment or dummying, for more or less addition agent, and other adjustments can be judged.

The interpretation of the appearance of a Hull cell cathode is largely a matter of experience, although much has been published concerning specific solutions. Vendors' technical brochures are good sources of information. But the best is a collection of Hull cell cathodes collected over time by the operator—cathodes representing satisfactory conditions, and cathodes representative of various specific troubles which have been cured in known ways.

If a Hull cell panel displays a particular difficulty, appropriate additions or adjustments are made to the sample solution and the test is repeated; when the panel is finally satisfactory, the necessary adjustments to the bath which cured the trouble are translated into those required for the working solution, and appropriate corrections are made from the formula already given.

BENT CATHODE

Other than the Hull cell, the bent-cathode test is probably the most useful of the plating tests. In this test, a 25- × 100-mm cathode is bent at right angles 25 mm from the lower end. This cathode is plated in a 1000-mL beaker at an appropriate current density. The cathode provides an area of low current density, a shelf area, and areas of high and medium current density. The distance into the base of the right angle which is covered by the deposit is often a good indication of the condition of the bath.

Both Hull cell and bent-cathode tests can be run in the production bath itself. A special Hull cell with an open bottom can be so used.

Various other plating tests, run on small samples of the solution, are useful in some circumstances. For instance, the current efficiency can be checked more easily on a small panel, plated in a beaker and weighed before and after plating, than in the production tank; sample panels are also useful for checking the performance of precious-metal plating solutions.

Solution Analysis

Ideally, plating solutions, once made up, would remain constant in composition and analysis would seldom be required. The anodes would replenish the metal deposited at the cathode; addition agents would be stable, and after a little experience any additions required could be scheduled on an ampere-hour basis. Such an ideal situation, like most ideals, is never realized: drag-out is unpredictable; acidity increases or decreases; addition agents and cyanides decompose; impurities enter the solution from anodes and the addition of salts (never 100 percent pure), from drag-in, from dissolution of the substrate metals, and from other sources; mistakes are made by operators in making additions. For all these reasons, plating solutions must be analyzed, occasionally or frequently depending on the type of solution, its sensitivity to compositional changes, the work load, and the experience of the operator. As a general rule, frequency of analytical control should usually be greatest when installing a new bath or an unfamiliar process, and can often be decreased when experience is gained and signs of trouble are more readily recognized. The reason for analyzing a plating bath is not to determine its exact composition, but only to know if it is within the limits fixed by experience with the bath. If it is outside these limits, it is desirable to know what corrections are required or what steps to take to remove deleterious impurities.

The tests already mentioned, such as the Hull cell and bent-cathode tests, can tell an experienced operator a great deal about the condition of the solution and may obviate the necessity for frequent analysis. pH determination should be routine for any solution that is sensitive to it, since it can change more rapidly than most other bath characteristics and is perhaps the easiest to adjust. Sooner or later, however, the solution must be analyzed—always for its content of principal constituents, sometimes for trace impurities that may be harmful to bath operation.

In this book we shall not attempt to present directions for determining the various constituents of plating or auxiliary solutions. There are innumerable textbooks on analytical chemistry, at all levels, and some are directed specifically to the analysis of plating solutions. The methods available to the plater will be outlined. More specialized texts should be consulted for details and for the many precautions required in good analytical technique.

Analytical procedures are of two general types: qualitative and quantitative. The first determines what is present; the second, how much. From the standpoint of the plater, qualitative analysis is necessary only for determining whether certain impurities or trace elements are present in the solution; obviously the plater does not require a qualitative test to tell that nickel is present in a nickel-plating bath.

Most qualitative analyses can be made at least semiquantitative; they can usually distinguish between major, minor, and trace amounts of elements and ions. Complete schemes for qualitative analyses for the metallic elements have been developed; most depend on the division of the metals into groups according to the acidities at which their sulfides precipitate, followed by tests for those not precipitated by H_2S. Within the groups more sensitive tests are carried out to distinguish the individual elements. A complete qualitative analysis by any of these schemes is a long and tedious affair, and in most cases can be replaced by spectrographic methods, which in a single determination can yield results about equivalent to a complete qualitative scheme. They require expensive equipment and are usually carried out by completely equipped analytical laboratories. Specific spot tests not requiring complete separations are available for some metals.

Quantitative analysis is designed to yield information on the exact amount of specific constituents present in a sample. Except for the simplest cases, most quantitative methods depend on the ability to separate the desired constituents from other materials in the sample that would interfere with the determination. These separations are often more time-consuming than the determination itself. Often the need for such separations can be obviated by newer instrumental techniques.

There are three basic techniques in quantitative analytical chemistry of interest to electroplaters; various instrumental techniques may constitute a fourth. These three are gravimetric, volumetric, and colorimetric. They are useful for different purposes and have varying applications to the usual determinations performed by metal finishers in analyzing their solutions.

GRAVIMETRIC

Perhaps the oldest technique in analytical chemistry, gravimetric determinations are usually avoided except when they are the only ones available or their better precision is really required. In addition, gravimetric methods possess the advantage that they require no standards, as do colorimetric and volumetric techniques; it may not be worthwhile to prepare a standard solution if only an occasional analysis is required. Therefore they are useful for occasional and nonrepetitive or nonroutine determinations.

They depend on the precipitation of some compound of the ingredient being determined; this compound must be "analytically insoluble" and of known composition, and methods must be available for separating the desired ingredient from others that might interfere with the determination, for example, by coprecipitating. After collecting and washing the precipitate free of the liquid from which it is derived, the weight of the precipitate is determined, and from the known composition of the precipitate the amount of the material actually sought is easily obtained by use of an appropriate factor.

Properly applied, gravimetric methods are capable of high precision and accuracy, but they can be time-consuming and do not lend themselves to even elementary automation. In metal finishing, the sulfate concentration of chromic acid baths may be determined in this way (usually as an occasional check on more rapid methods); and the gold content of gold-plating baths is determined by precipitating gold metal and weighing it directly. This has the further advantage that the expensive metal is recovered directly for sale or re-use. Many other gravimetric methods are described, but most can be replaced by volumetric ones if desired. Gravimetric methods are not usually recommended for determination of trace quantities, where colorimetric or instrumental methods are preferable.

VOLUMETRIC

In a volumetric determination, a solution of known concentration (standard solution) is added in known amounts to the test solution containing an ingredient with which the standard reacts; when this reaction is completed, another reaction takes over as soon as an excess of the standard solution has been added. This second reaction is signaled by a change in color of an indicator, or occasionally by the appearance of a precipitate. The amount of standard solution added is recorded; and from a knowledge of the concentration of the standard, the amount added, and the nature of the reaction, the amount of the unknown is calculated.

In a few cases, an indicator is not required since the standard solution provides its own indicator: when potassium permanganate solution is added to a solution of a reducing agent, its color is discharged (MnO_4^- is reduced to the practically colorless Mn^{++}). But as soon as one drop of excess permanganate ion has been added, the intense purple color of the permanganate ion signals the end of the reaction.

The operation is known as *titration;* the standard solution added is the titrant, and the point at which the two are exactly equivalent is the equivalence or end point. The titrant is added from a buret, which is a glass tube marked off in milliliters or fractions of milliliters, provided with a stopcock adjustable to permit rapid or dropwise addition of the titrant, and of a size appropriate to the amount to be added and the necessary precision of the determination. The unknown or test solution is contained in a beaker or Erlenmeyer flask; efficient stirring is required to make sure that reactants have reached equilibrium before more titrant is added. Usually the approximate answer is known and the titrant can be added rapidly at the beginning of the determination; when the end point is approached, the addition is slowed down and time is allowed for equilibrium to be established before the next drop is added.

In some cases the standard reactant is added in excess, and this excess then titrated with another standard solution to complete the determination. This is called *back-titration*.

Volumetric methods are of three general types: acid-alkali, reductant-oxidant, and chelate or complex-forming, known respectively as acidimetric (or alkalimetric), oxidimetric (or reductimetric), and chelatometric methods. There are also a few special cases that do not fall under any of these classifications.

Acid and alkali, when not in the range suitable for pH measurement, are almost universally determined by volumetric methods, which we shall discuss in some detail. Other volumetric methods will be considered more briefly.

Acidimetry-Alkalimetry

Assume that a 10-mL sample of 0.1 N HCl is the "unknown." Other compounds may be present, but it is assumed that they do not interfere with the reaction. This 10-mL sample is diluted to about 100 mL in a beaker or Erlenmeyer flask, a drop or two of indicator solution is added, and standard 0.1 N NaOH solution is added from a buret.

At the start, before any base has been added, the pH of the sample corresponds to 10^{-2} N H^+ (because 10 mL of a sample of 10^{-1} N HCl was diluted to 10 times its volume), so that the pH is 2.0. When 1 mL of 0.1 N NaOH has been added (neglecting the slight change in the volume in the beaker throughout), 1 mL of the sample has been neutralized. So now the beaker contains 9 mL of 0.1 *N* HCl in 100 mL, the H^+ concentration is 9×10^{-3} *N*, and the pH is 3 − log 9, or 2.05. When 2 mL of NaOH has been added, similarly, the pH is 3 − log 8, or 2.10. Let us skip to 9 mL of 0.1 *N* NaOH added; the pH, by analogous reasoning, is 3 − log 1, or 3.0. At 9.5 ml, the H^+ concentration is 0.5×10^{-3} or 5×10^{-4} *N*, and the pH is 4 − log 5 = 3.3. Continuing, at 9.9 mL of NaOH, the H^+ concentration is 1×10^{-4} *N* and the pH is 4.0; and at 9.99 mL of NaOH the H^+ concentration is 10^{-5} *N* and pH = 5.0. When 10 mL of NaOH is added, the solution is neutral and the pH is 7. When the titrant, 0.1 *N* NaOH, is 0.1 mL in excess, OH^- concentration is 1×10^{-4}; the pOH is 4, and therefore the pH is 10. When the NaOH is 1 mL in excess, the OH^- concentration is 10^{-3} *N*, pH is 11, etc. These results are plotted in Fig. 24-3.

The figure clearly shows that at the beginning of the titration the pH changes slowly. As the end point approaches, the curve becomes very steep, almost vertical; and as the end point is passed, it levels off again. Such a curve is typical of the titration of a strong acid with a strong base, or vice versa (strong means completely dissociated).

This titration could be followed with the aid of a pH meter, using no indicator; the end point is the midpoint of the steepest portion of the curve, in this case at 10.0 mL. For most purposes, it is much easier to add a few drops of a solution of an organic dye, an indicator, that changes color at or very near the end point. In titrating a strong acid with a strong base (understand *vice versa* throughout the discussion), the choice of indicators is wide, since any indicator that changes in the pH range from about 4.5 to 8.5 will indicate the end point with little error. Methyl orange, methyl red, bromcresol purple, and even phenolphthalein can be used. Some common indicators and their pH ranges are listed in Table 24-1.

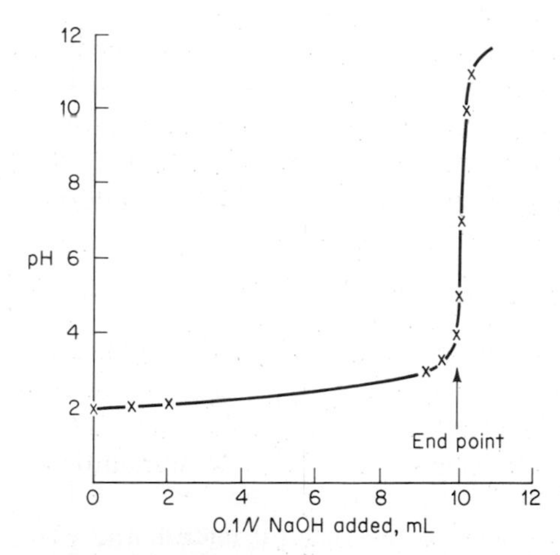

Fig. 24-3 Titration of 0.1 *N* HCl with 0.1 *N* NaOH.

Table 24-1 Typical Acidimetric Indicators

Indicator	*pH range*	*Acid–alkaline color*
Methyl yellow	2.9–4.0	Red–yellow
Bromphenol blue	3.0–4.6	Yellow–blue
Methyl orange	3.1–4.4	Red–orange
Methyl red	4.2–6.3	Red–yellow
Bromcresol purple	5.2–6.8	Yellow–purple
Phenol red	6.8–8.4	Yellow–red
Thymol blue	8.0–9.6	Yellow–blue
Phenolphthalein	8.2–10.0	Colorless–red
Thymolphthalein	9.3–10.5	Colorless–blue

Our example used 0.1 *N* HCl as the test solution. Similar calculations for 0.01 *N* HCl as the test solution and 0.01 *N* NaOH as the titrant would give comparable results, but the starting pH would be higher and the height of the steepest portion of the curve would be somewhat lower. As the concentration of the acid or base decreases, choice of appropriate indicators is narrowed accordingly.

When either the acid or the base is weak, i.e., not completely dissociated, the titration curve has a somewhat different form. If the acid is 0.1 *N* acetic acid, for example, the addition of NaOH yields what is in effect a buffer solution, which tends to resist pH change. At the equivalence point the solution will be distinctly alkaline, since it is a solution of sodium acetate, the salt of a weak acid and a strong base; and an indicator such as methyl orange will change color too soon. Phenol red or phenolphthalein is suitable. See Fig. 24-4. By similar arguments titration of ammonia (a weak base) with hydrochloric acid will be distinctly acidic at the equivalence point, and an indicator such as methyl red or methyl orange must be used. Phenolphthalein would not change color until the equivalence point had been passed.

When the acid is polybasic, dibasic such as sulfuric, or tribasic like phosphoric, the dissociation constant of each step in the ionization must be taken into account. Sulfuric acid yields two hydrogen ions, but even the second ionization, $HSO_4^- \rightarrow H^+ + SO_4^{--}$, is sufficiently strong that only the final end point, when both have been neutralized, will usually be noticed. The case of phosphoric acid is more complicated. Here the first and second ionizations

$$H_3PO_4 \rightarrow H_2PO_4^- + H^+ \qquad H_2PO_4^- \rightarrow HPO_4^{--} + H^+$$

can be observed at appropriate pH's, and suitable indicators are available; but the third yields hardly any inflection in the curve, and other methods must be used to determine total phosphoric acid if necessary.

Acid-base titrations, like all volumetric procedures, require that one of the solutions, the titrant, be of accurately known concentration. This in turn requires the availability of a *primary* standard. For standardizing sodium hydroxide solutions, either potassium acid phthalate or benzoic acid can be used; both are available in highly pure form, though the former is usually preferred. An appropriate quantity is accurately weighed, dissolved in a little distilled water, and titrated with the alkali to be standardized. Simple calculation yields the normality of the alkali (1.0211 g of potassium acid phthalate is equivalent to 50 mL of 0.1 *N* NaOH).

A suitable primary standard for hydrochloric acid is sodium carbonate or disodium phosphate, although it is often easier to use a standard NaOH solution as a secondary standard.

Oxidimetric and Reductimetric (Redox) Titrations

If the unknown to be determined exists in two or more valence states, such as Fe(II) and Fe(III) or Sn(II) and Sn(IV), it can be determined by oxidimetric

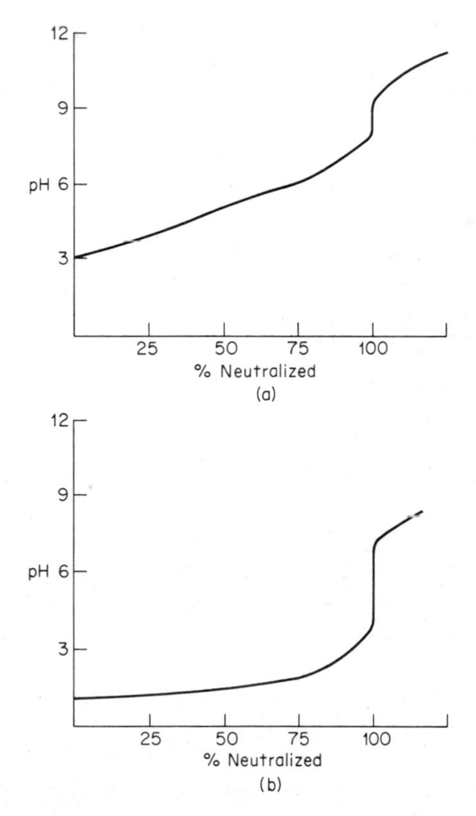

Fig. 24-4 Titrations: (*a*) acetic acid with sodium hydroxide; (*b*) hydrochloric acid with ammonium hydroxide.

titration. If one desires to determine iron, for example, one first makes sure that all of it is in the divalent form. Titration with an oxidant such as potassium permanganate oxidizes it to Fe(III):

$$5Fe^{++} + MnO_4^- + 8H^+ \rightarrow 5Fe^{3+} + Mn^{++} + 4\ H_2O$$

As soon as all the Fe^{++} has reacted, the next drop of permanganate solution will turn the solution pink; this is the end point. Similarly, stannous ions may be titrated with potassium iodate or iodine, using starch indicator:

$$Sn^{++} + I_2 \rightarrow Sn^{4+} + 2I^- \qquad \text{or} \qquad 3Sn^{++} + IO_3^- + 6H^+ \rightarrow 3Sn^{4+} + I^- + 3H_2O$$

A slight excess of iodine or potassium iodate will yield the familiar starch-iodine blue color, signaling the end point. Potassium iodate is preferred to iodine as a titrant because its solutions are more stable and do not require restandardization at intervals.

Cerate solutions [solutions of Ce(IV)] in various forms have many advantages over potassium permanganate for oxidimetry. The solutions are more stable, and the valence change of 1—Ce(IV) to Ce(III)—simplifies the stoichiometry. This oxidant became practical when suitable redox indicators such as the ferrous phenanthroline complex (ferroin) were developed.

In general, oxidimetric titrations are used more widely than reductimetric. In all cases, the stoichiometry of the reaction must be unambiguously known, and usually it is best to standardize the titrant against a known sample of the metal to be determined; for example, potassium iodate or iodine solutions should be standardized against known weights of tin if they are to be used for the determination of tin.

Methods for ensuring that the metal to be determined is entirely in the reduced form differ according to the individual metal.

The analogy between acid-base and redox titrations may be noted. The former depend on changes in pH, which can be measured either with indicators or with an electrode that is sensitive to changes in H^+ concentration. Similarly, the latter can be carried out with indicators whose color depends on an oxidation-reduction potential, or the potential can be determined by means of an appropriate electrode system. In the latter case, the operation is called a *potentiometric titration;* this technique is valuable in cases for which no suitable indicator is available and for research work.

Precipitation Titrations

The end points of several titrimetric determinations are signaled by the first appearance of a precipitate that persists instead of redissolving on stirring. That most important and familiar to electroplaters is the determination of cyanide by titration with standard silver nitrate solution, using iodide ion as indicator. The silver nitrate reacts with the cyanide to yield silver cyanide, but as long as any free cyanide ions remain, this precipitate redissolves to form $Ag(CN)_2^-$. As soon as $AgNO_3$ is in excess, a permanent, faintly yellow turbidity caused by the precipitation of silver iodide is observed.

Chelatometric (or Compleximetric) Titrations

EDTA (ethylenediamine tetraacetic acid, Fig. 24-5) forms complexes or chelates with many metals, and this property is useful for their volumetric determination. Standard solutions of EDTA are used as titrants; when all the metal has been complexed, an indicator sensitive to the metal changes color. This type of titration has become very popular for electroplating solutions; though accuracy often is not as high as that of other methods, it is sufficient for the purpose and the methods are readily adapted to routine use.

COLORIMETRY

Many reagents have been developed that react with one or more metals to produce highly colored complexes. In sufficiently dilute solutions, the intensity of this color is proportional to the concentration of the complex, and this property permits the determination of the metal by measuring the intensity of the color. In this context color means the absorption of light of a particular wavelength.

The colored solution is contained in a special cell through which light is passed. The light entering the cell is the incident light; that leaving the cell, having passed through a known length of solution, is the transmitted light. In sufficiently dilute solution, the relation between the intensity of the incident and the transmitted light of a specific wavelength (monochromatic light), in terms of the concentration of the colored complex in the solution and the depth of the solution, is found to be

$$I/I_0 = 10^{-kcl}$$

where I_0 = the intensity of the beam of incident light
I = the intensity of the transmitted light beam
k = a constant (the extinction coefficient)
c = the concentration of the colored substance
l = the depth of solution traversed by the light

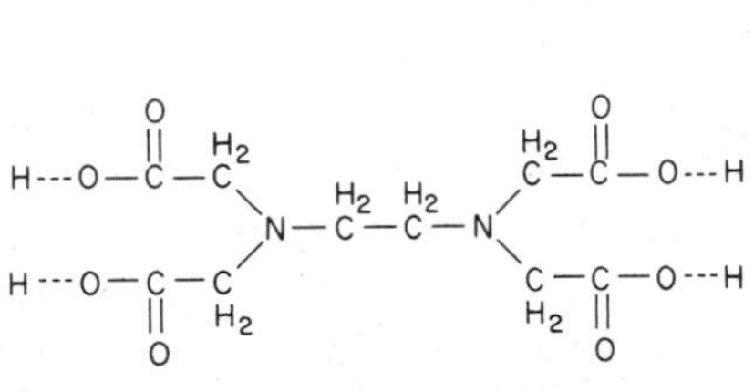

```
          O                               O
          ||  H2                      H2  ||
H---O—C—C                          C—C—O---H
               \      H2  H2      /
                N—C—C—N
               /                  \
H---O—C—C                          C—C—O---H
          ||  H2                      H2  ||
          O                               O
```

Fig. 24-5 EDTA (ethylenediamine tetraacetic acid).

This is known as the *Lambert-Beer law* (often called merely Beer's law); it is the fundamental law of colorimetry and spectrophotometry. If the law is not valid for a particular solution, colorimetric methods cannot be used. Most solutions follow this law within certain ranges of concentration of the colored substance, and the law breaks down if the concentration is too high. In most cases these concentrations are very dilute, so that colorimetric methods are most valuable for determining minor and trace amounts of metals. Although concentrated solutions could be diluted to bring them within the proper range, the act of dilution would introduce large errors, rendering the methods inappropriate for the determination of major constituents.

For each metal-reagent complex, the absorption of light is a maximum at a particular wavelength; at this point the method is most sensitive. Spectrophotometers are adjusted to provide light of the desired wavelength, and the transmitted light is measured by a photocell. A calibration curve, relating transmittance (or absorption) to concentration, must be constructed for each metal-reagent combination, using known solutions. The reading of the unknown is then simply compared to this calibration curve, and the result is read off.

White light may also be used, and either a photocell used to measure the intensity of the transmitted light or the reading may be by eye or comparison colorimeter. Most modern laboratories employ the spectrophotometric method.

Reagents may be highly specific or may form complexes with many metals. Naturally, the more specific the reagent, the fewer the preliminary separations required to prepare the sample for analysis. Colorimetric methods are available for almost all metals, and colorimetry (preferably spectrophotometry) is usually the method of choice for determining trace amounts of metals.*

INSTRUMENTAL METHODS

Instruments, of course, are used in all analytical procedures. Burets, pipets, and balances are instruments; so are pH meters and colorimeters. The word *instrumental,* however, usually connotes something more sophisticated (and expensive!) such as the spectrograph, nmr (nuclear magnetic resonance) instrument, GLC (gas-liquid chromatograph), and other machines routinely found in the modern analytical laboratory. It may be necessary to resort to one or more of these methods to solve specific problems in metal-finishing analysis, but we confine our mention to three that are most likely to be directly useful to the electroplater. As before, only a brief outline is presented, and more specialized texts must be consulted for details.

*In analytical texts, except those published recently, the wavelength of light used in colorimetric determinations is given in millimicrons, abbreviated mμ. This unit is obsolete, replaced by the SI unit nanometer, 1 nm = 10^{-9} m. Numerically they are identical.

Emission Spectrography

When the atoms in a sample are excited to high energy levels by means of an electric arc or spark, in falling back to their normal levels (ground states) they emit radiation of wavelengths that are characteristic of the elements that comprise the sample. By photographing the *spectrum* so obtained, the spectrographer is able to determine, from the frequency of the spectral lines, the elements present; and the amounts of these elements can be estimated from the intensity of the lines. Such estimation usually is accurate within somewhat less than an order of magnitude; spectrography is therefore a qualitative and semiquantitative tool.

In the cases of sodium and potassium ions, the high energy of an electric arc or spark is not necessary for excitation; these elements emit their characteristic radiation in an ordinary flame, and thus can be determined with the far less expensive spectrophotometer. In fact, flame emission spectrophotometry is the preferred method of determining sodium and potassium.

Atomic Absorption Spectrophotometry (AAS)

This technique has gained increasing popularity owing to its versatility, ease of operation, sensitivity, and (at least relative to most other instruments) economy. It is based on a principle in many ways the opposite of emission spectrography: atoms in their ground state will absorb radiation of specific wavelengths characteristic of the element, and the intensity of this absorption is a function of their concentration. The wavelengths absorbed are the same as those that would be emitted in exciting the element.

For atomic absorption the element does not need to be excited; it need only be dissociated from any chemical bonds. This requires a much lower temperature than excitation. Furthermore, only a small proportion of the atoms in a sample is excited and thereby produces an emission spectrum; a much larger proportion is dissociated from chemical bonds, so that the method is inherently more sensitive.

The essential elements of an atomic absorption spectrophotometer are shown schematically in Fig. 24-6. The radiation source is a hollow-cathode lamp, containing the element to be determined (a different lamp for each element) so that the radiation to be absorbed will be of the proper wavelength.

The atomic vapor is usually produced by atomizing the liquid sample into a flame. This gives sufficient energy to dissociate the compound and to yield an atomic vapor that absorbs the radiation.

The remainder of the instrument is concerned with selecting the appropriate wavelength and rejecting various extraneous radiations, as well as detectors and readout systems.

Almost all metals can be determined by AAS, although the sensitivity varies greatly. The method is convenient and precise, especially for impurities and trace

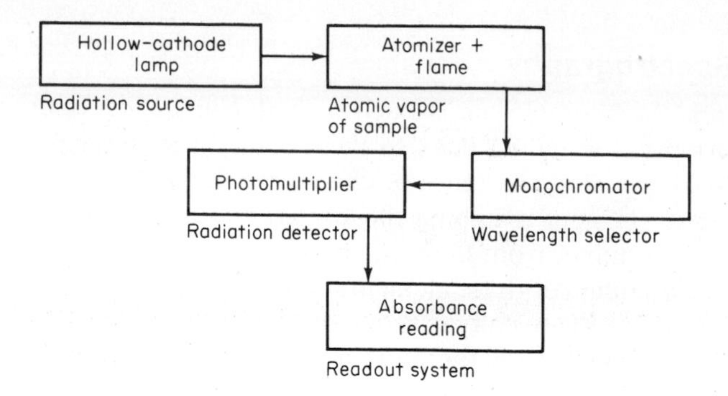

Fig. 24-6 Schematic diagram of atomic absorption spectrophotometer.

elements. It is applicable only to metals; anions of interest to platers, such as sulfate, chloride, and cyanide, are not amenable to this technique.

Specific (or Selective) Ion Electrodes

Just as the glass electrode is sensitive to the activity of hydrogen ions, certain electrode systems have been developed that are sensitive to other ions, so that a system much like a pH meter can be used to determine the activity of the ion to which the electrode responds. This field is in the process of development, and the number of ions measurable by the technique is accordingly increasing. Of particular interest to electroplaters is the ability to determine fluoride ions in this way, since ordinary methods for this determination are tedious. Manufacturers' literature should be consulted for details.

SUMMARY

Finally, to repeat what was said in introducing this chapter, the author's aim has been to alert the electroplater and metal finisher to the techniques of analytical chemistry most likely to be useful in controlling and maintaining solutions and products. In no respect is this discussion a substitute for the analytical literature. Analytical chemistry, not so long ago a neglected backwater of chemical science, is today among its most active areas. Much of this activity has been prompted by the needs for greater and greater sensitivity and precision in determining pollutants, sometimes in the parts per million or billion range, in the waters and the atmosphere, suspected carcinogens in foods, and similar problems of modern society. But whatever the reason, analytical techniques are being improved continually, and the metal finisher should remain alert to developments and adopt, or try out, those that appear to be relevant to the particular problems.

25 Waste Treatment and Metal Recovery

The subject of this chapter is basically a branch of chemical engineering rather than metal finishing; thus detailed consideration is not within our scope. Some awareness of this subject on the part of the metal finisher is, however, necessary. Before reading this chapter, the student should review Chap. 7 on Rinsing and Water. Efficient rinsing procedures will decrease the requirements for waste treatment, and the techniques for producing high-quality water—evaporation (distillation), ion exchange, and reverse osmosis—are used in systems for metal recovery also.

The present concern for the environment has spawned numerous government agencies that have issued regulations concerning the materials that can be legally discharged into the air and water in the effluent from manufacturing establishments. These agencies operate at all levels: federal, state, and municipal. Their regulations have the force of law, once they have undergone the usual formalities of tentative issuance, industry comment, possible compromise, and perhaps even tests in the courts. The electroplating trade is bound by such regulations, as is all of industry. Perhaps because the waste streams from a metal-finishing plant typically contain such obviously toxic materials as cyanides and chromates, the industry was one of the first to come to the notice of government regulatory bodies. The most important of these is the Environmental Protection Agency (EPA).

At this writing, the standards for effluents from finishing plants are still subject to negotiation or legal challenges; furthermore, because the whole field is so active and because new or revised methods of meeting the regulations are being brought forward almost continuously, we shall not consider the problems in detail. The finisher must remain alert to the current literature on the subject, which in addition to its technical aspects involves legal and economic considerations that are outside our scope.

Wastes from a metal-finishing plant originate in at least three principal ways: rinsewaters are discharged; spray from such operations as chromium plating may find its way into the atmosphere through exhaust fans; cleaning solutions regularly, and plating solutions occasionally, must be disposed of when it is no longer possible to adjust them to operable limits. A secondary source of effluent results when the finisher, in an attempt to meet regulations, precipitates the contaminants from the effluent; then there remains a sludge of one sort or another that must be disposed of.

Closely related to waste treatment is resource recovery: instead of merely treating the waste stream to bring it within legal limits, the finisher may (or perhaps must) choose to recover its valuable contents for re-use. Up to now, this has been principally a matter of economics: as long as it is cheaper to buy new chemicals than to recover them from waste, the operator has little incentive to recover and recycle. But at least two factors are rendering recovery attractive, if not mandatory: as the world's resources of high-grade ores are depleted, the price for all metals is bound to go up, bringing the cost of recovery more and more in line with that of purchase. More immediate is the strong probability that dumping of solid wastes will be forbidden by law, as land sites for disposal are filled up and regulations become stricter. The option of precipitating the regulated metals as sludge for subsequent land disposal may no longer be open, and the finisher will have no choice but to recover and re-use the metals in the effluent. This operation is referred to as *closed loop,* and it is being discussed ever more seriously.

The first step in waste treatment is to decrease, to the extent possible, the need for it. This entails efficient rinsing, installation of drain stations ahead of all tanks so that drag-out is returned to the preceding tank rather than carried over to the next, use of spray and fog rinses, and use of plating and other solutions in as dilute a form as is compatible with satisfactory performance.

In spite of precautions, some waste must be treated to reduce the pollutants in the effluent to legally mandated levels. Many schemes have been proposed; each has advantages, and the one most suitable for a given plant must be determined individually.

There are two approaches to compliance with government regulations (the third approach—ignoring them—is no longer possible). Plant managers may choose to treat effluents so as to render them fit for discharge by getting rid of the offending contaminants, or they may attempt to recover those same contaminants for sale or re-use in their own plants. For convenience, these approaches are termed *destruction* and *recovery*. They are related, but not identical.

DESTRUCTION

In a sense, this term is a misnomer, since metal-bearing wastes cannot be "destroyed." They can only be put in such form that their disposal is legal. This usually means that they are transformed from the soluble ionic state into insoluble precipitates; these precipitates, in the form of wet sludges, are then disposed of on land (or possibly dumped into the ocean). Advantage is taken of the insolubility, at some given pH, of most metal hydroxides; the waste effluent is adjusted to this pH, and the resulting hydroxide precipitate is either filtered or settled. The resulting sludge must usually be dewatered, to some extent, and hauled away, lagooned, buried in landfills, or otherwise disposed of.

Some wastes may be actually destroyed by burning: incineration may solve some of the waste-disposal problems connected with organic solvents and similar materials, but even this method is restricted by regulations limiting the amount of soot and ash that can be discharged, and other by-products of incineration such as hydrogen chloride and sulfur gases (from chlorine- and sulfur-containing organics) must be taken into account.

Of the toxic substances routinely handled by the electroplater, only cyanide is at present satisfactorily handled by truly destructive methods. Under appropriate conditions, cyanide ion may react with oxidizing agents to produce, finally, carbon dioxide and nitrogen, both innocuous.

Cyanides

The most commonly used oxidant is chlorine, either as such or in the form of hypochlorite ClO^-. Also used are peroxygen compounds in conjunction with formalin (an aqueous solution of formaldehyde HCHO); electrolytic oxidation has been suggested.

Oxidation of cyanide ion by chlorine is generally regarded as taking place in three steps, the first of which is a fast reaction taking place at almost any pH; the next two are slower, and depend on the pH of the solution.

1. $CN^- + Cl_2 \rightarrow CNCl + Cl^-$
2. $CNCl + 2OH^- \rightarrow CNO^- + Cl^- + H_2O$
3. $2CNO^- + 4OH^- + 3Cl_2 \rightarrow 6Cl^- + 2CO_2 + 2H_2O + N_2$

The first oxidation to cyanogen chloride is, as stated, a very fast reaction that occurs at all levels of pH. Reaction 2, hydrolysis of cyanogen chloride to cyanate ion, depends on pH and is fastest at a pH of 11.5 or higher, requiring a few minutes. At lower pH values the reaction is considerably slower, and since cyanogen chloride is toxic, low pH should be avoided. Step 3, oxidation of cyanate to nitrogen and carbon dioxide, is accelerated by decreasing pH; it takes 10 to 15 min at pH of 7.5 to 8, and at least 30 min at pH 9 to 9.5.

In some cases step 3 may not be required, since cyanate is about 1000 times less toxic than cyanide ion and may be tolerated by some municipal sewage treatment plants. For most purposes, however, the reaction must be taken all the way to carbon dioxide and nitrogen; this reaction at pH 7.5–8 is fairly slow and may require up to 1 h at room temperature unless excess chlorine is used.

The cyanide content of zinc- and cadmium-plating rinses, or of alkali cyanide solutions, reacts rapidly and efficiently according to these equations; but when cyanide is complexed with copper, nickel, or the precious metals, whose complexes are much more stable, excess chlorine over the stoichiometric amount must be used. In the extremely stable ferro- and ferricyanide complexes, the cyanide is hardly affected at all by the chlorination treatment.

Electrolytic decomposition has been suggested.*

Formalin plus a peroxygen compound has been recommended more recently as a replacement for chlorine in the oxidation of cyanide. The process uses a proprietary material, consisting principally of hydrogen peroxide with special additives to improve stability and flocculation of precipitates. The chemistry is complex; end products include ammonia, sodium cyanate, and glycolic acid amide, $CH_2OHCONH_2$. Advantages claimed include the elimination of chlorine (which may be deleterious in some forms of water treatment) and the biodegradability of the organic materials produced.

Oxidation with ozone has been suggested.

Heavy Metals

In "destructive" systems, heavy metals are usually precipitated as their hydroxides; since these precipitates are not usually stoichiometric hydroxides $M^n(OH)_n$ but oxides of rather indefinite composition with various amounts of more or less loosely bound water, the term *hydrous oxides* is more precise.

Most metal hydroxides are more or less insoluble, and therefore the metals can be precipitated by adjusting the pH to an appropriate value, which differs according to the metal and is most critical for amphoteric metals which dissolve at both low and high pH. Where the plating solution contains a cyanide complex, destruction of cyanide will free the metal from the complex and allow its precipitation. Some of the more recent noncyanide baths may pose a problem in this respect, in that the complexes may be so stable that precipitation with hydroxyl ion is possible only after special treatment, which will differ for each individual bath. Vendors of these almost universally proprietary solutions should be consulted concerning waste-disposal methods.

Chromium is a special case. In the form of Cr(VI) not only is it highly toxic and therefore stringently regulated, but no insoluble hydroxide is available. Some waste-treatment schemes have proposed precipitation as insoluble barium chromate without pretreatment, but such systems have not been notably successful. The usual scheme involves reduction of Cr(VI) to Cr(III) with a reducing agent such as sulfur dioxide or equivalent compounds such as sodium sulfite or bisulfite, followed by precipitation of Cr(III) hydroxide. Reactions may be formulated as

$$3SO_2 + 4H^+ + 2CrO_4^{--} \rightarrow 2Cr^{3+} + 3SO_4^{--} + 2H_2O$$

or

$$3SO_3^{--} + 10H^+ + 2CrO_4^{--} \rightarrow 3SO_4^{--} + 2Cr^{3+} + 5H_2O$$

Instead of the more usual sulfur dioxide or its derivatives, ferrous sulfate may

*See, for example, M. R. Hillis, *Trans. Inst. Metal Finishing,* **53,** 65 (1975).

be used to reduce Cr(VI) to Cr(III) for precipitation. Use of this compound results in the production of considerably more sludge, since both the iron and the chromium are precipitated. In some cases, however, the availability of cheap ferrous sulfate (or ferrous chloride) from spent pickling liquors may recommend its use.

Metal hydroxides (or hydrous oxides) are in general gelatinous, slow to settle, and difficult to filter. The sludges formed are very watery, often containing 5 percent solids or less, and their disposal presents problems.

Following precipitation of the various metals by pH adjustment, the precipitates must be settled or filtered, and the resulting sludges dewatered, at least to some extent. Where land disposal is permitted and sites are available, the sludge is then simply carted off to landfills. Lagooning—pumping of thin slurries to large settling ponds, where in time the solids settle to the bottom and the supernatant liquor either evaporates or can be allowed to drain away—is also used; it requires large amounts of land. Use of flocculating agents, filter aids, thickeners, and other expedients to speed settling or filtration may be indicated. How much dewatering is required will depend on the method of disposal: some dumps will accept sludges only if their water content is at a set maximum.

Alkali for precipitating metal hydroxides may be soda ash (sodium carbonate), slaked lime (calcium hydroxide), or caustic soda (sodium hydroxide). The choice is principally an economic matter, and may be influenced by the settling properties of the resulting gelatinous precipitates.

RECOVERY

Under present conditions, precipitation of metals and the disposal of their gelatinous hydroxides as landfill present problems in settling and disposal of the sludge. The availability of suitable landfill sites is becoming more and more questionable, and it seems certain that this method of disposal will not be tolerated in the future. If the various metals are segregated, sale of these sludges to a refiner or smelter for recovery of their values is a possibility, the practicality of which will vary according to geographic location, freight rates, intrinsic value of the metal, refining costs, and other factors. But it seems inevitable that the final answer will lie in so-called closed-loop operation: recovery on site of metals now being wasted, for re-use in the process or in some allied process.

Several methods of recovery and recycling have been suggested; all are being intensively researched, and, as with the other aspects of the subject of this chapter, current literature must be consulted for up-to-date information. Four processes have attracted the most attention; in many cases it may turn out that a combination of processes rather than just one will provide the best answer to the problem.

Reverse Osmosis

See also Chap. 7. The semipermeable membranes required for this process are being improved, and manufacturers continue to offer membranes with longer life and superior selectivity. Life of the membrane is one of the principal determinants of the economic feasibility of the method.

Reverse osmosis (RO) is often combined with evaporation to render the process adaptable to re-use of the effluents.*

Electrodialysis

Somewhat similar to reverse osmosis as a recovery technique is *electrodialysis;* electrodialysis cells are semipermeable membranes that contain ion-exchange sites, as do ordinary ion-exchange resins. An electrodialysis cell consists of a sandwich of alternating cation- and anion-permeable membranes. Application of electric current causes positively charged ions to pass through cation-permeable membranes; negatively charged ions move in the opposite direction and pass through anion-permeable membranes. The water in the center chamber of each membrane sandwich is thus depleted in salt, while water passing through the adjacent chambers is enriched in salt.

Ion Exchange

The principles of ion exchange (IX) have been described in Chap. 7, where the process was discussed in connection with water purification. The same principles can be applied to removing cations such as heavy-metal ions from wastewater, and subsequently eluting them from the IX resins in more concentrated form. The eluate still may not be sufficiently concentrated for direct use in a plating bath, so that the process is often used in connection with evaporation, which not only concentrates the solution to plating strength but yields pure water for plant use. Disposal of spent regenerating solutions may pose its own problems, but designers of ion-exchange systems can offer appropriate answers.

Evaporation

The principles of evaporation are in concept quite simple: the solution is simply raised to a temperature (which depends on the pressure of the system) at which the solvent water is boiled off, to be recondensed as pure water, leaving the solution sufficiently concentrated for efficient use of the metals or compounds contained. Simple in concept, the design and operation of evaporative recovery plants can be fairly complex in execution, and the economics can be quite as involved. Evaporation is a basic unit operation in chemical engineering, and as such its detailed discussion is not within the scope of this book; its application to recovery of metals in plating operations has been extensively reported in the current literature.

*R. G. Donnelly, R. G. Goldsmith, K. J. McNulty, and M. Tan, *Plating,* **61,** 432 (1974).

Contaminants

All closed-loop systems suffer, at least conceptually, from the disadvantage that recycling returns to the system not only the valuable constituents but also impurities and contaminants that would normally be kept under control by drag-out. In designing for recycle-recovery, such impurity buildup must be taken into account: recycle may be deliberately limited by allowing a certain proportion of the drag-out to go to waste (treated), or other methods of purification must be provided for.

DESIGN OF SYSTEMS

The treatments discussed in this chapter can be carried out in batches or continuously, with little or complete instrumentation, and either with constant supervision or almost completely automated. In general, metal finishers should not attempt to design their own waste-treatment plants; firms in this business, which has become highly specialized and is usually not handled by the regular plating supply houses, should be called upon for advice and actual installation, and the often conflicting claims for the superiority of one system or another must be carefully evaluated.

For orientation, Figs. 25-1 to 25-8 are offered as illustrations of the types of plant appropriate to several schemes for cyanide decomposition, chromate reduction, and recovery systems.

Pollutants other than those already mentioned also result from metal-finishing operations: shop dirt, oils and greases, buffing-compound residues, and the like; phosphates from cleaning baths must be controlled. Methods for removing such contaminants do not differ from those used by industry in general, and texts on water treatment should be consulted.

Many finishing shops are too small or too poorly capitalized to afford the sometimes expensive equipment mentioned here. There are springing up various contract disposal concerns who offer to cart away and dispose of waste material; their services may be of value to smaller establishments. These firms collect the

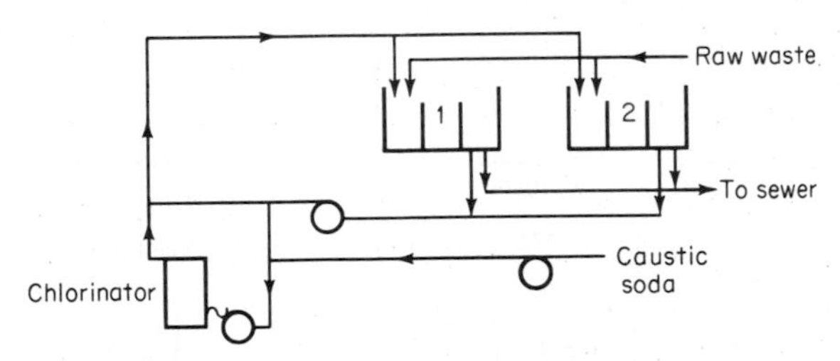

Fig. 25-1 Manual batch cyanide system. Key: 1,2—treatment tanks; 0—pumps.

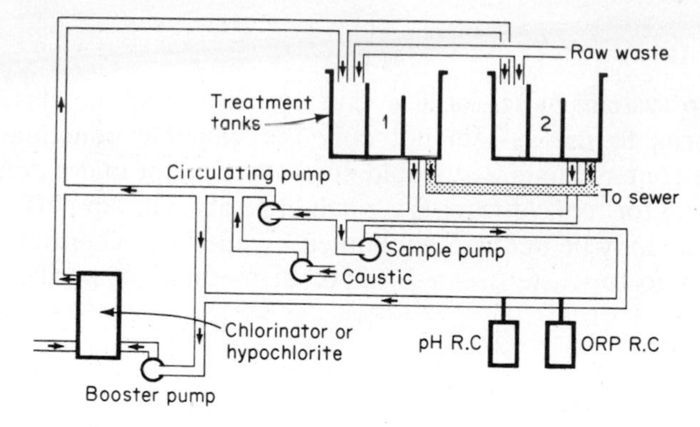

Fig. 25-2 Instrumented batch cyanide system.

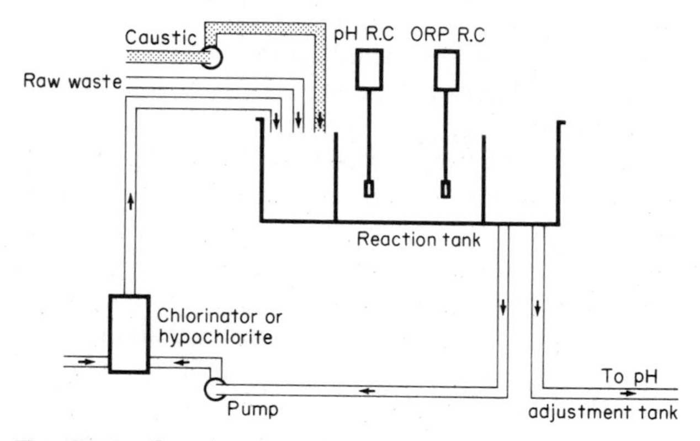

Fig. 25-3 Continuous cyanide treatment system.

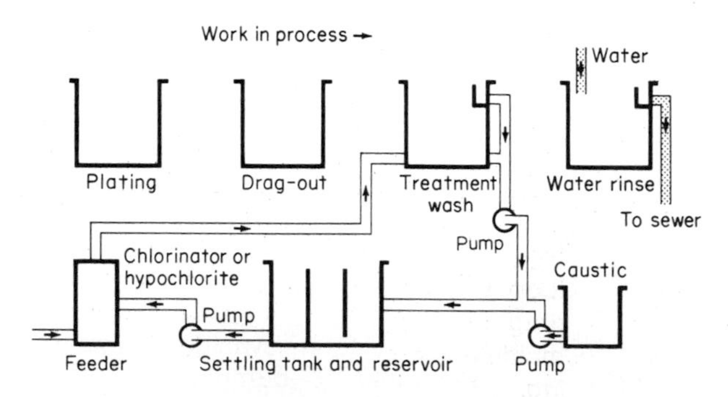

Fig. 25-4 "Integrated" cyanide treatment system.

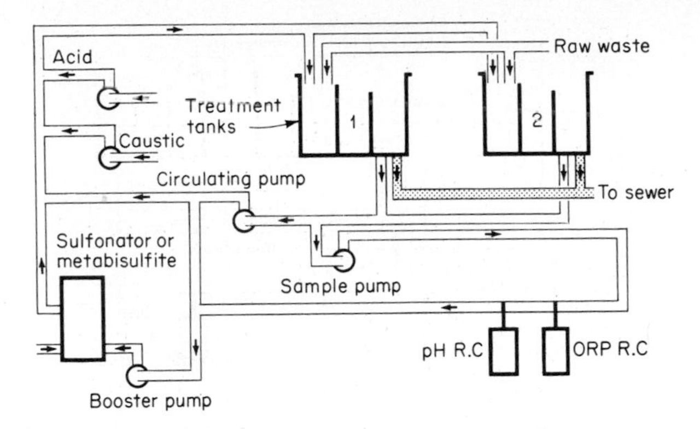

Fig. 25-5 Instrumented batch chromium waste-treatment system.

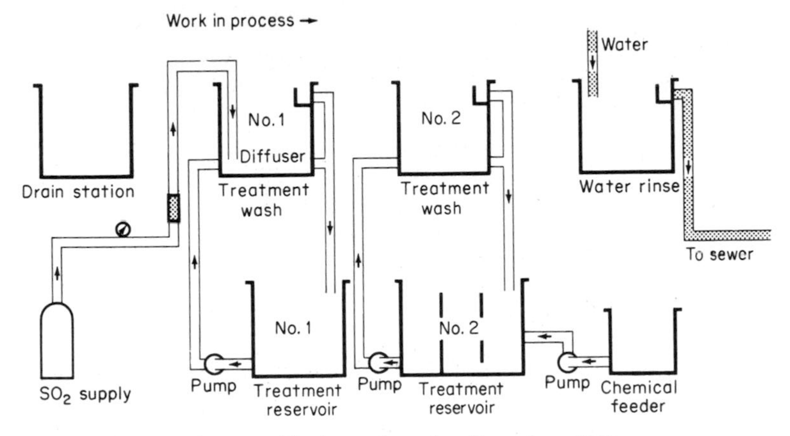

Fig. 25-6 "Integrated" chromium treatment system.

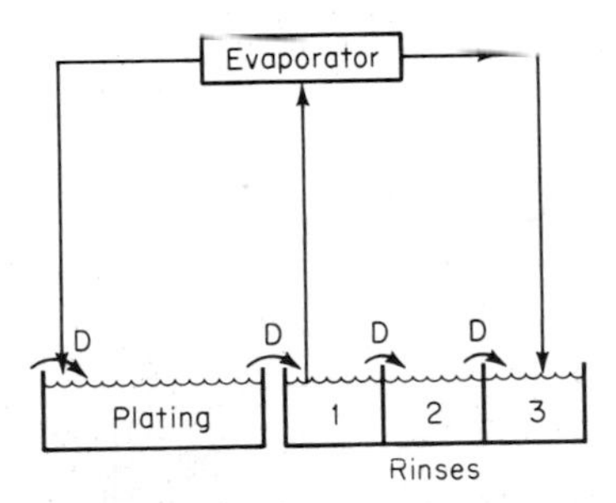

Fig. 25-7 Simple closed-loop evaporative system; D = drag-out; evaporator may be multiple-effect. [From *Plating,* **55**, 959 (1968). *Reproduced by permission.*]

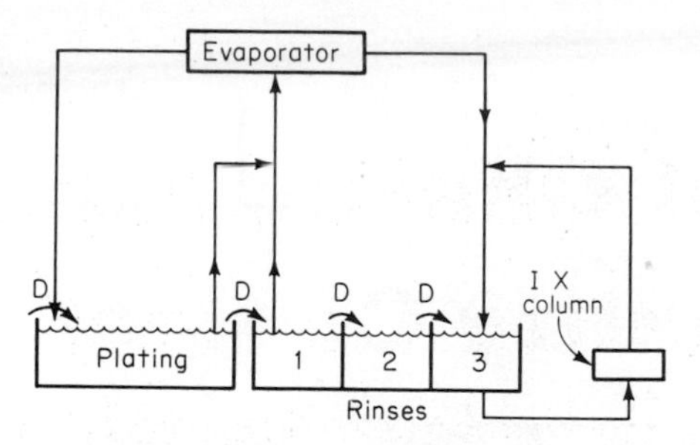

Fig. 25-8 Evaporative recovery and demineralization of final rinse.

effluent from many plants in their area and treat the combined wastes. (Some may even give the finisher credit for recoverable metals.) Although many offer certificates indicating that the wastes have been disposed of in accordance with applicable statutes, legal responsibility may still rest on the generator of the waste. No firm conclusions can be stated here, but the finisher should by all means retain expert legal advice.

Waste disposal and resource recovery are of tremendous importance and, as far as the business aspects of metal finishing are concerned, are paramount in the thoughts of many. The relatively short shrift given the topic here is due only to the fact that it is peripheral to the purpose of this book, which is the science of metal finishing, not its economics.

26
Control Charts, Sampling, and Quality Control

It is the purpose of this chapter only to call the attention of finishers to those aspects of their work that require the application of some kind of statistical knowledge. In this short space it is not possible to consider the mathematical background.

Statistics is a powerful tool for communicating and analyzing data, provided the data have resulted from appropriately chosen observations. The results of such statistical analysis can help manufacturing and purchasing departments:

To correct a process that is not producing up to its capabilities, i.e., is not in control

To improve an existing process or replace it with a better one

To leave a process alone if it is operating properly.

These ends are achieved through a branch of statistics called quality control, which is the technique of using numerical data to keep a process within reasonable bounds. This and other branches of statistics offer guidance in:

1. Process control: a system that sounds an alarm and detects when the process is deviating from its accepted limits and requires attention
2. Determination of process capability: what can be expected of a process if it is in control
3. Sampling inspection: a technique for avoiding 100 percent inspection of either production (by the manufacturer) or receipts (by the purchaser) in determining whether to accept or reject work coming off a production line

4. Experimental design: a technique for designing experiments so as to obtain the maximum information from the minimum amount of experimental data.

The need for considering statistics arises from the nature of almost all manufacturing operations, including metal finishing: a manufacturing process by its very nature does not produce items that are absolutely identical; they vary over wide or narrow ranges in their physical, mechanical, and dimensional characteristics. That is the way mass-production operations work, and except for extremely critical items, deviations from the design dimensions must be not only expected but tolerated. How large these deviations may be is set by the demands of users; they can tolerate a certain amount of variation without endangering the quality of their products, but pieces that fall outside their *tolerance* limits may not be suitable for the use and will lead to substandard performance in their product.

If the electroplater is required to deposit 25 μm of a metal, there is virtually no chance that all the parts plated will have exactly that thickness; some will have more, some less. As we have seen, even on a single part, the thickness of deposit will vary from place to place unless extremely stringent precautions are taken to ensure a uniform deposit. Some of these variations will be due to "chance," which is a way of saying that there are a number of variables inherent in any practical process over which the operator has no control. Some of the variation will be due to the inherent error in the method of measurement; we have seen that thickness-measuring techniques are at best accurate to only about ± 5 percent, and that figure is somewhat optimistic. But some of the variations may be due to errors, inadequate control, and other preventable causes. One function of statistical techniques is to separate the unavoidable variations from the avoidable ones, to ensure that the process is "in control." This aspect has to do with *control charts*.

Another important use of statistical techniques in electroplating, as in other manufacturing operations, relates to *sampling*. A sample is a set of elements drawn from a population in such a way that its characteristics will represent those of the entire population from which it is derived. Sampling is a means of selecting from this population a smaller group which, if proper procedures are followed, will represent the total population to any degree of accuracy required. The errors in assuming that a sample truly represents the whole can be as small as we care to make them, depending on how much trouble we are willing to take.

Sampling is carried out by both manufacturers and purchasers. Manufacturers—electroplaters in this case—sample to be sure a process is in control, to estimate costs, and to guard against shipping material that the purchaser will reject; this last may cost them a remunerative contract, and at best will cause unnecessary expense. It is better to discover rejects for oneself, and remedy the cause at the tank, than to have one's customer discover them, send back the material, and force the plater to rework it.

The purchasers (who may, or course, be simply another department of the same plant) carry out inspection for similar reasons: to ensure that the material they accept and pay for is indeed within their specifications and to guard the final quality of their products.

Both the manufacturer and the purchaser might ensure their aims by 100 percent inspection of the product. This is either impossible or impractical. It is impossible when inspection is destructive. A manufacturer of light bulbs wishes to represent them as being good for 1000 h of use; but if each bulb were tested to failure, to back up the claim, there would be no product to sell. Sampling is absolutely necessary in this case and in many similar ones. Tests for adhesion, ductility, corrosion resistance, and other properties of interest to metal finishers are destructive, and the only means of quality control and quality assurance is sampling inspection.

But 100 percent inspection is impractical in most cases even if the tests involved are nondestructive. Imagine testing the plate thickness on every zinc-plated screw in a barrel load! Inspection is usually regarded as an overhead cost, being nonproductive, and the more items that are inspected, the higher is this cost. The statistics of sampling, therefore, aims to arrive at a compromise between minimizing the cost of inspection and minimizing the chances of both accepting out-of-specification production and rejecting good material.

Sampling therefore always entails *some* risk—the purchaser's risk that some unacceptable lots will be accepted and the producer's risk that acceptable lots will be rejected. The aim of sampling plans is to equalize these risks in a fair manner, which is subject to negotiation between the manufacturer and the purchaser, or possibly set by the purchaser. In fact, even 100 percent inspection does not entirely eliminate risks, because even inspectors are human and make mistakes. Individual sampling plans are discussed later in this chapter.

The science of statistics is based on the mathematics of probability, first adumbrated some centuries ago by mathematicians to enable their noble patrons to judge their chances in gambling games. Building on these foundations, statisticians have developed powerful tools for judging the significance of experimental results, for designing experiments and sampling plans that will reduce, to any practical degree, the risks of judging a whole by its parts—under certain circumstances.

A third area in which a statistical approach to experimental results can be helpful is in the evaluation of new processes or changes in old ones. The metal finisher is bombarded with claims from vendors that their processes or products are better than their competitors', and indeed some of these claims are true, and some are partially true. In response to such claims, the finisher may decide to try out the new process and evaluate the results, perhaps by estimating the percentage of rejects from the new vs. the old or by other methods appropriate to the claims made. Then two questions must be answered: Is the difference between the new and the old process statistically significant? And is it important?

The statistical approach can help answer the first question but not the second. The matter of statistical significance has been studied most carefully in connection with the evaluation of new medical cures, but it is equally applicable to any question in which the results tend to scatter somewhat and are not clearly decisive on their face. It often turns out that the supposed differences between two competing processes are most probably due merely to chance and there is no significant difference between the two.

The fact that a difference between two products or processes turns out to be statistically significant does not prove that it is important. Process *A* may yield a cathode efficiency of 95 percent and process *B* 95.5 percent, and it is quite possible that this small difference is real—statistically significant—but it may not be very important: not worth changing one's manufacturing methods. That is a judgment factor which only the user can evaluate and which has nothing to do with statistics.

This chapter will not attempt a complete discussion of all the subjects that have been introduced; but it is necessary to consider two aspects that are fundamental to all statistical calculations and sampling plans: the normal distribution curve and the concept of randomness.

THE NORMAL DISTRIBUTION CURVE

Most experimental data and results are not pure numbers but approximations. If one measures the thickness of deposit on a substrate at exactly the same place several times, it is unlikely that the results will be exactly the same; and if one measures the thickness of a succession of pieces, all plated under the same conditions so far as it is possible to make them the same, the results will vary. If one analyzes the same sample of material for any given constituent, the analyses will not agree exactly. Most workers know intuitively that for reliance on an analysis, or a thickness reading or any other measurement, it is good practice to repeat the operation more than once, and to take an *average* (called the *arithmetic mean* in statistical work, to distinguish it from other types of central tendency such as the mode, the median, the harmonic mean, etc., which are not of primary interest here). The more one repeats the operation, the more faith one can place in this average result; on the other hand, repeating analytical determinations is expensive, and it is necessary to trade utmost precision vs. cost; this is determined by the use to be made of the results.

In any series of measurements—let us assume they are measurements of the thickness of zinc on steel parts, all alike and plated during one day in one tank under presumably constant conditions—the values will tend to cluster around a mean or average value. If the desired thickness is 25.0 μm, perhaps we shall obtain individual readings such as these (in μm): 25.1, 25.2, 25.0, 25.3, 24.9, 24.8, 24.3, 25.4, 25.0. The average, or arithmetic mean, is the total of these readings divided by their number: 225.0/9 = 25.0 μm. (This result is remarkably good; one should not expect such close agreement with specification in practice.) It is now necessary to calculate another important variable, the standard deviation. This is calculated as follows: Subtract the average from each individual value, neglecting sign. The individual deviations are 0.1, 0.2, 0, 0.3, 0.1, 0.2, 0.3, 0.4, 0. Now square each of these numbers: 0.01, 0.04, 0, 0.09, 0.01, 0.04, 0.09, 0.16, 0. Sum these squares, yielding 0.44; divide by the number of observations (some workers use one less than this number); this yields a number called the *variance*. Now take the square root of this number to derive the standard deviation:

$$\text{Variance} = 0.44/9 = 0.0489$$

$$\sqrt{0.0489} = 0.0699 \text{ (approx. 0.07)} = \text{standard deviation}$$

In symbolic terms:

$$\bar{x} = \Sigma x/n$$

where $\bar{x}$ = arithmetic mean or average (pronounced "x-bar")
Σ = the sum of
x = individual observations
n = number of observations

Also

$$s^2 = [(\bar{x} - x_1)^2 + (\bar{x} - x_2)^2 + (\bar{x} - x_3)^2 + \cdots + (\bar{x} - x_n)^2]/n$$

where s^2 = variance

and

$$\sigma = \sqrt{s^2}$$

where σ (sigma) = standard deviation

In summary,

$$\sigma = \sqrt{\frac{\Sigma(\bar{x} - x)^2}{n}}$$

Then we might express the result of our measurement as 25.0 ± 0.07 as giving not only a measure of the average result, but an approximation of how well the individual results cluster about that average.

For example, assume another set of measurements on a similar batch of zinc-plated steel: 25.9, 26.0, 24.0, 25.5, 23.9, 25.8, 23.8, 26.2, 23.9. The average is exactly the same as before, 25.0. But if we carry out the same operations on these figures, the separate deviations from the average are: 0.9, 1.0, 1.0, 0.5, 1.1, 0.8, 1.2, 1.2, 1.1. Squaring each, we get: 0.81, 1, 1, 0.25, 1.21, 0.64, 1.44, 1.44, 1.21. The sum of the squares is 9.0; dividing by the number of observations yields 1.0, the square root of which is 1.0. Therefore our new measurement is 25.0 ± 1.0 μm, the same average but a much wider spread or standard deviation.*

The significance of the average and the standard deviation for statistical

*For truly valid statistical treatment, nine observations are hardly sufficient for the laws of probability, upon which our statistics are based, to apply; probability is the mathematics of large populations. Our calculations are simplified for illustration.

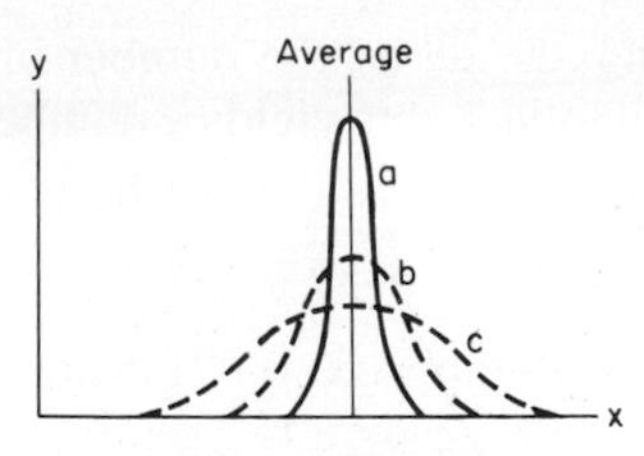

Fig. 26-1 The normal distribution curve: three curves with the same average but different standard deviations.

control and sampling is that most observations, if they conform to a few simple rules, follow what is called the *normal distribution;* that is, they tend to cluster about the mean or average. The number of observations on either side of this mean will be about the same; and the further away from the mean they are, the fewer the measurements will be. Graphically, they yield the familiar bell-shaped curve, often called the *error* or *Gaussian curve;** see Fig. 26-1. This figure shows three curves, all with the same mean value, but differing in height. The data for the steepest, *a,* cluster most closely around the average; the others, *b* and *c,* have the same average value but the observations are more scattered around it; their standard deviations are greater.

For statistical purposes, the important thing about this curve, and the standard deviation σ, is that under proper circumstances (which we come to shortly) 68 percent of all the observations will be found within one standard deviation from the mean (Fig. 26-2); 95 percent will be found within two standard deviations, and 99.7 percent within three standard deviations on either side of the mean. The remaining 0.3 percent will be what we ordinarily think of as "wild" points, constituting the "tails" of the curve. In fact they are not wild at all, and will be found 3 times out of every 1000 measurements, on the average.

The normal, error, or bell-shaped curve is not necessarily characteristic of all real distributions found in nature. The curve may be "skewed," i.e., one tail may be longer than the other, or perhaps missing entirely. For example, in a survey of the number of children in the "average" family, the lower limit is zero, while there is no upper limit; although families with more than ten children may be rare, they are not unknown. Such skewed distributions can be taken care of in statistical calculations, but will not be discussed here. Such a curve is illustrated in Fig. 26-3.

RANDOMNESS

The whole basis of statistics rests on the assumption that in any sampling of a population, as the word has been defined previously, every member of that

*The famous mathematician Gauss was not the first to propose it; it is usually credited to de Moivre.

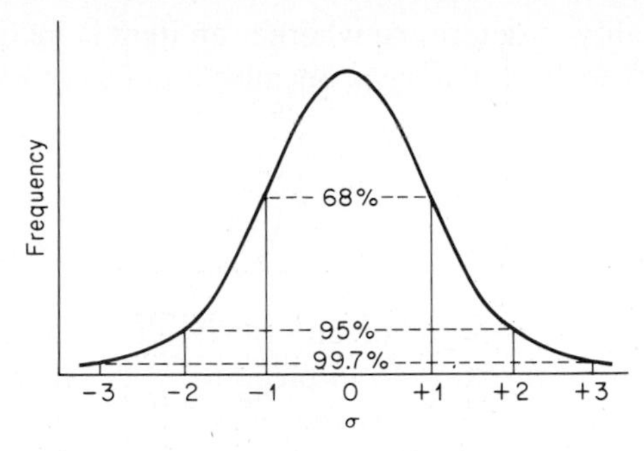

Fig. 26-2 Properties of the normal distribution curve.

population has an equal chance of being included in the sample; in other words, the sample is completely *random*. If this fundamental requirement is not met, the whole statistical treatment breaks down. In many aspects of statistical sampling, this randomness is one of the hardest things to achieve. In public opinion polls it is the main problem. Randomness in sampling of manufactured articles may be achieved in several ways.

As the articles are coming off the assembly line—in the case of rack-plating, for example—take every tenth (or some other number) item and choose it as a member of the sample. A danger here is that if the racks happen to hold ten pieces, every tenth item will come from the same position on the rack, and randomness will not be achieved. Tables of random numbers are available.* But what must be guarded against is that the person taking the sample does not use

*For example, in ASTM B 602, mentioned later in this chapter; directions for use of the table are also included.

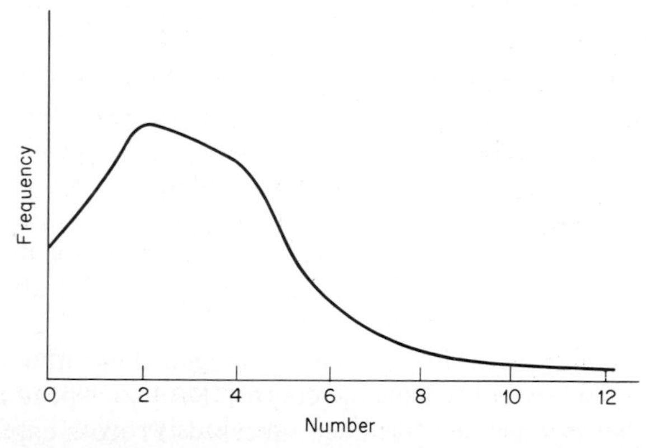

Fig. 26-3 Skewed distribution (number of children per family).

any judgment of quality to determine whether an item is included; as soon as the sample taker decides to discard a piece because it does not look right, the sample is biased.

CONTROL CHARTS

The data on thickness of deposit, or any other measurable characteristic of a plated article, will show variation from piece to piece; if these data are plotted with the individual measurements as the ordinate and time as the abscissa, a fluctuating line will be found such as could be drawn through the points of Fig. 26-4. (We could also plot the percentage of rejects from time to time and derive a similar chart.)

Any well-run, stable process will show such a normal fluctuation; no process is capable of producing a number of items all of which are *exactly* alike, unless the method of measurement is so crude as to hide these differences.

To construct a control chart, one simply draws a horizontal line representing the desired value, or the central line resulting from taking the mean value as previously calculated, as being the ideal value that is the aim of the process; such a central line may also represent the percentage of rejects one will tolerate from an operation, such as the percentage of items having a hazy chromium deposit or the number of cut-throughs from a buffing operation. Draw another line above this central line, representing the "upper control limit," and another horizontal line the same distance below, representing the "lower control limit." These are the lines between which almost all the measurements should fall if the process is in control.

We have seen that 95 percent of all the items in a distribution that follows a normal curve will fall within 2σ of the average, and 99.7 percent will fall within 3σ. If the process producing these items is in control (that is, producing only the normal variation characteristic of the process, which cannot be avoided by any known means), items coming off the line will give measurements as a rule between 2σ above and 2σ below the nominal or desired measurement—thickness of deposit, for example. A few more will be outside the 2σ limits but within 3σ of the nominal. And about 3 out of every 1000 will be outside even 3σ. With this knowledge, the operator can set up a control chart, with the nominal or average value of the measurement as the middle line and with lines 2σ and 3σ above and below it. Then the values of the samples, taken as indicated above, are plotted sequentially, with time as the abscissa. If the process is in control, these values will cluster about the average value, about as many above as below, mostly within 2σ and a few between 2σ and 3σ.

The natural capability of a process can be judged only after it has been in operation for some time under reasonable control. Realistic specifications can be written only after such capability has been assessed. Process capability may be expressed either as percentage of product out of specification or as a distribution

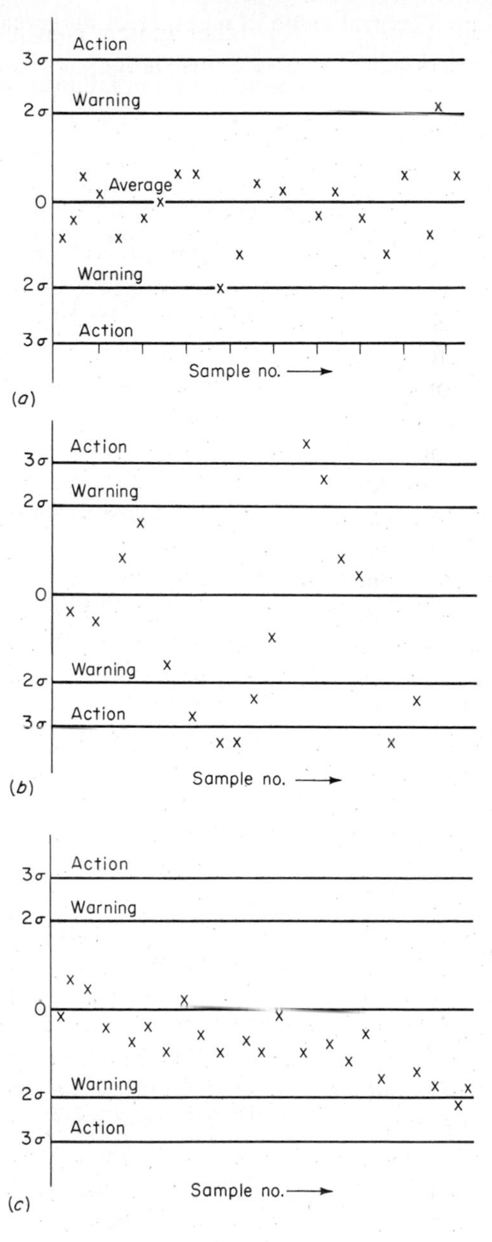

Fig. 26-4 Control charts: (*a*) normal; (*b*) out of control; (*c*) in control, but drifting.

of values measured by a central value, a measure of dispersion or spread, and a figure for the standard deviation.

Control charts, maintained over a period of time, fall into two main categories.

Normal Control Charts

Most of the points cluster about the central line; a few points spread out, and there are occasional ones at or near the control limits. Only a very occasional point is outside the control limits. And there are about an equal number above and below the central line. The process is in control: leave it alone.

Abnormal Control Charts

These are characterized by one or more characteristics:

1. Few points near the central line. Perhaps batches from two different plating tanks are being mixed.

2. Few points away from the central line and none near the limits. Perhaps the control limits are too wide: the process is capable of better precision than it is being given credit for. Other causes may include too crude a measuring device or, in the case of sampling for rejects, some bias in the sampling procedure.

3. Many points outside 2σ but within the control limits of 3σ. This is a warning that the process, although technically in control, is on dangerous ground. It is a signal that some action may be necessary.

4. Many points outside the control limits. The process is out of control, and immediate steps should be taken. Many causes are possible: temperature controllers out of order, careless operators, rectifier breakdown, etc. Whatever the reason, this is an alarm signal that must be answered. As soon as one point is seen to be out of the control limits, the rate of sampling should be increased. Perhaps it was just one of those 3 in 1000 that is to be expected even from a well-controlled process; if so, the next few points will be back in control. But if several points continue to fall outside the limits, remedial action must be taken.

5. Points tend to drift. Points remain within the control limits, but the average tends to drift up or down. The process is gradually changing, and the reason should be ascertained. It may be unavoidable, such as gradual wear of a tool, but this is not likely in electroplating operations.

6. A certain number of successive points are all on the same side of the central line. (Numbers commonly used are 7 in succession, 10 out of 11, and 12 out of 14). This is also a sign of gradual shift in the process, and the cause should be looked into.

In order to set the central line of the control chart, it is necessary to have some data to go on; generally these are gathered by setting up the process carefully, to ensure its best performance, and taking at least 20 samples, preferably more, to arrive at the central line, or average value. Further samples may cause one to alter this average, if it appears that the process is changing slightly. It is best to recompute the central line occasionally, to be sure that the process has not shifted abnormally.

SAMPLING FOR INSPECTION

An important application of statistical techniques, as mentioned previously, arises in dealings between the manufacturers (the metal finishers in this case) and the purchasers of their products or services. Presumably the purchasers have laid down certain requirements involving deposit thickness, appearance, solderability, corrosion resistance, or other properties. The finishers have produced the articles, and the problem now is to find a means whereby it can be determined whether the articles meet the specifications. For very critical or expensive items, this determination might be made by inspecting each item, provided that such inspection is nondestructive. We have already seen that for most transactions, 100 percent inspection is, to say the least, impractical. We must inspect a *sample,* with the hope that this sample tells us what we want to know about the whole shipment, or lot, of items. The problem then involves two questions: (1) How big a sample is needed? and (2) What percentage of substandard items may be permitted in the sample for a given quality level in the whole population?

This question has had the attention of statisticians for many years, and various types of sampling plans have been devised. Like other statistical problems, it is based, finally, on the laws of probability. Assume that we have a mixture of white and black balls, all alike except for color, in a bag. The white balls are "good" or acceptable items, and the black balls are rejects, or unacceptable items. The purchaser, being realistic, knows that in a large shipment there are bound to be a few "black balls;" a lot will be acceptable if it contains a given percentage, x percent, of rejects. In this case x is called the *acceptance quality level,* or AQL. Typical AQLs are 1.5, 4, and 5 percent.

This problem has been addressed in various ways, and the answer depends on several factors: what is the AQL, and what is the probability that a defective lot will actually be rejected and that an acceptable lot will actually be accepted?

Sampling plans may specify the sample size, based on lot size, as well as the number of defective items permissible in the sample for the lot to be considered acceptable. This is the approach taken by the ASTM method of sampling, B 602, summarized below.

It is not always necessary, in a sampling plan, to reject defective lots out of hand; in some plans, such lots are subjected to 100 percent inspection. Further discussion of the subject may be found in several ASTM documents: E 122, Choice of Sample Size to Estimate the Average Quality of a Lot or Process; E 105, Probability Sampling of Materials; and E 141, Acceptance of Evidence Based on the Results of Probability Sampling.

Still other sampling plans are documented in government specifications: MIL-STD-105D, Sampling Procedures and Tables for Inspection by Attributes; and General Services Administration Handbook FSS P 4440.1, Guide for the Use of MIL-STD-105D.

With particular reference to electroplated items, ASTM B 602, Sampling Procedures for Inspection of Electrodeposited Metallic Coatings and Related Finishes, specifies sample sizes and acceptance limits for destructive and non-destructive tests, as illustrated in Tables 26-1, 26-2, and 26-3, for, respectively, items produced by methods other than barrel plating; for barrel-plated items;* and for sampling for destructive tests. When tests must be destructive (such as

*The reason for different criteria is that barrel plating, in general, cannot be controlled as closely as rack plating and some additional leeway should be allowed. It should be reiterated that ASTM B 602 represents only one of many sampling procedures and criteria, and in its present form is somewhat controversial.

Table 26-1 Sampling of Items for Nondestructive Tests, Produced by Methods Other than Barrel Electroplating*

Number of product items in lot (lot size)	*Number of product items to be randomly selected for test (sample size)*	*Total number of defective product items for lot to be accepted (acceptance number)*	*Total number of defective product items for lot to be rejected (rejection number)*
2 to 8	2	0	1
9 to 15	3	0	1
16 to 25	5	0	1
26 to 50	8	0	1
51 to 90	13	0	1
91 to 150	20	1	2
151 to 280	32	1	2
281 to 500	50	2	3
501 to 1200	80	3	4
1201 to 3200	125	5	6
3201 to 10 000	200	7	8
10 001 and over	315	10	11

*Based upon MIL-STD-105, Level II, AQL of 1.5 percent, normal inspection.

Table 26-2 Sampling of Barrel Electroplated Items for Nondestructive Tests*

Number of product items in lot (lot size)	*Number of product items to be randomly selected for test (sample size)*	*Total number of defective product items for lot to be accepted (acceptance number)*	*Total number of defective product items for lot to be rejected (rejection number)*
2 to 15	2	0	1
16 to 25	3	0	1
26 to 90	5	0	1
91 to 150	8	1	2
151 to 500	13	1	2
501 to 1200	20	2	3
1201 to 10 000	32	3	4
10 001 and over	50	5	6

*Based upon MIL-STD-105, Level S-4, AQL of 4.0 percent, normal inspection.

corrosion, hydrogen embrittlement, solderability, or ductility), sample specimens may be used instead of the items themselves if the latter are of such shape, size, or value as to prohibit destructive testing or if the lots are too small (that is, too large a proportion of the items would be destroyed by the testing). Such sample specimens must, of course, be of the same basis metal and must be processed along with regular production.

In summary, sampling plans are compromises between the dangers of accepting unacceptable material and of rejecting acceptable material. Many such compromises are possible, depending on the degree of confidence required, the

Table 26-3 Sampling for Destructive Tests (Adhesion, Hydrogen Embrittlement, Corrosion Resistance, etc.)*

Number of product items in lot (lot size)	*Number of product items to be randomly selected for test (sample size)*	*Total number of defective product items for lot to be accepted (acceptance number)*	*Total number of defective product items for lot to be rejected (rejection number)*
2 to 25	2	0	1
26 to 150	3	0	1
151 to 1200	5	0	1
1201 to 35 000	8	0	1
35 001 and over	13	0	1

*Based upon MIL-STD-105, Level S-2, AQL of 1.5 percent, normal inspection.

nature of the items, the quality level required, and the degree of danger inherent in accepting a substandard item. This introduces a further complication, in that not all substandard items are unacceptable to the same degree: a critical defect, for example, is one that judgment and experience indicate is likely to result in hazardous or unsafe conditions, while a minor defect is one that is not likely to reduce the effective usefulness of the item but merely, e.g., detracts from its appearance. Inspection for critical defects may require 100 percent inspection.

Whatever plan is used, the manufacturer and the purchaser must agree on its parameters in advance. Here we have only introduced a complex problem; for details the specifications cited above and the Bibliography should be consulted.

APPENDIXES

Appendix A-1 Electrochemical Equivalents of Elements of Plating Interest

Table A-1

1	*2*
Element	*Symbol*
Aluminum	Al
Antimony	Sb
Bismuth	Bi
Cadmium	Cd
Chromium	Cr
Cobalt	Co
Copper	Cu
Gold	Au
Hydrogen	H
Indium	In
Iron	Fe
Lead	Pb
Manganese	Mn
Molybdenum	Mo
Nickel	Ni
Niobium	Nb
Osmium	Os
Oxygen	O
Palladium	Pd
Platinum	Pt
Rhenium	Re
Rhodium	Rh
Silver	Ag
Tantalum	Ta
Tin	Sn
Tungsten	W
Vanadium	V
Zinc	Zn

3	4	5	6	7	8	9
Atomic weight	Valence change	Electrochemical Equivalents			Density	
		$g/\mathfrak{F}$	mg/C	g/A·h	g/cm³	A·h/m²/μm
26.98	3	8.993	0.09319	0.3355	2.70	8.048
121.7	3	40.57	0.4204	1.513	6.684	4.418
209.0	3	69.67	0.7220	2.599	9.80	3.771
112.4	2	56.20	0.5824	2.097	8.642	4.121
52.00	6	8.667	0.08981	0.3233	7.20	22.27
	3	17.33	0.1796	0.6465		11.14
58.93	2	29.47	0.3054	1.099	8.9	8.098
63.55	1	63.55	0.6585	2.371	8.92	3.762
	2	31.78	0.3293	1.186		7.521
197.0	1	197.0	2.041	7.349	19.3	2.626
	3	65.67	0.6805	2.450		7.878
1.008	1	1.008	0.01045	0.03760	0.09×10^{-3}	—
114.8	3	38.27	0.3966	1.428	7.28	5.098
55.84	3	18.61	0.1928	0.6943	7.86	11.32
	2	27.92	0.2893	1.042		7.543
207.2	2	103.6	1.074	3.865	11.34	2.934
54.94	2	27.47	0.2847	1.025	7.30	7.122
96.94	3*	31.98	0.3314	1.193	10.2	8.550
58.71	2	29.36	0.3042	1.095	8.90	8.128
92.91	4*	23.23	0.2407	0.8666	8.57	9.889
190.2	3	63.40	0.6570	2.365	22.61	9.560
16.00	2	8.000	0.08290	0.2984	0.001331	—
106.4	2	53.20	0.5513	1.985	12.023	6.057
195.1	4	48.78	0.5055	1.820	21.45	11.79
186.2	7	26.60	0.2756	0.9923	21.04	21.20
102.9	3	34.30	0.3554	1.280	12.41	9.695
107.9	1	107.9	1.118	4.025	10.50	2.609
180.9	5*	36.18	0.3749	1.350	16.60	12.30
118.7	2	59.35	0.6150	2.214	7.28	3.288
	4	29.68	0.3076	1.107		6.576
183.9	4.5*	40.87	0.4325	1.525	19.35	12.69
50.94	3*	16.98	0.1760	0.6334	6.1	9.631
65.38	2	32.69	0.3388	1.220	7.14	5.852

*Atomic weights and electrochemical equivalents are to four significant figures only. Valence changes marked with an asterisk refer to conditions in fused-salt baths. Cathode efficiencies are assumed to be 100 percent.

Appendix A-2 Table of Conversion Factors: U.S. Customary Units ⇄ SI

Table A-2

A *U.S. Customary*
mil (0.001 in.)
microinch (μin.)
inch (in.)
foot (ft)
angstrom unit
millimicron (mμ)
square inch (in.^{2})
square foot (ft^2)
cubic inch (in.^{3})
cubic foot (ft^3)
gallon, U.S. liquid (gal)
1000 gallons
fluid ounce (fl oz)
grain (gr)
ounce, avoirdupois (oz)
ounce, troy (troy oz)
pound, avoirdupois (lb)
mg/in.^{2}
mg/in.^{2}
oz/in.^{2}
oz/in.^{2}
oz/ft^2
oz/ft^2
oz/gal
oz/gal
lb/100 gal
fl oz/gal
fl oz/gal

B SI	Multiply by factor to convert from: A to B		B to A	
Length or thickness				
micrometer (μm)	25.4	(25)*	0.03937	(0.04)
micrometer (μm)	0.0254	(0.025)	39.37	(40)
centimeter (cm)†	2.54	(2.5)	0.3937	(0.4)
meter (m)	0.3048	(0.3)	3.281	(3.3)
nanometer (nm)	0.1		10	
nanometer (nm)	1		1	
Area				
square centimeter (cm^2)	6.452	(6.5)	0.1550	(0.15)
square meter (m^2)	0.09290	(0.1)	10.76	(10)
Volume				
cubic centimeter (cm^3)	16.39	(16)	0.06102	(0.06)
cubic meter (m^3)	0.02832	(0.03)	35.31	(35)
liter (L)	3.785	(4)	0.2642	(0.25)
cubic meter (m^3)	3.785	(4)	0.2642	(0.25)
milliliter (mL)	29.57	(30)	0.03381	(0.034)
Mass (weight)				
milligram (mg)	64.80	(65)	0.01543	(0.015)
gram (g)	28.35	(28)	0.03527	(0.035)
gram (g)	31.10	(30)	0.03215	(0.03)
kilogram (kg)	0.4536	(0.45)	2.205	(2.2)
Weight (mass) per unit area				
g/m^2	1.550	(1.5)	0.6451	(0.65)
mg/cm^2†	0.1550	(0.15)	6.451	(6.5)
kg/m^2	43.95	(44)	0.02275	(0.023)
g/cm^2†	4.395	(4.4)	0.2275	(0.23)
kg/m^2	0.3052	(0.3)	3.277	(3.3)
mg/cm^2†	305.2	(300)	0.003277	(0.003)
Weight (mass) per unit volume				
g/L	7.490	(7.5)	0.1335	(0.13)
kg/m^3	7.490	(7.5)	0.1335	(0.13)
kg/m^3	1.198	(1.2)	0.8347	(0.8)
cm^3/m^3†	7812	(8000)	0.0001280	(0.00013)
mL/L	7.812	(8)	0.1280	(0.13)

Table A-2 (*Continued*)

A *U.S. Customary*
A/ft^2
A/ft^2
A/ft^2
$A/in.^2$
$A/in.^2$
calorie (cal)
kilowatt-hour (kWh)
$lb/in.^2$ (psi)
1000 psi
kilogram-force/mm^2 (kgf/mm^2)
$A \cdot h/ft^2$ for 1 mil
$A \cdot h/ft^2$ for 1 mil
$A \cdot h/ft^2$ for 1 mil
$°F = 9/5\ °C + 32$
$°C = 5/9\ (°F - 32)$
$t\ °C = t_K - 273.15$

B SI	Multiply by factor to convert from: A to B		B to A	
Current density				
A/m²	10.76	(10)	0.09290	(0.1)
mA/cm²†	1.076	(1)	0.9290	(1)
A/dm²†	0.1076	(0.1)	9.290	(10)
kA/m²	1.550	(1.5)	0.6451	(0.65)
mA/cm²†	155.0	(150)	0.006451	(0.0065)
Energy				
joule (J)	4.19	(4)	0.239	(0.25)
megajoule (MJ)	3.600	(exact)	0.2778	(0.3)
Pressure				
pascal (Pa)	6895	(7000)	0.0001450	(0.00015)
megapascal (MPa)	6.895	(7)	0.145	(0.15)
megapascal (MPa)	9.806	(10)	0.1020	(0.1)
Quantity of electricity to plate unit thickness				
C/m² for 1 μm	1530	(1500)	0.0006535	(0.00065)
A·h/m² for 1 μm	0.4236	(0.4)	2.361	(2.4)
mA·h/cm² for 1 μm	0.04236	(0.04)	23.61	(24)
Temperature				
°F = Fahrenheit temperature				
°C = Celsius (centigrade) temperature				
t_K = kelvins				

*Numbers in parentheses may be used for rough or mental calculation; they are accurate to ± 10 percent.

†Use of these units, though permitted in SI, is discouraged because they violate the rule that units should differ by multiples of 1000.

Appendix A-3
Glossary of Terms

Many terms are defined in the text, and for the most part their definitions are not repeated here; refer to the Index. See also the special glossary of metallurgical terms appended to Chap. 4.

In general, words to be found in any good dictionary are not repeated here, except when they have special meanings applicable to the subject of this book. Many of the words defined below also have other meanings not relevant to our subject.

Definitions marked with an asterisk are those approved by ASTM in B 374, Standard Definitions of Terms Relating to Electroplating.

Absorption (light). Retention of waves of light by a solid, liquid, or gas. The absorption coefficient is the constant k in the equation $I = I_0e^{-kcd}$, where I_0 and I are the intensities of the incident and transmitted light, respectively, c is the concentration, d is the thickness of solution (or length of light path), and e is the base of natural logarithms.

***Activation.** Elimination of a passive condition on a surface.

***Activity (ion).** The ion concentration corrected for deviations from ideal behavior. Concentration multiplied by *activity coefficient.* (The *effective* concentration of an ion.)

***Addition agent.** A material added in small quantities to a solution to modify its characteristics. It is usually added to a plating solution for the purpose of modifying the character of the deposit.

***Adhesion.** The attractive force that exists between an electrodeposit and its substrate, that can be measured as the force required to separate an electrodeposit and its substrate.

Alcohol. R—OH; alkyl compound containing a hydroxyl group.

Allotrope. One of two or more isomeric forms of an element.

***Amorphous.** Noncrystalline, or devoid of regular structure.

Ampere (A). That current which, if maintained in two straight parallel conductors of infinite length, of negligible cross section, and placed 1 m apart in vacuum, would produce between these conductors a force equal to 2×10^{-7} N/m (SI definition). Practically, the current that will deposit silver at the rate of 0.0011180 g/s, or current flowing at the rate of 1 C/s.

Amphoteric. Describing substances having both acid and basic properties.

***Back-emf.** The potential set up in an electrolytic cell that opposes the flow of current, caused by such factors as concentration polarization and electrode films.

***Barrel plating (or cleaning).** Plating or cleaning in which the work is processed in bulk in a rotating container.

***Base metal.** A metal that readily oxidizes or dissolves to form ions. The opposite of noble metal.

***Basis metal (or material).** Material upon which coatings are deposited; substrate. This term is preferred over base metal, to avoid confusion with the preceding.

***Bipolar electrode.** An electrode that is not directly connected to the power supply but is placed in the solution between the anode and the cathode so that the part nearest the cathode becomes anodic.

***Black oxide.** A finish on metal produced by immersing a metal in hot oxidizing salts or salt solutions.

***Blister.** A dome-shaped imperfection or defect, resulting from loss of adhesion between a metallic deposit and the substrate.

***Brightener.** An addition agent that leads to the formation of a bright plate or that improves the brightness of the deposit.

***Bright plating.** A process that produces an electrodeposit having a high degree of specular reflection in the as-plated condition.

Buffer. A substance or combination of substances which when dissolved in a solution produces a solution that resists a change in pH on the addition of acid or alkali. It may also be used as a verb: to buffer a solution, to add a buffer to it for the above purpose.

***Buffing.** The smoothing of a surface by means of a rotating flexible wheel to the surface to which fine abrasive particles are applied in liquid suspension, paste, or grease-stick form.

***Burnt deposit.** A rough, nonadherent, or otherwise unsatisfactory deposit produced by the application of an excessive current density and usually containing oxides or other inclusions.

***Bus (bus bar).** A rigid conducting section, for carrying current to the anode and cathode bars.

***Butler finish.** A finish composed of fine, uniformly distributed parallel lines, having a characteristic luster usually produced with rotating wire brushes or cloth wheels with applied abrasives.

***Calomel half-cell (calomel electrode).** A half-cell containing a mercury electrode in contact with a solution of potassium chloride of specified concentration that is saturated with mercurous chloride (calomel).

The three common calomel electrodes are the saturated, decinormal, and normal; these terms refer to the concentration of the potassium chloride in the cell. The potentials vs. the standard hydrogen electrode (SHE) at 20°C are, respectively, +0.2492, 0.3379, and 0.2860 V.

Cavitation. The production of emulsions by disruption of a liquid into a two-phase system of liquid and gas, when the hydrodynamic pressure in the liquid is reduced to the vapor pressure.

Chelate. Pertaining to a molecular structure in which a ring can be formed by the residual valences (unshared electrons) of neighboring atoms. ***Chelate compound.** A compound in which the metal is contained as an integral part of a ring structure and is not readily ionized. ***Chelating agent.** A compound capable of forming a chelate compound with a metal ion.

Colloid. A state of subdivision of matter which comprises either single large molecules or aggregations of smaller molecules. The particles of ultramicroscopic size (dispersed phase) are surrounded by different matter (external phase). Both phases may be solid, liquid, or gaseous. ***Colloidal particle.** An electrically charged particle, generally smaller than 200 nm, dispersed in a second molecular phase.

***Complex ion.** An ion composed of two or more ions or radicals, both of which are capable of independent existence. ***Complexing agent.** A compound that will combine with metallic ions to form complex ions.

***Conditioning (in plating on plastics).** The conversion of a surface to a suitable state for successful treatment in succeeding steps.

Conductance (unit). The siemens (S) is the electric conductance of a conductor in which a current of one ampere is produced by an electric potential difference of one volt. Identical to the older unit mho, or reciprocal ohm.

Coordinate covalent bond. A bond based on a shared pair of electrons both of which originate from the same atom.

Coordination. The formulation of covalent bonds.

Coordination compound. A compound whose molecular structure contains a central atom bonded to other atoms by coordinate covalent bonds. **Coordination number.** A number indicating the number of molecules or atoms linked to the central atom by coordinate bonds.

Coulomb (C). One ampere flowing for one second; a unit of quantity of electricity.

Covalency. The number of electron pairs that an atom can share with other atoms.

Covalent compound. A compound in which electron pairs are shared by the elements constituting the compound. See also *Coordinate covalent bond.*

Curie point. The temperature above which the molecular forces of magnetism of paramagnetic bodies cease to exist.

***Diaphragm.** A porous or permeable membrane separating anode and cathode compartments of an electrolytic cell from each other or from an intermediate compartment.

Dielectric constant (permittivity, inductivity). ϵ in the equation $F = QQ'/\epsilon_r^2$ where F is the force of attraction between two charges Q and Q' separated by a distance r in a uniform medium. (In general, in a solvent the greater ϵ is, the less the force between two charged ions Q and Q' and thus the better ionizing solvent it would be. This, however, is a rough guide, not a rule.)

Dipole. A combination of two electrically or magnetically charged particles separated by a very small distance.

Donor. An atom or radical that furnishes an electron pair to form a covalent bond with another atom called the acceptor.

***Dummy (or dummy cathode).** A cathode in a plating solution that is not used after plating. Often used for removal or decomposition of impurities. **Dummying** is the use of dummy cathodes for this purpose or for "breaking in" a new bath.

***Electrode.** A conductor through which current enters or leaves an electrolytic cell, at which there is a change from conduction by electrons to conduction by charged particles, or vice versa.

***Electrode potential.** The difference in potential between an electrode and the immediately adjacent electrolyte referred to some standard electrode potential as zero.

Dynamic. The electrode potential (EP) measured when current is passing between the electrode and the electrolyte.

Equilibrium. A static EP when the electrode and the electrolyte are in equilibrium with respect to a specified chemical reaction.

Static. The EP measured when no net current is flowing between the electrode and the electrolyte.

Standard. An equilibrium EP for an electrode in contact with an electrolyte in which all the components of a specified chemical reaction are in their standard states. The standard states for an ionic constituent are unit ionic activity.

***Electrophoresis.** The movement of colloidal particles produced by the application of an electrode potential.

Emissivity. The ratio of radiation intensity from a surface to the radiation intensity at the same wavelength of a black body at the same temperature.

Emulsion. A fluid consisting of a microscopically heterogeneous mixture of two normally immiscible phases in which one liquid forms minute droplets suspended in the other liquid. An invert emulsion is water in oil, whereas a normal emulsion is oil in water. See *Colloid*.

Equivalent conductivity. In an electrolyte, the conductivity of the solution divided by the number of equivalents of conducting solute per unit volume; i.e., the conductivity divided by the normality of the solution.

Ester. An organic salt formed from an alcohol (base) and an organic acid by elimination of water.

$$\underset{\text{alcohol}}{\mathrm{ROH}} + \underset{\text{acid}}{\mathrm{R'COOH}} \rightarrow \underset{\text{ester}}{\mathrm{R'COOR}} + \mathrm{H_2O}$$

Eutrophication. Process by which a body of water in which the increase of mineral and organic nutrients has reduced the dissolved oxygen, producing an environment that favors plant life over animal life.

***Flash plate (flash).** A thin electrodeposit, less than 2.5 μm. See *Strike*.

Flocculation. Coagulation of a finely divided precipitate so that it settles more readily.

***Free cyanide. True.** The actual concentration of cyanide radical or equivalent alkali cyanide, not combined in complex ions with metals in solution.

Calculated. The concentration of cyanide or alkali cyanide, present in solution in excess of that calculated as necessary to form a specified complex ion with a metal or metals present in solution.

Analytical. The free cyanide content of a solution as determined by a specified analytical method.

Since the value of true free cyanide is seldom known, the calculated or analytical value usually is used in practice.

***Grinding.** The removal of metal by means of rotating rigid wheels containing abrasive.

Half-cell. A setup consisting of a metal in contact with a solution of its ions, thereby setting up an electrode potential. Usually referring to a standardized half-cell whose potential is known so that the electrode potential of another similar system can be determined. See, e.g., *Calomel electrode*.

Hertz (Hz). A frequency of one cycle per second.

Hot lead (hot contact). To enter a bath with a hot contact is to have the electrolytic potential already applied so that current flows as soon as the work enters the bath.

***Hydrogen embrittlement.** Embrittlement of a metal or alloy caused by absorption of hydrogen during a pickling, cleaning, or plating process.

***Inhibitor.** A substance used to reduce the rate of a chemical or electrochemical reaction, usually corrosion or pickling.

Isomers. Compounds composed of the same number of the same atoms but differing in molecular structure.

Joule (J). A unit of energy, work, or quantity of heat equal to 1 N·m.

Kinetics. The study of all aspects of motion. Specifically, in electrochemistry, the study of the mechanism rather than the total result of an electrochemical process.

Ligand. A group of atoms around a central metallic ion in a complex compound.

***Limiting current density. Cathodic.** The maximum current density at which satisfactory deposits can be obtained.

Anodic. The maximum current density at which the anode behaves normally, without excessive polarization.

Lithosphere. The solid part of the earth, as distinguished from the hydrosphere and atmosphere.

***Mechanical plating.** A process whereby hard, small spherical objects (such as glass shot) are tumbled against a metallic surface in the presence of finely divided metal powder (such as zinc dust) and appropriate chemicals for the purpose of covering such surfaces with metal. (Peen plating is a trade mark for such a process.)

Newton (N). The force required to give an acceleration of 1 m/s^2 to 1 kg of mass.

***Noble metal.** A metal that does not tend to furnish ions readily and therefore does not dissolve readily, nor easily enter into such reactions as oxidations. The opposite of base metal.

***Nodule.** A rounded projection formed on a cathode during electrodeposition.

***Nucleation (in plating on plastics).** The preplating step in which a catalytic material, often a palladium or gold compound, is adsorbed on the surface. The catalyst is not necessarily in its final form.

Ohm (Ω). The electric resistance between two points of a conductor when a constant difference of potential of one volt, applied between these two points, produces in this conductor a current of one ampere, this conductor not being the source of any electromotive force.

Paramagnetic. Of a substance that has magnetic properties stronger than those of air.

Pascal (Pa.) A pressure or stress of 1 N/m^2.

Passivity. The inertness of certain substances under conditions in which chemical activity is expected.

Peen plating™. See *Mechanical plating.*

Peptize. To change from a gel to a liquid other than by melting.

***Pickle.** An acid solution used to remove oxides or other compounds from the surface of a metal by chemical or electrochemical action.

***Pickling.** The removal of oxides or other compounds from a metal surface by means of a pickle.

Polar. Characterized by electrical attraction. A polar compound is one in which the atoms are held by electrostatic attraction.

Polarization. The change in the potential of an electrode during electrolysis, such that the potential of an anode always becomes more noble and that of a cathode less noble than their respective static potentials. Equal to the difference between the static potential and the dynamic potential.

Polarograph. An instrument that records minute changes in the intensity of a current resulting from a gradually increasing applied voltage, in electrolysis with a dropping mercury cathode.

Polarography. Use of a polarograph.

Polymer. A substance composed of very large molecules which consist essentially of recurring structural units. (*Dimer*: a polymer consisting of two such units; similarly *trimer, tetramer,* etc.)

***Post-nucleation (in plating on plastics).** The step in which, if necessary, the catalyst is converted to its final form. This is the final step before autocatalytic (electroless) plating.

***Primary current distribution.** The distribution of the current over the surface of an electrode in the absence of polarization.

***Rack (plating rack).** A frame for suspending and carrying current to articles during plating and related operations. **Racking.** Placing articles to be processed on racks.

Reference electrode. A half-cell whose electrode potential vs. the standard hydrogen electrode (SHE) is known and which consequently can be used as a secondary standard electrode.

***Resist.** A material applied to a part of a cathode or plating rack to render the surface nonconductive. A material applied to a part of the surface of an article to prevent reaction of metal from that area during chemical or electrochemical processes.

Resistance. See *Ohm.*

***Ripple (dc).** Regular modulations in the dc output wave of a rectifier unit or a motor-generator set, originating from the harmonics of the ac input system in the case of a rectifier or from the harmonics of the induced voltage of a motor-generator set.

Robber. See *Thief.*

***Saponification.** The alkaline hydrolysis of fats whereby a soap is formed; more generally, the hydrolysis of an ester by an alkali with the formation of an alcohol and a salt of the acid portion.

***Satin finish.** A surface finish that behaves as a diffuse reflector and which is lustrous but not mirror-like.

***Scale.** An adherent oxide coating that is thicker than the superficial film referred to as tarnish.

Scratch finish. See *Butler finish.*

***Sensitization (in plating on plastic).** The adsorption of a reducing agent, often a stannous compound, on a surface.

Sequestering. The removal of a metal ion from a system by forming a complex ion which does not have the chemical reactions of the ion removed.

***Sequestering agent.** An agent that forms soluble complex compounds with, or sequesters, a simple ion, thereby suppressing the activity of that ion. See also *Chelating agent.*

Shield. (n.) A nonconducting medium for altering the current distribution on an anode or cathode. (vb.) To alter the normal current distribution on an anode or cathode by the interposition of a nonconductor.

Silver–silver chloride electrode. The half-cell Ag/AgCl-KCl, used as a reference electrode; its potential vs. the SHE is +0.2222 V at 25°C.

Soap. A salt of a higher fatty acid with an alkali or metal.

Solubility product (K_{sp}). $(M^+)(X^-)/MX$ where ( ) indicates the concentrations (or activities) of the components of the equilibrium. $MX \rightarrow M^+ + X^-$. If $(M^+)(X^-) > K_{sp}$, MX will precipitate. Usually applied only to compounds of limited solubility.

***Spotting out.** The delayed appearance of spots and blemishes on plated or finished surfaces.

Standard hydrogen electrode (SHE). See text or index.

Stoichiometry. Numerical relationships between elements or compounds; determination of the proportions in which the elements combine and the mass relations of reactions.

***Stopping off.** The application of a resist to any part of an electrode: cathode, anode, or rack. The use of a resist or stopoff.

***Strike.** (n.) A thin film of metal to be followed by other coatings. A solution used to deposit a strike. (vb.) To plate for a short time, usually at a high initial current density.

Substrate. The material upon which deposits are applied; basis metal or material.

***Surface-active agent; surfactant.** A substance that affects markedly the interfacial or surface tension of solutions even when present in very low concentrations.

***Surface tension.** That property, due to molecular forces, that exists in the surface film of all liquids and tends to prevent the liquid from spreading.

***Tarnish.** The dulling, staining, or discoloration of metals due to superficial corrosion. The film so formed.

Thermodynamics. The study of the empirical relations between heat energy and other forms of energy.

***Thief.** An auxiliary cathode so placed as to divert to itself some current from portions of the work which would otherwise receive too high a current density.

***Total cyanide.** The total content of cyanide expressed as the radical CN^- or alkali cyanide whether present as simple or complex ions. The sum of the combined and free cyanide content of a solution.

***Trees.** Branched or irregular projections formed on a cathode during electrodeposition especially at edges and other high-current-density areas.

Transducer. A device used to convert a pulsating electric potential into periodic vibrations at ultrasonic frequencies.

Volt (V). A unit of electric potential difference and electromotive force, being the difference of electric potential between two points of a conductor carrying a constant current of one ampere, when the power dissipated between these points is equal to one watt.

Watt (W). The power which gives rise to the production of energy at the rate of one joule per second.

***Wetting agent.** A substance that reduces the surface tension of a liquid, thereby causing it to spread more readily on a solid surface.

***Whiskers.** Metallic filamentary growths, often microscopic, sometimes formed during electrodeposition and sometimes forming spontaneously during storage or service, after finishing.

***Work.** The material being plated or otherwise finished.

BIBLIOGRAPHY

Suggestions for Further Reading

Relatively few references to journal articles are cited in the text because books cited below will generally provide needed references to the older literature (the chapter on Anodizing is an exception). These always may be superseded by more recent publications, and awareness of current literature is the only way to keep abreast of new developments. Recommended for this purpose are:

Plating and Surface Finishing (through 1975, simply *Plating*), the official journal of the American Electroplaters' Society.

Transactions of the Institute of Metal Finishing (British) and the trade journals *Metal Finishing* and *Products Finishing*.

Surface Technology (Elsevier Sequoia, Switzerland) publishes in English, French, and German.

For the most part, the following suggestions for additional reading are books, though a few journal articles are cited.

The following general references are useful for individual articles relevant to some aspects of metal finishing, such as ion-exchange resins and the individual elements.

A. Standen (ed.), "Kirk-Othmer Encyclopedia of Chemical Technology," 2d ed., 22 vols. plus supplement and index, Interscience-Wiley, New York, 1963–1971.

One-volume encyclopedias of interest include:

D. M. Considine (ed.), "Chemical and Process Technology Encyclopedia," McGraw-Hill Book Company, New York, 1974.

C. A. Hampel (ed.), "Encyclopedia of the Chemical Elements," Reinhold Publishing Co., New York, 1968.

C. A. Hampel and G. G. Hawley (eds.), "The Encyclopedia of Chemistry," 3d ed., Van Nostrand Reinhold Co., New York, 1973.

General Chemistry. The number of texts on general and inorganic chemistry is too large to list. Almost any modern text should be satisfactory at an elementary or intermediate level. For more advanced students, an outstanding text is F. A. Cotton and G. Wilkinson, "Advanced Inorganic Chemistry," 3d ed., Interscience-Wiley, New York, 1972.

References for Specific Chapters

Chapter 1:

G. R. Palin, "Electrochemistry for Technologists," Pergamon Press, New York, 1969 (paper).

Chapters 2 and 8. *Elementary:*

A. R. Denaro, "Elementary Electrochemistry," 2d ed., Butterworths, London, 1971 (paper).

Intermediate:

E. C. Potter, "Electrochemistry," The Macmillan Co., New York, 1956.

J. Robbins, "Ions in Solution" (2) and G. Pass, "Ions in Solution" (3), Oxford University Press, New York, 1972, 1973 (paper).

Advanced:

J. O'M. Bockris and A. K. N. Reddy, "Modern Electrochemistry," (2 vols.,) Plenum Press, New York, 1970 (available in paper).

I. Fried, "The Chemistry of Electrode Processes," Academic Press, New York, 1973.

H. R. Thirsk and J. A. Harrison, "A Guide to the Study of Electrode Kinetics," Academic Press, New York, 1972.

G. Milazzo, "Electrochemistry," American Elsevier, New York, 1963.

Chapter 3:

H. H. Uhlig, "Corrosion and Corrosion Control," 2d ed., John Wiley & Sons, Inc., New York, 1971.

N. D. Tomashov, "Theory of Corrosion and Protection of Metals," The Macmillan Co., New York, 1966.

L. L. Shrier (ed.), "Corrosion," 2 vols., Halsted Press (John Wiley & Sons), New York, 1963.

U. R. Evans, "Corrosion and Oxidation of Metals," plus supplementary volume, St. Martin's Press, New York, 1960, 1969.

A more practical approach than the above can be found in:

F. L. LaQue and H. R. Copson, "Corrosion Resistance of Metals and Alloys," 2d ed., Reinhold Publishing Corp., New York, 1963 (ACS Monograph series).

R. M. Burns and W. W. Bradley, "Protective Coatings for Metals," 3d ed., Reinhold Publishing Corp., New York, 1967 (ACS Monograph series). Also useful for several chapters on organic finishing.

Chapter 4: *Relatively elementary:*

B. A. Rogers, "The Nature of Metals," American Society for Metals, Novelty, Ohio, 1964.

A.H. Cottrell, "An Introduction to Metallurgy," St. Martin's Press, New York, 1970.

More advanced:

C. S. Barrett and T. B. Massalski, "Structure of Metals," 3d ed., McGraw-Hill Book Company, New York, 1966.

B. Chalmers, "Physical Metallurgy ," John Wiley & Sons, Inc., New York, 1959.

A classic in the field is W. Hume-Rothery, "Electrons, Atoms, Metals, and Alloys," 3d ed., Dover Publications, Inc., New York, 1963 (paper).

With particular reference to the metallurgical and physical properties of electrodeposits:

W. H. Safranek, "The Properties of Electrodeposited Metals and Alloys," American Elsevier, New York, 1974.

Chapter 8. See also Chap. 2. The theory of leveling is well discussed, with copious references, in an annual William Blum lecture:

O. Kardos, "Current Distribution on Microprofiles," *Plating and Surface Finishing,* **61,** 129, 229, 316 (1974).

Chapters 10 to 16. Several books are available on electroplating practice; the most important are the following. These books supplement the present volume in various ways.

F. A. Lowenheim (ed.), "Modern Electroplating," 3d ed., John Wiley & Sons, Inc., New York, 1974. Offers more detail on the plating of the individual metals and chapters on theory and special topics more detailed than the present book can offer. Also valuable as a source of references to the literature. A volume in the Electrochemical Society Series.

A. K. Graham (ed.), "Electroplating Engineering Handbook," 3d ed.; Van Nostrand Reinhold Co., New York, 1971. Covers many aspects of metal finishing not considered in the present book, such as engineering, generators and rectifiers, automatic machines, design of facilities, etc.

J. A. Murphy (ed.), "Surface Preparation and Finishes for Metals," McGraw-Hill Book Company, New York, 1971. Includes organic finishing.

N. Hall (ed.), "Metal Finishing Guidebook-Directory," Metals and Plastics Publications, Inc., Hackensack, N.J. Published annually and received with subscription to *Metal Finishing*. Good for operating directions for the individual baths; has sections on organic finishing, buffing, etc.

W. Blum and G. B. Hogaboom, "Principles of Electroplating and Electroforming," 3d ed., McGraw-Hill Book Company, New York, 1949. A classic in the field, obsolete in some respects but still useful; deservedly famous for being the first to apply scientific principles to the "art" of electroplating.

Chapters 11 to 16. The American Electroplaters' Society (AES) has prepared a series (being added to at irregular intervals) of Illustrated Lectures on many aspects of finishing technology. For a complete listing, consult the Society's Headquarters. Intended for oral presentation with accompanying slides, they are also valuable for individual study.

Few monographs on plating the individual metals are available. These few include:

Chromium: J. D. Greenwood, "Hard Chromium Plating," 2d ed., International Publications Services (Portcullis), New York, 1971.

Nickel: R. Brugger, "Nickel Plating," International Publications Services (Portcullis), New York, 1970.

J. K. Dennis and T. E. Such, "Nickel and Chromium Plating," Halsted Press (John Wiley & Sons), New York, 1972.

Gold: F. H. Reid and W. Goldie (eds.), "Gold Plating Technology," Electrochemical Publications Ltd., Ayr, Scotland, 1974.

Chapter 19:

W. Goldie, "Metallic Coating of Plastics," 2 vols., Electrochemical Publications Ltd., Ayr, Scotland, 1968/69.

G. D. R. Jarrett, C. R. Draper, G. Muller, and D. W. Baudrand, "Plating on Plastics," 2d ed., International Publications Services (Portcullis), New York, 1971.

Also of interest: C. F. Coombs, Jr. (ed.), "Printed Circuits Handbook," McGraw-Hill Book Company, New York, 1967.

Chapter 20:

P. Spiro, "Electroforming," 2d ed., International Publications Services, New York, 1971.

Chapter 22:

J. Alexander, trans. W. Lewis, "The Anodic Oxidation of Aluminum and Its Alloys," Charles Griffin & Co., Ltd., London, 1940.

"Anodized Aluminum," STP 388; Philadelphia, ASTM, 1965.

W. E. Hubner and A. Schiltknecht, "The Practical Anodizing of Aluminium," McDonald and Evans, London, 1960.

A. W. Brace, "The Technology of Anodizing Aluminium," Robert Draper Ltd., Teddington, England, 1968.

J. W. Diggle, T. C. Downie, and C. W. Goulding, *Chem. Rev.*, **69,** 365 (1969).

S. Tajima, in "Advances in Corrosion Sciences and Technology," vol. 1, pp. 229, 362, Plenum Press, New York, 1970.

T. P. Hoar and N. F. Mott, *J. Physics & Chem. of Solids,* **9,** 97 (1959).

G. E. Best, J. G. Hecker, J. W. McGrew, and R. V. Vanden Berg, *Proc. ASTM,* **59,** 277 (1959).

R. W. Franklin, *Proc. Conf. on Anodizing Aluminum,* University of Nottingham, Aluminium Development Association, London, 1962.

K. Wafers and W. T. Evans, *Plating and Surface Finishing,* **62,** 951 (1975).

K. Wafers, *Aluminium,* **49,** nos. 8, 9 (1973).

Aluminum Finishing Seminar, Chicago, March 1973, The Aluminum Association.

"Aluminum Association Designation System for Aluminum Finishes," 6th ed., The Aluminum Association, New York, 1973.

See also references cited in the chapter.

Chapter 23. The indispensable references are the ASTM documents referring to electroplated and related finishes, many of which are cited in the text. They are the responsibility of ASTM Committee B.08 (or B-8); they are reconsidered at frequent intervals, and by ASTM regulations must be reexamined and either reaffirmed, revised, or withdrawn at stated intervals. The "Annual Book of Standards" of ASTM is published in many parts; at present the electroplating standards appear in Part 9, issued about July of each year. Anyone using a specification, test, or recommended practice should be sure to use the latest revision.

Also valuable is the record of an Electrochemical Society symposium:

R. Sard, H. Leidheiser, Jr., and F. Ogburn (eds.), "Properties of Electrodeposits, Their Measurement and Significance," The Electrochemical Society, Inc., Princeton, N.J., 1975 (paper).

Chapter 24:

D. G. Foulke (ed.), "Electroplaters' Process Control Handbook," rev. ed., Robert E. Krieger Publishing Co., Huntington, N.Y., 1975.

Analytical: especially oriented toward plating solution analysis:

K. E. Langford and J. E. Parker, "Analysis of Electroplating and Related Solutions," 4th ed., Robert Draper, Ltd., Teddington, England, 1971.

The book by Foulke also contains analytical methods, as do the "Electroplating Engineering Handbook'' (Graham) and the ''Metal Finishing Guidebook-Directory.'' General analytical texts are plentiful; for reference, see:

F. D. Snell and L. S. Ettre (eds.), ''Encyclopedia of Industrial Chemical Analysis,'' 19 vols. plus index, Interscience-Wiley, New York, 1969–1974.

Atomic absorption spectrophotometry: manufacturers' literature; see also:

W. Slavin, ''Atomic Absorption Spectroscopy,'' Interscience Publishers, New York, 1968.

Ion Selective Electrodes:

R. A. Durst, 'Ion Selective Electrodes,'' National Bureau of Standards special publication 314, Government Printing Office, Washington, D.C.

Chapter 26. There are too many advanced texts to list. The following books employ a minimum of advanced mathematics and serve as good introductions to the field.

W. J. Reichman, ''Use and Abuse of Statistics,'' Pelican (Penguin) Books, Baltimore, Md., 1944.

M. J. Monroney, ''Facts from Figures,'' Penguin Books, Baltimore, Md., 1953.

R. Langley, ''Practical Statistics Simply Explained,'' Dover Publications, Inc., New York, 1970.

E. L. Bauer, ''A Statistical Manual for Chemists,'' 2d ed., Academic Press, New York, 1971.

The books and journals cited are in English. This fact implies no lack of merit in the many excellent publications in German, Russian, and other languages, and students acquainted with those languages certainly should take advantage of them. It is believed that for most of the intended audience for this book, the choice referred to is appropriate.

Index

NOTE: Words found in the Glossary, on pp. 569–576, are not separately indexed unless they are also discussed in the text.

Numbered ASTM specifications mentioned or discussed in the text are not separately indexed since they are too numerous. See Part 9 of the ASTM "Annual Book of Standards."